Biomaterials: From Theory to Applications

Biomaterials: From Theory to Applications

Editor: Shay Fisher

NYRESEARCH
P R E S S

New York

Published by NY Research Press
118-35 Queens Blvd., Suite 400,
Forest Hills, NY 11375, USA
www.nyresearchpress.com

Biomaterials: From Theory to Applications
Edited by Shay Fisher

International Standard Book Number: 978-1-63238-752-3 (Hardback)

Trademark Notice: Registered trademark of products or corporate names are used only for explanation and identification without intent to infringe.

Cataloging-in-Publication Data

Biomaterials : from theory to applications / edited by Shay Fisher.
     p. cm.
Includes bibliographical references and index.
ISBN 978-1-63238-752-3
1. Biomedical materials. 2. Biomedical engineering. 3. Biocompatibility. I. Fisher, Shay.
R857.M3 B56 2020
610.28--dc23

Contents

Permissions

List of Contributors

Index

Preface

This book has been a concerted effort by a group of academicians, researchers and scientists, who have contributed their research works for the realization of the book. This book has materialized in the wake of emerging advancements and innovations in this field. Therefore, the need of the hour was to compile all the required researches and disseminate the knowledge to a broad spectrum of people comprising of students, researchers and specialists of the field.

Biomaterial is a substance that is engineered to interact with biological systems for medical purposes. These purposes can either be therapeutic or diagnostic. The study of biomaterials is called biomaterial science or biomaterial engineering. Biomaterial science makes use of elements from medicine, biology, chemistry, tissue engineering and material science. Biomaterials can be derived from nature. They can also be synthesized in the laboratory by using varieties of chemical approaches such as utilizing metallic components, ceramics or composite materials and polymers. Biomaterials comprise a part or a whole of living structure or biochemical device which performs, augments or replaces a natural function. They are used for various medical purposes such as joint replacements, bone plates, intraocular lenses, bone cement, artificial ligaments and tendons, blood vessel prostheses, dental implants, heart valves, etc. The ever growing need of advanced technology is the reason that has fueled the research in the field of biomaterials in recent times. Most of the topics introduced in this book cover new techniques and the applications of biomaterials. The extensive content of this book provides the readers with a thorough understanding of the subject.

At the end of the preface, I would like to thank the authors for their brilliant chapters and the publisher for guiding us all-through the making of the book till its final stage. Also, I would like to thank my family for providing the support and encouragement throughout my academic career and research projects.

Editor

Effect of Injection Molding Melt Temperatures on PLGA Craniofacial Plate Properties during *In Vitro* Degradation

Liliane Pimenta de Melo,[1] **Gean Vitor Salmoria,**[1,2]
Eduardo Alberto Fancello,[1,3] **and Carlos Rodrigo de Mello Roesler**[1]

[1]*LEBm Biomechanics Engineering Laboratory, University Hospital (HU), Federal University of Santa Catarina,
88040-900 Florianópolis, SC, Brazil*
[2]*NIMMA Laboratory of Innovation on Additive Manufacturing and Molding, Federal University of Santa Catarina,
88040-900 Florianópolis, SC, Brazil*
[3]*GRANTE, Department of Mechanical Engineering, Federal University of Santa Catarina, 88040-900 Florianópolis, SC, Brazil*

Correspondence should be addressed to Liliane Pimenta de Melo; liliane.eng@gmail.com

Academic Editor: Junling Guo

The purpose of this article is to present mechanical and physicochemical properties during *in vitro* degradation of PLGA material as craniofacial plates based on different values of injection molded temperatures. Injection molded plates were submitted to *in vitro* degradation in a thermostat bath at $37 \pm 1°C$ by 16 weeks. The material was removed after 15, 30, 60, and 120 days; then bending stiffness, crystallinity, molecular weights, and viscoelasticity were studied. A significant decrease of molecular weight and mechanical properties over time and a difference in FT-IR after 60 days showed faster degradation of the material in the geometry studied. DSC analysis confirmed that the crystallization occurred, especially in higher melt temperature condition. DMA analysis suggests a greater contribution of the viscous component of higher temperature than lower temperature in thermomechanical behavior. The results suggest that physical-mechanical properties of PLGA plates among degradation differ per injection molding temperatures.

1. Introduction

Poly(lactic-co-glycolic acid), PLGA, is a biocompatible, biodegradable, and FDA-approved polymer. In the last two decades, PLGA has been considered as one of the most promising polymers for biomedical engineering applications, such as plates and screws in craniofacial surgery [1–4]. One of the techniques used in the manufacturing of medical devices is injection molding, which allows the development of complex mold geometries. However, mechanical properties stability and physicochemical properties of the resorbable material can also be strongly influenced by manufacturing process and design of the devices [5–8].

PLGA plates and screws must have suitable strength and ductility for biomechanical function, biocompatibility, and degradation. In particular, melt processing temperatures during injection molding could develop different microstructures of the manufactured device, including other operative

parameters, such as mold temperature, injection flow rate, and holding pressure [9]. As semicrystalline, PLGA devices are heterogeneous systems comprised of highly anisotropic crystallites, a phase in which the chains show long-range 3D order. The size and distribution of these crystals and the viscoelasticity are extremely dependent on the molecular weight distribution and the conditions under which the material is processed [10]. In manufacturing, the parameters can affect viscosity and chain orientation during the process of molded devices, where crystallinity is an important parameter because it can increase flexural stiffness and decrease the impact properties of the final product [10, 11].

Once properties of resorbable polymers devices are established, degradation rate must be evaluated. Many factors could influence degradation rate, such as implant site, mechanical stress, molar mass distribution, chemical/stereoisometric composition, crystallinity, morphology, size and geometry of the carrier, and surface roughness

[10, 12–17]. Biodegradation and reabsorption process of poly(α-hydroxy acids) is a succession of events. Material is initially hydrated exposed to the body's aqueous fluids. With water molecules presence, the degradation process occurs through the hydrolysis of the ester linkages, resulting in products in the form of soluble and nontoxic oligomers (or monomers). The degradation proceeds passive hydrolytic cleavage, characterized by changes in molecular weight, glass transition temperature (T_g), moisture content, and mechanical properties, such as tensile and compressive strength [13, 18, 19]. Therefore, time for osteosynthesis must be less than the time of the mechanical properties retention.

In this work, PLGA plates were designed and manufactured by injection molding as craniofacial bioresorbable medical devices. Two different melt temperatures were tested for the injection molding process (i.e., 240 and 280°C). The high melt temperature condition was defined as the upper work limit temperature of injection molding manufacturing, while lower temperature was defined by minimum processability temperature. For both conditions, PLGA craniofacial plates were evaluated by mechanical properties (bending stiffness, flexural maximum strength, and storage modulus), physicochemical properties (crystallinity and transitions temperatures), and morphology during *in vitro* degradation.

2. Materials and Methods

2.1. Material. Poly(L-lactic-co-glycolic acid) 85/15 granules (PURASORB PLG 8531) were purchased by PURAC Biomaterials (Netherlands). The PLGA 85/15 showed an average molecular weight, Mn = 224.27 g/mol, and a polydispersity index of 1.87 (Gel Permeation Chromatography, Viscotek VE 2001, Viscotek Detector TDA 302, USA, 2008). The transition temperatures declared by the manufacturer are Tg = 57 ± 1°C and Tm = 140°C and intrinsic viscosity was 3.04 dl/g (chloroform, 25°C, c = 0.1 g/dl) (PURAC, 2012).

2.2. Plates Design and Processing. The implant was designed by 3D CAD SolidWorks 2014 software (Concord, MA). The design of the craniofacial plate device is characterized by 2 mm of thickness, 5.8 mm of width, and 38.7 mm of length and by 8 aligned holes of 2 mm semicircle screw thread. The pellets of PLGA were processed using an injection molding machine (ARBURG 270S, 250-70 model). To investigate processing influence on PLGA plates properties, two different melt injection temperatures were considered: low temperature (T = 240°C, PLGA_lowT) and high temperature (T = 280°C, PLGA_highT). The other processing parameters were kept constant, as summarized in Table 1.

2.3. In Vitro Hydrolytic Degradation. Injection molded plates (0.25 ± 0.01 g per sample/plate) were desinfected by immersion in 70% v/v ethanol. Samples were immersed (n = 16, per each condition, per time point) in the phosphate-buffered saline (PBS) solution (0.25 g/30 ml, pH = 7.4) by storing them in the thermostatic bath (37 ± 1°C) for 15, 30, 60, and 120 days. At each time point, the degradation solution pH (pH = 7.4) was recorded and the samples were removed from the buffer solution, washed, and held in distilled water for 1 hour

TABLE 1: Injection molding parameters used to produce PLGA plates. All the production parameters were kept constant while varying the melt injection temperature (240°C for PLGA_lowT and 280°C for PLGA_highT).

	PLGA_lowT	PLGA_highT
Melt injection temperatures	240°C	280°C
Mold temperature	25°C	
Injection pressure	1500 MPa	
Holding pressure	25 MPa	
Injection time	2 s	
Cooling time	90 s	
Screw speed	100 rpm	

to remove as much buffer solution as possible. They were weighed in the wet condition and then dried in a vacuum oven at 23°C for 8 h. The samples were kept under vacuum prior to the characterization tests.

2.4. Dynamic Mechanical Analyses. PLGA_lowT and PLGA_highT specimens were submitted to the mechanical characterization, performed by a Dynamic Mechanical Analyzer (DMA Q800, TA Instruments) at the different degradation points (i.e., 0, 15, 30, 60, and 120 days). For viscoelasticity analysis, samples (n = 6, gauge length = 16 mm) were tested by setting 1 Hz frequency and 0.1% relative strain of area. DMA tests were performed in the temperature range of 30–80°C at a temperature rate of 3°C min^{-1}. From each test, storage modulus (E'), loss modulus (E''), and tan $\delta(E''/E')$ trends in function of the temperature were obtained and the transition temperatures were determined as peak in tan δ trends.

For three-point bending tests, samples (n = 5, gauge length = 10 mm) were tested in three-point flexural mode (ASTM D790) [20], kept at 37°C, using a test speed of 2 N min^{-1}. The considered mechanical parameters were bending stiffness (E_f), maximum bending deformation (ε_f), and flexural stress (σ_f), calculated according to the following equation:

$$\sigma_f = \frac{3PL}{2bd^2}, \qquad (1)$$

where σ_f is stress in the outer fibers at midpoint (MPa); P is load at a given point on the load-deflection curve (N); L is support span (mm); b is width of beam tested (mm); and d is depth of beam tested (mm).

2.5. Differential Scanning Calorimeter Analysis. The crystallinity and thermal properties (melting point, T_m, glass transition temperature, T_g, enthalpy of cold crystallization, ΔH_c, and enthalpy of melting, ΔH_m) were obtained by using a DSC (Shimadzu DSC-6000) in a nitrogen atmosphere of 19 cm^3 m^{-1}, using aluminum oxide as standard. The applied heating rate was 10°C min^{-1}, from 10 to 250°C, using an average sample size of 7 mg taken from the central region of

TABLE 2: Mechanical properties of injection molded PLGA 85/15 plates (mean ± standard deviation, $n = 3$): flexural stiffness, E; flexural strength, σ_r; and maximum flexural strain, ε_r ($^*p < 0.05$).

		E (GPa)*	σ_r (MPa)*	ε_r (%)*
PLGA_lowT	0 days	2.2 ± 0.1	41.4 ± 11.8	2.6 ± 0.8
	15 days	2.1 ± 0.2	54.6 ± 3.9	3.5 ± 0.5
	30 days	1.5 ± 0.1	12.5 ± 7.7	1.3 ± 0.7
	60 days	1.2 ± 0.1	19.05 ± 0.8	2.8 ± 0.4
PLGA_highT	0 days	1.9 ± 0.1	30.1 ± 3.1	2.3 ± 0.2
	15 days	2.1 ± 0.07	42.3 ± 7.2	2.9 ± 0.6
	30 days	1.9 ± 0.3	24.5 ± 15.3	1.8 ± 0.7
	60 days	0.5 ± 0.1	4.1 ± 0.8	1.0 ± 0.1

molded specimens ($n = 3$). Degree of crystallinity (X_c) was determined by using the following formula:

$$X_c = 100 \times \left(\frac{\Delta H_m - \Delta H_{cc}}{\Delta H_m^c} \right) \times \frac{1}{1 - m_f}, \qquad (2)$$

where ΔH_m is enthalpy of fusion, ΔH_{cc} is enthalpy of cool crystallization, ΔH_m^c is heat of melt of purely crystalline PLA, taken as 93 J/g [10, 19], and $(1 - m_f)$ is weight fraction of the polymer in the sample. Crystallization temperature (T_{cc}) was obtained from the second heating.

2.6. Fourier Transform Infrared Spectroscopy. Fourier transform infrared spectroscopy (FT-IR) was performed using attenuated total reflection (ATR) mode on a Shimadzu spectrometer, model TENSOR 27. The spectra of the samples ($n = 3$) were obtained in 4000 to 600 cm^{-1} wavenumbers by 4 cm^{-1} resolution. FTIR analysis identified bioresorbable copolymer functional groups and the possible changes due to the degradation.

2.7. Gel Permeation Chromatography. Molar mass distribution of the PLGA copolymer was verified by a high-performance liquid chromatography (GPC) Viscotek VE 2001 coupled to the Viscotek Detector TDA 302, Houston, Texas, USA (2008). THF solvent was used as the mobile phase and the parameters included flow rate at 1000 ml/min, injection volume of 100 ul, increment volume of 0.00333 ml, and detector and column temperature of 45°C. The injected volume was always 100 μL and flow velocity was 1 cm^3 min^{-1}. PLA samples were used as standard.

2.8. Scanning Electron Microscopy. SEM was used to observe the surface morphology of PLGA plates during the *in vitro* degradation. At each time point (i.e., 0, 15, 30, 60, and 120 days), PBS was removed from the samples and immersed in distilled water for 2 h. After that, the samples were kept under vacuum for SEM observation. The samples ($n = 2$) were covered with a thin layer of gold/palladium using a cathodic spray (Diode Sputtering System, International Scientific Instruments) and observed at different magnifications (original ×13) with an acceleration voltage of 10 kV in a scanning electron microscope (SEM) Jeol JSM-6390L model.

2.9. Statistical Data Treatment. Analysis of variance (ANOVA) was performed considering a statistical significance set at 0.05; p value was investigated for significance of the factors among melt temperatures. All data are reported as mean ± standard deviation.

3. Results

3.1. Mechanical Tests Analyses. The flexural stiffness, flexural strength, and maximum flexural strain of PLGA_ low and PLGA_highT are summarized in Table 2.

Flexural stiffness values, E, showed 2.2 ± 0.1 and 1.9 ± 0.1 GPa for plates without degradation (0 days) for PLGA_lowT and PLGA_highT, respectively, and the flexural strength, σ_r, was 41.4±11.8 and 30.1±3.1 MPa, respectively. In the flexural test, PLGA plates presented break under all conditions. This is probably due to the presence of crystallinity and chain organization during the solidification of the material in the injection molding process. PLGA_lowT plates support additional load, exhibiting greater flexural strength at 0 days of degradation. Flexural curves for PLGA_lowT and PLGA_highT during degradation are shown in Figure 1.

Mechanical properties decrease especially after 30 days studied. DMA curves of PLGA_lowT and PLGA_highT during the degradation study are shown in Figure 2. Storage modulus (E') showed comparable trends and values for the two injection molding temperatures. The storage modulus E' (T_g at 37°C) was 1.2 ± 0.2 GPa for both conditions in plates without degradation. The behavior of tan δ was different for the two conditions processed. In particular, the highest values achieved for PLGA_highT are related to a higher loss modulus. This behavior suggests a greater contribution of the viscous component of PLGA_highT than PLGA_lowT in thermomechanical behavior. T_g detected as peak in the tan δ trend was 57.4 ± 1.8°C for both conditions.

Storage modulus E' is the component related to the elastic energy stored; and E'' is related to dissipated viscous energy. Both properties decreased significantly at point 3 (60 days).

3.2. Differential Scanning Calorimeter Analysis. Values of crystallinity (X_c), transition temperatures (T_g, T_{m1}, T_{m2}, T_c), and enthalpies detected by DSC (ΔH_{cc}, ΔH_m, ΔH_g) are shown in Table 3. The values correspond to PLGA_lowT and

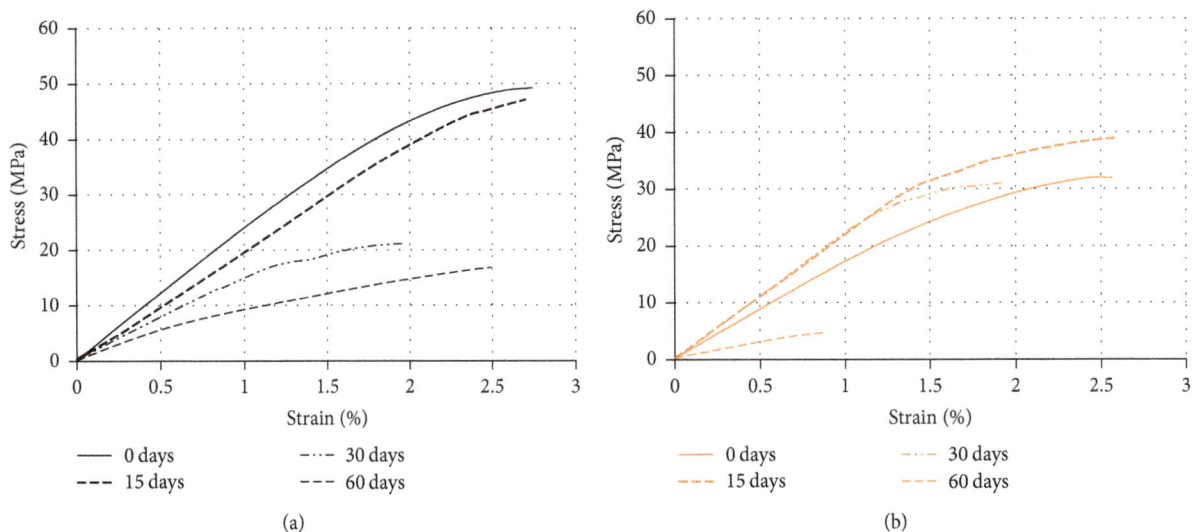

FIGURE 1: Representative stress-strain curves of PLGA_lowT (a) and PLGA_highT (b) at all degradation time points obtained by flexural tests.

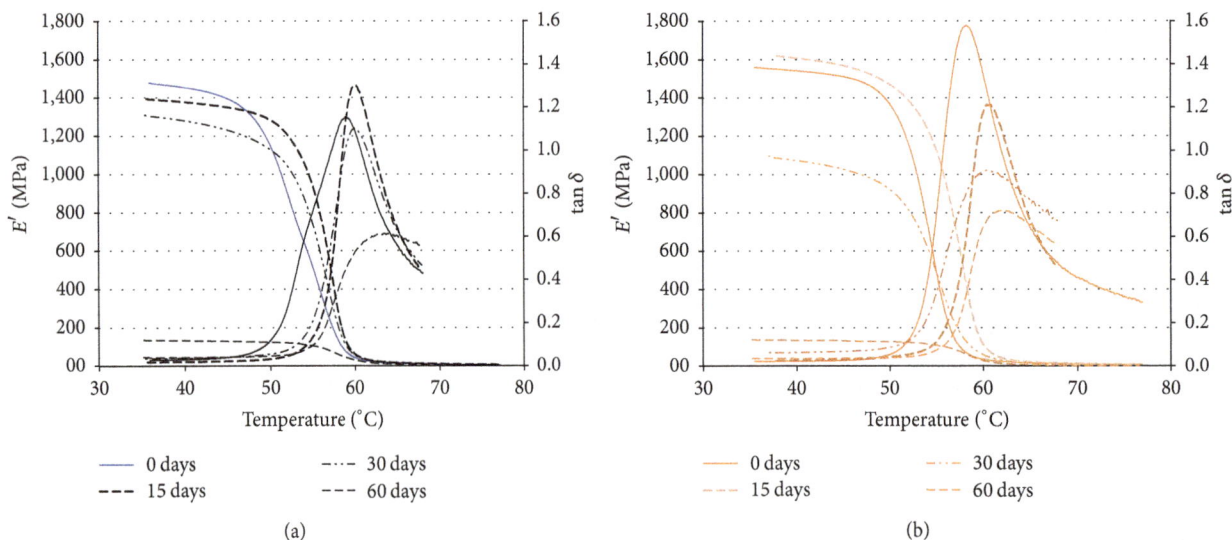

FIGURE 2: Representative curves of storage modulus (E') and tan $\delta(E''/E')$ as a function of the temperature obtained by DMA of the different degradation points (e.g., 0, 15, 30, 60, and 120 days) for PLGA_lowT (a) and PLGA_highT (b).

PLGA_highT, (i.e., 0, 15, 30, 60, and 120 days) and for PLGA_pellet as a control.

The glass transition temperature occurs at 56.0 ± 0.5°C for PLGA_lowT and at 52.7 ± 0.9°C for PLGA_highT at the first point studied without degradation (i.e., 0 days). The glass transition temperature of the preprocessed PLGA samples (pellet) was detected at 60.4 ± 1.8°C. Changes in transition temperatures for PLGA_lowT and PLGA_highT, compared to PLGA_pellet, may be referred to changes related to the molecular chain during the injection molding processing. The endothermic melting peak of PLGA_lowT appeared at 156.6 ± 0.2°C. Considering PLGA_highT, this peak occurs twice in one shoulder at 151.0 ± 2.2°C and 157.9 ± 1.1°C, due to differences in crystallinity. Percent crystallinity was measured by the calculation of (2).

The DSC curves show the difference in the endothermic peaks and the presence of high crystallinity only for PLGA_highT. In addition, there is shoulder presence indicating two melt temperatures (146 and 157°C), related to the PGA (15%) and PLA (85%) fractions, respectively. For PLGA_lowT, a very mild melting endotherm occurs at 153.1 ± 0.3°C, which could indicate a second T_m, even if this value is very close to the detected T_m.

The decrease presented by Tg of the copolymer (Figure 3), as a function of the degradation time, in the period from 0 to 120 days (from 56.0 ± 0.5°C with 0 days to 41.7 ± 6.6°C for PLGA_lowT in 120 days and from 52.7 ± 0.9 to 41.6 ± 0.8°C for PLGA_highT) indicates rapid hydrolytic degradation of plates in PBS medium. In fact, this is characteristic of PLGA copolymers. At the beginning of the degradation process, T_g

TABLE 3: Degree of crystallinity (X_c) and thermal properties measured by DSC: glass-transition temperature (T_g), melting point (T_m), enthalpy of cold crystallization (ΔH_c), enthalpy of glass (ΔH_g), and enthalpy of melting (ΔH_m) ($^*p < 0.05$).

		T_g [°C]*	T_{m1} [°C]	T_{m2} [°C]	T_c [°C]*	ΔH_{cc} [J/g]*	ΔH_m [J/g]*	ΔH_g [J/g]*	X_c [%]*
	Pellet	60.4 ± 1.8	—	146.1 ± 0.1	—	—	33.4 ± 2.4	1.7 ± 0.4	35.7
PLGA_lowT	0 days	56.0 ± 0.5	153.1 ± 0.3	156.6 ± 0.2	130.7 ± 2.4	−2.3 ± 0.6	4.3 ± 1.5	5.1 ± 0.5	7.1
	15 days	53.0 ± 3.2	153.8 ± 2.2	157.5 ± 1.8	131.0 ± 0.1	−8.9 ± 2.2	5.0 ± 2.6	5.9 ± 1.3	14.8
	30 days	51.8 ± 1.2	153.6 ± 0.9	156.4 ± 1.0	99.9 ± 7.7	−17.5 ± 3.1	14.5 ± 6.9	9.6 ± 4.0	34.1
	60 days	48.3 ± 0.3	138.7 ± 4.2	147.3 ± 5.4	87.4 ± 4.1	−35.3 ± 7.8	28.6 ± 6.8	7.1 ± 12.0	68.2
	120 days	41.7 ± 6.6	120.4 ± 0.2	133.5 ± 1.0	95.5 ± 0.1	−30.2 ± 0.9	19.7 ± 8.7	14.1 ± 3.8	53.3
PLGA_highT	0 days	52.7 ± 0.9	151.0 ± 2.2	157.9 ± 1.0	124.5 ± 4.2	−24.0 ± 8.1	17.9 ± 5.1	3.1 ± 0.3	44.7
	15 days	52.5 ± 3.1	154.1 ± 0.1	157.2 ± 2.0	127.1 ± 14.5	−9.9 ± 4.3	7.9 ± 3.0	9.1 ± 0.4	19.0
	30 days	52.6 ± 0.8	152.0 ± 0.1	156.6 ± 0.4	102.1 ± 8.3	−68.6 ± 3.6	19.1 ± 7.1	10.5 ± 0.5	93.7
	60 days	52.0 ± 1.1	153.7 ± 0.7	157.7 ± 1.7	89.0 ± 3.9	−47.9 ± 2.2	7.9 ± 7.0	11.8 ± 6.5	59.6
	120 days	41.6 ± 0.8	144.4 ± 2.8	152.4 ± 3.2	96.6 ± 1.3	−28.9 ± 6.9	23.2 ± 4.1	0.4 ± 0.1	55.7

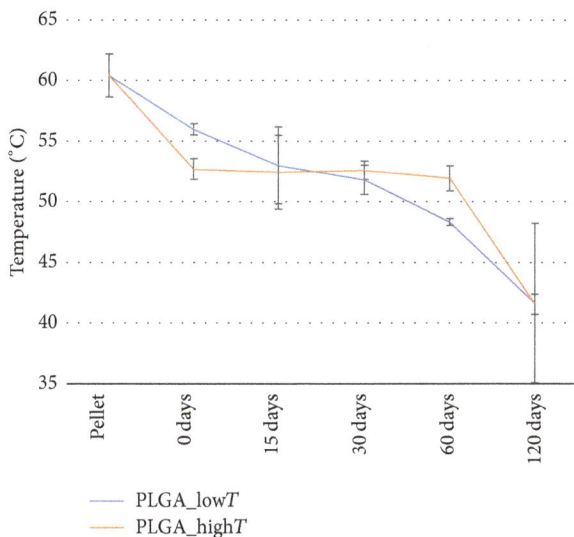

FIGURE 3: Glass transition temperatures (Tg) measured by the DSC for PLGA_lowT and PLGA_highT.

TABLE 4: Absorption bands identified in the FT-IR assay characteristic of the functional groups present in the PLGA copolymer.

Absorption bands (cm^{-1})	Groups
3000–2850	CH, CH$_3$ e CH$_2$
1760–1745	C=O
1600–1500	O-C=O (oligomer)
1450–1370	CH$_3$ and CH$_2$
1350–1150	CH$_2$ and CH
1300–1150	C-O
800–750	CH

decrease could be associated with the plasticizing effect of the plates by the H$_2$O absorption [17]. Moreover, there is an advanced degree of degradation of plates from 60 days.

3.3. Fourier Transform Infrared Spectroscopy.

Absorption bands were identified in the spectrum of the PLGA_lowT and PLGA_highT plates at the different degradation periods: an intense band between 1760 and 1750 cm^{-1}, characteristic of carbonyl (C=O), present in the two monomers, and a bonding band (C-O) between 1300 cm^{-1} and 1150 cm^{-1}, characteristic of the ester groups. The absorption bands that are characteristic of the functional groups present in the PLGA copolymer can be observed in the spectra (Figure 4) and are shown in Table 4.

The presence of the O-C=O group near the absorption band 1600–1500 cm^{-1} indicates signs of degradation; note that the presence of this band occurs in 60 and 120 days of degradation in PBS solution. Usually the appearance of

a band at 3400 cm^{-1} relating to vibration of OH bands in groups -COO-H and -CO-H at the ends of strings is observed, indicating a reduction in their size, and also the occurrence of the peak at 1605 cm^{-1} corresponds to the asymmetric stretch of the -COO- group in oligomers.

3.4. Gel Permeation Chromatography.

Figure 5 shows the molar mass distribution (e.g., M_n, M_w, M_z, and M_p) of the PLGA studied for the pellet and for the other times of degradation.

Based on the results presented, PLGA plates have a narrow (low polydispersity) and monomodal (only one peak) molar mass distribution. The pellet represents unprocessed PLGA, in which high molecular mass (e.g., 183407 ± 28895 Da) was identified. As shown in Figure 5, values of molar mass M_n decrease from pellet to 0 days of degradation, which can be explained by material processing during injection molding.

PLGA plates' degradation can be observed by mass loss during the time of contact with the phosphate buffer solution; however, the degradation speed between PLGA_lowT and PLGA_highT was different. This behavior probably occurred due to the reduction of the amorphous phase of the polymers, since the molecular interaction of the solution with the copolymer is more propitious in the amorphous phase. According to Middleton and Tipton (2000) [12], in the first stage of degradation, there is diffusion of water in the amorphous regions of the polymer and hydrolytic cleavage

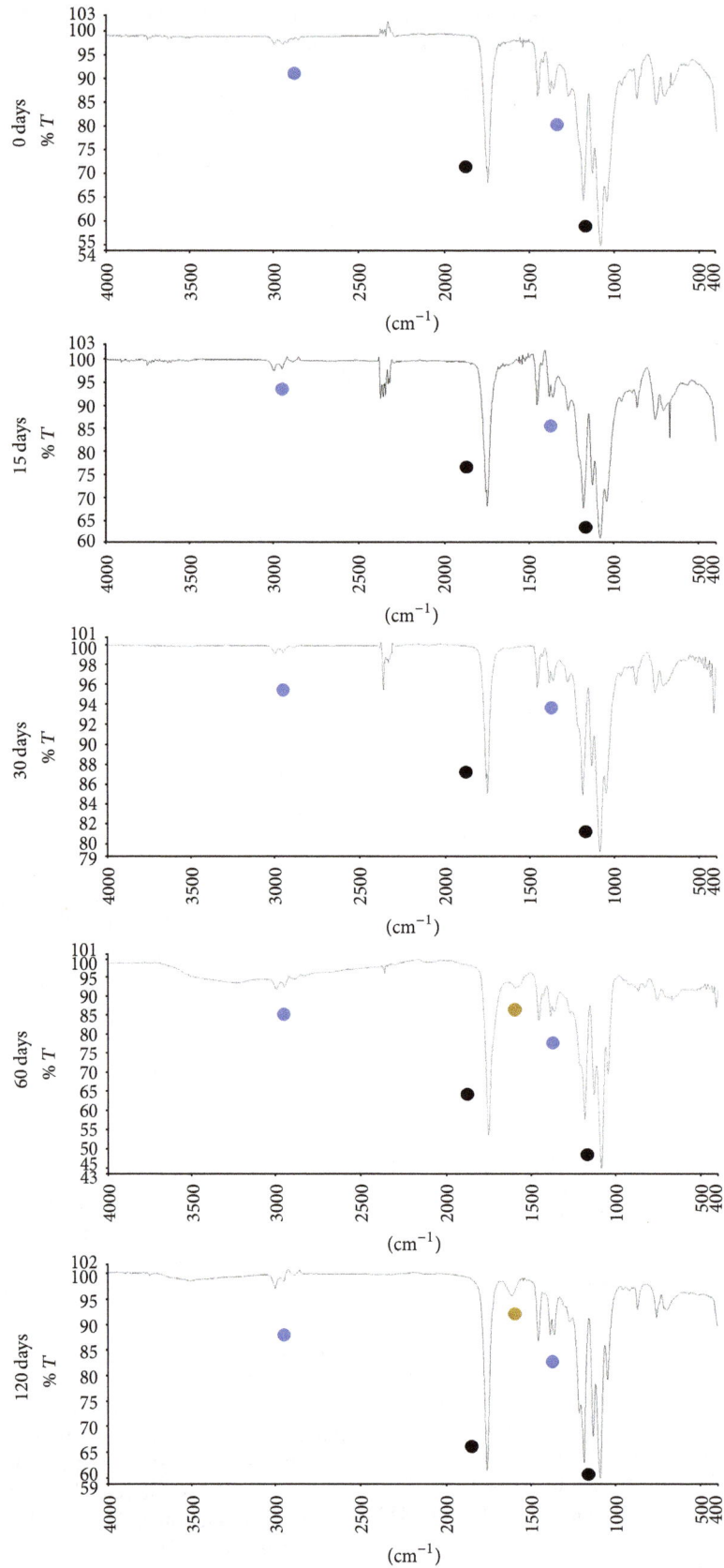

FIGURE 4: Spectrum related to degradation points 0 to 4 of PLGA plates for the period of 0, 15, 30, 60, and 120 days, respectively. Ester groups = red: 1760–1745 cm^{-1} and 1300–1150 cm^{-1}; alkanes groups = blue: 3000–2800 cm^{-1} and 1450–1370 cm^{-1}; oligomer groups = yellow: 1600–1500 cm^{-1}.

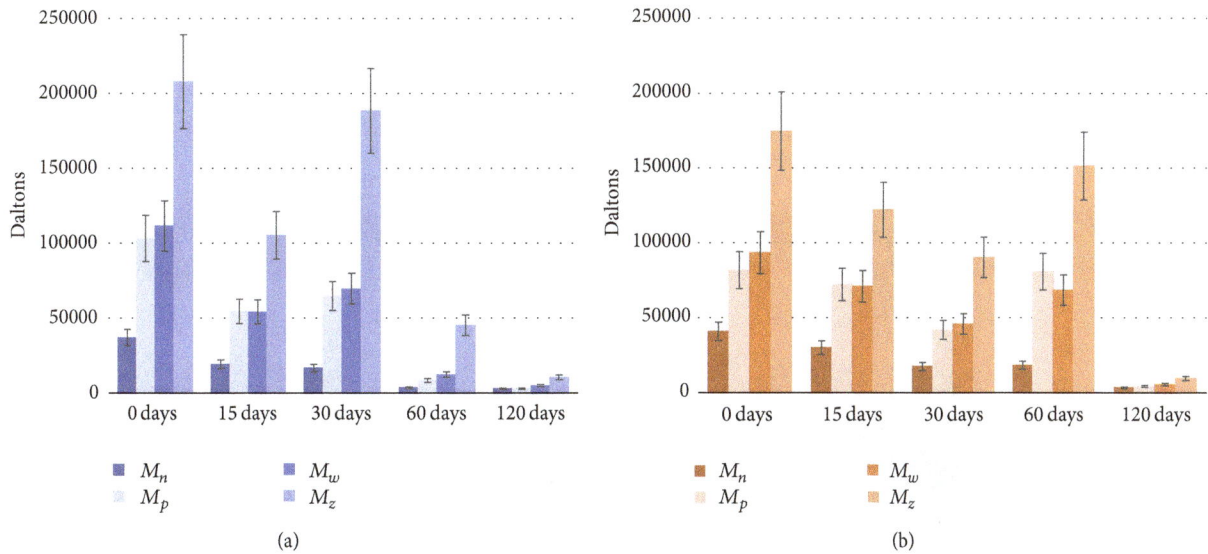

FIGURE 5: Molar mass (mean ± deviation) of the PLGA plates in the 0, 15, 30, 60, and 120 days of degradation in PBS solution for PLGA_low$T$ (a) and PLGA_high$T$ (b).

of the ester bonds of the polymer chains. After much of the amorphous phase undergoing degradation begins the second stage in the crystalline phase; therefore, there is a percentage increase in the degree of crystallinity.

3.5. Scanning Electron Microscopy. PLGA plates' degradation during the immersion period in phosphate buffer solution (PBS, 37°C) is qualitatively confirmed by the SEM images (magnification ×13), as shown in Figure 6.

Figure 7 shows the material degradation process for PLGA_low$T$ and PLGA_high$T$, which coincides with the results obtained from the SEM analyses.

Initially transparent PLGA plates become opaque when degradation process begins. The whitish aspect of the device is clear in the first 15 days in phosphate buffer solution, which is an indication of the influence of the degradation process, as a function of the organization of the chains during the degradation. During the process, the deformation and brittle feature of the material can be noted. After 120 days, it was possible to notice the crumbling of the material and the absence of mechanical properties. The whitish form can be noted in point 1 for PLGA_high$T$, which is less evident in PLGA_low$T$. That could indicate that PLGA_high$T$ is more sensitive to hydrolytic degradation in less time.

4. Discussion

PLGA copolymer is a promising material for medical devices applications as craniofacial plates. Medical devices manufactured from aliphatic polyesters degrade through hydrolysis of the polymer backbone primarily through a bulk degradation process that includes decline of molecular weight, reduction in mechanical properties, and loss of mass. Hydrolytic degradation can be evaluated through molar mass (GPC), presence of polar groups or oligomers and monomers (FTIR), and changes in mechanical properties (DMA), transition

temperature changes (DSC), and surface and geometry (SEM) [21–23]. This study reported the *in vitro* degradation of PLGA craniofacial plates tested in two different melt temperatures of the injection molding process (i.e., 240 and 280°C), corresponding to the processable limits of PLGA, at different time points. Physicochemical properties such as molecular weight and mechanical properties were monitored by FTIR, DSC, and DMA analysis. The results suggested that the property changes differ according to the injection molding temperature.

The flexural strengths for the PLGA plates studied in this work ranged from 1.9 ± 0.1 to 2.2 ± 0.1 GPa, which compared with the stiffness (E) of bone (E_{bone} ~ 6–20 GPa), metal (E_{metal} = 100–200 GPa), and poly(lactic acid) (11–72 MPa) [24], indicating a possible use of these plates under investigation in non-load-bearing body sites, such as for craniofacial bone fractures. However, bioresorbable plates PGA-based copolymers have higher degradation rate than other bioresorbable polymers, which limits the useful time of the devices [25, 26]. Both PLGA plates conditions (i.e., PLGA_low$T$ and PLGA_high$T$) showed rapid degradation, regardless of the different characteristics of the microstructure during the degradation.

The evaluated mechanical properties of PLGA_low$T$ and PLGA_high$T$ showed suitable values at the beginning of degradation [27–30]. However, PLGA_high$T$ plates reached flexural strength peak and maximum flexural strain after 15 days, as observed in Figure 1. A possible reason for this result is related to the diffusion effect order from PBS solution to PLGA plates: first, the diffusion occurs to lower molecular weight chains of PLGA and then to higher molecular chains. In addition, degradation of resorbable polymers begins from higher molecular weight to lower molecular weight chains over time on PBS solution. In addition, smaller polymer chains can rearrange and relax over time before larger chains degrade. Thus, the diffusion to the solution of lower

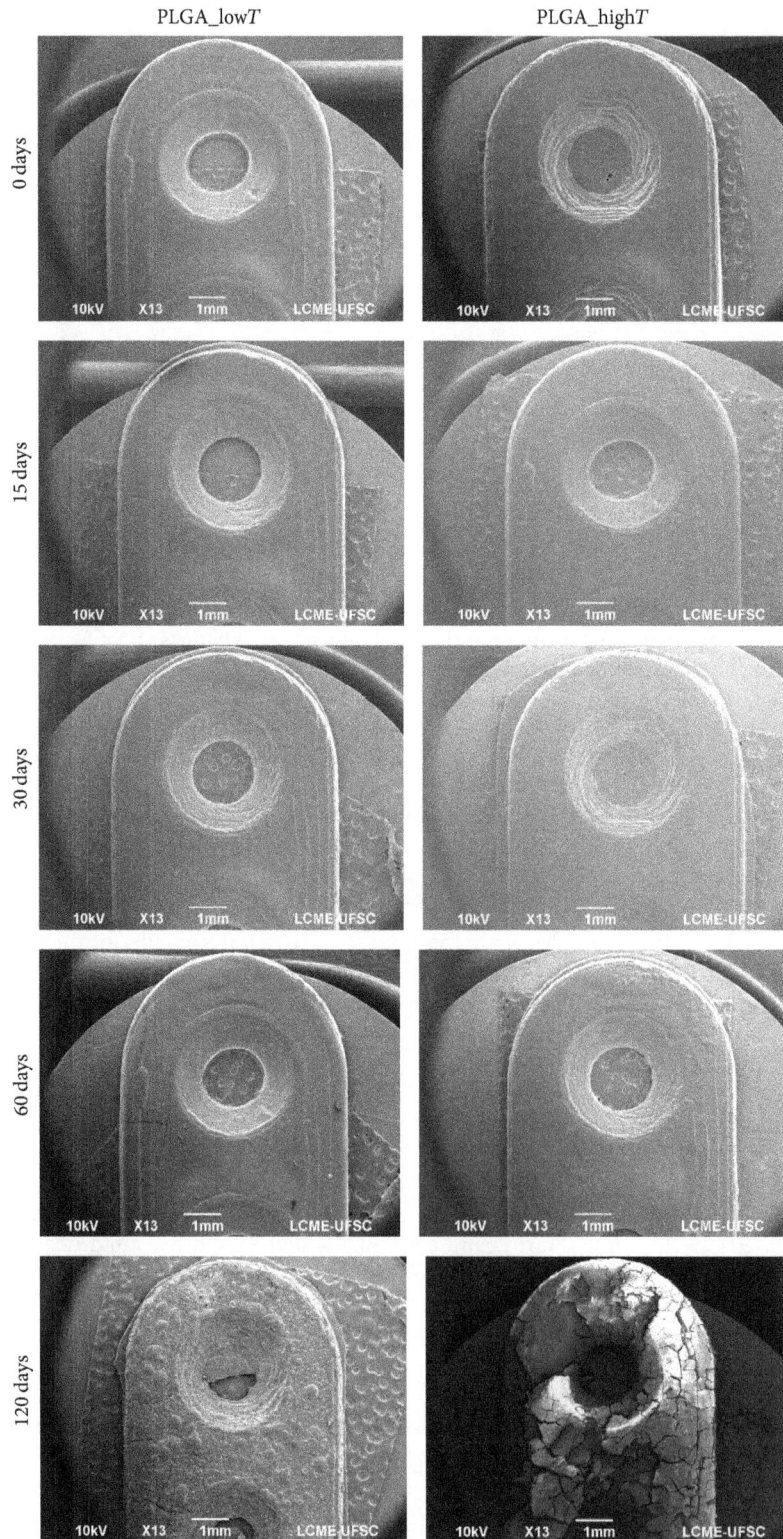

FIGURE 6: SEM images of the PLGA_lowT and PLGA_highT plates at different degradation time (0, 15, 30, 60, and 120 days) (bar scale = 1 mm).

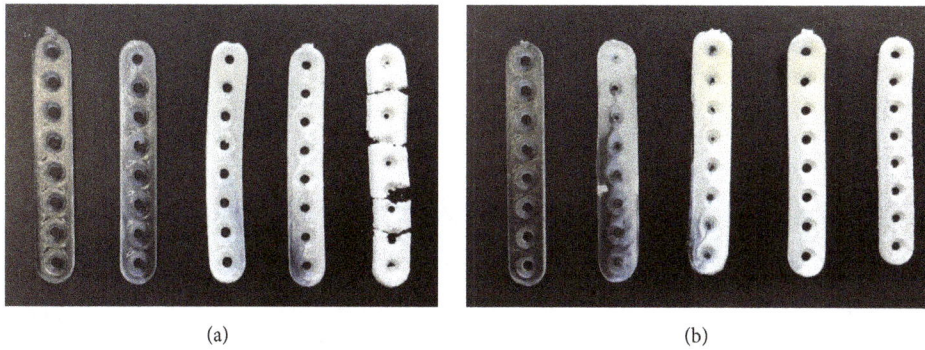

(a)   (b)

FIGURE 7: Visual appearance images of the degradation of the PLGA_low$T$ (a) and PLGA_high$T$ (b) plates at the different degradation points: 0, 15, 30, 60, and 120 days from left to right.

molecular weight chains of PLGA prior to degradation of the higher molecular weight fractions renders the material stiffer after 15 days but with continuing decrease after degradation of the material.

Moreover, Table 3 shows crystallinity and thermal properties on degradation of the plates which could be associated with chains hydrolysis, diffusion, and erosion. At the beginning of the degradation, the phenomenon of surface erosion can be observed, where molar weight loss is exclusively from the outside to the inside of the material, where diffusion of the water molecules, for example, is slower than the release of fragments from the surface. In another case, the volumetric degradation occurs when water penetrates the polymer matrix homogeneously, causing hydrolysis throughout. In this event, there is a relationship between hydrolysis of the chains and their diffusion and erosion. If any disturbance occurs, this equilibrium may be undone and a variation of the mechanism known as autocatalysis via carboxylic and hydroxyl groups may occur. This autocatalysis in volumetric degradation causes an acid gradient in the inner part of the body, causing accelerated degradation to occur at this site compared to the surface. The oligomers generated in the central regions can easily diffuse to the surface. This effect, accompanied by the presence of acid products, may result in inflammatory reactions in *in vivo* cases. It is worth mentioning that the degradation of devices implanted in the human body, an object of interest in this work, tends to present an increase in the rate of diffusion and consequent degradation due to the body temperature around 37°C, variations in pH, and eventual efforts which may increase the probability of breaking connections [13, 18, 19, 31]. An increase in molecular weight will result in more covalent bonds and thus an increased number of entanglements and thereby increasing resorption/degradation time [32].

The decrease of properties (Figure 8) was evident at each time point of the degradation for PLGA_low$T$; however, viscoelasticity properties possibly influenced the dispersion for PLGA_high$T$ in the first 60 days of degradation, which showed stable values of the properties found, with drop in the last two points.

In DSC curves, it was evident that the PLGA_high$T$ plates would have greater propensity to crystallize than PLGA_low$T$ plates due to their greater steric regularity

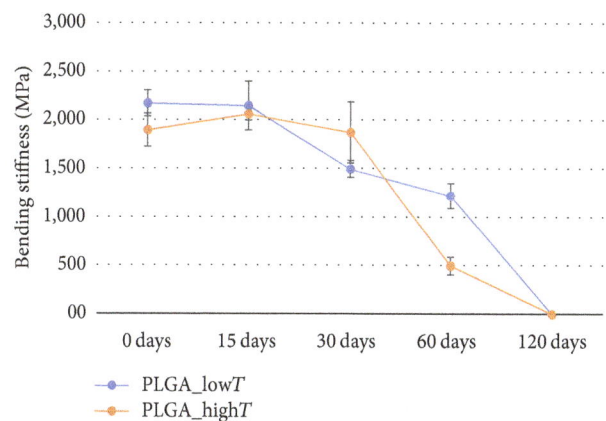

FIGURE 8: Bending stiffness along the hydrolytic degradation of PLGA_low$T$ and PLGA_high$T$ plates.

along the polymer chain during processability [33]. Injection molded PLGA_high$T$ and rapid cooling had the effect to reduce T_g by about 4°C. Furthermore, the cooling time (i.e., 90 s) of injection molding was the same for both plates' conditions. It means that the cooling rate to reach the temperature of the mold (23°C) was faster for PLGA_high$T$ plates. Because of this difference, different crystalline phases were formed, which may have resulted in the formation of different sizes of spherulites and irregular crystals in polymer structure [26, 34, 35].

Moreover, several polymers properties that are important in terms of their processability and applications are directly related to the specific molar mass. That could be related to the fact that mechanical, chemical, and physical properties are drastically affected by the crystallinity and especially by the low and high molar mass fractions. Devices for this application need to be deeper investigated to overcome complications on manufacturing and designs that could influence the degradation rate after placement such as properties stability.

5. Conclusion

We have proposed a work limit of temperatures (low and high) on medical devices as PLGA craniofacial plates manufactured by injection molding and tested biomechanical function degradation of the two conditions of melt process temperature. Both working temperatures allowed producing craniofacial plates devices. At low and high temperature conditions (i.e., 240 and 280°C, resp.), the PLGA plates were evaluated for mechanical properties (apparent elastic modulus, maximum stress, and storage modulus) and crystallinity. The mechanical properties (i.e., 2.2 ± 0.1 and 1.9 ± 0.1 GPa of flexural stiffness) of the plate are suitable for osteosynthesis in non-load-bearing anatomical sites (e.g., craniofacial applications). The differences in crystallinity showed that we can choose the plate with degradation kinetics more suitable for the application. Based on all these results, we can conclude that the proposed process temperatures are adequate for the manufacture of PLGA craniofacial plates. In addition, the knowledge presented is useful to better understand the working limits of bioresorbable implants and the development of implant geometries with property control.

Conflicts of Interest

The authors declare that there are no conflicts of interest regarding the publication of this paper.

Acknowledgments

The authors would like to acknowledge PRONEX/FAPESC, CNPQ, and FINEP for financial support.

References

[1] R. Suuronen, P. E. Haers, C. Lindqvist, and H. F. Sailer, "Update on bioresorbable plates in maxillofacial surgery," *Facial Plastic Surgery*, vol. 15, no. 1, pp. 61–72, 1999.

[2] B. L. Eppley, "Use of resorbable plates and screws in pediatric facial fractures," *Journal of Oral and Maxillofacial Surgery*, vol. 63, no. 3, pp. 385–391, 2005.

[3] H. Peltoniemi, N. Ashammakhi, R. Kontio et al., "The use of bioabsorbable osteofixation devices in craniomaxillofacial surgery," *Oral Surgery, Oral Medicine, Oral Pathology, Oral Radiology, and Endodontics*, vol. 94, no. 1, pp. 5–14, 2002.

[4] R. E. Holmes, S. R. Cohen, G. B. Cornwall, K. A. Thomas, K. K. Kleinhenz, and M. Z. Beckett, "MacroPore resorbable devices in craniofacial surgery," *Clinics in Plastic Surgery*, vol. 31, no. 3, pp. 393–406, 2004.

[5] C. S. Leiggener, R. Curtis, and B. A. Rahn, "Effects of chemical composition and design of poly (L/DLLactide) implants on the healing of cranial defects," *Journal of Cranio-Maxillo-Facial Surgery*, vol. 26, p. 151, 1998.

[6] C. Schiller, C. Rasche, M. Wehmöller et al., "Geometrically structured implants for cranial reconstruction made of biodegradable polyesters and calcium phosphate/calcium carbonate," *Biomaterials*, vol. 25, no. 7-8, pp. 1239–1247, 2004.

[7] S.-H. Hyon, K. Jamshidi, and Y. Ikada, "Effects of Residual Monomer on the Degradation of DL-Lactide Polymer," *Polymer International*, vol. 46, no. 3, pp. 196–202, 1998.

[8] S. Ghosh, J. C. Viana, R. L. Reis, and J. F. Mano, "Effect of processing conditions on morphology and mechanical properties of injection-molded poly(L-lactic acid)," *Polymer Engineering and Science*, vol. 46, no. 7, pp. 1141–1147, 2007.

[9] J. C. Viana, N. M. Alves, and J. F. Mano, "Morphology and mechanical properties of injection molded poly(ethylene terephthalate)," *Polymer Engineering and Science*, vol. 44, no. 12, pp. 2174–2184, 2004.

[10] A. M. Harris and E. C. Lee, "Improving mechanical performance of injection molded PLA by controlling crystallinity," *Journal of Applied Polymer Science*, vol. 107, no. 4, pp. 2246–2255, 2008.

[11] H. Zhao and G. Zhao, "Mechanical and thermal properties of conventional and microcellular injection molded poly (lactic acid)/poly (ε-caprolactone) blends," *Journal of the Mechanical Behavior of Biomedical Materials*, vol. 53, pp. 59–67, 2016.

[12] J. C. Middleton and A. J. Tipton, "Synthetic biodegradable polymers as orthopedic devices," *Biomaterials*, vol. 21, no. 23, pp. 2335–2346, 2000.

[13] S. H. Barbanti, C. A. Zavaglia, and E. A. Duek, "Degradação acelerada de suportes de poli(épsilon-caprolactona) e poli(D,L-ácido láctico-co-ácido glicólico) em meio alcalino," *Polímeros*, vol. 16, no. 2, pp. 141–148, 2006.

[14] L. Lu, S. J. Peter, M. D. Lyman et al., "In vitro degradation of porous poly(L-lactic acid) foams," *Biomaterials*, vol. 21, no. 15, pp. 1595–1605, 2000.

[15] D. Bendix, "Chemical synthesis of polylactide and its copolymers for medical applications," *Polymer Degradation and Stability*, vol. 59, no. 1-3, pp. 129–135, 1998.

[16] S. Li, "Hydrolytic degradation characteristics of aliphatic polyesters derived from lactic and glycolic acids," *Journal of Biomedical Materials Research*, vol. 48, no. 3, pp. 342–353, 1999.

[17] E. W. Fischer, H. Goddar, and G. F. Schmidt, "Determination of degree of crystallinity of drawn polymers by means of density measurements," *Journal of Polymer Science Part A-2: Polymer Physics*, vol. 7, no. 1, pp. 37–45, 1969.

[18] R. T. MacDonald, S. P. McCarthy, and R. A. Gross, "Enzymatic degradability of poly(lactide): Effects of chain stereochemistry and material crystallinity," *Macromolecules*, vol. 29, no. 23, pp. 7356–7361, 1996.

[19] L.-T. Lim, R. Auras, and M. Rubino, "Processing technologies for poly(lactic acid)," *Progress in Polymer Science*, vol. 33, no. 8, pp. 820–852, 2008.

[20] D790 ASTM, *Standard Test Methods for Flexural Properties of Unreinforced and Reinforced Plastics and Electrical Insulating Materials*, ASTM International, West Conshohocken, PA, USA, 2015, https://www.astm.org/.

[21] S. Lyu and D. Untereker, "Degradability of polymers for implantable biomedical devices," *International Journal of Molecular Sciences*, vol. 10, no. 9, pp. 4033–4065, 2009.

[22] D. Zuchowska, D. Hlavatá, R. Steller, W. Adamiak, and W. Meissner, "Physical structure of polyolefin - starch blends after ageing," *Polymer Degradation and Stability*, vol. 64, no. 2, pp. 339–346, 1999.

[23] H. Essig, D. Lindhorst, T. Gander et al., "Patient-specific biodegradable implant in pediatric craniofacial surgery," *Journal of Cranio-Maxillofacial Surgery*, vol. 45, no. 2, pp. 216–222, 2017.

[24] A. U. Daniels, M. K. Chang, K. P. Andriano, and J. Heller, "Mechanical properties of biodegradable polymers and composites proposed for internal fixation of bone," *Journal of Applied Biomaterials*, vol. 1, no. 1, pp. 57–78, 1990.

[25] N. J. Ostrowski, B. Lee, A. Roy, M. Ramanathan, and P. N. Kumta, "Biodegradable poly(lactide-co-glycolide) coatings on magnesium alloys for orthopedic applications," *Journal of Materials Science: Materials in Medicine*, vol. 24, no. 1, pp. 85–96, 2013.

[26] P. Gentile, V. Chiono, I. Carmagnola, and P. V. Hatton, "An overview of poly(lactic-co-glycolic) Acid (PLGA)-based bio-materials for bone tissue engineering," *International Journal of Molecular Sciences*, vol. 15, no. 3, pp. 3640–3659, 2014.

[27] V. Hasirci, K. U. Lewandrowski, S. P. Bondre, J. D. Gresser, D. J. Trantolo, and D. L. Wise, "High strength bioresorbable bone plates: Preparation, Mechanical properties and in vitro analysis," *Bio-Medical Materials and Engineering*, vol. 10, no. 1, pp. 19–29, 2000.

[28] N. Ashammakhi, H. Peltoniemi, E. Waris et al., "Developments in craniomaxillofacial surgery: Use of self-reinforced bioabsorbable osteofixation devices," *Plastic and Reconstructive Surgery*, vol. 108, no. 1, pp. 167–180, 2001.

[29] R. B. Bell and C. S. Kindsfater, "The use of biodegradable plates and screws to stabilize facial fractures," *Journal of Oral and Maxillofacial Surgery*, vol. 64, no. 1, pp. 31–39, 2006.

[30] R. E. Lins, B. S. Myers, R. J. Spinner, and L. S. Levin, "A comparative mechanical analysis of plate fixation in a proximal phalangeal fracture model," *Journal of Hand Surgery*, vol. 21, no. 6, pp. 1059–1064, 1996.

[31] M. A. Woodruff and D. W. Hutmacher, "The return of a forgotten polymer—polycaprolactone in the 21st century," *Progress in Polymer Science*, vol. 35, no. 10, pp. 1217–1256, 2010.

[32] D. D. Wright, *Degradable Polymer Composites. In Encyclopedia of Biomaterials and Biomedical Engineering*, Marcel Dekker Inc, New York, NY, USA, 2004.

[33] W. S. Pietrzak, "Rapid cooling through the glass transition transiently increases ductility of PGA/PLLA copolymers: A proposed mechanism and implications for devices," *Journal of Materials Science: Materials in Medicine*, vol. 18, no. 9, pp. 1753–1763, 2007.

[34] J. Y. Nam, S. Sinha Ray, and M. Okamoto, "Crystallization behavior and morphology of biodegradable polylactide/layered silicate nanocomposite," *Macromolecules*, vol. 36, no. 19, pp. 7126–7131, 2003.

[35] S. Farè, P. Torricelli, G. Giavaresi et al., "In vitro study on silk fibroin textile structure for Anterior Cruciate Ligament regeneration," *Materials Science and Engineering C*, vol. 33, no. 7, pp. 3601–3608, 2013.

2

New Biofunctional Loading of Natural Antimicrobial Agent in Biodegradable Polymeric Films for Biomedical Applications

Bakhtawar Ghafoor, Murtaza Najabat Ali, Umar Ansari, Muhammad Faraz Bhatti, Mariam Mir, Hafsah Akhtar, and Fatima Darakhshan

Biomedical Engineering and Sciences Department, School of Mechanical and Manufacturing Engineering (SMME), National University of Sciences and Technology (NUST), Islamabad, Pakistan

Correspondence should be addressed to Murtaza Najabat Ali; murtaza_bme@hotmail.com

Academic Editor: Rosalind Labow

The study focuses on the development of novel *Aloe vera* based polymeric composite films and antimicrobial suture coatings. Polyvinyl alcohol (PVA), a synthetic biocompatible and biodegradable polymer, was combined with *Aloe vera*, a natural herb used for soothing burning effects and cosmetic purposes. The properties of these two materials were combined together to get additional benefits such as wound healing and prevention of surgical site infections. PVA and *Aloe vera* were mixed in a fixed quantity to produce polymer based films. The films were screened for antibacterial and antifungal activity against bacterial (*E. coli, P. aeruginosa*) and fungal strains (*Aspergillus flavus* and *Aspergillus tubingensis*) screened. *Aloe vera* based PVA films showed antimicrobial activity against all the strains; the lowest *Aloe vera* concentration (5%) showed the highest activity against all the strains. *In vitro* degradation and release profile of these films was also evaluated. The coating for sutures was prepared, *in vitro* antibacterial tests of these coated sutures were carried out, and later on *in vivo* studies of these coated sutures were also performed. The results showed that sutures coated with *Aloe vera*/PVA coating solution have antibacterial effects and thus have the potential to be used in the prevention of surgical site infections and *Aloe vera*/PVA based films have the potential to be used for wound healing purposes.

1. Introduction

Nosocomial infections are hospital-acquired infections (HAI) that usually develop in patients during their hospital stay, affecting the health expenditure of the patient [1]. The main factors that make patients prone to nosocomial infection include concurrent infections, medical devices, surgery, immunosuppressive agents, and emergence of multidrug resistant pathogens. Pathogens are responsible for such infections known as nosocomial pathogens. Among them 90% bacterial pathogens are involved; however mycobacterial, viral, fungal, or protozoal agents are less commonly involved [2]. According to the data, *Escherichia coli, Staphylococcus aureus*, enterococci, and Pseudomonas *aeruginosa* are the most common nosocomial pathogens [3]. Among the fungal pathogens, *Candida albicans* [4], *Aspergillus* spp., and especially *Aspergillus fumigatus, A. flavus*, and *A. terreus*

have also been reported as the common cause of nosocomial infection in highly immunocompromised patients. These pathogens can be transmitted through either inhalation or direct contact with occlusive materials [5, 6].

One of the reasons of nosocomial infections is surgical site infections (SSIs) mainly caused due to infected suture materials used in surgery and medical implants [7]. These infections are usually difficult to resolve and may cause complications in extreme cases. In order to prevent surgical site infections, scientists have been using several natural and synthetic materials like plant extracts and polymers which may be used as coating materials on surface of medical devices such as surgical implants or sutures [8]. The addition of antibiotics to these coating biomaterials can provide the local delivery of antibiotic directly at implantation or suture site, thereby decreasing the onset of infection [9]. Synthetic and natural biomaterials have also been used in other biomedical applications such as drug delivery systems, wound infections, and antitumor and anti-inflammatory agents [10].

Among synthetic biomaterials, one of the extensively used polymers is poly(vinyl alcohol) (PVA). Due to its suitable chemical and physical properties, biocompatibility, biodegradability, easy preparation with excellent film forming properties, and nontoxic nature, PVA has been studied intensively in different biomedical applications including wound dressings, contact lenses, coatings for sutures, and catheters [11, 12].

Aloe vera, as a natural source of bioactive compounds, is widely studied for biomedical applications. *Aloe vera* belongs to the Liliaceae family and is known as the oldest therapeutic herb. It has the ability to promote wound healing as well as treat burnt areas on the skin [13, 14]. Due to its properties, many researchers have shown the antibacterial, antiviral, antitumor, and anti-inflammatory activity of different parts of *Aloe vera* such as its stem, root, and leaf extracts [15–17]. The chemical composition of *Aloe vera* has also proved its potential use in cosmetic formulations, food supplements, and medical devices [15, 18, 19].

The inner part of *Aloe vera* contains a clear mucilaginous tissue commonly known as *Aloe* gel. The major portion of *Aloe* gel contains water while almost 1% contains bioactive compounds such asaloin, emodin (anthraquinones), flavonoids, saponin, and *Aloe*-mannan along with many different amino acids and vitamins. These bioactive compounds play a major role for antibacterial activity of *Aloe* gel [15, 20, 21].

The present work focuses on the antibacterial and antifungal activity of *Aloe vera*/PVA composite membranes and the application of these blends in the prevention of nosocomial infections; for the specific purpose of investing this, sutures coated with the PVA/*Aloe* gel blend have been used for both *in vitro* and *in vivo analysis*. *Aloe vera*/PVA films have been characterized through SEM and FTIR analysis. The results of *in vitro* and *in vivo* analysis proposed that this composition can be used as a coating for the prevention of surgical site infections that are caused by infections through sutures. To our knowledge, such biofunctional *Aloe* loaded PVA coatings have not been investigated against surgical infection causing bacteria in recent studies. The results of our study indicate the potential of such coatings as part of larger preventive measures against surgical infections.

2. Materials and Methods

2.1. Collection of Plant Material. Fresh *Aloe vera* plants were collected from local nurseries and the leaves washed well with distilled water to remove all contaminants present at the surface. The gel was harvested from the leaves in an autoclaved container and kept at room temperature for further use; the storage time of the gels at room temperature was one minute. During this time period, any residual solid leaf particles were mechanically separated from the gel.

2.2. Test Organisms for In Vitro Studies. In order to investigate antimicrobial and antifungal activity (*in vitro* studies),

pure cultures of bacterial and fungal strains including *Pseudomonas aeruginosa (P. aeruginosa)*, *Escherichia coli (E. coli)*, *Aspergillus tubingensis,* and *Aspergillus flavus* were obtained from Mycovirus Research Lab, National University of Sciences and Technology (NUST) H-12, Islamabad. The pure bacterial and fungal cultures were stored in nutrient agar at 4°C.

2.3. Suture Materials. Commercially available silk braided black surgical sutures, nonabsorbable (1.5 metric, size 4-0) supplied by Foosin Medical supplies Inc., Ltd., Shandong, China and manufactured by WEGOSUTURES, were used to carry out *in vitro* and *in vivo* studies. The suture material was delivered in sterile single peelable foil packages and stored at room temperature. For investigation, the sutures were cut into defined lengths (1 cm) under aseptic conditions.

2.4. Preparation of Aloe vera Based PVA Films. Polyvinyl alcohol (PVA) supplied by AppliChem, Germany, a biocompatible polymer, was used for the formation of polymer/*Aloe vera* films [22]. Dimethyl Formamide (DMF) manufactured by TEDIA Company Inc, USA, was selected as solvent for the formation of PVA-*Aloe vera* films, due to its high volatility.

Solvent-casting method was used for the fabrication of *Aloe vera* gel/PVA films. 1 g of PVA was dissolved in 40 mL of DMF. The solution was stirred with a constant RPM of 580 at 60°C until PVA was completely dissolved, and a clear solution was obtained. This was followed by the addition of different amounts of *Aloe* gel. *Aloe* gel was added in the amounts of 5%, 10%, 15%, and 20%, respectively, for the fabrication of *Aloe vera*/PVA films with varying *Aloe* gel compositions. The heating was turned off while constant magnetic stirring was continued to obtain a homogenized mixture of *Aloe vera* gel and PVA in DMF. The mixture was poured into Petri dishes and placed in oven at 37°C for 20 h to evaporate the solvent completely and dry films were harvested for further testing. A solution of PVA in DMF was also prepared by the same procedure to obtain PVA films that were used as control for antimicrobial activity.

2.5. Antimicrobial Testing of Aloe vera Based PVA Films. The antifungal and antibacterial activities of films were evaluated using standard procedure of disc diffusion [23]. For antibacterial activity sterile nutrient agar (pH: 7.4) was prepared using Tryptone 10 g supplied by BioWorld, USA, yeast extract 5 g, supplied by MERCK, Germany, Sodium Chloride 10 g supplied by AnalaR, England, and nutrient agar 12 g supplied by MERCK, Germany, dissolved in 1000 mL of distilled water. After autoclaving the nutrient agar was poured in Petri dishes which were inoculated with the 0.1 mL of bacterial inoculum from preculture of test bacterial strains.

For antifungal investigation, sterile potato dextrose agar was prepared and poured onto the Petri plates and pure fungal cultures were obtained from test fungal strains.

For disc diffusion test, films were cut into discs of about 7 mm in diameter and placed on the bacterial and fungal inoculated plates with certain distances. Each Petri plate contained six discs one of which included the control sterile Whatman filter paper number 1, PVA film, and other four

Aloe vera/polymer based films with varying concentrations of *Aloe vera* (5%, 10%, 15%, and 20%). Antimicrobial activity of pure *Aloe vera* gel was also recorded using well diffusion method. 10 mm of diameter of well was made in solid agar medium in which 0.1 mL of pure *Aloe vera* gel was delivered into the well after incubating the plate with the bacterial strains.

For antibacterial testing a positive control (Tetracycline disc) was used. All plates were incubated at 37°C for 24 h. The zone of inhibition diameter in millimeter (mm) was measured. The study was performed in triplicate and mean was calculated.

2.6. Characterization of Aloe vera Based PVA Films

2.6.1. Fourier Transform Infrared (FTIR) Analysis. Fourier transform infrared (FTIR) spectroscopy (Perkin Elmer, spectrum 100 FTIR spectrophotometer) of *Aloe vera*/PVA films was carried out (at 256 scans, 8 cm^{-1} resolution) to investigate the presence of functional groups and types of interaction between the *Aloe vera* and PVA components.

2.6.2. Morphological Analysis: SEM. Scanning Electron Microscopy (SEM) was performed to find out the surface morphology of the casted films. The assessment of the surface morphology of the *Aloe vera*/PVA based films was done using JSM-6490A Analytical scanning electron microscope (JEOL, Tokyo, Japan). SEM images were collected at an activation voltage of 20 kV.

2.7. In Vitro Degradation and Aloe Release Profile Testing of Aloe vera Based PVA Film. The degradation profile was assessed by recording weight differences after regular time intervals while *Aloe* release profile of *Aloe vera*/PVA films was assessed through UV-Vis spectrophotometry. A portion of *Aloe vera*/PVA films with measurable size (1″ by 1″) were cut and placed in 3 mL of PBS (pH 7.4) at 37°C. The remaining PBS was removed after every 10-minute interval and replaced with fresh 3 mL of PBS. The films were weighed before addition of PBS and afterwards they were taken out of the PBS solution, in wet state; the weights were subtracted and recorded. Moreover the drained PBS solutions were evaluated for *Aloe* release profile by UV-VIS spectrophotometer (Systronics 2202) absorbance at $\lambda_{max} = 301$ nm. The degradation and release tests were carried out in triplicate and an average value was calculated.

2.8. Coating for the Sutures. Dip coating method was used to coat the sutures. For dip coating, the solution was prepared by mixing 2 g of *Aloe vera* and 1 g of PVA in 40 mL of DMF. The sutures (30 cm length) were first sterilized and then dipped in the dip coating solution (for 60 minutes) followed by removal and air drying of suture for 24 h. The confirmation of coating of the suture was done by measuring the weight before coating and after coating.

2.9. In Vitro Evaluation of Coated and Uncoated Sutures. The silk sutures (with and without *Aloe vera*/PVA coating) were evaluated *in vitro* for antibacterial activity against two bacterial strains, that is, *E. coli* and *P. aeruginosa*. Nutrient agar media (pH: 7.4) plates were prepared and the coated suture of the size 4 cm was placed over agar. The plates were then inoculated with bacterial strains (*E. coli* and *P. aeruginosa*) and antibacterial activity was recorded.

2.10. In Vivo Evaluation of Coated Sutures. BALB/c mice were purchased from National Institute of Health (NIH) for the *in vivo* analysis of coated sutures. To check the antimicrobial activity of the coated sutures *in vivo*, mice were given an incision of about 2 cm on both sides of the spine. The incision was inoculated with *E. coli* 30×10^6 colony forming unit (CFU) of 100 μL with the help of a syringe. A coated suture was then placed in one incision, whereas an uncoated suture was placed in the other incision. A discontinuous suturing was done to close the incision site. The same procedure was carried with the mice using *P. aeruginosa* (50×10^6 CFU) of 100 μL for inoculation. The entire experiment was performed in triplicate using sterilized instruments. The sutured incision sites were covered with surgical tape for two days. After two days, sutures from both sides of mice were taken out and placed in separate 1.5 mL centrifuge tubes containing 100 μL PBS solution, the sutures were placed on the Petri dishes containing nutrient agar and placed in an incubator at 37°C overnight.

2.11. Statistical Analysis. All the quantitative data were expressed as mean value with standard deviation. The statistical analyses of the results were done by using *t*-test in Graph Pad Prism 6.0 software. The values that were $p < 0.05$ were considered statistically significant value.

3. Results and Discussion

3.1. Scanning Electron Microscopy (SEM). The surface morphology of different films was assessed by SEM which has been demonstrated in Figure 1. The SEM images showed the aggregates of *Aloe vera* dispersed on the surface of films which contributed to the film surface roughness. Similar results have been reported by Pereira et al., while studying the properties of alginate based *Aloe vera* films [8].

3.2. Fourier Transform Infrared (FTIR). FTIR analysis was performed to identify the nature of linkages between PVA and *Aloe vera*. The FTIR spectra of pure *Aloe vera*, PVA, *Aloe vera* in DMF, and *Aloe vera*/PVA films with varying concentrations have been shown in Figure 2. The peak that appeared between 3500 cm^{-1} and 3200 cm^{-1} in all films indicates the presence of hydroxyl group (OH) [24]. The absorption band between 3000 cm^{-1} and 2800 cm^{-1} centered at 2932.68 cm^{-1} in 5% *Aloe vera*/PVA and 2926 cm^{-1} in 20% *Aloe vera*/PVA. Both peaks had shifted from 2922 cm^{-1}; this was a characteristic of asymmetric stretching of CH$_2$ groups [25]. The shift indicated the intermolecular interactions at these functional groups in *Aloe vera* and PVA. The peaks obtained at the range of 1720 cm^{-1} to 1710 cm^{-1} correspond to the stretching of C=O group which indicated the presence of carbonyl compounds in *Aloe vera*. The presence of C-O-C (phenol ether) group was

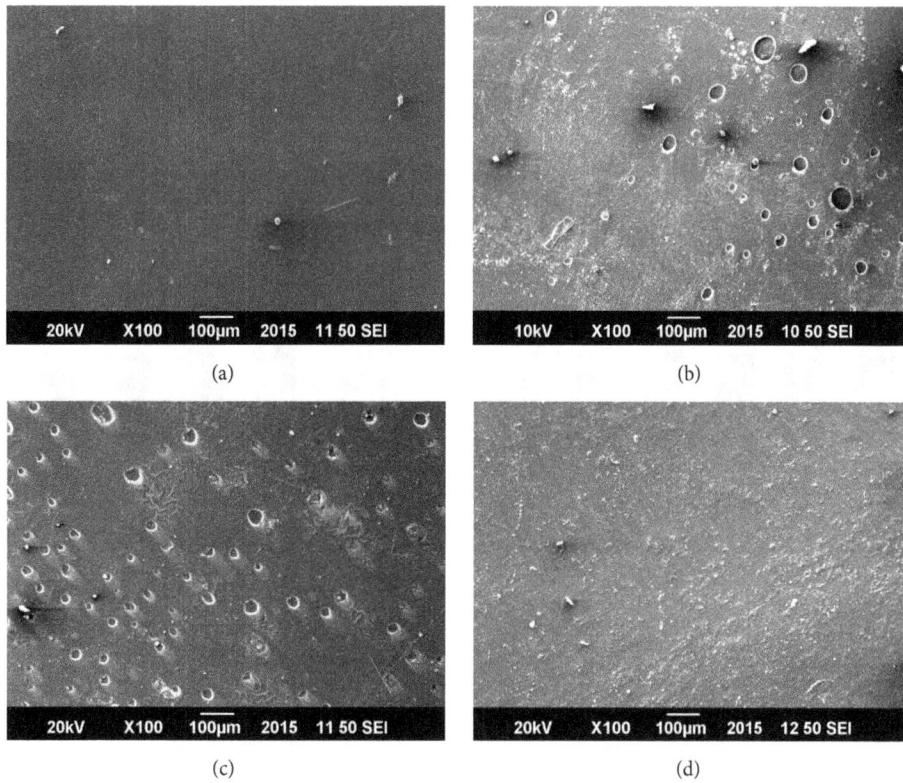

FIGURE 1: SEM images of *Aloe vera*/polymer films. (a) Film with 5% *Aloe vera* concentration at 20 kV and ×100 magnification. (b) Film with 10% concentration at 10 kV and ×100 magnification. (c) Film with 15% at 20 kV and ×100 magnification and (d) with 20% concentration at 20 kV and ×100 magnification.

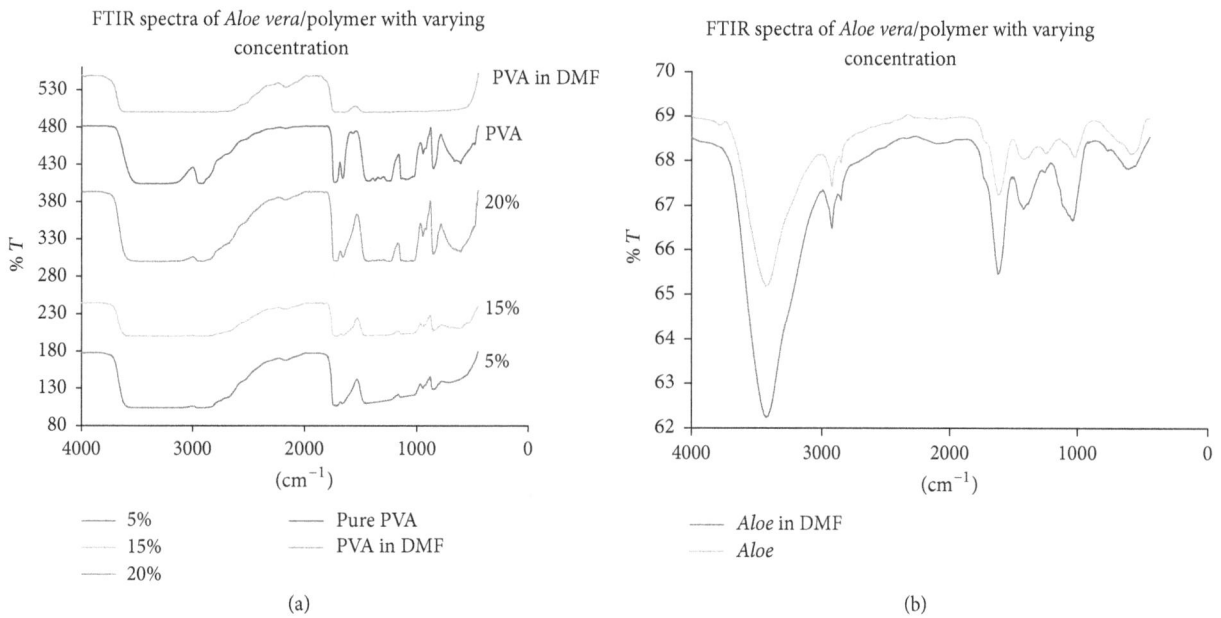

FIGURE 2: FTIR results of (a) PVA film and *Aloe vera*/polymer films with 5%, 15%, and 20% and (b) results of pure *Aloe* and *Aloe* incubated in DMF.

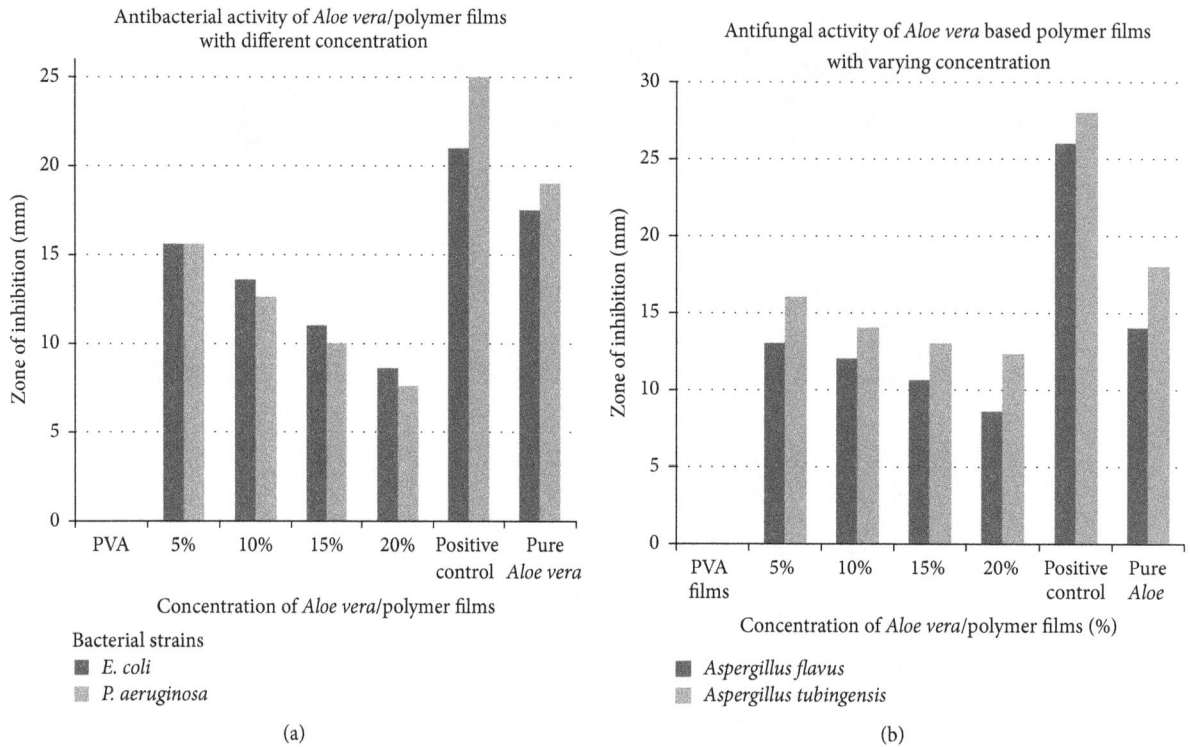

FIGURE 3: Graphical representation of antibacterial (a) and antifungal (b) activity of different concentrations of *Aloe vera*/polymer films. *y*-axis shows zones in mm while *x*-axis shows varying concentration of *Aloe vera*/polymer films.

indicated by the bands located at 1036 cm^{-1} in films having 20% *Aloe vera*/PVA concentration. The peak in pure *Aloe vera* at 1075 cm^{-1} [25] was shifted to 1036 cm^{-1} indicating the presence of C-N functional groups in the films; the shift observed in the peak can be attributed to interactions between amine groups and hydroxyl groups of *Aloe vera* and PVA, respectively [26]. The absorption band 1460 cm^{-1} to 1410 cm^{-1} appeared in all concentrations of *Aloe vera*/PVA films, hence representing symmetric stretching vibrations of COOH groups in films [26]. The broad peak at 1150 cm^{-1} to 1130 cm^{-1} could indicate either (C-O) stretching vibrations in films with concentrations of 5% and 15%. The absorption peaks obtained at 860 cm^{-1} to 840 cm^{-1} correspond to rocking vibrations of CH$_2$ bonds in PVA [27]. The bending of C-H alkyl groups present in *Aloe vera* and PVA at a peak range of 950 cm^{-1} to 940 cm^{-1} can easily be seen in FTIR results. A new peak at 2171.18 cm^{-1} in 5%, 2167.69 cm^{-1} in 15%, and 2168 cm^{-1} in 20% *Aloe vera*/PVA film indicates the occurrence of interactions between CH group of PVA with CH group of *Aloe vera*. The band at 1660 cm^{-1} and 1264 cm^{-1} in 20% *Aloe vera*/PVA film demonstrated the interaction between hydrogen groups and C-O-C of PVA and C=O and C-O-C groups of *Aloe vera* [25, 28, 29].

The occurrence of peaks of COOH, C-H, C-O-C, NH$_2$, and OH shows that the pharmacologically active compounds of *Aloe vera* such as anthraquinones, saponins, and polysaccharides are still in their active form. This can be further correlated with antibacterial activity which confirmed that

the active components are still intact and are not affected through interactions with PVA. Thus, the stability of pharmacologically active moiety of *Aloe* after preparation in DMF and loading in PVA films has been confirmed.

3.3. Antimicrobial Testing Results of Films. The *Aloe vera*/PVA films when positioned on bacterial and fungal inoculated plates gave zones of inhibition which were recorded after 24 h of positioning the films (Figure 3). All films demonstrated the antimicrobial activity due to the release of *Aloe vera* from the surface of the films. The maximum activity was indicated by 5% *Aloe vera*/PVA combination. The potential reason could be the presence of a lower number of interactions between *Aloe vera* and PVA; because of lower concentrations of *Aloe* gel, they were not chemically bound to each other thus keeping the components and their respective functional groups of *Aloe vera* chemically active against microbial activity as shown by FTIR results. Increased levels (10%, 15%, and 20%) of *Aloe vera* in the PVA blend lead to the increased interactions between pharmacologically active components of *Aloe vera* and PVA which causes the shift in FTIR peak (Figure 2); thus such interactions of *Aloe vera*/PVA may influence antimicrobial activity. Renisheya Joy Jeba Malar et al., demonstrated the antimicrobial activity of DMSO extracts of *Aloe vera* gel against human pathogens and highest zone of inhibition (13 mm) against *E. coli* was recorded [30]. In another study, the zone of inhibition against *E. coli*, *P. aeruginosa*, and *Aspergillus flavus* was recorded

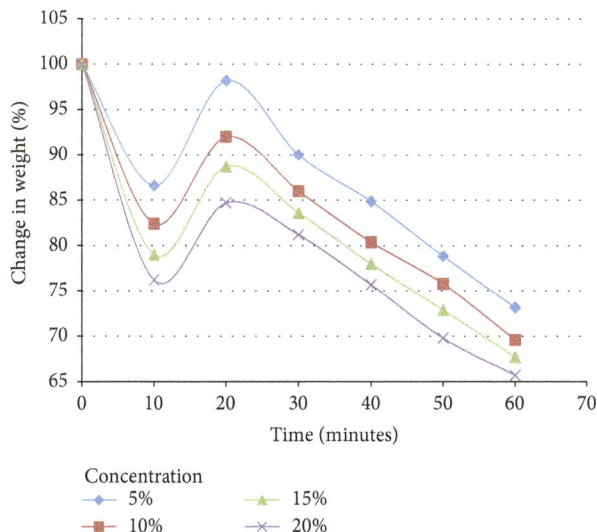

FIGURE 4: Degradation profile of *Aloe vera*/polymer films with varying concentration. Time in minutes is shown on *x*-axis and % change in weight is shown on *y*-axis.

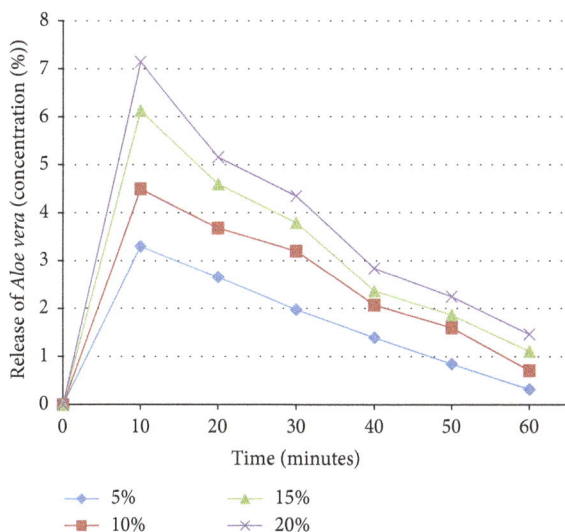

FIGURE 5: *Aloe vera* release profile of *Aloe vera*/polymer films with different concentrations. *x*-axis shows time in minutes and *y*-axis shows release of *Aloe vera* concentration in %.

as 15 mm, 20 mm, and 15 mm, respectively [31]. In current research, the mean zone of inhibition is 15 mm for both *E. coli* and *P. aeruginosa* and 16 mm for *Aspergillus tubingensis* (Figures 4 and 5) while for pure *Aloe vera* the zone was 19 mm and no zone was recorded against PVA films. Thus it can be concluded that blend of *Aloe vera* and PVA has not much affected the antimicrobial activity of *Aloe vera* and antimicrobial activity of *Aloe vera* was maintained in blend form.

3.4. Degradation of Aloe vera Release Profile Test Results. The degradation profile of *Aloe vera*/PVA composite was evaluated by recording weight loss at predetermined time points (Figure 4). The degradation profile was divided into three stages; during first 10 minutes a sudden loss of weight was observed due to the initial burst release of *Aloe vera*, followed by sudden increase in the weight of the films (Figure 4) because of the absorption of buffer solution by the PVA. PVA, when exposed to aqueous media, absorbs the liquid and swells, resulting in an increase in weight; later it becomes solvated and starts losing mass [32]. However, after 30 minutes, the weight loss by the films became linear. The rate of swelling of the PVA films after the initial burst decreased with the increase in the ratio of the *Aloe* component of the films. This was due to the fact that, with the increase in the *Aloe vera* concentration of PVA based films, absorption of the liquid medium by the PVA decreased [33].

The initial burst release followed by slow surface release of *Aloe vera* from the polymer based *Aloe vera* films was observed (Figure 5). An initial burst release of *Aloe vera* from the surface of the films was detected during the first 10 minutes. This may be attributed to the presence of aggregates of *Aloe vera* components on the film surface (verified later by SEM images of the films). The aggregates of *Aloe vera* over the surface of the films were observed causing the initial burst release. Later, the amount of *Aloe vera* released from the surface decreased because of entrapment of *Aloe vera* in PVA mass. During first 10 minutes *Aloe vera* was released only by diffusion from the surface while after 20 minutes the degradation of *Aloe vera*/polymer film also contributes to the release of *Aloe vera* [34]. The release profile of all the concentrations, that is, 5%, 10%, 15%, and 20%, showed the same behavior but with the increase in concentration from 5% to 20% greater initial burst release was observed which is due to the increased amount of *Aloe vera*. Moreover, increased concentration resulted in decreased *Aloe* release from the surface in the later stages because increasing the *Aloe* amount lowers the rate of diffusion of *Aloe* from the surface [35].

An accelerated degradation and release study is performed for short period of time because, keeping the application in mind, the chance of infections occurring is greater at initial stages. The *Aloe vera*/polymer composite film is flexible and can easily be placed on body surfaces, hence making it an ideal candidate for wound healing devices. The initial *Aloe vera* release from the surface is intended to be used as an antimicrobial so as to prevent the entry and proliferation of the microbes into the wound area [36]. Also, slow release marks the potential for an ideal microbial-free environment for wound healing [36].

3.5. Coating of the Sutures. The dry weight of the sutures before dipping into the coating solution was 0.045 g and after dipping into the coating solution was increased to 0.075 g. This increase in weight demonstrated the coating of the suture with the coating material.

3.6. In Vitro Testing of Coated Suture. The zones of inhibitions against both the bacterial strains (*E. coli* and *P. aeruginosa*) were evaluated with coated and uncoated sutures (Figure 6). The results were compared with uncoated sutures

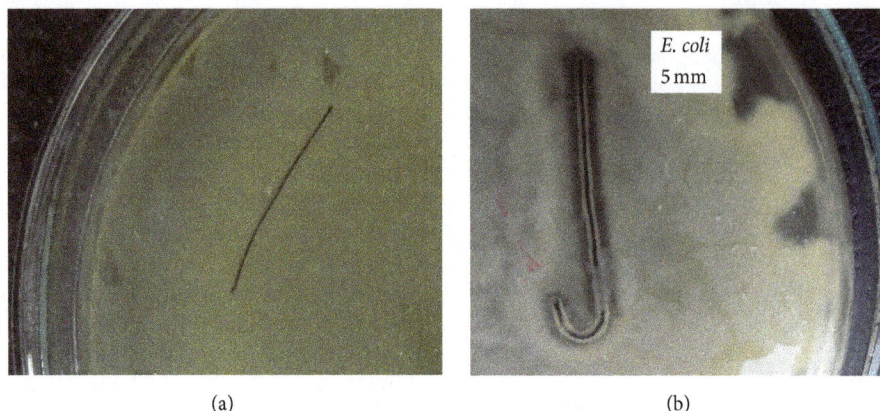

(a) (b)

FIGURE 6: *In vitro* testing against *E. coli* of (a) uncoated suture and (b) coated suture.

TABLE 1: *In vitro* testing of coated and uncoated sutures against *E. coli* and *P. aeruginosa*.

Bacteria	Zone of inhibition (mm)	
	Coated suture	Uncoated suture
E. coli	4.6 ± 0.577	0 ± 0
P. aeruginosa	3.16 ± 0.288	0 ± 0

which demonstrated no zone of inhibition, using a paired t-test (Table 1).

The zone of inhibition with *E. coli* was 4.6 ± 0.577 mm (mean of three triplicates), p value = 0.0051, while with *P. aeruginosa* it was 3.16 ± 0.28 mm (mean of three triplicates), p value < 0.0028.

3.7. In Vivo Testing of Coated Sutures. The silk sutures coated with *Aloe vera*/polymer coatings showed significant reduction in microbial colonization by *E. coli* and *P. aeruginosa* in mice models (Table 2). The coated sutures demonstrated reduction in *E. coli* to about 97% (p < 0.0001) and 80% with *P. aeruginosa* (p < 0.0001) (Figure 7).

In this present study, silk sutures coated with *Aloe vera*/polymer coating exhibited substantial zone of inhibitions against *E. coli* and *P. aeruginosa* in vitro because of the pharmacologically active components present in *Aloe vera* such as anthraquinones which remain active after being blended with PVA. Coated sutures showed results against bacterial strains while no inhibition zones were observed with uncoated sutures. For *in vivo* studies, mice models were used in which control and test sutures were used in the same animal and the incision site was inoculated with a known number of bacteria to evaluate the effectiveness of the coated sutures. The results of both *in vivo* and *in vitro* studies along with FTIR confirmed that the active components that are responsible for antimicrobial activity present in *Aloe vera* still remain in their active form.

Sutures with *Aloe vera*/polymer coating illustrated noticeable reduction in the growth of the *P. aeruginosa* and even greater reduction against *E. coli*. The test results of the *in vivo* and *in vitro* investigations suggested that sutures with

TABLE 2: *In vivo* bacterial colonization of suture with coated material.

Bacterial strains	Log CFU/explanted[a]	% kill bacteria relative to inoculum introduced	p value[b]
E. coli			
With coated material	03	97	<0.0001
Without coated material	>300	NA	
P. aeruginosa			
With coated material	11	80	<0.0001
Without coated material	>300	NA	

[a] Average of three animals.
[b] Paired t-test.
NA: not applicable.

Aloe vera/polymer coating are bactericidal. It was verified by calculating the bacterial colony count at incision site which was reduced in case of coated sutures; this shows that the *Aloe vera*/PVA composite may be used as a suture coating that has the potential to prevent the spread of infections during surgical procedures.

4. Conclusion

The biocompatibility and biodegradative properties of PVA have been combined with the intrinsic bactericidal properties of *Aloe vera*. The composition was screened for antimicrobial activity against bacterial and fungal strains, that is, *E. coli*, *P. aeruginosa*, *Aspergillus flavus*, and *Aspergillus tubingensis*, respectively. The polymeric films with lowest concentration (5%) of *Aloe vera* illustrated the best results with regard to antimicrobial activity against all the strains. Commercially available sutures were coated with *Aloe vera*/PVA solution and tested for antimicrobial activity in *in vitro* and *in vivo* systems. These coated sutures illustrated a potential

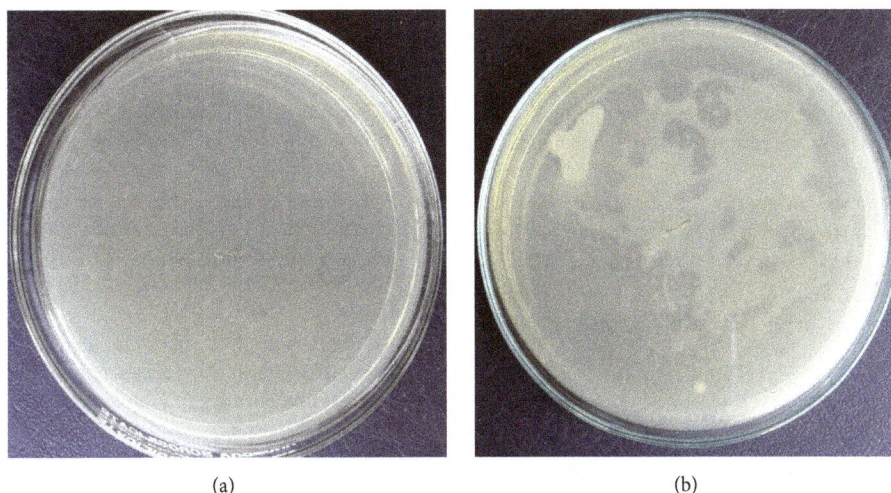

(a)   (b)

FIGURE 7: *In vivo* colonization of *E. coli* with coated and uncoated suture (a) shows the results of *in vivo* antibacterial activity with coated suture while (b) shows the *in vivo* antibacterial results with uncoated sutures.

for antibacterial/antifungal coatings in commercial surgical sutures that can play a role in preventing infections at surgical sites. Biocompatibility tests and clinical trials need to be conducted to better ascertain the potential of this *Aloe*/polymer composite as an option for use in surgical procedures as a suture coating, as part of prophylactic measures to prevent surgical infections.

Competing Interests

The authors declare that they have no competing interests.

References

[1] A. Nautiyal, N. V. Satheesh, Madhav et al., "Review on nosocomial infections," *Caribbean Journal of Science and Technology*, vol. 3, pp. 781–788, 2015.

[2] G. D. Taylor, M. Buchanan-Chell, T. Kirkland, M. McKenzie, and R. Wiens, "Nosocomial gram-negative bacteremia," *International Journal of Infectious Diseases*, vol. 1, no. 4, pp. 202–205, 1997.

[3] T. C. Horan, M. Andrus, and M. A. Dudeck, "CDC/NHSN surveillance definition of health care–associated infection and criteria for specific types of infections in the acute care setting," *American Journal of Infection Control*, vol. 36, no. 5, pp. 309–332, 2008.

[4] S. N. Banerjee, T. G. Emori, D. H. Culver et al., "Secular trends in nosocomial primary bloodstream infections in the United States, 1980–1989. National Nosocomial Infections Surveillance System," *The American Journal of Medicine*, vol. 91, no. 3, pp. 86S–89S, 1991.

[5] G. P. Bodey, "The emergence of fungi as major hospital pathogens," *Journal of Hospital Infection*, vol. 11, pp. 411–426, 1988.

[6] S. K. Fridkin and W. R. Jarvis, "Epidemiology of nosocomial fungal infections," *Clinical Microbiology Reviews*, vol. 9, no. 4, pp. 499–511, 1996.

[7] C. D. Owens and K. Stoessel, "Surgical site infections: epidemiology, microbiology and prevention," *Journal of Hospital Infection*, vol. 70, supplement 2, pp. 3–10, 2008.

[8] R. Pereira, A. Tojeira, D. C. Vaz, A. Mendes, and P. Bártolo, "Preparation and characterization of films based on alginate and aloe vera," *International Journal of Polymer Analysis and Characterization*, vol. 16, no. 7, pp. 449–464, 2011.

[9] S. Goldstein, R. J. Levy, V. Labhasetwar, and J. F. Bonadio, "Compositions and methods for coating medical devices," US Patent 6143037 A, 1996.

[10] M. G. Cascone, B. Sim, and D. Sandra, "Blends of synthetic and natural polymers as drug delivery systems for growth hormone," *Biomaterials*, vol. 16, no. 7, pp. 569–574, 1995.

[11] J. Walker, G. Young, C. Hunt, and T. Henderson, "Multi-centre evaluation of two daily disposable contact lenses," *Contact Lens and Anterior Eye*, vol. 30, no. 2, pp. 125–133, 2007.

[12] S.-H. Yang, Y.-S. J. Lee, F.-H. Lin, J.-M. Yang, and K.-S. Chen, "Chitosan/poly(vinyl alcohol) blending hydrogel coating improves the surface characteristics of segmented polyurethane urethral catheters," *Journal of Biomedical Materials Research—Part B Applied Biomaterials*, vol. 83, no. 2, pp. 304–313, 2007.

[13] J. M. Schmidt and J. S. Greenspoon, "Aloe vera dermal wound gel is associated with a delay in wound healing," *Obstetrics & Gynecology*, vol. 78, no. 1, pp. 115–117, 1991.

[14] M. Y. Wani, N. Hasan, and M. A. Malik, "Chitosan and *Aloe vera*: two gifts of nature," *Journal of Dispersion Science and Technology*, vol. 31, no. 6, pp. 799–811, 2010.

[15] J. H. Hamman, "Composition and applications of *Aloe vera* leaf gel," *Molecules*, vol. 13, no. 8, pp. 1599–1616, 2008.

[16] R. Pandey and A. Mishra, "Antibacterial activities of crude extract of *Aloe barbadensis* to clinically isolated bacterial pathogens," *Applied Biochemistry and Biotechnology*, vol. 160, no. 5, pp. 1356–1361, 2010.

[17] T. Reynolds and A. C. Dweck, "Aloe vera leaf gel: a review update," *Journal of Ethnopharmacology*, vol. 68, no. 1–3, pp. 3–37, 1999.

[18] M. H. Radha and N. P. Laxmipriya, "Evaluation of biological properties and clinical effectiveness of *Aloe vera*: a systematic

review," *Journal of Traditional and Complementary Medicine*, vol. 5, no. 1, pp. 21–26, 2015.

[19] S. S. Silva, S. G. Caridade, J. F. Mano, and R. L. Reis, "Effect of crosslinking in chitosan/aloe vera-based membranes for biomedical applications," *Carbohydrate Polymers*, vol. 98, no. 1, pp. 581–588, 2013.

[20] M. Fani and J. Kohanteb, "Inhibitory activity of *Aloe vera* gel on some clinically isolated cariogenic and periodontopathic bacteria," *Journal of Oral Science*, vol. 54, no. 1, pp. 15–21, 2012.

[21] A. Surjushe, R. Vasani, and D. G. Saple, "*Aloe vera*: a short review," *Indian Journal of Dermatology*, vol. 53, no. 4, pp. 163–166, 2008.

[22] A. Aytimur, S. Koçyiğit, and İ. Uslu, "Synthesis and characterization of poly(vinyl alcohol)/poly(vinyl pyrrolidone)-iodine nanofibers with poloxamer 188 and chitosan," *Polymer-Plastics Technology and Engineering*, vol. 52, no. 7, pp. 661–666, 2013.

[23] A. W. Bauer, W. M. Kirby, J. C. Sherris, and M. Turck, "Antibiotic susceptibility testing by a standardized single disk method," *American Journal of Clinical Pathology*, vol. 45, no. 4, pp. 493–496, 1966.

[24] R. A. Kumar, M. Gayathiri, S. Ravi, P. Kabilar, and S. Velmurugan, "Spectroscopy studies on the status of aloin in Aloe vera and commercial samples," *Journal of Experimental Sciences*, vol. 2, no. 8, pp. 10–13, 2011.

[25] Z. X. Lim and K. Y. Cheong, "Effects of drying temperature and ethanol concentration on bipolar switching characteristics of natural *Aloe vera*-based memory devices," *Physical Chemistry Chemical Physics*, vol. 17, no. 40, pp. 26833–26853, 2015.

[26] K. S. Venkatesh, S. R. Krishnamoorthi, N. S. Palani et al., "Facile one step synthesis of novel TiO_2 nanocoral by sol–gel method using Aloe vera plant extract," *Indian Journal of Physics*, vol. 89, no. 5, pp. 445–452, 2015.

[27] G.-M. Kim, *Fabrication of Bio-Nanocomposite Nanofibers Mimicking the Mineralized Hard Tissues via Electrospinning Process*, INTECH Open Access Publisher, 2010.

[28] G.-M. Kim, "Fabrication of bio-nanocomposite nanofibers mimicking the mineralized hard tissues via electrospinning process," in *Nanofibers*, A. Kumar, Ed., chapter 4, pp. 69–88, InTech, Rijeka, Croatia, 2010.

[29] N. A. Abdullah Shukry, K. Ahmad Sekak, M. R. Ahmad, and T. J. Bustami Effendi, "Characteristics of electrospun PVA-*Aloe vera* nanofibres produced via electrospinning," in *Proceedings of the International Colloquium in Textile Engineering, Fashion, Apparel and Design 2014 (ICTEFAD 2014)*, pp. 7–10, Springer, 2014.

[30] T. Renisheya Joy Jeba Malar, M. Johnson, S. Nancy Beaulah, R. S. Laju, G. Anupriya, and T. Renola Joy Jeba Ethal, "Anti-bacterial and antifungal activity of Aloe vera gel extract," *International Journal of Biomedical and Advance Research*, vol. 3, no. 3, pp. 184–187, 2012.

[31] S. Arunkumar and M. Muthuselvam, "Analysis of phytochemical constituents and antimicrobial activities of *Aloe vera* L. against clinical pathogens," *World Journal of Agricultural Sciences*, vol. 5, no. 5, pp. 572–576, 2009.

[32] E.-R. Kenawy, F. I. Abdel-Hay, M. H. El-Newehy, and G. E. Wnek, "Controlled release of ketoprofen from electrospun poly(vinyl alcohol) nanofibers," *Materials Science and Engineering A*, vol. 459, no. 1-2, pp. 390–396, 2007.

[33] M. Jannesari, J. Varshosaz, M. Morshed, and M. Zamani, "Composite poly (vinyl alcohol)/poly (vinyl acetate) electrospun nanofibrous mats as a novel wound dressing matrix for controlled release of drugs," *International Journal of Nanomedicine*, vol. 6, pp. 993–1003, 2011.

[34] R. Rosenberg, W. Devenney, S. Siegel, and N. Dan, "Anomalous release of hydrophilic drugs from poly(ε-caprolactone) matrices," *Molecular Pharmaceutics*, vol. 4, no. 6, pp. 943–948, 2007.

[35] S. G. Kumbar and T. M. Aminabhavi, "Synthesis and characterization of modified chitosan microspheres: effect of the grafting ratio on the controlled release of nifedipine through microspheres," *Journal of Applied Polymer Science*, vol. 89, no. 11, pp. 2940–2949, 2003.

[36] S.-J. Park and K.-S. Kim, "Influence of hydrophobe on the release behavior of vinyl acetate miniemulsion polymerization," *Colloids and Surfaces B: Biointerfaces*, vol. 46, no. 1, pp. 52–56, 2005.

3

Preparation and Characterization of Polyelectrolyte Complexes of *Hibiscus esculentus* (Okra) Gum and Chitosan

Vivekjot Brar and Gurpreet Kaur ⓘ

Department of Pharmaceutical Sciences and Drug Research, Punjabi University, Patiala, Punjab 147002, India

Correspondence should be addressed to Gurpreet Kaur; kaurgpt@gmail.com

Academic Editor: Carlo Galli

Polyelectrolyte complexes (PECs) of Okra gum (OKG) extracted from fruits of *Hibiscus esculentus* (Malvaceae) and chitosan (CH) were prepared using ionic gelation technique. The PECs were insoluble and maximum yield was obtained at weight ratio of $7:3$. The supernatant obtained after extracting PECs was clearly representing complete conversion of polysaccharides into PECs. Complexation was also evaluated by measuring the viscosity of supernatant after precipitation of PECs. The dried PECs were characterized using FTIR, DSC, zeta potential, water uptake, and SEM studies. Thermal analysis of PECs prepared at all ratios $(10:90, 20:80, 30:70, 40:60, 50:50, 60:40, 70:30, 80:20,$ and $90:10$; OKG:CH) depicted an endothermic peak at approximately $240°C$ representing cleavage of electrostatic bond between OKG and CH. The optimized ratio $(7:3)$ exhibited a zeta potential of -0.434 mV and displayed a porous structure in SEM analysis. These OKG-CH PECs can be further employed as promising carrier for drug delivery.

1. Introduction

Chitosan (CH) is among the most commonly employed natural polymers for drug delivery systems due to its biocompatible and biodegradable nature. It is a naturally occurring cationic polysaccharide consisting of glucosamine and N-acetyl-glucosamine obtained by partial deacetylation of chitin [1]. It is degraded in vertebrates by lysozyme and by bacterial enzymes in the colon [2]. These properties of CH make it desirable candidate for use as excipient in drug formulations. However, high solubility of CH in acidic conditions limits its use in sustained release oral preparations. The solubility of CH is attributed to the protonation of the free amine group. If this amine group is not free then there might be a possibility of using this biodegradable polymer in oral sustained release drug delivery systems. A large number of polyelectrolyte complexes (PECs) have been reported in the literature for controlling the release of drugs. Chitosan is cationic in nature, this property can be employed in the formation of PECs using ionic cross linking with poly anions like tripolyphosphate (TPP) [3], other anionic natural polysaccharides such as alginate [4, 5], pectin [6], carrageenan [7], xanthan gum [8], and gum kondagogu [9], synthetic anionic polymers such as poly acrylic acids [10], and semisynthetic polymers such as carboxymethylcellulose [11].

Okra gum (OKG) is anionic polymer obtained from the fruits of *Hibiscus esculentus* (family Malvaceae) consisting of galactose, rhamnose, and galacturonic/glucuronic acid as monomers [12]. It is used in pharmaceutical preparations as binder [13] and film coating agent [14], in colon targeting [15], and in buccal delivery [16].

CH and OKG both cannot be used alone in the formulation of sustained release dosage forms as they are limited by their high solubility. Complexation of CH with anionic polymers has been reported to decrease its solubility. This property is useful in sustaining the drug release. The interaction of polymer with an oppositely charged polymer results in an exchange of counterions and it is accompanied with an increase in entropy [17]. This increase in entropy suggests that the complexation reaction is spontaneous. The formation and properties of PECs depend on molecular weight, density of charge, and degree of neutralization of the polymers employed [18]. The PECs between oppositely charged polymers impart characteristic final properties to polymers (solubility, swelling, and rheology). The PECs when used in drug delivery may control the release of drug from the dosage form. Also, depending on different polymeric factors and formulation conditions the PECs can be formulated as different dosage forms such as microcapsules [19], gel nanoparticles [20], nanoparticles [21], controlled release membranes [22], gels [8], inserts [23], films [24], and carriers for oral administration [25].

There are no published reports of formation of polyelectrolyte complexes of Okra gum so the research in hand was designed to prepare and characterize the polyelectrolyte complexes between OKG and CH. The carboxylic acid groups present in OKG may react with amine groups present in chitosan resulting in formation of polyelectrolyte complexes. These complexes may be further evaluated for their potential as carrier for drug delivery systems.

2. Material and Methods

2.1. Material. Chitosan (CH), degree of deacetylation > 80%, was purchased from India Sea Foods, Cochin, India. Acetic acid, acetone, barium hydroxide ($Ba(OH)_2$), and zinc sulfate ($ZnSO_4$) used in the study were purchased from Loba Chemie and were of analytical grade.

2.2. Extraction of Okra Gum. Okra gum (OKG) was extracted using method described by Kaur et al. [16]. Briefly, crushed Okra fruits were allowed to swell in water for 8–12 h. This was followed by filtration through muslin cloth and deproteinization with barium hydroxide ($Ba(OH)_2$) and zinc sulfate ($ZnSO_4$). The supernatant was then precipitated with acetone. The powdered gum was dried by freeze drying (Allied Frost, Delhi) and stored in air tight container.

2.3. Preparation of Polyelectrolyte Complexes. OKG and CH were dissolved separately in distilled water and acetate buffer (pH 5.0), respectively, producing a 1% w/v of polymer solutions. The OKG solution was then added to CH solution dropwise (10 : 90, 20 : 80, 30 : 70, 40 : 60, 50 : 50, 60 : 40, 70 : 30, 80 : 20, and 90 : 10; OKG : CH). The suspensions were stirred at room temperature for one hour. The suspension was then incubated for 24 h in shaking incubator (Remi shaking incubator CIS 24, Mumbai, India) at temperature of 37°C and the precipitates were separated by centrifugation at 10,000 rpm (6708 g) for 10 min (Remi cooling centrifuge, Mumbai, India), washed with distilled water, and freeze dried. The practical yield of the PECs was calculated employing the following equation:

$$\text{Percent yield (\%)} = \frac{W_0}{W_t} \times 100, \qquad (1)$$

where W_0 is weight of PEC obtained and W_t is total weight of polymers taken.

2.4. Viscosity Measurements. The viscosity of the supernatants obtained after mixing OKG solutions with CH solution in different ratios was determined with Brookfield viscometer spindle S18 using small volume adapter (Brookfield viscometer LVDV1, Bruker, UK). Briefly, 7-8 mL sample was poured in adapter. The samples were equilibrated to a temperature of 25°C using a circulatory water bath for 15 minutes. Thereafter the viscosity measurements were carried at suitable rpm.

2.5. Fourier Transform Infrared (FTIR) Measurements. The spectral characteristics of different ratios of PECs were determined employing FT Infrared Spectrophotometer Model RZX (Perkin Elmer). The dried samples of PECs were compressed with KBr to form pellets for FT-IR measurement. Sixty-four scans were single-averaged at a resolution of $4 \, \text{cm}^{-1}$.

2.6. Thermal Analysis. The thermal properties of the PECs were evaluated using differential scanning calorimeter (EVO 131, SETARAM Instrumentation, France). The samples were prepared by weighing accurately 5-6 mg of sample and crimped in aluminum crucibles with a pin holed lid and heated in the range of 40–400°C at a heating rate of $10°C \, \text{min}^{-1}$ under a purge of nitrogen at rate of 30 mL/min.

2.7. Zeta Potential Measurements. Zeta potential measurements of different ratio of PECs were carried out using a Zetasizer (Malvern Instruments Ltd., Malvern, UK). The zeta potential measurements were performed using an aqueous dip cell in an automatic mode. Samples were diluted in triple distilled water and placed in a clear disposable zeta cell, with the cell position being adjusted.

2.8. Swelling Measurements. Accurately weighed amounts of PECs prepared employing different ratios of CH and OKG were compressed into pellets (6 mm, radial diameter) using a tablet compression machine. These pellets were positioned on top of a sponge. The sponge was previously soaked in hydrochloric acid buffer (0.2 M) solution of pH 1.2 and phosphate buffer pH 6.8 (0.2 M) in a Petri plate. The water uptake by the pellets was calculated by weighing the initial weight of the pellet before keeping it on hydration medium and after eight hours [23]. The swelling behavior of the PECs was calculated using the following equation:

$$P_s = \frac{(W_s - W_i)}{W_i} \times 100, \qquad (2)$$

where P_s is percent swelling, W_s is weight of the swollen pellet at time "t," and W_i is initial weight of the pellet.

2.9. SEM Analysis. Sample of powdered PECs was sprinkled on the sample stub with double-sided tape, extra powder was removed and coated for 70 s under an argon atmosphere with gold coating, and then the powdered samples on the SEM grid were allowed to air dry for 10 min. Images were captured at different magnifications using SEM (Jeol JSM-6610 LV, USA) machine.

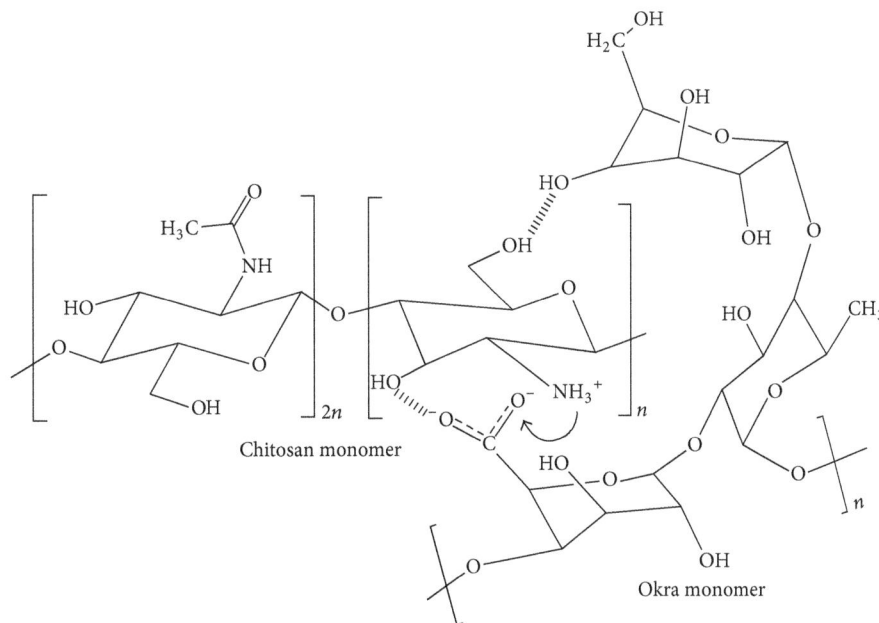

FIGURE 1: Electrostatic interactions between the monomers of OKG and CH.

TABLE 1: Effect of different ratios of OKG and CH on percentage yield and thermal properties of polyelectrolyte complexes.

OKG : CH	Percent yield (in %)	Thermal properties	
		Peak maximum (°C)	Heat (J/g)
1 : 9	0	-	-
2 : 8	36	231.241	122.989
3 : 7	40	237.622	88.628
4 : 6	44	243.356	110.485
5 : 5	54	239.529	63.955
6 : 4	58	242.384	84.341
7 : 3	60	242.042	60.377
8 : 2	54	240.543	84.087
9 : 1	20	239.475	68.245

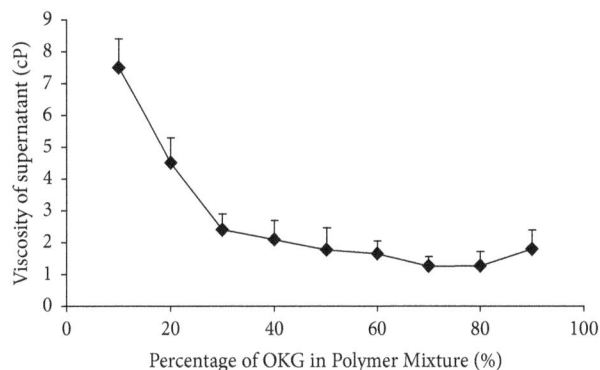

FIGURE 2: Effect of increasing the percentage of OKG in polymer mixture on viscosity of supernatant.

3. Results and Discussions

3.1. Extraction of OKG. The extraction technique used for the isolation of OKG from fresh pod of *Hibiscus esculentus* resulted in a yield of 0.5% w/w. The gum was pale green in color with a characteristic odor [16].

3.2. Yield of Complexes. The percent yield of the PECs obtained at different ratios is depicted in Table 1. The PEC formation occurs when protonated amine group of CH interacts electrostatically with the negatively charged carboxylic group of OKG as depicted in Figure 1.

No complexation was observed on addition of OKG solution to CH solution at ratio of 1 : 9. As the concentration of OKG increased there was a steady increase in yield till ratio of 7 : 3; after that the yield started to decrease suggesting that at this ratio maximum amount of OKG interacts with the CH

and beyond that point increasing the amount of OKG will only result in solubilization of gum in the solution [10].

3.3. Viscosity Measurements. The graph shows the viscosity of the supernatant obtained after PECs were removed from the solution. OKG is anionic polymer due to presence of $-COO^-$ groups while CH is cationic due to presence of NH_3^+ when ionized. A decline in the viscosity of the supernatant was observed as the concentration of OKG in the PECs increases (Figure 2).

This decrease in viscosity is observed till the concentration of 8 : 2 is reached. A further increase in OKG leads to slight increase in the viscosity. Previous studies have also reported a drop in viscosity to minimum when oppositely charged polymers like CH and carrageenan [26] or CH and chondroitin [27] are mixed. A slight increase in the viscosity of solution could be attributed to the presence of free OKG in the solution after the maximum interaction occurs as explained earlier by Chavasit and Torres [10].

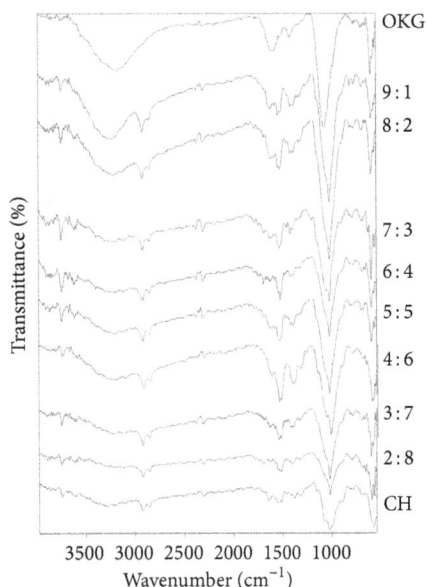

FIGURE 3: The FTIR spectra of OKG, PECs of ratios 2:8, 3:7, 4:6, 5:5, 6:4, 7:3, 8:2, and 9:1, and CH showing characteristic bands.

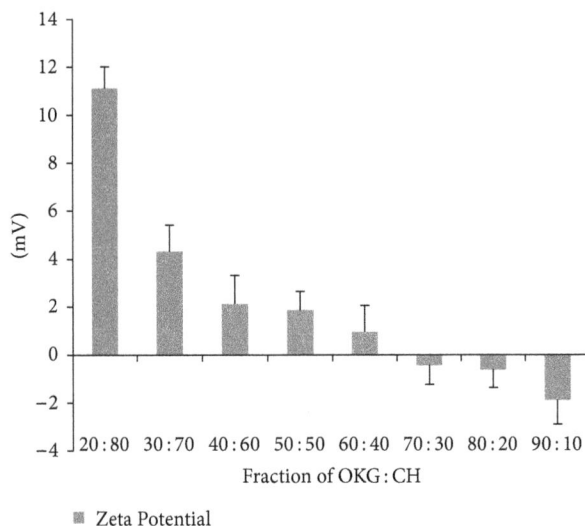

FIGURE 4: Effect of different ratios on zeta potential.

3.4. FTIR Measurements. The FTIR spectra of OKG, CH, and PECs are depicted in Figure 3.

The presence of absorption band at 3180.14 cm^{-1} in the FTIR spectra of OKG represents the vibrational stretches due to hydroxyl groups participating in hydrogen bonding that correspond to the basic carbohydrate structure of the polysaccharides. The FTIR spectra of OKG show an absorption peak at 1585.05 cm^{-1} which indicates the presence of carboxylic acid group in the form of –COO^{-} corresponding to antisym stretch and at 1419.42 cm^{-1} due to sym stretch. The absorption spectra of CH show two distinct peaks at 3736.65 cm^{-1} and 3606.38 cm^{-1} which also correspond to the hydroxyl groups. Two characteristic peaks at 2922.56 cm^{-1} and 2857.65 cm^{-1} are observed in the FTIR spectra of CH which corresponds to –CH$_3$ and –CH$_2$– due to antisym and sym stretching. These peaks are not present in OKG but in the PECs these two peaks are very prominent [28]. As the concentration of OKG increases in the PECs from 2:8 to 9:1, the broad hydrogen bonding peak becomes more prominent but the single distinct peaks are also visible. As the PECs are formed it was noticed that the peak of carboxylate ion shifts to the peak of acid alone and appears in the range of 1690–1710 cm^{-1} which indicates that the –COO^{-} is getting converted to –COOH. This conversion is indicative of formation of electrostatic bond which requires the transfer of counterion from one polymer to another. CH FTIR spectra show characteristic peak of primary amine at 3410.05 cm^{-1} and 1646.94 cm^{-1}. Both are distinctively visible in the PECs of all concentrations. FTIR spectra of both CH and OKG show sharp peaks in the region of 1015–1065 cm^{-1} which correspond to the cyclic alcohols which are present in both the polymers. These peaks are also seen in all the PECs [28].

3.5. Thermal Analysis. The DSC of the pure OKG and its complexes with CH is tabulated in Table 1.

The OKG polymer thermogram depicts a broad endothermic peak starting from 45°C and shouldering up to 105.2°C with peak maxima of 55.062°C which may be due to close endothermic changes of glass transition (T_g) and water loss [29]. The peak that represents removal of water present in the bound state is higher than the boiling point of water as more amount of heat is required to break the ionic bonds that water has made with the polymer [30]. CH showed an endothermic peak at 80.22°C and an endothermic peak at 311.30°C. The DSC thermograms of PECs show endotherms ranging from peak maxima from 231.241°C to 243.356°C with varying amounts of heat transfer for different concentration ratios. These transitions could be associated with cleavage of electrostatic interactions between oppositely charged constituents of OKG and CH since it is not observed in pure components [31].

3.6. Zeta Potential Measurements. Zeta potential of pure CH is 12 mV and OKG is –11.47 mV which is due to presence of free amino groups in CH and free carboxylic groups of OKG giving it positive and negative net charge, respectively. The decrease in the zeta potential values was observed when the concentration of OKG increases in the PECs (Figure 4).

This decrease suggested the neutralization of the free positive charge associated with ionized CH upon the addition of OKG. The zeta value was reduced to –0.434 mV at the ratio of 7:3. This indicated a complete neutralization of all the free charges associated with CH and a maximum interaction. Zeta potential values drop further to negative values as after that the increase in OKG concentration only leads to increase in free negative charge that is associated with OKG [32].

3.7. Swelling Measurements. The water uptake capacity of the polymers was determined in pH 1.2 and phosphate buffer pH 6.8. Figure 5 represents the water uptake as it is represented as percentage weight gain of total weight of the PECs. As the concentration of OKG increases in the PECs a decrease

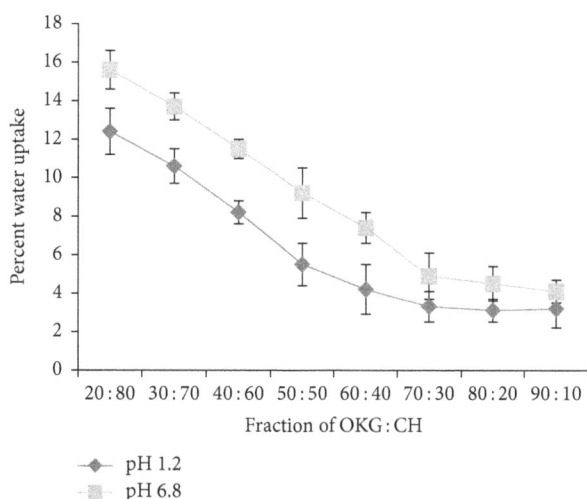

FIGURE 5: Effect of different ratios on percent water uptake by the PECs.

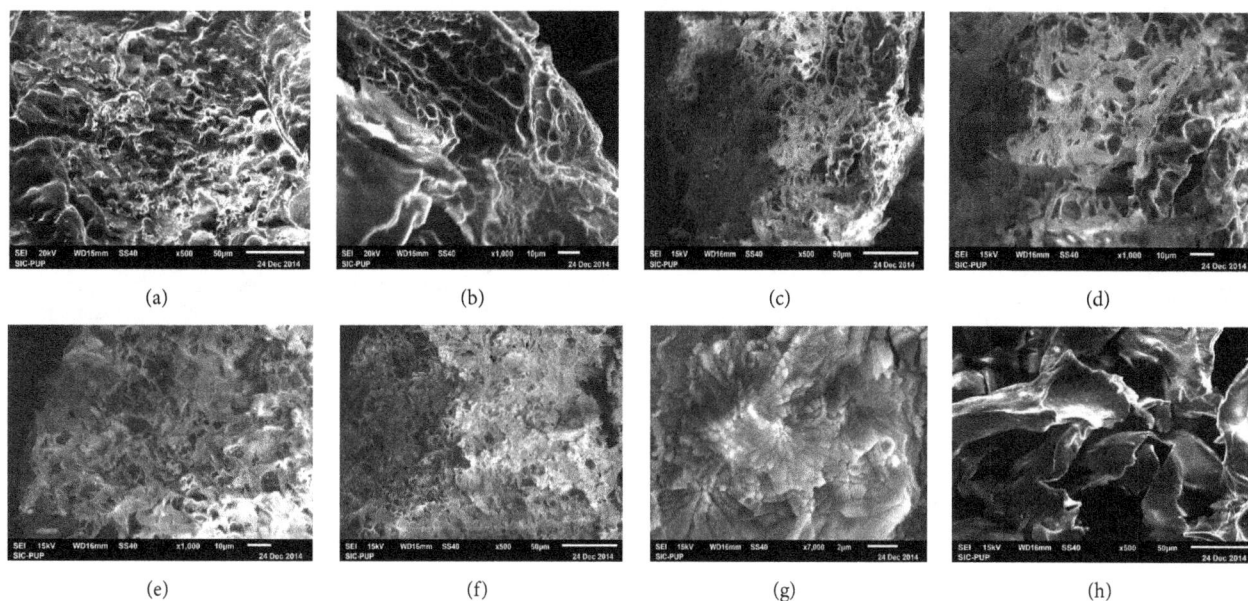

(a)   (b)   (c)   (d)

(e)   (f)   (g)   (h)

FIGURE 6: SEM images of different PECs between OKG and CH at concentration ratios 2 : 8 (a), 3 : 7 (b), 4 : 6 (c), 5 : 5 (d), 6 : 4 (e), 7 : 3 (f), 8 : 2 (g), and 9 : 1 (h).

in the water uptake in both the pH media was observed till a ratio of 7 : 3 after which no significant change in the weight gain was observed. Uptake of water is associated with the ability of unionized groups present in polysaccharides to form ionic bonds with water. An increase in the concentration of OKG decreases the number of free ionic groups; as a result there is less water uptake. After 7 : 3 slight increase in the water uptake was observed. However, it is not significant and may be due to free ionic groups of OKG which may uptake the water molecules forming hydrogen bonding [24].

Another noteworthy observation is that percent water uptake of PECs at pH 1.2 is less as compared to pH 6.8. The reason for this observation may be the difference in the ionization of the two polymers at different pH. In acidic conditions the amine group of CH is protonated which causes electrostatic interactions of carboxylic group of OKG with protonated CH to produce a tight network which may lead to lower water uptake at lower pH. As the pH is higher the protonation of the CH decreases which leads to decrease in electrostatic interaction and hence the network of the matrix is loose which may lead to higher water uptake in the free space. This phenomenon was explained by Fahmy and Fouda in PECs of alginate and CH [33] and gum kondagogu and CH [34].

3.8. SEM Analysis. The morphology of the PECs between OKG and CH at concentration ratios ranging from ratios 2 : 8 to 9 : 1 is shown in the Figure 6 as observed by SEM.

The PECs were dried using freeze drying which formed highly porous matrixes of the complexes as water from the surface sublimes damping the surface from which it sublimes. The cross linking between the oppositely charged polymers may also be responsible for the formation of sponge like matrix structure. The porosity of the PECs seems to decrease with an increase in OKG levels which is in consistence with water retention studies where structures with higher porosity in three-dimensional structures of the PECs tend to retain more water [23]. After the ratio where the maximum interaction is achieved, the additional OKG tends to reside at the interstitial surfaces of complex forming a smoother surface.

4. Conclusion

The PECs based on electrostatic interaction between OKG and CH were prepared by varying the ratios of two polymers using ionic gelation technique. The insoluble PECs obtained had a porous matrix and were characterized using FTIR and DSC studies. Further investigations suggest that the ratio of 7 : 3 leads to maximum interaction as implied by highest yield, lowest viscosity of supernatant, and lowest zeta potential at this ratio. Swelling studies show that swelling of PECs occurs more in pH 6.8 as compared to pH 1.2 which may in future be used in pH dependent sustained drug delivery system.

Disclosure

Portions of this study have been presented in abstract form at 69th Indian Pharmaceutical Conference, Chitkara University, Rajpura, Punjab, India, on December 22, 2017.

Conflicts of Interest

The authors declare that they have no conflicts of interest.

References

[1] S. V. Madihally and H. W. T. Matthew, "Porous chitosan scaffolds for tissue engineering," *Biomaterials*, vol. 20, no. 12, pp. 1133–1142, 1999.

[2] T. Kean and M. Thanou, "Biodegradation, biodistribution and toxicity of chitosan," *Advanced Drug Delivery Reviews*, vol. 62, no. 1, pp. 3–11, 2010.

[3] A. R. Dudhani and S. L. Kosaraju, "Bioadhesive chitosan nanoparticles: Preparation and characterization," *Carbohydrate Polymers*, vol. 81, no. 2, pp. 243–251, 2010.

[4] M. G. Sankalia, R. C. Mashru, J. M. Sankalia, and V. B. Sutariya, "Reversed chitosan-alginate polyelectrolyte complex for stability improvement of alpha-amylase: optimization and physicochemical characterization," *European Journal of Pharmaceutics and Biopharmaceutics*, vol. 65, no. 2, pp. 215–232, 2007.

[5] H. V. Sæther, H. K. Holme, G. Maurstad, O. Smidsrød, and B. T. Stokke, "Polyelectrolyte complex formation using alginate and chitosan," *Carbohydrate Polymers*, vol. 74, no. 4, pp. 813–821, 2008.

[6] F. Bigucci, B. Luppi, T. Cerchiara et al., "Chitosan/pectin polyelectrolyte complexes: Selection of suitable preparative conditions for colon-specific delivery of vancomycin," *European Journal of Pharmaceutical Sciences*, vol. 35, no. 5, pp. 435–441, 2008.

[7] A. V. Briones and T. Sato, "Encapsulation of glucose oxidase (GOD) in polyelectrolyte complexes of chitosan-carrageenan," *Reactive and Functional Polymers*, vol. 70, no. 1, pp. 19–27, 2010.

[8] S. Argin-Soysal, P. Kofinas, and Y. M. Lo, "Effect of complexation conditions on xanthan-chitosan polyelectrolyte complex gels," *Food Hydrocolloids*, vol. 23, no. 1, pp. 202–209, 2009.

[9] V. T. P. Vinod and R. B. Sashidhar, "Solution and conformational properties of gum kondagogu (Cochlospermum gossypium) - A natural product with immense potential as a food additive," *Food Chemistry*, vol. 116, no. 3, pp. 686–692, 2009.

[10] V. Chavasit and J. A. Torres, "Chitosan-Poly(acrylic acid): Mechanism of Complex Formation and Potential Industrial Applications," *Biotechnology Progress*, vol. 6, no. 1, pp. 2–6, 1990.

[11] T. Cerchiara, A. Abruzzo, C. Parolin et al., "Microparticles based on chitosan/carboxymethylcellulose polyelectrolyte complexes for colon delivery of vancomycin," *Carbohydrate Polymers*, vol. 143, pp. 124–130, 2016.

[12] R. L. Whistler and H. E. Conrad, "A Crystalline Galactobiose from Acid Hydrolysis of Okra Mucilage," *Journal of the American Chemical Society*, vol. 76, no. 6, pp. 1673-1674, 1954.

[13] K. Ameena, C. Dilip, R. Saraswathi, P. N. Krishnan, C. Sankar, and S. P. Simi, "Isolation of the mucilages from *Hibiscus rosasinensis* linn. and Okra (*Abelmoschus esculentus* linn.) and studies of the binding effects of the mucilages," *Asian Pacific Journal of Tropical Medicine*, vol. 3, no. 7, pp. 539–543, 2010.

[14] I. Ogaji and O. Nnoli, "Film coating potential of okra gum using paracetamol tablets as a model drug," *Asian Journal of Pharmaceutics*, vol. 4, no. 2, pp. 130–134, 2010.

[15] A. Rajkumari, M. S. Kataki, K. B. Ilango, S. D. Devi, and P. Rajak, "Studies on the development of colon specific drug delivery system of ibuprofen using polysaccharide extracted from Abelmoschus esculentus L. (Moench.)," *Asian Journal of Pharmaceutical Sciences*, vol. 7, no. 1, pp. 67–74, 2012.

[16] G. Kaur, D. Singh, and V. Brar, "Bioadhesive okra polymer based buccal patches as platform for controlled drug delivery," *International Journal of Biological Macromolecules*, vol. 70, pp. 408–419, 2014.

[17] A. F. Thünemann, M. Müller, H. Dautzenberg, J. Joanny, and H. Löwen, "Polyelectrolyte Complexes," in *Polyelectrolytes with Defined Molecular Architecture II*, vol. 166 of *Advances in Polymer Science*, pp. 113–171, Springer Berlin Heidelberg, Berlin, Heidelberg, 2004.

[18] A. I. Gamzazade and S. M. Nasibov, "Formation and properties of polyelectrolyte complexes of chitosan hydrochloride and sodium dextransulfate," *Carbohydrate Polymers*, vol. 50, no. 4, pp. 339–343, 2002.

[19] N. Devi and T. K. Maji, "A novel microencapsulation of neem (Azadirachta Indica A. Juss.) seed oil (NSO) in polyelectrolyte complex of k-carrageenan and chitosan," *Journal of Applied Polymer Science*, vol. 113, no. 3, pp. 1576–1583, 2009.

[20] O. Masalova, V. Kulikouskaya, T. Shutava, and V. Agabekov, "Alginate and chitosan gel nanoparticles for efficient protein

entrapment," in *Proceedings of the 2nd European Conference on Nano Films, ECNF 2012*, pp. 69–75, Italy, June 2012.

[21] A. Kumar and M. Ahuja, "Carboxymethyl gum kondagogu-chitosan polyelectrolyte complex nanoparticles: Preparation and characterization," *International Journal of Biological Macromolecules*, vol. 62, pp. 80–84, 2013.

[22] M. R. El-Aassar, G. F. El Fawal, E. A. Kamoun, and M. M. G. Fouda, "Controlled drug release from cross-linked κ-carrageenan/hyaluronic acid membranes," *International Journal of Biological Macromolecules*, vol. 77, pp. 322–329, 2015.

[23] B. Luppi, F. Bigucci, A. Abruzzo, G. Corace, T. Cerchiara, and V. Zecchi, "Freeze-dried chitosan/pectin nasal inserts for antipsychotic drug delivery," *European Journal of Pharmaceutics and Biopharmaceutics*, vol. 75, no. 3, pp. 381–387, 2010.

[24] M. Jindal, V. Kumar, V. Rana, and A. K. Tiwary, "Physicochemical, mechanical and electrical performance of bael fruit gum-chitosan IPN films," *Food Hydrocolloids*, vol. 30, no. 1, pp. 192–199, 2013.

[25] E. Assaad, Y. J. Wang, X. X. Zhu, and M. A. Mateescu, "Polyelectrolyte complex of carboxymethyl starch and chitosan as drug carrier for oral administration," *Carbohydrate Polymers*, vol. 84, no. 4, pp. 1399–1407, 2011.

[26] C. Tapia, Z. Escobar, E. Costa et al., "Comparative studies on polyelectrolyte complexes and mixtures of chitosan-alginate and chitosan-carrageenan as prolonged diltiazem clorhydrate release systems," *European Journal of Pharmaceutics and Biopharmaceutics*, vol. 57, no. 1, pp. 65–75, 2004.

[27] K. Kaur and G. Kaur, "Formulation and evaluation of chitosan-chondroitin sulphate based nasal inserts for zolmitriptan," *BioMed Research International*, vol. 2013, Article ID 958465, 2013.

[28] J. B. Lambert, H. F. Shurvell, D. A. Lightner, and R. G. Cooks, *Introduction to Organic Spectroscopy*, Macmillan, New York, 1987.

[29] M. Emeje, P. Nwabunike, C. Isimi et al., "Isolation, characterization and formulation properties of a new plant gum obtained from *Cissus refescence*," *International Journal of Green Pharmacy*, vol. 3, no. 1, pp. 16–23, 2009.

[30] N. D. Zaharuddin, M. I. Noordin, and A. Kadivar, "The use of hibiscus esculentus (Okra) gum in sustaining the release of propranolol hydrochloride in a solid oral dosage form," *BioMed Research International*, vol. 2014, Article ID 735891, 2014.

[31] C. L. Silva, J. C. Pereira, A. Ramalho, A. A. C. C. Pais, and J. J. S. Sousa, "Films based on chitosan polyelectrolyte complexes for skin drug delivery: development and characterization," *Journal of Membrane Science*, vol. 320, no. 1-2, pp. 268–279, 2008.

[32] W.-B. Chen, L.-F. Wang, J.-S. Chen, and S.-Y. Fan, "Characterization of polyelectrolyte complexes between chondroitin sulfate and chitosan in the solid state," *Journal of Biomedical Materials Research Part A*, vol. 75, no. 1, pp. 128–137, 2005.

[33] H. M. Fahmy and M. M. G. Fouda, "Crosslinking of alginic acid/chitosan matrices using polycarboxylic acids and their utilization for sodium diclofenac release," *Carbohydrate Polymers*, vol. 73, no. 4, pp. 606–611, 2008.

[34] V. G. M. Naidu, K. Madhusudhana, R. B. Sashidhar et al., "Polyelectrolyte complexes of gum kondagogu and chitosan, as diclofenac carriers," *Carbohydrate Polymers*, vol. 76, no. 3, pp. 464–471, 2009.

4

Silver Nanoparticles: Biosynthesis Using an ATCC Reference Strain of *Pseudomonas aeruginosa* and Activity as Broad Spectrum Clinical Antibacterial Agents

Melisa A. Quinteros,[1] **Ivana M. Aiassa Martínez,**[2] **Pablo R. Dalmasso,**[3] **and Paulina L. Páez**[2]

[1]IMBIV, CONICET, Departamento de Farmacia, Facultad de Ciencias Químicas, Universidad Nacional de Córdoba, Ciudad Universitaria, 5000 Córdoba, Argentina
[2]UNITEFA, CONICET, Departamento de Farmacia, Facultad de Ciencias Químicas, Universidad Nacional de Córdoba, Ciudad Universitaria, 5000 Córdoba, Argentina
[3]CITSE, CONICET, Universidad Nacional de Santiago del Estero, RN 9, Km 1125, 4206 Santiago del Estero, Argentina

Correspondence should be addressed to Pablo R. Dalmasso; p-dalmasso@hotmail.com

Academic Editor: Vijaya Kumar Rangari

Currently, the biosynthesis of silver-based nanomaterials attracts enormous attention owing to the documented antimicrobial properties of these ones. This study reports the extracellular biosynthesis of silver nanoparticles (Ag-NPs) using a *Pseudomonas aeruginosa* strain from a reference culture collection. A greenish culture supernatant of *P. aeruginosa* incubated at 37°C with a silver nitrate solution for 24 h changed to a yellowish brown color, indicating the formation of Ag-NPs, which was confirmed by UV-vis spectroscopy, transmission electron microscopy, and X-ray diffraction. TEM analysis showed spherical and pseudospherical nanoparticles with a distributed size mainly between 25 and 45 nm, and the XRD pattern revealed the crystalline nature of Ag-NPs. Also it provides an evaluation of the antimicrobial activity of the biosynthesized Ag-NPs against human pathogenic and opportunistic microorganisms, namely, *Staphylococcus aureus*, *Staphylococcus epidermidis*, *Enterococcus faecalis*, *Proteus mirabilis*, *Acinetobacter baumannii*, *Escherichia coli*, *P. aeruginosa*, and *Klebsiella pneumonia*. Ag-NPs were found to be bioactive at picomolar concentration levels showing bactericidal effects against both Gram-positive and Gram-negative bacterial strains. This work demonstrates the first helpful use of biosynthesized Ag-NPs as broad spectrum bactericidal agents for clinical strains of pathogenic multidrug-resistant bacteria such as methicillin-resistant *S. aureus*, *A. baumannii*, and *E. coli*. In addition, these Ag-NPs showed negligible cytotoxic effect in human neutrophils suggesting low toxicity to the host.

1. Introduction

The continuing appearance of antibiotic resistance in pathogenic and opportunistic microorganisms obliges the scientific community to constantly develop new drugs and drug targets. The costs of healthcare-associated infections are clearly high and increasing as the number of infections that are caused by multiple drug-resistant microorganisms increases [1]. More than 70% of bacterial nosocomial infections are resistant to one or more of the antibiotics traditionally used to treat them, and people infected with drug-resistant microorganisms usually spend more time in the hospital and require a treatment that uses two or three different antibiotics which is less effective, more toxic, and more expensive [2].

Even though the goal of many scientists is designing drugs acting *via* novel mechanisms of action, few new antibiotics have been introduced by the pharmaceutical industry in the last decade, and none of them have improved the activity against multidrug-resistant bacteria [3]. In the current scenario, nanotechnology offers opportunities to reexplore the biological properties of already known antimicrobial materials by manipulating their size to alter the effect [4].

Recently, the application of nanoparticles in various fields has expanded considerably. Nanoparticles possess unique physicochemical characteristics, such as a high ratio of surface area to mass, high reactivity, and sizes in the range of nanometers (10^{-9} m). Nanoparticles have been successfully used in nanochemistry to enhance the immobilization and activity of catalysts, in sensors, in medical and pharmaceutical nanoengineering for delivery of therapeutic agents, and in the food industry to limit bacterial growth [5–8]. Due to nanoparticles which have also demonstrated antimicrobial activities, the development of novel applications in this field makes them an attractive alternative to antibiotics.

In recent years, there has been growing interest in the synthesis and study of silver nanoparticles (Ag-NPs), because silver has long been known for its antimicrobial properties and the Ag-NPs are considered as nontoxic and environmentally friendly antibacterial materials that may be linked to broad spectrum activity and far lower propensity to induce microbial resistance compared to antibiotics [8, 9]. Currently, many methods have been reported for the synthesis of Ag-NPs by using chemical, physical, and biological routes [10]. The latter has emerged as a green alternative and it is highly advantageous for it is eco-friendly, cost-effective, and easily scaled up. The biosynthesis of Ag-NPs has great potential with natural reducing agents and/or stabilizing compounds from bacteria, fungi, yeasts, algae, or plants [10, 11].

In this work, we provide a simple and eco-friendly strategy for the green synthesis of Ag-NPs using the metal-reducing culture supernatant of *Pseudomonas aeruginosa* ATCC 27853. UV-vis spectroscopy and transmission electron microscopy were used to characterize the Ag-NPs biosynthesized. While a similar strategy has been used previously by Kumar and Mamidyala [12], this work provides the first extracellular biosynthesis of Ag-NPs using a *P. aeruginosa* strain from a reference culture collection. Also we evaluated the *in vitro* antimicrobial efficacy of the Ag-NPs against representative Gram-positive and Gram-negative bacteria such as *Staphylococcus aureus*, *Staphylococcus epidermidis*, *Enterococcus faecalis*, *Proteus mirabilis*, *Acinetobacter baumannii*, *Escherichia coli*, *P. aeruginosa*, and *Klebsiella pneumoniae*. To the best of our knowledge, this is the first work reporting the helpful use of the biosynthesized Ag-NPs as bactericidal agents for clinical strains of multiresistant human pathogenic microorganisms, namely, methicillin-resistant *S. aureus*, *A. baumannii*, and *E. coli*. In addition, we are submitting the preliminary results of cell viability assays of biosynthesized Ag-NPs-treated human neutrophils.

2. Materials and Methods

2.1. Reagents. Tryptic soy broth (TSB) and Mueller Hinton broth (MHB) were obtained from BritaniaLab and prepared according to manufacturer's recommendations. Silver nitrate (>99% purity) was purchased from Cicarelli, Argentina, and employed to prepare fresh silver solutions (10 mM) in sterile distilled water for each experiment. Dextran from *Leuconostoc mesenteroides* (average molecular weight 78,000), Ficoll-Hypaque (Histopaque-1077), and Trypan blue solution were obtained from Sigma. Hank's balanced salt solution (HBSS) was prepared with sterile distilled water.

2.2. Biosynthesis of Ag-NPs. TSB medium was prepared, sterilized, and inoculated with a fresh growth of *P. aeruginosa* ATCC 27853, being incubated at 37°C for 24 h. After the incubation time, the culture was centrifuged at 10,000 rpm and the culture supernatant was used for the synthesis of Ag-NPs. Different concentrations of *P. aeruginosa* culture supernatant (10, 30, and 50% by volume) were separately added to the reaction vessels containing silver nitrate at different concentrations (1, 5, and 10 mM).

2.3. Characterization of Ag-NPs. The bioreduction of the Ag^+ ions was monitored at regular intervals by sampling aliquots (2 mL) of the reaction mixture and measuring the UV-vis spectrum of the mixture. UV-vis spectra of these samples aliquots were recorded from 200 to 800 nm on a Shimadzu UV-vis spectrophotometer at room temperature. The colloidal stability of Ag-NPs was evaluated by zeta potential measurements using a Delsa™Nano C instrument (Beckman Coulter). Furthermore, the biosynthesized nanoparticles were characterized using transmission electron microscopy (TEM). Morphological analysis of Ag-NPs was carried out using TEM images acquired with a JEM-JEOL 1120 EXII model microscope operating at 80 kV. Samples were prepared by adding one drop of the reaction mixture onto a holey carbon-coated copper TEM grid and allowing it to dry in air. The crystal structure and chemical composition of Ag-NPs were determined by X-ray diffraction (XRD) analysis using an X-ray diffractometer (PANalytical X-Pert Pro) with Cu K-alpha radiation that was operated at 40 kV and 40 mA at 2θ range of 30–70°.

2.4. Bacterial Strains. The antimicrobial activity of biosynthesized Ag-NPs was examined in several representative Gram-positive and Gram-negative bacterial strains. The following Gram-positive microorganisms were evaluated: *S. aureus* ATCC 29213, methicillin-sensitive *S. aureus* (MSSA) clinical strain 1, MSSA clinical strain 2, MSSA clinical strain 3, methicillin-resistant *S. aureus* (MRSA), *S. epidermidis* ATCC 12228, and *E. faecalis* ATCC 29212. Among Gram-negative microorganisms were tested *P. mirabilis* clinical strain, *A. baumannii* clinical strain, *E. coli* ATCC 25922, *E. coli* clinical strain 1, *E. coli* clinical strain 2, *P. aeruginosa* ATCC 27853, and *K. pneumoniae* ATCC 700603. All bacterial strains were grown aerobically in MHB for 24 h at 37°C.

2.5. Determination of Minimum Inhibitory Concentration and Minimum Bactericidal Concentration of the Ag-NPs and Time-Death Assays. The standard tube dilution method on MHB was used to evaluate the antimicrobial efficacy of the Ag-NPs. Strains coming from cultures of 24 h in MHB medium were diluted to 10^6 CFU/mL and incubated for 10 min at 37°C. The Ag-NPs concentrations added to bacterial suspensions were ranged from 0.025 to 51.2 pM. Bacterial growth was observed at 18 h of incubation following the indications of the Clinical and Laboratory Standards Institute (CLSI). The lowest concentration of the Ag-NPs that inhibited bacterial growth was considered to be the minimum inhibitory concentration (MIC). Minimum bactericidal concentration

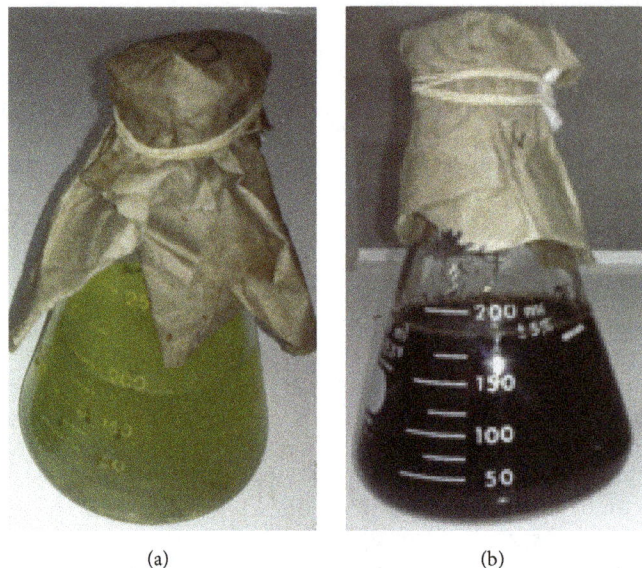

(a)   (b)

Figure 1: Biosynthesis of Ag-NPs. (a) Culture supernatant of *P. aeruginosa* without Ag⁺ ions after 24 h of incubation. (b) Culture supernatant of *P. aeruginosa* with AgNO₃ 10 mM after 24 h of incubation.

(MBC) measured was the lowest concentration that reduced initial inoculums to 99.9%. Time-death assays were conducted in the *S. aureus* and *E. coli* reference strains in the presence of 0.6 pM Ag-NPs biosynthesized. Both strains at a starting inoculum of 10^7 CFU/mL in 2 mL of MHB were incubated for 2.5 h at 37°C with constant agitation and then they were spiked with the nanoparticles. In different times, an aliquot of the bacterial suspension was collected, diluted in phosphate buffer solution, and plated on Mueller Hinton agar plates in the absence of Ag-NPs. Colonies were counted after 24 hours at 37°C.

2.6. Neutrophils Preparation from Human Blood and Cell Viability Assay. Human neutrophils were isolated by a combined dextran/Ficoll-Hypaque sedimentation procedure. Sedimentation in dextran 6% was performed before gradient centrifugation. A mixture of Ficoll-Hypaque was then used to isolate the mononuclear cells from the remaining haematic cells. After sedimentation, hypotonic lysis of the erythrocytes was carried out. The neutrophil layer was washed twice and suspended in HBSS. Cell preparations were adjusted to ~10^6 cells/mL for the assay.

The Trypan blue exclusion test was used to determine the number of viable cells present in a cell suspension exposed to Ag-NPs at 40 pM. In this test, a cell suspension is simply mixed with Trypan blue 0.02% and then examined to determine whether cells take up or exclude dye. In the protocol presented here, a viable cell will have a clear cytoplasm whereas a nonviable cell will show blue cytoplasm. Values of viability of treated cells were expressed as percentage of that from corresponding control cells.

2.7. Ethics Statement. Healthy volunteers were involved in this study for the human blood donation and all participants signed written informed consent before participation. This study was approved by the Chemical School Institutional Review Board and complies with the Argentinean (ANMAT 5330/97) and international (Declaration of Helsinki) principles and bioethical codes.

3. Results and Discussion

Addition of different concentrations of *P. aeruginosa* culture supernatant (10, 30, and 50% by volume) to aqueous AgNO₃ solution at different concentrations (1, 5, and 10 mM) resulted in the biosynthesis of Ag-NPs. However, the best compromise to generate higher amount of Ag-NPs with lower polydispersity was reached with a 10 mM AgNO₃ solution and a *P. aeruginosa* culture supernatant concentration at 30% by volume. Figures 1(a) and 1(b) display the visual change in color from greenish to yellowish brown of the culture supernatant incubated at 37°C with Ag⁺ ions after 24 h of reaction, whereas no color change could be observed in culture supernatant without AgNO₃. The bioreduction of the Ag⁺ ions was confirmed by UV-vis spectroscopy as shown in Figure 2. Among the UV-vis spectra, a strong-broad absorption band centered at about 420 nm is observed and assigned to a surface plasmon [13], indicating the presence of Ag-NPs biosynthesized using the *P. aeruginosa* culture supernatant, while the absorption peak centered at around 300 nm is attributed to the silver ions. The zeta potential of Ag-NPs in the present study was found to be −36.0 mV suggesting that the repulsive forces between the nanoparticles would be responsible for electrostatic stability. This proves evidence that Ag-NPs were dispersed in the medium. Morphology and size distribution of Ag-NPs obtained were examined by transmission electron microscopy (TEM). A representative TEM image and a particle size histogram of the biosynthesized nanoparticles by extracellular matrix from *P. aeruginosa* are

TABLE 1: Minimum inhibitory concentration (MIC) of Ag-NPs and ciprofloxacin and minimum bactericidal concentration (MBC) of Ag-NPs for different bacterial species.

Bacterial strain	Ag-NPs			Ciprofloxacin
	MIC (pM)	MBC (pM)	MBC/MIC	MIC (μM)
Gram-positive bacteria				
S. aureus ATCC 29213	0.8	0.8	1.0	1.6
MSSA clinical strain 1	0.8	0.8	1.0	0.8
MSSA clinical strain 2	0.4	0.4	1.0	0.8
MSSA clinical strain 3	0.4	0.8	2.0	0.4
MRSA clinical strain	3.2	3.2	1.0	99.1
S. epidermidis ATCC 12228	3.2	6.2	1.9	3.1
E. faecalis ATCC 29212	0.8	0.8	1.0	0.8
Gram-negative bacteria				
P. mirabilis clinical strain	0.4	0.4	1.0	0.4
A. baumannii clinical strain	0.8	0.8	1.0	1.6
E. coli ATCC 25922	1.6	3.2	2.0	0.4
E. coli clinical strain 1	1.6	1.6	1.0	0.8
E. coli clinical strain 2	3.2	3.2	1.0	1.0
P. aeruginosa ATCC 27853	6.4	6.4	1.0	3.1
K. pneumoniae ATCC 700603	0.8	1.6	2.0	0.4

MSSA: methicillin-sensitive S. aureus; MRSA: methicillin-resistant S. aureus.

shown in Figures 3(a) and 3(b), respectively. It can be seen that the nanoparticles are spherical and roughly spherical and relatively uniform in diameter between 25 and 45 nm. A possible mechanism that may explain the biosynthesis of Ag-NPs is considering that the NADH-dependent nitrate reductase, which is an enzyme secreted by P. aeruginosa, may be responsible for the reduction of Ag^+ to Ag^0 and the subsequent Ag-NPs formation. The bioreduction may occur by means of the electrons from NADH where the NADH-dependent reductase can act as a carrier [9, 11]. An X-ray diffraction pattern of the biosynthesized Ag-NPs is shown in Figure 4. Three peaks at 38.1°, 44.2°, and 64.5° corresponding to the (111), (200), and (220) planes of silver were confirmed using standard powder diffraction data of JCPDS number 04-0783. All peaks corresponded to a face centered cubic (fcc) symmetry. In addition to these representative peaks of fcc silver nanocrystal, other peaks can be observed in Figure 4 suggesting the crystallization of a bioorganic phase on the surface of nanoparticles and Ag-NPs stabilization [14].

The continuous selection of bacteria that are resistant to a wide range of antibiotics has led to the resurgence in the research of novel unconventional sources of antibiotics. Accordingly, the antimicrobial properties of the biosynthesized Ag-NPs against representative Gram-positive and Gram-negative bacterial pathogens were explored in this work. We challenged clinical and reference strains of S. aureus, S. epidermidis, E. faecalis, P. mirabilis, A. baumannii, E. coli, P. aeruginosa, and K. pneumonia with different concentrations of Ag-NPs (from 0.1 to 51.2 pM) using the conventional tube macrodilution method to determine MIC and MBC of the Ag-NPs (see Table 1).

It can be observed in Table 1 that the biosynthesized Ag-NPs were effective agaynst all the bacterial species studied

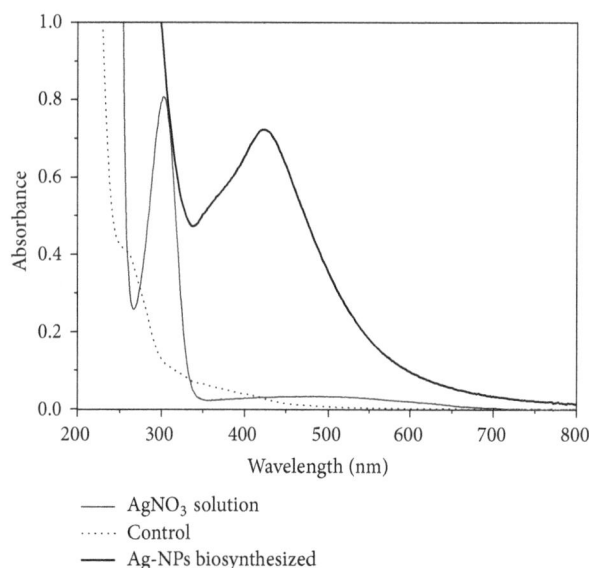

FIGURE 2: UV-visible spectra of Ag-NPs biosynthesized (black line), AgNO$_3$ solution (gray line), and P. aeruginosa culture supernatant (control, dotted line). The absorption of Ag-NPs was recorded after the addition of a culture supernatant of P. aeruginosa at 30% by volume to 10 mM AgNO$_3$ solution. The curve was recorded after 24 h of incubation.

and notable for their MIC at picomolar levels estimated between 0.4 and 6.4 pM. Comparing with a conventional clinical antibiotic, such as ciprofloxacin, the Ag-NPs obtained showed the higher growth inhibition effect against all of the tested bacterial species and significantly lower levels of concentration (μM and pM for ciprofloxacin and Ag-NPs, resp.).

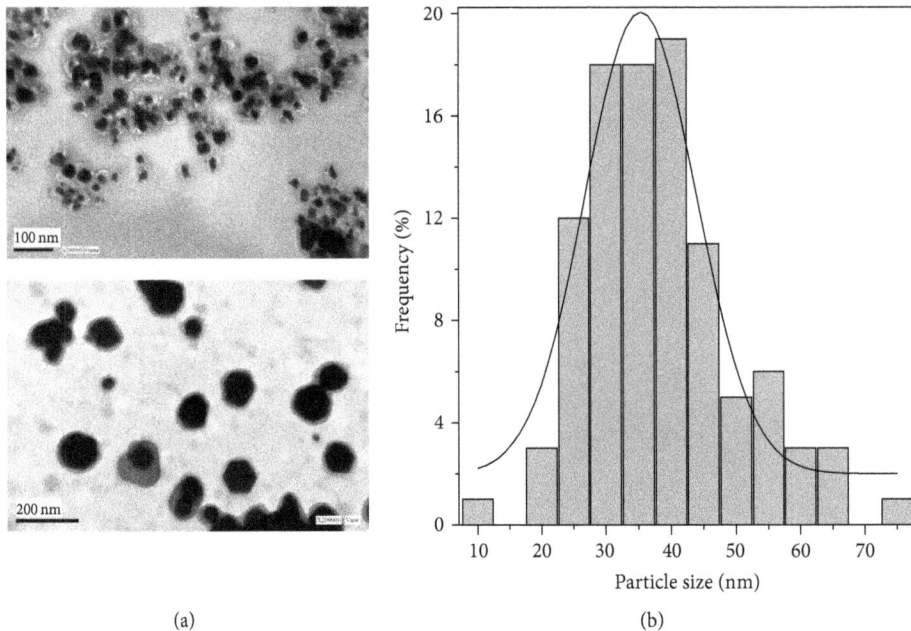

(a)   (b)

FIGURE 3: (a) Representative TEM images of Ag-NPs biosynthesized by reducing Ag^+ ions using a culture supernatant of *P. aeruginosa*. (b) Particle size histogram of Ag-NPs from TEM image showing the distribution of nanoparticles.

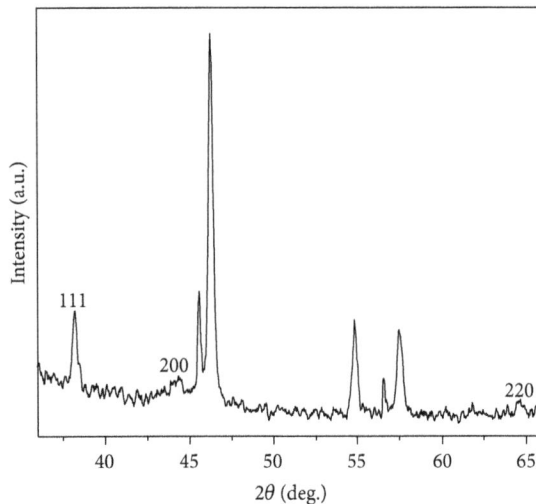

FIGURE 4: XRD spectrum of the biosynthesized Ag-NPs.

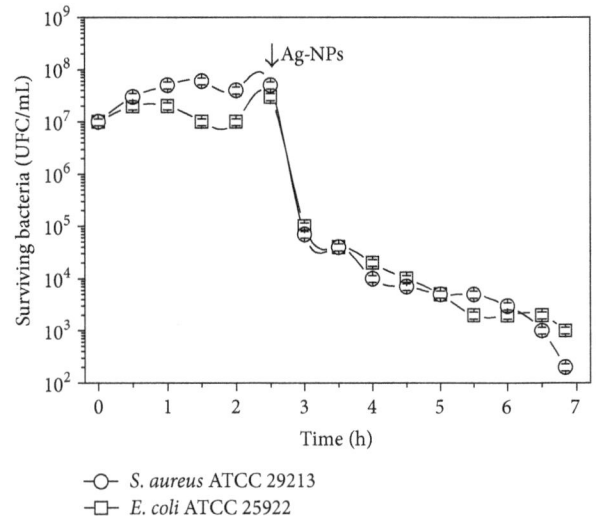

FIGURE 5: Time-death curves for *S. aureus* ATCC 29213 (-O-) and *E. coli* ATCC 25922 (-□-) using 0.6 pM Ag-NPs biosynthesized.

These results demonstrated that Ag-NPs may be used as potential antimicrobial agents and suggest the broad spectrum nature of their antimicrobial activity. The MIC values observed for *P. aeruginosa* and *S. epidermidis* were higher than for other bacterial strains, which could be explained for their capacity to form biofilm [15] and then to reduce the Ag-NPs-mediated antimicrobial action. Considering the MBC/MIC ratio as a measure of the bactericidal power of an antimicrobial agent (bactericidal agent: MBC/MIC ≤ 2; bacteriostatic agent: MBC/MIC > 2), the results listed in Table 1 allow pointing out a bactericidal activity of Ag-NPs in the bacterial species tested. Additionally, the

bactericidal kinetics of Ag-NPs biosynthesized were analyzed from time-death curve experiments using *S. aureus* ATCC 29213 and *E. coli* ATCC 25922, as models for Gram-positive and Gram-negative bacteria, respectively. The results obtained showed a reduction of $3 \log_{10}$ after 4 h of incubation with an Ag-NPs concentration at 0.6 pM (see Figure 5). Ag-NPs were powerful bactericidal agents against clinical pathogenic strains of *methicillin-resistant S. aureus*, *A. baumannii*, and *E. coli*, which have been considered some of the most virulent multidrug-resistant microorganisms for the human population [16]. This is a markedly promising result since the use of the biosynthesized Ag-NPs may be one of the approaches for overcoming bacterial resistance and playing an advanced key role in pharmacotherapeutics.

The mechanism of the Ag-NPs-mediated bactericidal effect remains to be understood. Several studies propose that Ag-NPs attach to the cell wall affecting its membrane integrity, thus disturbing permeability and respiration functions of the cell [9]. Likewise, the antibacterial activity of Ag-NPs is size dependent, and smaller Ag-NPs having the large surface area available for interaction are more effective antimicrobial agents than larger ones. Then, it is possible that Ag-NPs not only interact with the cell membrane, but can also penetrate inside the bacteria [8]. Another possible mechanism involved in the antimicrobial activity of Ag-NPs is the release of Ag^+ ions that play a partial but important role in their bactericidal effect [9].

Cell viability in response to Ag-NPs was estimated by Trypan blue exclusion test for cells in contact with much higher Ag-NPs concentrations than the MIC/MBC determined. After 30 min and 3 h incubation, the cell viability was greater than 80% and 50%, respectively. These preliminary results demonstrated that the biosynthesized Ag-NPs have a negligible cytotoxic effect in human neutrophils even after 3 h of exposure to nanoparticles, suggesting low toxicity to the host. Thus, the unconventional antimicrobial agent obtained may be used in patients without side effect, being an alternative to control the infectious diseases caused by different pathogenic bacteria.

4. Conclusion

We reported a simple and green chemistry approach for the biological synthesis of Ag-NPs using the culture supernatant of a *P. aeruginosa* reference strain at 37°C and without any harmful reducing agents. The nanoparticles were characterized by means of UV-vis spectroscopy and transmission electron microscopy. TEM analysis confirmed the relatively uniform distribution of Ag-NPs and their roughly spherical shapes. The antimicrobial activity of the biosynthesized Ag-NPs was evaluated and it was found that this nanomaterial at picomolar concentration levels has bactericidal activity against representative human Gram-positive and Gram-negative pathogens including clinically isolated multidrug-resistant bacteria such as methicillin-resistant *S. aureus*, *A. baumannii*, and *E. coli*. This is notable since Ag-NPs have proved to be effective antibacterial agents regardless of the drug-resistance mechanisms that exist in human pathogenic microorganisms and may be a potential candidate as effective broad spectrum bactericidal agents and nontoxic to the host.

Competing Interests

The authors declare no competing interests regarding the publication of this paper.

Authors' Contributions

Melisa A. Quinteros and Ivana M. Aiassa Martínez contributed equally to this work.

Acknowledgments

The authors wish to acknowledge the financial support of CONICET (PIP 11220130100702CO, PIO 14520140100013CO) and ANPCyT-FONCyT (PICT 2014 N° 1663) from Argentina and SECyT-UNC from Córdoba, Argentina. Melisa A. Quinteros and Ivana M. Aiassa Martínez thank CONICET for the doctoral and postdoctoral fellowships, respectively. The authors would also like to thank Juan Carlos Fraire for assistance in TEM experiments.

References

[1] B. Spellberg, R. Guidos, D. Gilbert et al., "The epidemic of antibiotic-resistant infections: a call to action for the medical community from the infectious diseases society of America," *Clinical Infectious Diseases*, vol. 46, no. 2, pp. 155–164, 2008.

[2] D. J. Diekema, K. J. Dodgson, B. Sigurdardottir, and M. A. Pfaller, "Rapid detection of antimicrobial-resistant organism carriage: an unmet clinical need," *Journal of Clinical Microbiology*, vol. 42, no. 7, pp. 2879–2883, 2004.

[3] A. R. M. Coates and Y. Hu, "Novel approaches to developing new antibiotics for bacterial infections," *British Journal of Pharmacology*, vol. 152, no. 8, pp. 1147–1154, 2007.

[4] M. J. Hajipour, K. M. Fromm, A. Akbar Ashkarran et al., "Antibacterial properties of nanoparticles," *Trends in Biotechnology*, vol. 30, no. 10, pp. 499–511, 2012.

[5] M.-A. Neouze, "Nanoparticle assemblies: main synthesis pathways and brief overview on some important applications," *Journal of Materials Science*, vol. 48, no. 21, pp. 7321–7349, 2013.

[6] L. Zhang, F. X. Gu, J. M. Chan, A. Z. Wang, R. S. Langer, and O. C. Farokhzad, "Nanoparticles in medicine: therapeutic applications and developments," *Clinical Pharmacology and Therapeutics*, vol. 83, no. 5, pp. 761–769, 2008.

[7] H. M. C. De Azeredo, "Antimicrobial nanostructures in food packaging," *Trends in Food Science and Technology*, vol. 30, no. 1, pp. 56–69, 2013.

[8] M. Rai, A. Yadav, and A. Gade, "Silver nanoparticles as a new generation of antimicrobials," *Biotechnology Advances*, vol. 27, no. 1, pp. 76–83, 2009.

[9] S. Eckhardt, P. S. Brunetto, J. Gagnon, M. Priebe, B. Giese, and K. M. Fromm, "Nanobio silver: its interactions with peptides and bacteria, and its uses in medicine," *Chemical Reviews*, vol. 113, no. 7, pp. 4708–4754, 2013.

[10] K. N. Thakkar, S. S. Mhatre, and R. Y. Parikh, "Biological synthesis of metallic nanoparticles," *Nanomedicine: Nanotechnology, Biology, and Medicine*, vol. 6, no. 2, pp. 257–262, 2010.

[11] N. I. Hulkoti and T. C. Taranath, "Biosynthesis of nanoparticles using microbes—a review," *Colloids and Surfaces B: Biointerfaces*, vol. 121, pp. 474–483, 2014.

[12] C. G. Kumar and S. K. Mamidyala, "Extracellular synthesis of silver nanoparticles using culture supernatant of *Pseudomonas aeruginosa*," *Colloids and Surfaces B: Biointerfaces*, vol. 84, no. 2, pp. 462–466, 2011.

[13] P. Mulvaney, "Surface plasmon spectroscopy of nanosized metal particles," *Langmuir*, vol. 12, no. 3, pp. 788–800, 1996.

[14] N. Basavegowda and Y. Rok Lee, "Synthesis of silver nanoparticles using Satsuma mandarin (*Citrus unshiu*) peel extract: a novel approach towards waste utilization," *Materials Letters*, vol. 109, pp. 31–33, 2013.

[15] K. Kalishwaralal, S. BarathManiKanth, S. R. K. Pandian, V. Deepak, and S. Gurunathan, "Silver nanoparticles impede the biofilm formation by *Pseudomonas aeruginosa* and *Staphylococcus epidermidis*," *Colloids and Surfaces B: Biointerfaces*, vol. 79, no. 2, pp. 340–344, 2010.

[16] A. L. Cohen, D. Calfee, S. K. Fridkin et al., "Recommendations for metrics for multidrug-resistant organisms in healthcare settings: SHEA/HICPAC position paper," *Infection Control and Hospital Epidemiology*, vol. 29, no. 10, pp. 901–913, 2008.

5

Influence of Processing Conditions on the Mechanical Behavior and Morphology of Injection Molded Poly(lactic-co-glycolic acid) 85:15

Liliane Pimenta de Melo,[1,2] **Gean Vitor Salmoria,**[1,2] **Eduardo Alberto Fancello,**[1,3] **and Carlos Rodrigo de Mello Roesler**[1]

[1]*Biomechanical Engineering Laboratory (LEBm), University Hospital (HU), Federal University of Santa Catarina, 88040-900 Florianópolis, SC, Brazil*

[2]*Laboratory of Innovation on Additive Manufacturing and Molding (NIMMA), Federal University of Santa Catarina, 88040-900 Florianópolis, SC, Brazil*

[3]*GRANTE, Department of Mechanical Engineering, Federal University of Santa Catarina, 88040-900 Florianópolis, SC, Brazil*

Correspondence should be addressed to Liliane Pimenta de Melo; liliane.eng@gmail.com

Academic Editor: Jie Deng

Two groups of PLGA specimens with different geometries (notched and unnotched) were injection molded under two melting temperatures and flow rates. The mechanical properties, morphology at the fracture surface, and residual stresses were evaluated for both processing conditions. The morphology of the fractured surfaces for both specimens showed brittle and smooth fracture features for the majority of the specimens. Fracture images of the notched specimens suggest that the surface failure mechanisms are different from the core failure. Polarized light techniques indicated birefringence in all specimens, especially those molded with lower temperature, which suggests residual stress due to rapid solidification. DSC analysis confirmed the existence of residual stress in all PLGA specimens. The specimens molded using the lower injection temperature and the low flow rate presented lower loss tangent values according to the DMA and higher residual stress as shown by DSC, and the photoelastic analysis showed extensive birefringence.

1. Introduction

Implants for medical applications using resorbable polymers derived from a class of aliphatic polyesters, polyhydroxy acids, are widely used for internal fracture fixation, wound closure, sutures, small vessel ligation, and drug delivery [1, 2]. During the injection molding process, polymeric materials undergo complex thermomechanical histories and significant changes in their rheological, mechanical, and thermochemical properties due to the large pressure variations, cooling times, mold geometry, and the manufacturing process [3–7]. The polymer's mechanical properties (apparent elastic modulus, maximum strength), morphology, crystallinity, and frozen layer thickness are also influenced by injection molding parameters, such as the melting process temperature, injection flow rate, holding pressure, mold temperature,

and average bulk temperature [2, 8–11]. Poly(glycolide) and poly(L-lactide-co-glycolide), which are the synthetic copolymers of lactic acid (α-hydroxypropionic acid) and glycolic acid (hydroxyacetic acid), respectively, have good fiber-forming properties but their thermomechanical histories influence the ductility and degradability of the corresponding manufactured devices [12, 13].

The crystallinity and frozen layer thickness are controlled by the combined effect of the cooling rate and the stress fields imposed during the melting process [14–16]. Viana and collaborators [5] concluded that the thickness of the PLLA frozen layer increases with the stress level and decreases with temperature, while its degree of crystallinity increases with both shear stresses and temperature. On the other hand, Pantani et al. (2005) [17] indicated that the poly(acid

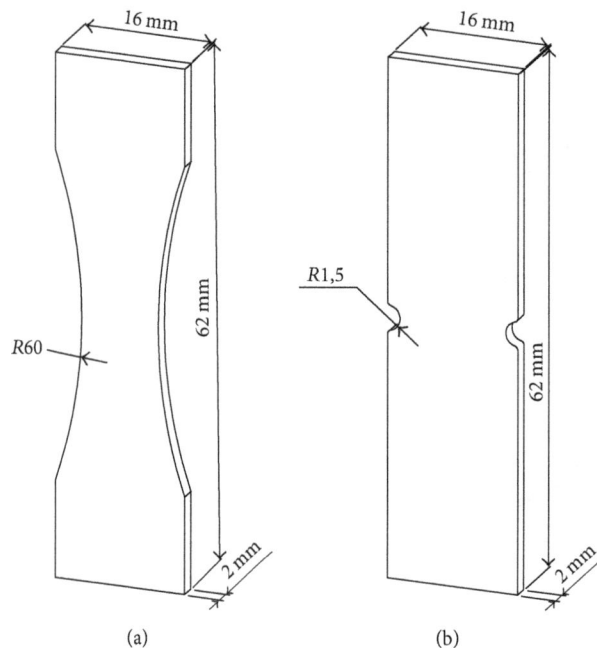

FIGURE 1: Illustration of the specimen's geometries used for injection molding: unnotched specimen (a) and notched specimen (b).

lactide) frozen layer thickness increased when either the flow rate or the mold temperature decreased and that a correlation existed between the two parameters. Residual stresses and molecular orientations throughout a product provide important information about how that product will perform. Residual stresses are introduced by nearly all techniques used for polymer manufacturing and they can also be introduced by nonuniform flow, differential packing, or cooling. Therefore, an assessment of the mechanical behavior and structural characteristics of PLGA resulting from distinct injection molding parameters of absorbable polymers can provide valuable information.

This study provided an overview among processing conditions, morphology, and mechanical property relationship of injection molded PLGA. Two specimen groups with different geometries (notched and unnotched) were injection molded using two melting temperatures and flow rates (low and high). These choices generated four different processing conditions for both groups. For each processing condition, the mechanical properties (apparent elastic modulus, ultimate strength, elongation at failure, storage modulus, and loss tangent), morphology at fracture surface, and residual stress were evaluated.

2. Experimental

2.1. Materials. Poly(lactic-co-glycolic acid) 85/15 granules, commercially available as Purasorb PLG 8531, purchased from Corbion Purac Biomaterials (Holland), were used in this study. The PLGA 85/15 average molecular weight of Mn = 224,271 g/mol and polydispersity of 1.87 were determined using Gel Permeation Chromatography (GPC) (Viscotek

TABLE 1: Injection molding conditions for notched and unnotched PLGA specimens.

Injection condition	T (°C)	Q (cm^3 s^{-1})
I	240	25
II	240	10
III	210	25
IV	210	10

VE 2001, Viscotek detector TDA 302, USA, 2008). This copolymer has T_g of 57 ± 1°C, T_m = 140°C, and 3.04 dL/g of intrinsic viscosity (chloroform, 25°C, c = 0.1 g/dL).

2.2. Injection Molding Specimens. PLGA pellets were injection molded with an ARBURG 270S/250-70 machine into two groups of specimen geometries, notched and unnotched, adapted from ASTM D1822 type S and ASTM D638 type V. Both groups had a rectangular format of 62 × 16 mm of length and a cross section of 10 × 2 mm. Notched specimens had a 1.5 mm notch radius (stress concentration factor of 2.4), while unnotched specimens had a narrower section with a radius of 60 mm (see Figure 1).

Two (low and high) melt injection temperatures and two injection (low and high) flow rates were investigated, generating four injection conditions shown in Table 1.

The other processing parameters had the following fixed values: mold temperature: 25°C, injection pressure: 1500 MPa, holding pressure: 25 MPa, injection time: 2 s, cooling time: 90 s, and screw speed: 100 rpm.

TABLE 2: Mechanical properties of notched and unnotched PLGA specimens injected with different molding conditions.

	Injection condition	T_{inj} (°C)	Q_{inj} (cm^3 s^{-1})	E (GPa)	σ_u (MPa)	ε_f (%)
Notched	I	240	25	5.6 ± 0.4	63.1 ± 1.2	2.7 ± 0.3
	II	240	10	4.8 ± 0.2	65.5 ± 1.4	3.3 ± 0.5
	III	210	25	4.8 ± 0.4	54.0 ± 11.0	1.9 ± 1.0
	IV	210	10	4.8 ± 0.3	67.6 ± 0.7	4.5 ± 0.3
Unnotched	I	240	25	3.5 ± 0.1	63.4 ± 1.1	4.34 ± 1.9
	II	240	10	3.4 ± 0.1	62.3 ± 2.3	7.1 ± 4.3
	III	210	25	3.7 ± 0.2	62.3 ± 3.0	3.4 ± 2.0
	IV	210	10	4.0 ± 0.3	64.9 ± 1.1	4.9 ± 1.1

2.3. Mechanical Characterization

2.3.1. Tensile Test.
The two different specimens were tested in an EMIC testing machine, model DL-3000, in the tensile mode as per ISO 527-1. The elongation of the specimens was measured using an extensometer, Instron/EMIC 2630-107. The tests were performed using a moving grip speed of 1 mm min^{-1} at a controlled room temperature of 23°C. Six specimens (n = 6) for each condition for each group were tested. The mechanical properties investigated were the apparent elastic modulus (taken as the initial slope of the engineering stress-strain curve) E, ultimate strength (maximum stress value of the engineering stress-strain curve) σ_u, and engineering strain at failure ε_f.

2.3.2. Dynamical Mechanical Analysis.
A DMA-Q800 analyzer (TA Instruments) with a single cantilever clamp was used for the viscoelastic tests. Dynamic mechanical analysis (DMA) provided the storage modulus E' and tan δ values at a frequency of 1 Hz within the temperature range of 30°C to 120°C using a heating rate of 3°C/min and a transversal displacement amplitude of 0.3% of the effective length of the specimen.

2.4. Scanning Electronic Microscopy.
Scanning Electronic Microscopy (SEM) analysis was used to evaluate the fractured surface of the PLGA (85/15) specimens submitted to the tensile test and also to observe the frozen layer thickness and other morphological characteristics such as molecular orientation of shear force caused by molding injection. The analysis was conducted on all conditions of injection molding for the two groups of specimens.

In order to obtain good quality PLGA images, the specimens were fixed to a support with a double-sided carbon tape. For electronic conductivity, the specimens were covered with a thin layer of gold in a sputter model D2 Diode Sputtering System, manufactured by ISI (International Scientific Instruments). The fractured surfaces and thicknesses were observed using a JEOL JSM-6390LV (FEI Company, Japan) scanning electron microscope with an accelerating voltage of 5 kV.

2.5. Differential Scanning Calorimetry and Residual Stress Analysis.
Differential scanning calorimetry (DSC) was used to determine the thermal transitions and residual stress enthalpy of the injection molded PLGA specimens in a Shimadzu DSC-6000 with nitrogen atmosphere (19 cm^3 m^{-1}), using aluminum oxide as standard. The heating rate was from 10°C to 250°C at 10°C m^{-1}, using an average sample weight of 7 mg taken from the central region of the molded specimens. The residual stresses of the manufactured specimens were also evaluated by the polarized light technique using a polariscope with polarizing and quarter-wave lenses of 250 mm diameters, following the ASTM D4093 [18].

2.6. Data Analysis.
Analysis of variance (ANOVA) was performed considering statistical significance set at 0.05; the p value was investigated for significance of the factors among injection molding conditions. All data are reported as mean ± standard deviation.

3. Results and Discussion

The processing conditions were systematically varied following a DOE array involving notch presence on geometry, melt temperature, and flow rate. Mechanical properties, as apparent elastic modulus, ultimate strength, elongation at failure, storage modulus, and loss tangent, were estimated by tensile test and DMA. The two morphological parameters, morphology at fracture surface and residual stress, were interpreted by the thermomechanical parameters. ANOVA was performed to measure statistically the significant response. The relationships between the morphology and mechanical properties were then established.

3.1. Tensile Test.
Figure 2 shows the curves for stress versus strain for the notched and unnotched molded specimens using low and high values of melt injection temperatures and the two injection flow rates.

Table 2 contains the average values of E, σ_u, and ε_f at each injection condition for both notched and unnotched groups. Similar mechanical properties were found for the injected PLGA specimens that were molded under different processing parameters (notched or unnotched samples), as shown in Table 2. But there is an evident difference between notched and unnotched samples relative to the apparent elastic modulus E mechanical property. The apparent elastic modulus E and ultimate strength σ_u show low sensitivity to the injection conditions for both the notched and the

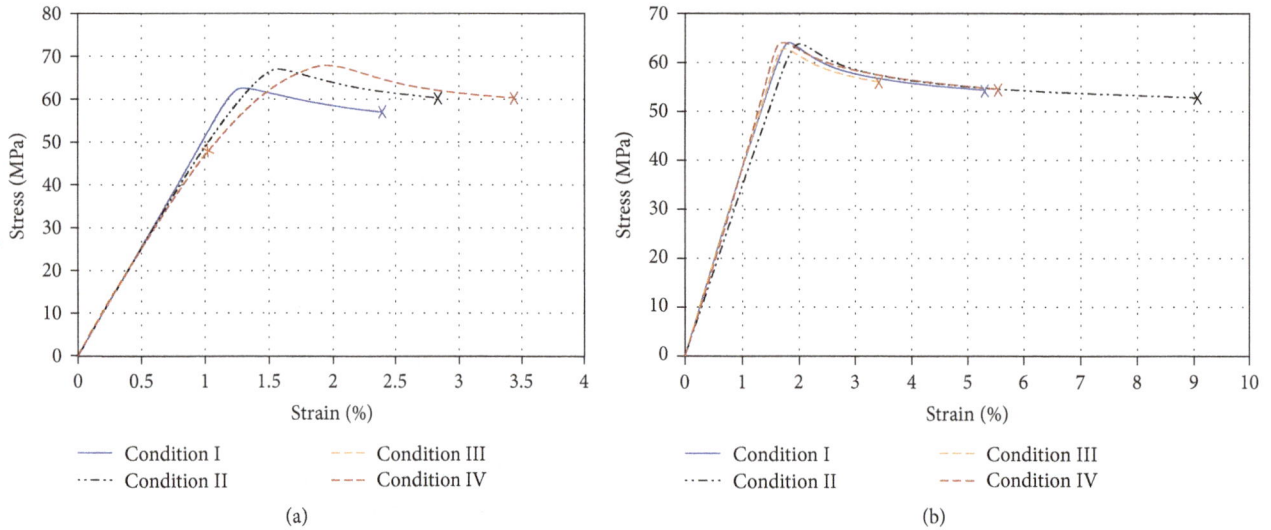

FIGURE 2: Stress-strain curves (means) of notched (a) and unnotched (b) specimens of injection molded PLGA.

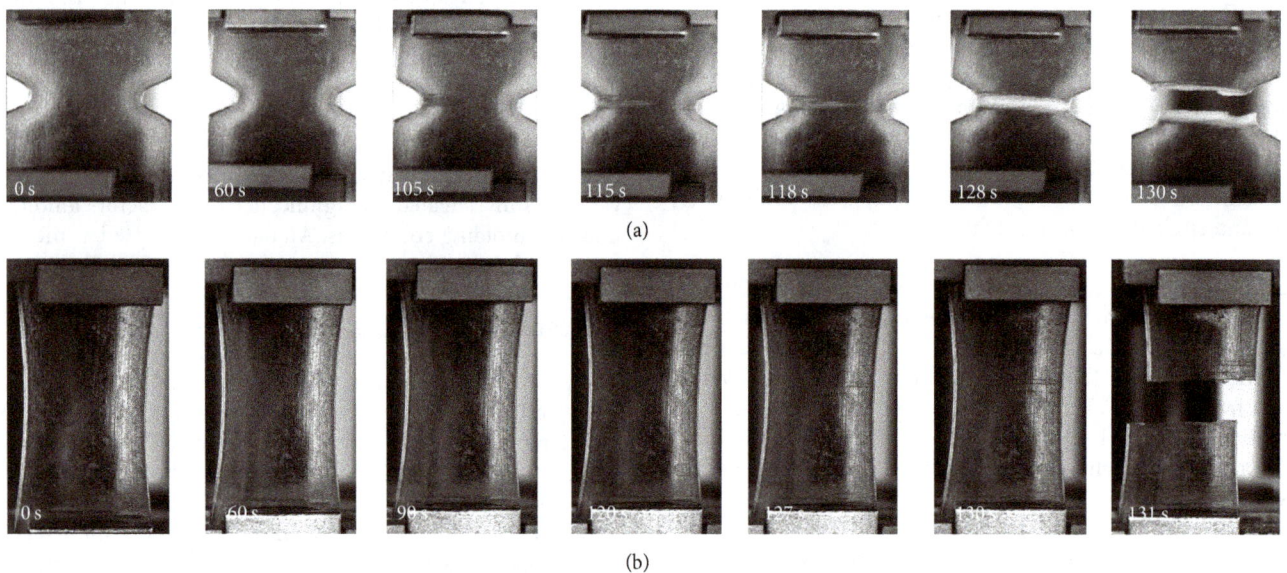

FIGURE 3: Sequence of images of tensile test for notched (a) and unnotched (b) specimens injected under Condition I.

unnotched groups of specimens. However, there was a high standard deviation found for σ_u in the notched group injected under Condition III (see Table 2).

This low sensitivity reveals a certain level of material toughness. In order to achieve nearly the same value of σ_u for both geometries, the material localized near the notch valley seems to allow plastic deformation during loading to finally reach an almost constant stress distribution prior to the occurrence of a complete cross section (plastic) collapse. On the other hand, it is important to note that specimens, even those injected under the same conditions, showed different macroscopic behaviors at failure. While some showed a clear necking formation, others fractured without this formation. This observation is consistent with the large standard deviation found for ε_f.

Sensitivity to the injection conditions showed that strains at failure presented slightly higher mean values for Conditions II and IV (low injection flow rates) than for Conditions I and III (high injection flow rates), for both notched and unnotched groups.

3.2. Scanning Electronic Microscopy. Figure 3 presents a sequence of images that illustrate the progressive localization (necking) of strains prior to total failure of one of the notched specimens injected under Condition I. In these pictures, the capacity of the material to withstand plastic strain is macroscopically visible, as was already mentioned when discussing ultimate strength σ_u.

Unnotched specimens, as Figure 3 shows, present elongation along the specimen with longitudinal spread failure

Condition   Unnotched   Notched

(I)

(II)

(III)

(IV)

(a)   (b)

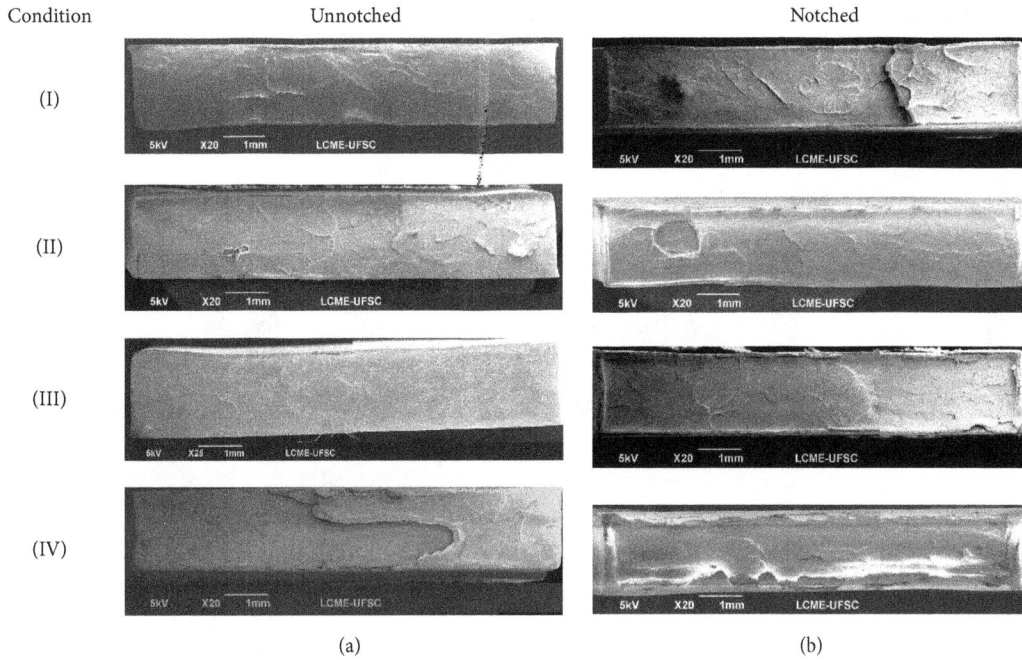

FIGURE 4: SEM of fractured surface unnotched (a) and notched (b) PLGA specimens molded under different processing conditions.

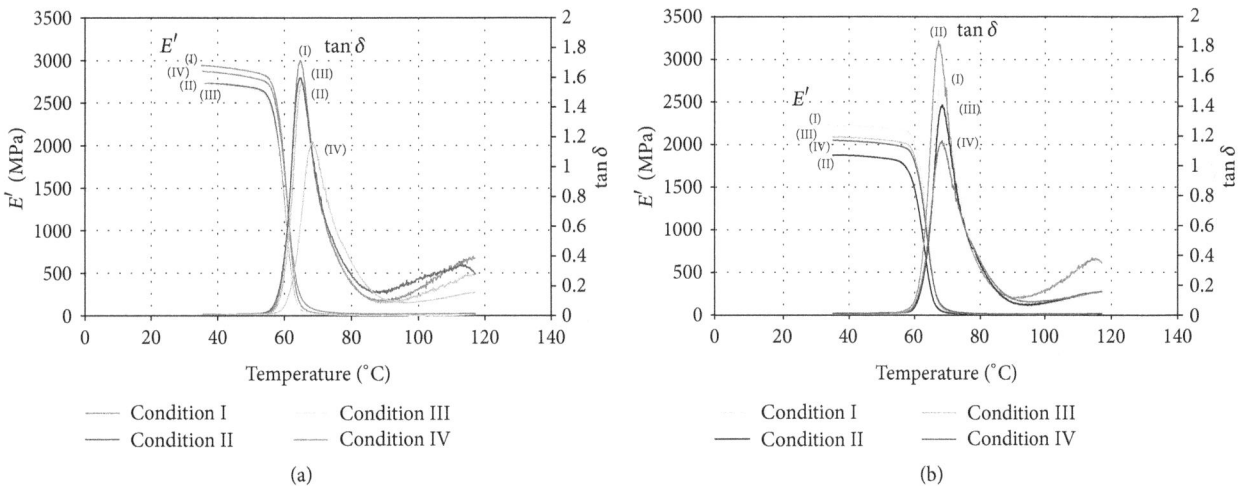

(a)   (b)

FIGURE 5: Storage modulus and loss tangent as a function of temperature for PLGA specimen injection molded under different conditions (I, II, III, and IV). Curve for unnotched (a) and notched (b) specimens.

after 60 s, while notched specimens present stretching in the center region due to the stress concentrator (e.g., notch). Although there were different behaviors, the brittle feature of the material occurs in both geometries (see Figure 4).

It is worth emphasizing once more that a different macroscopic behavior at failure was observed even in specimens injected under the same condition; while some specimens presented clear necking formation, others failed without this formation showing flatter fracture surfaces (Figure 3).

SEM images of fractured surfaces are shown in Figure 4. These images are representative of those samples that did not show a clear necking formation. In these figures, it is possible to see flat fractured surfaces with some evidence of

localized plastic flow, mainly in the notched specimens. The plastic fractures along the borders of the notched specimens are clear. This can be related to a different behavior between the core and borders and is due to the existence of the frozen layer. The existence of a thicker frozen layer in this group of specimens seems to be consistent with the fact that thicker frozen layers are related to higher flow rates as seen in the notched region of this group.

3.3. Dynamical Mechanical Analysis. Figure 5 shows representative curves of storage modulus E' and tan δ (tan $\delta = E''/E'$) as functions of temperature for notched and unnotched specimens.

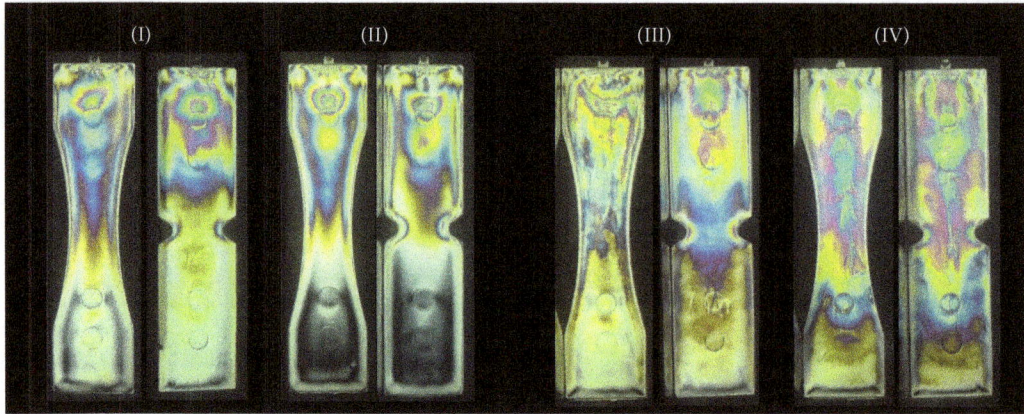

FIGURE 6: Photoelastic fringes of notched and unnotched specimens for Conditions I, II, III, and IV.

TABLE 3: Relaxation enthalpy values for PLGA specimens determined by DSC.

PLGA specimens	Condition I	Condition II	Condition III	Condition IV
Unnotched	2.8 (±0.3) J/g	4.2 (±0.6) J/g	3.7 (±0.3) J/g	3.8 (±0.5) J/g
Notched	3.6 (±0.2) J/g	3.7 (±0.5) J/g	5.2 (±0.3) J/g	4.5 (±0.4) J/g

Sensitivity of the storage modulus to the injection conditions did not show a clear trend in the notched and unnotched specimens. Remember that due to the differences in geometry the storage moduli of the notched and unnotched specimens are not comparable. The only visible response is the lowest value of the loss tangent (tan δ) reached by Condition IV (lower temperature and flow rate) for both notched and unnotched geometries, characterizing lower dissipation due to viscous micromechanisms.

Moreover, DMA was carried out to characterize the resulting copolymer in the four conditions. As shown in Figure 5, one obvious transition behavior was observed, designated as relaxation. It is well known that the glass transition temperature (T_g) of a polymer can be determined by relaxation, as it is usually related to the segment movements in the noncrystalline area. This behavior, associated with broad peaks, nonexisting in a second heating scan, is characteristic of an incomplete crosslink process of copolymers formulations. In fact, tan δ also indicates the composite damping capacity, which has a maximum value at the amorphous transition. As a PLGA copolymer, a possible reason why the curves went up above 100°C is the presence of residual monomers.

3.4. Differential Scanning Calorimetry and Residual Stress Analysis. The injection molding of transparent polymers can induce a peculiar stress field that is clearly detected by photoelastic stress analysis [18, 19]. In Figure 6, the isochromatic maps for notched and unnotched specimens are shown. Notably, the isochromatic fringes have an asymmetric pattern distribution. This means that injection molding imposed an asymmetric thermomechanical environment onto the injected polymer that is related to the concentration of residual stresses near the injection gate.

The stress concentration decreases uniformly on the opposite side of the specimen at a different rate for each molding condition. The residual stresses arise during the filling and the packing processes. The wide distribution of residual stresses present in the specimens molded with the lower temperature is probably due to the nonuniform shear stress during cavity filling and rapid solidification. On the other hand, the concentrated residual stresses near the gate present in the specimens molded with the higher temperature are due to the compressive force caused by the holding pressure during the slower solidification.

In terms of stress shielding, Condition IV is the most propulsive to show a different behavior, since Condition 4 presents the lower temperature and lower flow rate. Then, the mold filling during injection molding along the specimen exhibited different characteristics due to the high shear stress between the polymer and the wall of the mold. The shear stress causes different lines of residual stress between notched and unnotched specimens.

DSC curves (Figure 7) show the transitions for the PLGA pellets and for the PLGA molded specimens under different conditions. The molded specimen curves present a clear endothermic peak together with the glass transition related to the stress relaxation enthalpy of PLGA [20]. The residual stress was determined by measuring this relaxation enthalpy peak area at glass transition and is presented in Table 3. The relaxation enthalpy values were higher for the notched specimens molded under Conditions III and IV; that is, the lower temperature resulted in higher residual stresses at the center of the specimens in the notched region.

The DSC curves evidence the difference in the endothermic peaks and the presence of crystallinity only for Condition II. For this condition, a shoulder indicating two melting temperatures is present, related to PGA (15%) and PLA (85%)

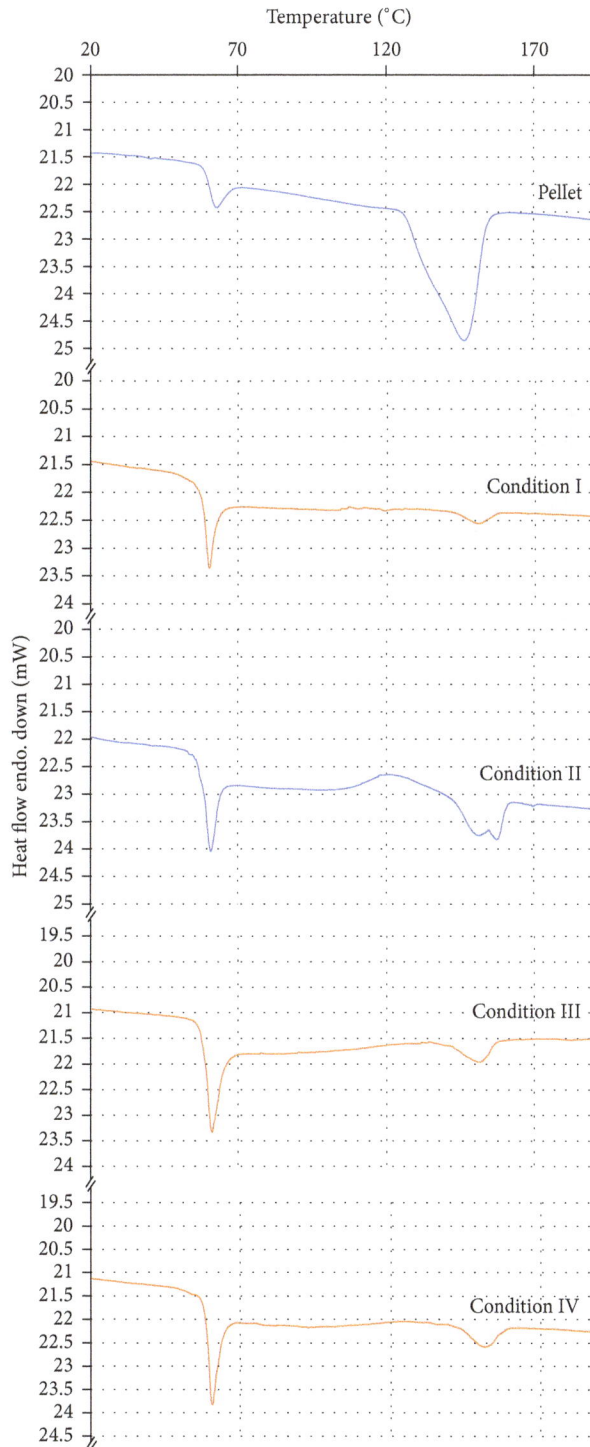

FIGURE 7: Calorimetry curves for PLGA pellets and molded specimens under different process conditions (I, II, III, and IV).

fractions, respectively, even though this value is very close to the detected T_m.

Moreover, it is possible for residual stress to be reduced or eliminated with heat treatment to avoid influencing mechanical properties and fracture surface morphology of PLGA specimens. The present DSC curves are about the processed specimens without any treatment to show the results of the injection molding process. Heat treatment after the manufacturing of the devices in medical applications could possibly cause degradation of the material.

4. Conclusion

Similar mechanical properties were found for the injected PLGA specimens that were molded under different processing parameters. The melt temperature can influence the injection molded device and can be influenced by other parameters of injection molding. Flow rate is strongly associated with the shear rate and therefore has effects on melt temperature, molecular orientation, strains at failure, and residual stresses. The morphology of fractured surfaces of the notched and unnotched specimens showed flat and smooth fractures for the majority of the specimens. The macroscopic mechanical behavior of the injected specimens presented low notch sensitivity, suggesting the existence of a certain level of material toughness. The strains at failure presented slightly higher mean values for Conditions II and IV (low injection flow rate) than for Conditions I and III (high injection flow rate), for both notched and unnotched specimens. There were localized deformations near the specimen surface different from the core region. This can be related to the orientation of the skin layer, especially in the notched specimens. Polarized light techniques indicated birefringence throughout all specimens, especially in those molded under lower temperature, which suggests residual stress due to rapid solidification. DSC analysis confirmed the existence of residual stress in all PLGA specimens. The specimens molded using the lower injection temperature and lower flow rate (Condition IV) presented lower loss tangent values according to DMA and higher residual stress as shown by DSC, and photoelastic analysis demonstrated extensive birefringence along the specimen. Molecular restriction in the chain rotation and conformation due to the thick oriented skin layer can explain the less viscous behavior observed.

Conflicts of Interest

The authors declare that there are no conflicts of interest regarding the publication of this paper.

Acknowledgments

The authors would like to thank PRONEX/FAPESC, CNPQ, and FINEP for financial support and the Center of Microscopy, UFSC, for providing the micrographs.

References

[1] L. Fambri, C. Migliaresi, K. Kesenci, and E. Piskin, "Biodegradable polymers," in *Integrated Biomaterials Science*, R. Barbucci, Ed., pp. 119–187, Kluwer Academic/Plenum Publishers, New York, USA, 2002.

[2] J. W. Leenslag, A. J. Pennings, R. R. M. Bos, F. R. Rozema, and G. Boering, "Resorbable materials of poly(l-lactide). VI. Plates and screws for internal fracture fixation," *Biomaterials*, vol. 8, no. 1, pp. 70–73, 1987.

[3] S. Ghosh, J. C. Viana, R. L. Reis, and J. F. Mano, "Effect of processing conditions on morphology and mechanical properties of injection-molded poly(L-lactic acid)," *Polymer Engineering and Science*, vol. 46, no. 7, pp. 1141–1147, 2007.

[4] D. Cardozo, "Three Models of the 3D Filling Simulation for Injection Molding: A Brief Review," *Journal of Reinforced Plastics and Composites*, vol. 27, pp. 1963–1974, 2008.

[5] J. C. Viana, A. M. Cunha, and N. Billon, "The thermomechanical environment and the microstructure of an injection moulded polypropylene copolymer," *Polymer*, vol. 43, no. 15, pp. 4185–4196, 2002.

[6] S.-L. Yang, Z.-H. Wu, W. Yang, and M.-B. Yang, "Thermal and mechanical properties of chemical crosslinked polylactide (PLA)," *Polymer Testing*, vol. 27, no. 8, pp. 957–963, 2008.

[7] L.-T. Lim, R. Auras, and M. Rubino, "Processing technologies for poly(lactic acid)," *Progress in Polymer Science*, vol. 33, no. 8, pp. 820–852, 2008.

[8] C. D. Han, *Rheology and Processing of Polymeric Materials*, vol. 1, Polymer Technology Oxford University Press, New York, USA, 2007.

[9] D. F. Gibbons, "Tissue response to resorbable synthetic polymers," in *Degradation Phenomena on Polymeric Biomaterials*, H. Planck, M. Dauner, and M. Renardy, Eds., pp. 97–104, Springer, New York, USA, 1992.

[10] S. Ghosh, J. C. Viana, R. L. Reis, and J. F. Mano, "Bi-layered constructs based on poly(l-lactic acid) and starch for tissue engineering of osteochondral defects," *Materials Science and Engineering C*, vol. 28, no. 1, pp. 80–86, 2008.

[11] H. Ben Daly, B. Sanschagrin, K. T. Nguyen, and K. C. Cole, "Effect of polymer properties on the structure of injection-molded parts," *Polymer Engineering & Science*, vol. 39, no. 9, pp. 1736–1751, 1999.

[12] T. Kijchavengkul, R. Auras, M. Rubino, S. Selke, M. Ngouajio, and R. T. Fernandez, "Biodegradation and hydrolysis rate of aliphatic aromatic polyester," *Polymer Degradation and Stability*, vol. 95, no. 12, pp. 2641–2647, 2010.

[13] G. L. Racey, W. R. Wallace, C. J. Cavalaris, and J. V. Marguard, "Comparison of a polyglycolic-polylactic acid suture to black silk and plain catgut in human oral tissues," *Journal of Oral Surgery*, vol. 36, no. 10, pp. 766–770, 1978.

[14] S.-H. Hyon, K. Jamshidi, and Y. Ikada, "Synthesis of polylactides with different molecular weights," *Biomaterials*, vol. 18, no. 22, pp. 1503–1508, 1997.

[15] A. M. Brito, A. M. Cunha, A. S. Pouzada, and R. J. Crawford, "Predicting the Skin-Core Boundary Location in Injection Moldings," *International Polymer Processing*, vol. 4, pp. 307–404, 1991.

[16] H.-C. Kuo and M.-C. Jeng, "Effects of part geometry and injection molding conditions on the tensile properties of ultra-high molecular weight polyethylene polymer," *Materials and Design*, vol. 31, no. 2, pp. 884–893, 2010.

[17] R. Pantani, I. Coccorullo, V. Speranza, and G. Titomanlio, "Modeling of morphology evolution in the injection molding process of thermoplastic polymers," *Progress in Polymer Science*, vol. 30, no. 12, pp. 1185–1222, 2005.

[18] ASTM, "Standard test method for photoelastic measurements of birefringence and residual strains in transparent or translucent plastic materials," ASTM D4093, ASTMA, Consshohocken, PA, USA, 2014.

[19] W. Dally and F. R. William, *Experimental Stress Analysis*, College House Enterprises, LLC, New York, third edition, 1991.

[20] W. S. Pietrzak, "Rapid cooling through the glass transition transiently increases ductility of PGA/PLLA copolymers: A proposed mechanism and implications for devices," *Journal of Materials Science: Materials in Medicine*, vol. 18, no. 9, pp. 1753–1763, 2007.

6

Metabolic Engineered Biocatalyst: A Solution for PLA based Problems

Sundus Riaz[ID],[1,2] **Nosheen Fatima,**[1] **Ahmed Rasheed,**[3] **Mehvish Riaz,**[4] **Faiza Anwar,**[2] **and Yamna Khatoon**[5]

[1]*Department of Biomedical Engineering and Sciences, National University of Sciences & Technology, Islamabad, Pakistan*
[2]*Pakistan Agricultural Research Council, FQSRI, SARC, Karachi, Pakistan*
[3]*PhD. Scholar, Sun Yat-Sen University (East Campus), Higher Education Mega Centre North, Guangzhou, China*
[4]*MPH, London South Bank University, UK*
[5]*Postgraduate Scholar, Department of Agriculture and Agribusiness Management, University of Karachi, Karachi, Pakistan*

Correspondence should be addressed to Sundus Riaz; sundusriaz_fuuast@yahoo.com

Academic Editor: Wen-Cheng Chen

Polylactic acid (PLA) is a biodegradable thermoplastic polyester. In 2010, PLA became the second highest consumed bioplastic in the world due to its wide application. Conventionally, PLA is produced by direct condensation of lactic acid monomer and ring opening polymerization of lactide, resulting in lower molecular weight and lesser strength of polymer. Furthermore, conventional methods of PLA production require a catalyst which makes it inappropriate for biomedical applications. Newer method utilizes metabolic engineering of microorganism for direct production of PLA through fermentation which produces good quality and high molecular weight and yield as compared to conventional methods. PLA is used as decomposing packaging material, sheet casting, medical implants in the form of screw, plate, and rod pin, etc. The main focus of the review is to highlight the synthesis of PLA by various polymerization methods that mainly include metabolic engineering fermentation as well as salient biomedical applications of PLA.

1. Introduction

Polylactic acid (PLA) is a rigid thermoplastic polymer that has semicrystalline or amorphous geometry, depending on the optical purity of the polymer backbone [1]. Lactic acid has two optically active forms out of which L-lactic acid is the natural and most common form, whereas D-lactic acid is produced either by microorganisms or racemization. Furthermore, it acts much like comonomers which optimize the kinetics of crystallization for specific fabrication processes and applications [1]. Properties of PLA are similar to polyethylene terephthalate (PET) and polypropylene; these are petrochemical based polymer used for packaging applications [1]. PLA is a polymer which has wide range of applications in both biomedical and packaging industry, because it has ability to be stress crystallized, impact modified, filled, thermally crystallized, copolymerized, and processed in most polymer processing equipment [1]. It is unique in many ways and behaves like PET but also performs a lot like polypropylene. PLA has better organoleptic characteristics which makes it excellent for food contact and their related packaging applications [1].

Polylactic acid is produced by polymerization of lactic acid, and that is produced by two methods, i.e., chemical method and fermentation method [2]. Chemical method utilizes petrochemical resources followed by addition of HCN and specific catalyst to synthesized lactic acid [2]. On the other hand, fermentation method utilizes renewable resources, such as carbohydrate (monosaccharide and disaccharides) in the fermented broth to obtain lactic acid [3]. Optical purity of lactic acid is very important and hence is of major concern in production of PLA. Chemical method produces racemic mixture of both D (-) and L (+) lactic acid while fermentation method produces only one optically pure form of D (-) or L (+) lactic acid, respectively [2].

Production of lactic acid

Chemical Method   Fermentation Method

Utilizes petrochemical   Utilizes renewable Resources
resources and catalyst   and microbial fermentation
for lactic acid Production   process for lactic acid production

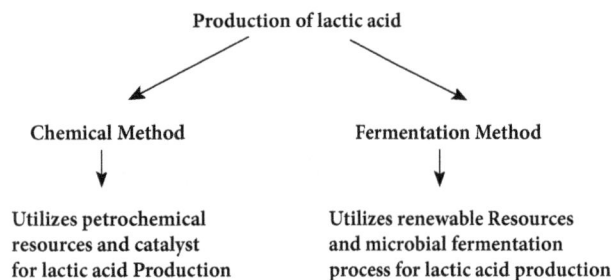

FIGURE 1: Methods of production of lactic acid.

The main advantage of PLA that has encouraged its use in packaging industry is its high strength, biodegradability, antimicrobial, and antioxidant properties [3]. Although there is still a big market of petrochemical based polymer, these polymers have many disadvantages, because they adversely affect oil and gas resources which make them harmful as far as environment is concerned [3]. PLA is environmentally friendly because of its biodegradable properties and it can rapidly be degraded into less toxic byproducts like CO_2 and H_2O which saves environment from hazardous effects [3].

2. Synthesis of Lactic Acid Monomer for PLA Production

The cost factor for synthesis of lactic acid is raw material which is used in fermentation medium [4]. Production of lactic acid by fastidious lactic acid bacteria is usually a costly procedure [4]. Raw materials for lactic acid production are usually based on cheap polymeric waste and side stream materials [4]. These cheap materials are widely studied for high yield lactic acid [4]. For quality production of PLA, both optically and chemically pure lactic acid are required. Lactic acid synthesized from microbial strains produces optical pure lactic acid under optimized fermentation conditions [4].

Lactic acid yield from fermentation of monosaccharide usually has a very high molecular weight (> 90 %) [4]. Main impurity in the fermentation medium is being the cell mass itself that can be easily separated from the product [4]. Figure 1 illustrates the methods by which lactic acid is produced.

3. Chemical Method of Lactic Acid Production

This method utilizes acetaldehyde reaction with hydrogen cyanide in the presence of catalyst to produce lactonitrile [19]. This reaction occurs in liquid phase and at high atmospheric temperature [19]. After completion of reaction, lactonitrile is purified and hydrolyzed to produce lactic acid [19]. This method produces racemic mixture of both D (-) and L (+) lactic acid [2]. Furthermore, the metal catalyst employed in this process is difficult to remove which makes it unfit in many applications. Figure 2 presents the schematic illustration of chemical method.

4. Fermentation Method of Lactic Acid Production

Fermentation is an energy yielding process and is a characteristic of anaerobic bacteria [19]. Bacteria produce lactic acid by utilizing simple sugars like glucose, lactose, and galactose, without any requirement of heating process [19]. There are three types of fermentation process: (1) batch fermentation, (2) fed batch fermentation, and (3) continuous fermentation [19]. Batch and fed batch fermentation produce high concentration of lactic acid, whereas continuous fermentation produces higher productivity. Fermentation is usually carried out in controlled temperature and pH condition [19].

Both bacteria and fungi can produce lactic acid through fermentation but the yield of lactic acid by fungi is very low as compared to yield of lactic acid through bacteria [19]. To decrease cost for lactic acid production by fermentation, cheap raw material like lignocellulosic biomass is employed which is a promising feedstock due to its great availability, sustainability, and low cost as compared to refined sugars [20]. But the commercial use of lignocellulose for lactic acid production is still problematic because extensive pretreatment of enzyme is required to obtain fermentable sugars from lignocelluloses biomass [20].

Microorganism chosen for production of lactic acid should have high yield factor along with low cell mass at the expense of low cost raw material in low pH and at high temperature, along with negligible byproducts [21]. Continual improvements have been carried out in production and purification of lactic acid and are summarized in Table 1 [1].

5. Role of Bacterial Cultures in Lactic Acid Production

Lactic acid producing bacteria serve as starting material for production of lactic acid [3]. Both bacteria and fungi produce lactic acid but, for fungal production of lactic acid, aerial condition is required because it also shows low reaction rate [3].

Bacteria that carry out fermentation are divided into two groups, namely, homofermentative and heterofermentative, respectively. A homofermentative bacterium produces one product only at one time so production of side products can be minimized [3]. Example of such lactic acid bacteria by homofermentation are Lactococcus, Enterococcus, Streptococcus, and some Lactobacilli [4]. Homofermentative lactic acid bacteria metabolize hexose sugar entirely by Embden-Meyerhof pathway [4]. Industries are using homofermentative procedure for L-lactic acid production through specie of Lactobacillus genus, specifically with Lactobacillus delbrueckii, L. amylophilus, L. bulgaricus, and L. leichmannii [22]. Other than lactic acid bacteria, there are two other bacteria that produce lactic acid by fermentation. Thermotolerant B. coagulans utilizes glucose and xylose to produce yield of 96 % and 88 %, respectively; this is achieved at R_p (2.5 g/h) and product concentration (100 g/l) [4]. Yeast-like Candida utilis has been metabolically engineered by pyruvate decarboxylase deletion and L-lactate dehydrogenase expression to produce lactic acid from glucose with yield of 95% [4].

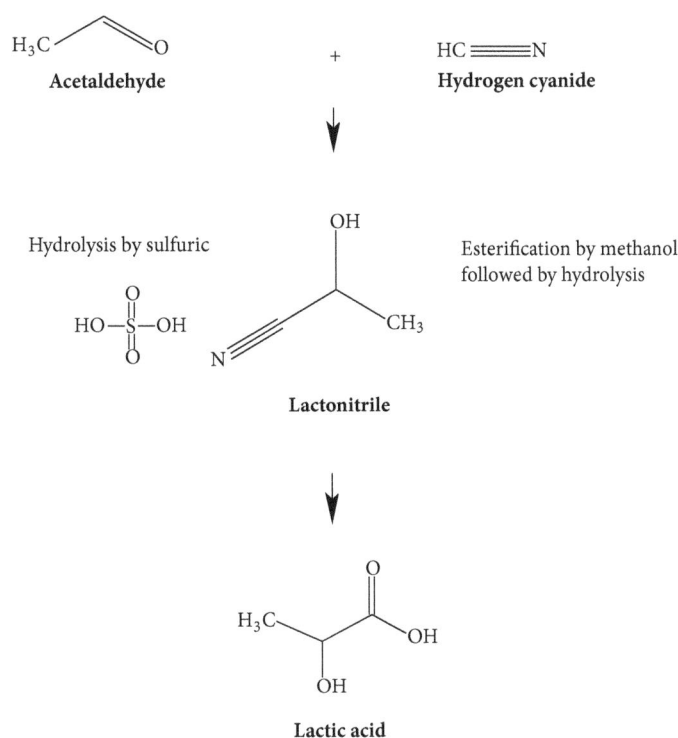

FIGURE 2: A schematic presentation of production of Lactic Acid by chemical process.

TABLE 1: Microorganism along with their yield of production of lactic acid by fermentation.

Sr. No.	Microorganism producing lactic acid	Substrate involved	Genetic modification	Yield of lactic acid
1	*T. aotearoense* SCUT27[5]	Lignocellulosic biomass[5]	Engineered to block the acetic acid formation pathway[5]	0.93 g/g glucose with an optical purity of 99.3%[5]
2	*Lactobacillus amylovorus* ATCC 33622[6]	Liquefied corn starch[6]	Nil	$20 \, g \, l^{-1} \, h^{-1}$[6]
3	*L. helveticus* [6]	Whey[6]	Nil	$35 \, g \, l^{-1} \, h^{-1}$[6]
4	*Enterococcus faecalis* CBRD01[7]	Glucose[7]	Nil	5 g l–1 h–1[7]
5	*L. delbrueckii* NCIM 2025[8]	Cane molasses concentration of 150 g/L (equivalent to 78 g total sugar). [8]	*adh-ve* mutant by UV radiation[8]	78±1.2 (g/g) [8]
6	*L. plantarum* LMISM6 [9]	Molasses 193.50 g L-1[9]	NIL	94.8 g L-1[9]
7	Thermophilic *Bacillus* sp. XZL4[10]	Corn stover hydrolyzate 162.5 g L-1[10]	NIL	1.86 g L-1 h-1[10]
8	*Lactococcus lactis*[11]	Glucose 60 gl-1 [11]	Nil	35 gl-1[11]
9	Escherichia coliBAD-ldh[12]	1g l-1 of fructose[12]	Overexpression of L-ldh gene derivative[12]	0.62 g l-1[12]
10	Escherichia coli[13]	56 g/L of crude glycerol[13]	Overexpression of GlpK/GlpD gene[13]	50 g/L of L-lactic acid[13]

FERMENTATION

↓

CELL MASS AND PROTEIN REMOVAL

↓

RECOVERY AND PURIFICATION OF
LACTIC ACID

↓

CONCENTRATION OF LACTIC ACID

↓

COLOR REMOVAL

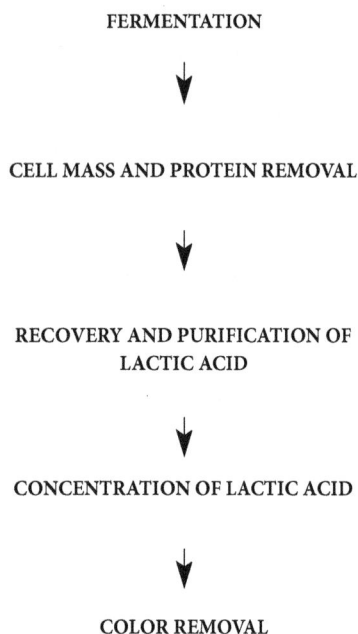

FIGURE 3: A schematic presentation of steps involved in production and purification of lactic acid by fermentation.

Fermentation media containing lactic acid

↓ Calcium Hydroxide

Coagulation of proteins, inactivation of microorganism, solubilize Ca lactate and degrade residual sugars

↓ Carbon treatment

Colored components in broth is removed

↓ Acidification with 63% Sulfuric acid

Precipitation of calcium sulphate

↓

Bleaching, concentration and evaporation

↓

Lactic acid 82% concentration

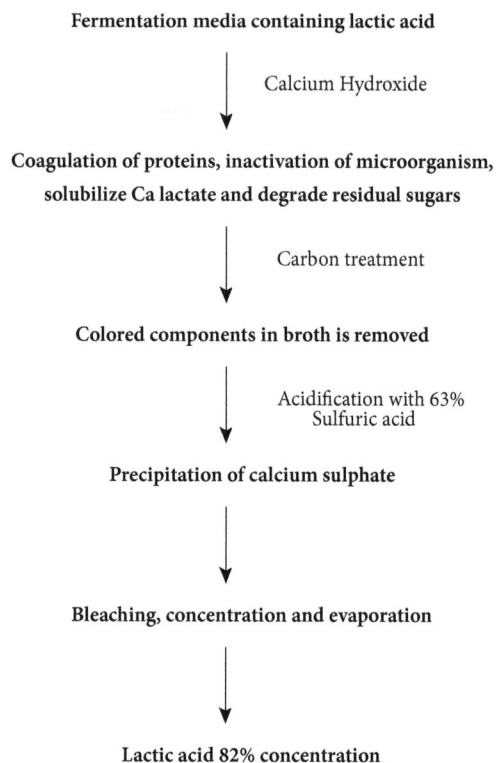

FIGURE 4: A schematic presentation of recovery and purification of lactic acid from fermentation of broth by adsorption.

A heterofermentative bacterium produces ethanol and CO_2 along with lactic acid [3]. Examples of organism that are heterofermentative are *Leuconostoc, Weissella,* and *Lactobacillus brevis* [4]. Heterofermentative bacteria utilize both hexose and pentose sugars to produce lactic acid. Heterofermentative bacteria are also employed to produce polyols such as mannitol, erythritol, ethanol, and acetic acid [4]. Recently, *Lactobacillus* strains are being utilized in fermentation for production of lactic acid [3].

New biotechnological improvements have been carried out so as to increase the yield of lactic acid production and to reduce side products. Metabolic engineering has been carried out in *Lactobacillus* strains so as to increase flux of lactic acid production [4]. In some metabolic engineering experiments, overexpression of genes does not cause increase in lactic acid yield. Such experiments are performed on *Lactobacillus plantarum* and *L. lactis*. *Lactobacillus plantarum* are metabolically engineered by overexpression of L-LDH but still show no increase in yield of lactic acid [4]. Similarly, overexpression of glyceraldehyde-3-P dehydrogenase (GAPDH) in *L. lactis* strain does not limit the glycolytic flux either in growing or resting cells [4].

Metabolic engineering is also performed to obtain optically pure lactic acid. For example, altering activity of L-LDH in *Lactobacillus helveticus* is used for production of optically pure L-lactic acid [4]. Figure 3 shows steps involved in purification of lactic acid by fermentation.

6. Isolation and Purification of Lactic Acid from Fermentation Medium

For recovery and purification, lactic acid is adsorbed in suitable polymeric adsorbents [23]. Since polymeric adsorbents are nontoxic to fermentation broth, they can be used directly in fermentation medium [23]. In this process, strong alkali adsorbent is added which converts lactic acid into its basic salt [23]. The adsorbed lactic acid is then desorbed by adding strong acid like H_2SO_4 [23]. Figure 4 details the isolation scheme of lactic acid from medium.

Lactic acid can also be isolated from fermentation medium by reactive extraction [23]. This process requires liquid-liquid extraction along with reversible chemical complexion [23]. In this method, tertiary amine is used as extractant, L-decanol is used as diluent, and trimethylamine is used as stripping solution. Aqueous phase comprises fermentation medium [24]. To carry out extraction process, equal amounts of aqueous and organic phases are added and shaken for a definite period of time [24]. After shaking, aqueous phase is decanted and concentration of acid is determined on organic phase.

7. Production of Polylactic Acid

The following are two methods that are conventionally being used in polymerization of lactic acid.

(i) Direct polymerization [25]

(ii) Ring opening polymerization of cyclic diester lactide [26]

7.1. Direct Polymerization

7.1.1. Polycondensation of Lactic Acid. In this method, lactic acid (either produced chemically or by fermentation) is subjected to heat under vacuum at 50°C [25]. Disadvantage of this process is that it produces many side products of distillation which can contaminate the reaction mixture such as lactyl lactic acid [24]. Product of polycondensation yields low-molecular-weight polymer with low mechanical properties along with higher reaction times [26].

7.1.2. Melt Condensation.
This type of condensation is possible only if the temperature of the reaction remains above the melting temperature of the polymer [27]. This process produces high molecular weight polymer in short period of time. Reaction time for melt polymerization is ≤15 h. This method is cost-effective because of its simplified procedure but it requires sensitive reaction condition. To avoid these limitations, melt/solid polycondensation technique is developed that uses a binary catalyst which is tin dichloride hydrate and p-toluene sulfonic acid. Process involves thermal oligocondensates of lactic acid which were first subjected to melt polycondensation and later to solid state polycondensation [26]. So, after reaction, high molecular weight PLA is obtained having molecular weight around 600,000. Melt condensation is a relatively economical and easy to control process. Melt condensation is affected by factors such as temperature, reaction time catalyst, and pressure [27]. So consideration must be employed to these factors to obtain high molecular weight PLA [27]. Metal catalyst used in this process is difficult to remove and makes resulting polymer unfit for biomedical applications.

7.2. Ring Opening Polymerization of Cyclic Diester Lactide

7.2.1. Formation of Lactide.
Lactide is oligomer of polylactic acid and for its formation lactic acid is added to a reactor containing vacuum and a stirrer, and zinc oxide or $Sn(OEt)_2$ is added in to the reaction mixture as catalyst [26]. As water is removed at high temperature the oligomerization is promoted; after that, temperature is quickly increased and yellow liquid is distilled which on cooling converts into needle-like crystals. These crystals are recrystallized at least 4 times to obtain pure colorless crystals of lactide [26].

7.2.2. Ring Opening Polymerization.
Ring opening polymerization (ROP) requires reaction initiator (Tetraphenyltin) for polymerization [28]. In this method, pure lactide is placed in clean, dried polymerization tube; the appropriate amount of initiator is dissolved in benzene and placed in polymerization tube [28]. The whole system is kept in oil bath under vacuum or nitrogen at 60-100°C. The reaction mixture is allowed to sublime [28]. Following sublimation, the tube is again immersed in oil bath, and after predetermined interval the contents of polymerization are removed and kept at -15°C. The extent of polymerization is determined by gel permeation chromatography [28]. Disadvantage of ROP processes is that it requires high temperature which initiates side reactions that hinder its propagation [15]. ROP is divided into two categories, namely, cationic ring opening polymerization and anionic ring opening polymerization [15]. The process of cationic ROP can produce low-molecular-weight poly lactic acid while anionic ROP process can lead to racemization [15]. Furthermore, ROP uses tin as catalyst which is incorporated into polymer system, making the resulting polymer unfit for biomedical applications.

8. Bioproduction of Polylactic Acid

Polylactic acid and its copolymer can be prepared via genetic manipulation of microorganism following the process of fermentation [14]. Metabolically engineered E. coli with propionate CoA transferase and polyhydroxyalkanoate (PHA) synthase can be used for production of polylactic acid and its copolymers [14]. This metabolically engineered bacterium utilizes glucose as a substrate for the production of polylactic acid and its copolymers [14]. To enhance biosynthesis of polylactic acid and its copolymers, MBEL 6–19 PHA synthase (PhaC1Ps6–19) is engineered with in vitro mutagenesis to generate lactyl CoA, which enhance the PLA production, which is analyzed by gas chromatography [14]. Metabolic pathways of E. coli for production of PLA are further modified by deletion of genes, which are ackA, ppc, and adhE genes, respectively [16]. Promoters' genes are also replaced from ldhA and acs to trc promoter based on in silico genome-scale metabolic flux. Recombinant strain of E. coli can be used for making homopolymer lactic acid [16].

Chemical medium having pH 7.0 is used for production of recombinant strains of E. coli [17]. This medium contains potassium dihydrogen phosphate, ammonium phosphate, magnesium sulphate hepta hydrate, citric acid, and trace metal solution per liter, respectively [17]. Seed cultures of E. coli are prepared in Luria-Bertani medium. After incubation, seed culture is inoculated to MR medium containing glucose [17]. Flask cultures are kept under temperature of 30°C at rotary shaker.

Polymer produced by bacteria is analyzed by gas chromatography, flame ionization detector [17]. To determine amount of PLA formed by fermentation, dried pellets of PLA are subjected to methanolysis with 15% sulphuric acid with internal standard as benzoic acid [17]. Products of methanolysis, such as carboxylic acid and lactate, are analyzed by gas chromatography. Methyl lactate was analyzed on mass spectrometric detector. For determination of molecular weights, gel permeation chromatography is used. PLA is subjected to differential scanning calorimetry (DSC) [17]. Figure 5 summarizes production of polylactic acid through metabolic engineering and Table 2 summarizes genetically manipulated microorganisms that are used in PLA production.

9. Source and Cloning of Gene Containing Propionate CoA Transferases

Propionate CoA transferase (PCT) gene is present in Clostridium propionicum which is regarded as alanine fermenting organism [29]. This organism is found in black mud of San Francisco bay [29]. Other organisms that produce Propionate CoA transferase are Megasphaera elsdenii, Bacteroides ruminicola, and Clostridium homopropionicum [29]. When PCT enzyme gene is overexpressed in E. coli, a

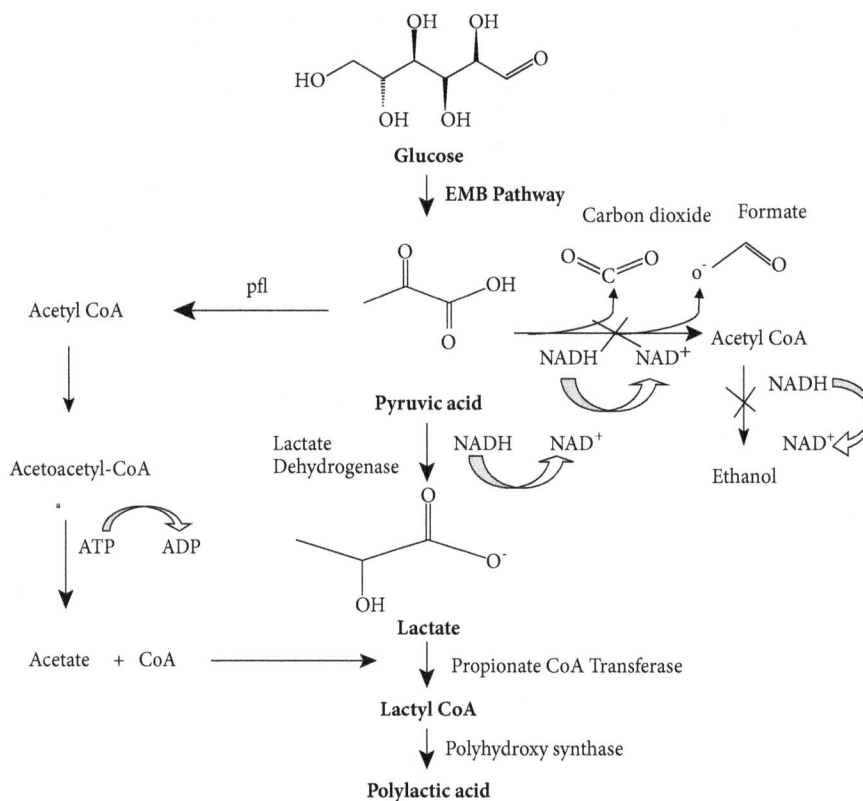

FIGURE 5: A schematic presentation of production of polylactic acid by metabolic engineering.

serious metabolic disorder is observed, which causes death of all recombinant E. coli, when an inducer is added in an isopropyl-.beta.-D-thio-galactoside (IPTG)-inducible protein expression system having a T7 promoter [30]. Due to this, constitutive expression system which expresses gene weakly but continuously with growth of a microorganism is induced [30].

For cloning of Propionate CoA transferase gene, a degenerated primer pair is introduced [29]. This primer is used to amplify a 300-bp fragment of genomic DNA obtained from C. propionicum using PCR [29]. Labelled PCR product is used for screening a library of genomic DNA from C. propionicum using k-ZAP-Express phage vector [29].

10. Source and Cloning of Gene Containing PHA Synthase

Polyhydroxyalkanoates synthase enzyme is mainly found in most genera of bacterium and members of the family Halobacteriaceae of the Archaea [31]. This enzyme utilizes thioesters of hydroxyalkanoic acids as substrate and converts them into polyhydroxyalkanoic acids [31]. For cloning of PHA gene, 8 different strategies are employed [31]. Strategy A has enzymatic approach which is employed to screen clones for functional expression of PHA gene [31]. In strategy B, after transposon mutagenesis, homologous gene probes have been obtained [31]. This strategy is used to identify the respective gene intact within the same genome [31]. In strategy C a well characterized R. eutropha PHA synthase gene is used which is used to identify corresponding genes from genomic libraries [31]. Strategy D focuses on design of short oligonucleotides with short highly conserved stretches of PHA synthases [31]. In strategy E, PHA synthase protein is purified and their oligonucleotide is designed from its N-terminal. This strategy is used to identify gene from a genomic library [31]. The most successful strategy for cloning of PHA synthase is strategy F, in which genomic libraries are screened to obtain PHA-negative wild-type organism [31]. Strategy G aimed to clone heterologous phaC genes to a PhaC-negative mutated organism [31]. In strategy H, homologous proteins encoding PHA are cloned subsequently [31]. In strategy I, bacteria is allowed to grow in a medium without carbon for storing polymer [31]. Bacteria from which PHA genes are cloned are *Paracoccus denitrificans, Rhodobacter capsulatus, Chromobacterium violaceum, Pseudomonas putida* BM01, *Methylobacterium extorquens, Comamonas acidovorans, Ectothiorhodospira shaposhnikovi, Synechocystis* sp., and *Zoogloea ramigera*, respectively [31].

11. Deletion of Other Pathways

Red recombinase expression plasmid is constructed for one-step inactivation of gene encoding pyruvate formate lyase, fumarate reductase, and LacI transcriptional repressor, respectively [14]. This Recombinant E. coli containing Red recombinase expression plasmid was cultivated at 30°C [14].

TABLE 2: Table summarizes genetically manipulated microorganisms used in PLA production.

Sr. No.	Name of Microorganism	Genetic Manipulation	Substrate Utilized	Yield of PLA	Analytical Technique
1	Escherichia coli[14]	Insertion of propionate CoA-transferase and polyhydroxyalkanoate (PHA) synthase gene[15]	Glucose [14]	43 wt% [14]	Gas Chromatography[14]
2	Escherichia coli[16]	Insertion of propionate CoA-transferase and polyhydroxyalkanoate (PHA) synthase gene [16] Knocking out the ackA, ppc, and adhE genes[16] Replacing the promoters of the ldhA and acs genes with the trc promoter[16]	56 wt% from glucose[16]	55-86 mol%[16]	Gas Chromatography[16]
3.	Escherichia coli[17]	Introduction of propionate CoA transferase (PctCp) gene from Clostridium propionicum And 19 polyhydroxyalkanoate (PHA) synthase 1 (PhaC1Ps6-19) from Pseudomonas sp. into Escherichia coli for the generation of lactyl-CoA endogenously and incorporation of lactyl-CoA[17]	62wt% glucose[17]	20–49 mol%[17]	Gas Chromatography[17]
4.	Escherichia coli[18]	introduction of heterologous pathways having engineered propionate CoA-transferase and polyhydroxyalkanoate (PHA) synthase in to E. coli for generation of lactyl-CoA [18]	46 wt% glucose[18]	70mol% [18]	Gas Chromatography[17]

The expression of Red recombinase is induced by adding 10 mM L-arabinose [14]. Pyruvate formate lyase gene is deleted by homologous recombination in two steps [14]. Firstly, 1234bp DNA fragment which contains lox71 site, chloramphenicol resistance gene, and lox66 site fused together to be obtained by PCR product [14]. Primers utilized are FDpflB1 and RDpflB1, respectively [12]. The final pcr product introduced to E. coli has pKD46 gene [14]. Screening of colonies is done on Luria-Bertani (LB) agar plate containing chloramphenicol and subsequently by direct colony PCR [14]. AdhE gene is deleted by using FDfrd1, RDfrd1, FDfrd2, and RDfrd2 for frdABCD, FDadhE1, RDadhE1, FDadhE2, and RDadhE2 primers, while lacI gene is deleted by FDlacI1, RDlacI1, FDlacI2, and RDlacI2 primers, respectively [14]. E. coli harboring chloramphenicol resistant mutants is transformed with pJW168 primer and ampicillin-resistant gene [14]. Screening is done on Luria-Bertani agar containing 100 gml^{-1} ampicillin and 1 mM IPTG. Positive colonies are cultivated and screed by PCR [14].

12. Biomedical Application of PLA

Polylactic acid is a group of bioresorbable polymers that show higher tensile strength and it erodes into harmless components when interacting with physiological fluids [32].

Due to its high tensile strength, it can be braided into sutures, stents, and scaffolds. PLA based biomaterials can be manufactured by injection molding, extrusion, spinning film, and casting process [18]. Rate of absorption of PLA depends upon molecular weight, morphology, and enantiomeric purity of PLA as PLA with high molecular weight is used to absorb it completely [18].

Nowadays, collagen and hyaluronan-based matrices are among the most popular scaffolds in clinical use, because their substrate is essential for cartilage support [18]. Furthermore PLA based scaffolds are being extensively used in tissue engineering [32].

PLA based drug delivery systems can be used in the form of pellets, nanoparticles, microcapsules, microparticles, and sustained release dosage forms. In addition, PLA based devices and drug delivery systems have been extensively used in tumors.

13. Conclusion

PLA is a biodegradable polymer used in manufacturing many biomedical devices and packaging applications. It either can be found as a sole polymer or can copolymerize with other polymers. PLA is produced from polymerization of lactic acid, and monomer lactic acid synthesized by chemical or fermentation method. Chemical method produces racemic mixture of both D(-) and L(+) forms of lactic acid while fermentation method only produces optically active L(+) form of lactic acid, which is required for PLA production. Lactic acid, whether produced from fermentation or chemical method, is further polymerized by two methods: direct polymerization of lactic acid and ring opening polymerization. These polymerization methods have many drawbacks. Firstly, the polymer synthesized from direct polymerization method produces low mechanical strength polymer; secondly, ring opening polymerization uses catalyst that makes polymer unsuitable for biomedical applications. Synthetic formation of lactic acid contains many limitations such as inability to form required L- lactic acid isomer and low product yield because it utilizes by-product as a reactant, thus making it a high cost procedure.

Recent methods of PLA utilize genetic manipulation of microorganisms that can produce polylactic acid directly by fermentation. In this context, E. coli seemed suitable for genetic modification with the insertion of propionate CoA transferase and polyhydroxyalkanoate (PHA) synthase gene that have ability to produce lactyl CoA and PLA directly by fermentation. Industries are using homofermentative procedure for L-lactic acid production because it leads to greater yield and lower amount of by-products. The PLA produced by fermentation is found to be mechanically fit for biomedical applications along high molecular weight, strength, and yield as compared to conventional methods.

Conflicts of Interest

We confirmed that there are no known conflicts of interest associated with this publication and there has been no significant financial support for this work that could have influenced its outcome.

References

[1] D. E. Henton, P. Gruber, J. Lunt, and J. Randall, "Polylactic acid technology," *Natural Fibers, Biopolymers, and Biocomposites*, vol. 16, pp. 527–577, 2005.

[2] Y.-J. Wee, J.-N. Kim, and H.-W. Ryu, "Biotechnological production of lactic acid and its recent applications," *Food Technology and Biotechnology*, vol. 44, no. 2, pp. 163–172, 2006.

[3] M. Jamshidian, E. A. Tehrany, M. Imran, M. Jacquot, and S. Desobry, "Poly-Lactic Acid: production, applications, nanocomposites, and release studies," *Comprehensive Reviews in Food Science and Food Safety*, vol. 9, no. 5, pp. 552–571, 2010.

[4] S. Taskila and H. Ojamo, "The current status and future expectations in industrial production of lactic acid by lactic acid bacteria," in *Lactic Acid Bacteria-R & D for Food, Health And Livestock Purposes*, InTech, 2013.

[5] X. Yang, Z. Lai, C. Lai et al., "Efficient production of L-lactic acid by an engineered Thermoanaerobacterium aotearoense with broad substrate specificity," *Biotechnology for Biofuels*, vol. 6, no. 1, Article ID 124, 2013.

[6] R. P. John, K. M. Nampoothiri, and A. Pandey, "Fermentative production of lactic acid from biomass: an overview on process developments and future perspectives," *Applied Microbiology and Biotechnology*, vol. 74, no. 3, pp. 524–534, 2007.

[7] M. R. Subramanian, S. Talluri, and L. P. Christopher, "Production of lactic acid using a new homofermentative Enterococcus faecalis isolate," *Microbial Biotechnology*, vol. 8, no. 2, pp. 221–229, 2015.

[8] B. Sheelendra Mangal and S. K. Srivastava, "High yield of Lactic acid production by Mutant Strain of L. delbrueckii U12-1 and Parameter Optimization by Taguchi methodology," *Annals of Biological Research*, vol. 3, pp. 2579–2592, 2012.

[9] L. F. Coelho, C. J. B. De Lima, C. M. Rodovalho, M. P. Bernardo, and J. Contiero, "Lactic acid production by new lactobacillus plantarum LMISM6 grown in molasses: Optimization of medium composition," *Brazilian Journal of Chemical Engineering*, vol. 28, no. 1, pp. 27–36, 2011.

[10] Z. Xue, L. Wang, J. Ju, B. Yu, P. Xu, and Y. Ma, "Efficient production of polymer-grade L-lactic acid from corn stover hydrolyzate by thermophilic bacillus sp. strain XZL4," *SpringerPlus*, vol. 1, no. 1, Article ID 43, 2012.

[11] L. S. Cock and A. R. de Stouvenel, "Lactic acid production from a mixture of cultures of Lactococcus lactis and Streptococcus salivarius using batch fermentation," *Revista Colombiana De Biotecnologia*, vol. 7, no. 1, 32 pages, 2005.

[12] T. E. T. Z. Mulok, M.-L. Chong, Y. Shirai, R. A. Rahim, and M. A. Hassan, "Engineering of E. coli for increased production of L-lactic acid," *African Journal of Biotechnology*, vol. 8, no. 18, 2009.

[13] S. Mazumdar, M. D. Blankschien, J. M. Clomburg, and R. Gonzalez, "Efficient synthesis of L-lactic acid from glycerol by metabolically engineered Escherichia coli," *Microbial Cell Factories*, vol. 12, no. 1, article no. 7, 2013.

[14] Y. K. Jung and S. Y. Lee, "Efficient production of polylactic acid and its copolymers by metabolically engineered Escherichia coli," *Journal of Biotechnology*, vol. 151, no. 1, pp. 94–101, 2011.

[15] J. E. Yang, S. Y. Choi, J. H. Shin, S. J. Park, and S. Y. Lee, "Microbial production of lactate-containing polyesters," *Microbial Biotechnology*, vol. 6, no. 6, pp. 621–636, 2013.

[16] Y. K. Jung, T. Y. Kim, S. J. Park, and S. Y. Lee, "Metabolic engineering of Escherichia coli for the production of polylactic acid and its copolymers," *Biotechnology and Bioengineering*, vol. 105, no. 1, pp. 161–171, 2010.

[17] T. H. Yang, T. W. Kim, H. O. Kang et al., "Biosynthesis of polylactic acid and its copolymers using evolved propionate CoA transferase and PHA synthase," *Biotechnology and Bioengineering*, vol. 105, no. 1, pp. 150–160, 2010.

[18] R. P. Pawar, S. U. Tekale, S. U. Shisodia, J. T. Totre, and A. J. Domb, "Biomedical applications of poly(lactic acid)," *Recent Patents on Regenerative Medicine*, vol. 4, no. 1, pp. 40–51, 2014.

[19] T. Ghaffar, M. Irshad, Z. Anwar et al., "Recent trends in lactic acid biotechnology: a brief review on production to purification," *Journal of Radiation Research and Applied Sciences*, vol. 7, no. 2, pp. 222–229, 2014.

[20] M. A. Abdel-Rahman, Y. Tashiro, and K. Sonomoto, "Lactic acid production from lignocellulose-derived sugars using lactic acid bacteria: Overview and limits," *Journal of Biotechnology*, vol. 156, no. 4, pp. 286–301, 2011.

[21] N. Narayanan, P. K. Roychoudhury, and A. Srivastava, "L (+) lactic acid fermentation and its product polymerization," *Electronic Journal of Biotechnology*, vol. 7, no. 2, pp. 167–178, 2004.

[22] R. Mehta, V. Kumar, H. Bhunia, and S. N. Upadhyay, "Synthesis of poly(lactic acid): A review," *Journal of Macromolecular Science - Polymer Reviews*, vol. 45, no. 4, pp. 325–349, 2005.

[23] Evangelista R. L. and Z. L. Nikolov, "Recovery and purification of lactic acid from fermentation broth by adsorption," in *Seventeenth Symposium on Biotechnology for Fuels and Chemicals*, pp. 471–480, Humana Press, Totowa, NJ, USA.

[24] M. Jarvinen, L. Myllykoski, R. Keiski, and J. Sohlo, "Separation of lactic acid from fermented broth by reactive extraction," *Bioseparation*, vol. 9, no. 3, pp. 163–166, 2000.

[25] P. Laonuad, N. Chaiyut, and B. Ksapabutr, "Poly(lactic acid) preparation by polycondensation method," *Optoelectronics and Advanced Materials – Rapid Communications* , vol. 4, no. 8, pp. 1200–1202, 2010.

[26] H. Chae Hwan, S. Hwan Kim, J.-Y. Seo, and D. S. Han, "Development of four unit processes for biobased PLA manufacturing," *ISRN Polymer Science*, vol. 2012, Article ID 938261, 6 pages, 2012.

[27] X. Lin, B. Wang, G. Yang, and M. Gauthier, "Poly (lactic acid)-based biomaterials: synthesis, modification and applications," in *Biomedical Science, Engineering And Technology*, InTech, 2012.

[28] F. E. Kohn, J. W. A. Van Den Berg, G. Van De Ridder, and J. Feijen, "The ring-opening polymerization of D,L-lactide in the melt initiated with tetraphenyltin," *Journal of Applied Polymer Science*, vol. 29, no. 12, pp. 4265–4277, 1984.

[29] T. Selmer, A. Willanzheimer, and M. Hetzel, "Propionate CoA-transferase from Clostridium propionicum," *The FEBS Journal*, vol. 269, no. 1, pp. 372–380, 2002.

[30] S.-J. Park, T.-W. Kim, H.-O. Kang, T.-H. Yang, and S.-Y. Lee., "Recombinant microorganism capable of producing polylactate or polylactate copolymer from sucrose and method for producing polylactate or polylactate copolymer from sucrose using the same," U.S. Patent 8,420,357, 2013.

[31] B. H. A. Rehm and A. Steinbüchel, "Biochemical and genetic analysis of PHA synthases and other proteins required for PHA synthesis," *International Journal of Biological Macromolecules*, vol. 25, no. 1-3, pp. 3–19, 1999.

[32] A. J. R. Lasprilla, G. A. R. Martinez, B. H. Lunelli, A. L. Jardini, and R. M. Filho, "Poly-lactic acid synthesis for application in biomedical devices - A review," *Biotechnology Advances*, vol. 30, no. 1, pp. 321–328, 2012.

7

Cuspal Displacement Induced by Bulk Fill Resin Composite Polymerization: Biomechanical Evaluation using Fiber Bragg Grating Sensors

Alexandra Vinagre,[1] João Ramos,[1] Sofia Alves,[1] Ana Messias,[1] Nélia Alberto,[2] and Rogério Nogueira[2]

[1] Faculty of Medicine, University of Coimbra, Avenida Bissaya Barreto, Blocos de Celas, 3000-075 Coimbra, Portugal
[2] Instituto de Telecomunicações (IT), Campus Universitário de Santiago, 3810-193 Aveiro, Portugal

Correspondence should be addressed to Alexandra Vinagre; vinagrealexandra@gmail.com

Academic Editor: Feng-Huei Lin

Polymerization shrinkage is a major concern to the clinical success of direct composite resin restorations. The aim of this study was to compare the effect of polymerization shrinkage strain of two resin composites on cuspal movement based on the use of fiber Bragg grating (FBG) sensors. Twenty standardized Class II cavities prepared in upper third molars were allocated into two groups ($n = 10$). Restorations involved the bulk fill placement of conventional microhybrid resin composite (Esthet•X® HD, Dentsply DeTrey) (Group 1) or flowable "low-shrinkage" resin composite (SDR™, Dentsply DeTrey) (Group 2). Two FBG sensors were used per restoration for real-time measurement of cuspal linear deformation and temperature variation. Group comparisons were determined using ANCOVA ($\alpha = 0.05$) considering temperature as the covariate. A statistically significant correlation between cuspal deflection, time, and material was observed ($p < 0.01$). Cuspal deflection reached 8.8 μm (0.23%) and 7.8 μm (0.20%) in Groups 1 and 2, respectively. When used with bulk fill technique, flowable resin composite SDR™ induced significantly less cuspal deflection than the conventional resin composite Esthet•X® HD ($p = 0.015$) and presented a smoother curve slope during the polymerization. FBG sensors appear to be a valid tool for accurate real-time monitoring of cuspal deformation.

1. Introduction

Volumetric shrinkage remains a major drawback to the clinical performance of the resin composite restorations. Shrinkage leads to deformation of the resin composites and generates stress due to the confinement of the resin to the cavity walls generated by the bonding procedure. This shrinkage stress is transferred to the tooth and may lead to cuspal deflection or enamel microcracks, whereas stress at the tooth-composite interface increases the likelihood of interfacial adhesive failures [1].

Cuspal deflection occurs due to the interaction between the polymerization shrinkage stress of the resin composite, the adhesive interface, and the compliance of the cavity wall [2]. Compliance is defined as the change in dimension per unit of force applied or generated, being essentially the inverse of stiffness [1]. Several studies have described it as a valuable method to assess the effects of polymerization shrinkage stress [3–7] and dimensional changes have been reported to range from 4 to 25 μm [4, 6, 8, 9]. The amplitude of this inward cuspal movement can depend on several factors, namely, the size and configuration of the cavity [2, 3, 10]; the properties of the resin composite [2, 4, 5, 9]; the bonding system [3, 5]; the hydration condition of the teeth [2]; and the experimental conditions [4]. Even though different model designs have been used for cusp deflection assessment, such as glass rods, aluminum blocks, or tooth structure, all inherently present with distinct compliance behaviors [4, 11, 12]. In order to overcome this limitation, system compliance similar to that of teeth is necessary to accurately detect stress [4, 11, 12]. Considering substrate structural deformability, both C-factor and resin composite volume seem to have an impact on the substrate compliance. When the substrate is only slightly deformable, the increase of the stress correlates better with the C-factor but if the compliance is higher, the resin composite volume would correlate better with stress development [10]. These findings demand careful data interpretation across studies concerning different methodologies for cuspal deflection assessment.

Additionally, the development of inward cuspal deflection can also be related to the strategies employed for managing shrinkage stress of resin composites [1]. These clinical approaches to reduce polymerization shrinkage include incremental placement techniques [2, 4, 11], the use of low-modulus intermediate liner materials as stress absorbers [4, 7], and modification of the light application methods to reduce curing speed [13]. Also, factors related to resin composite formulations like changes in filler amount, shape or surface treatment, variations in monomer structure or chemistry, and modification of polymerization resin kinetics have been more recently introduced aiming to reduce the polymerization shrinkage [1, 4, 5, 13]. All these strategies encompass a new class of resin composites known as "low-shrinkage resin based composites" that are generally allowed to be placed in a bulk fill mode due to the increased depth of cure, probably related to higher translucency [14]. Bulk filling techniques are undoubtedly more user friendly than the necessary meticulous incremental layering techniques advocated for conventional resin based composites (RBCs) [8], which justifies the growing interest in these so-called "low-shrinkage" RBCs and raises the need for exhaustive studies to clarify their potentialities [5, 8, 14–16].

Many methods have been used to evaluate cuspal deflection, involving technologies that go from linear variable differential transformers (LVDT) [2, 4, 11], strain gauges [9], profilometry [3], or twin channel deflection measuring gauge [6–8], among others. Fiber Bragg grating (FBG) sensors can be used to perform real-time local temperature and strain measurements [17–20]. Fiber optical sensors have the advantage of presenting immunity to electromagnetic interference [21], small dimensions [17–20], high resolution and sensibility, chemical inertness [17–19], biocompatibility [17], long-term stability [20], multiplexing capability, possibility to be embedded in different structures [17, 22], and ability to perform remote measurements [21].

The aim of this study was to compare the cuspal displacement induced by the polymerization shrinkage of a bulk fill resin composite (SDR™) and a conventional microhybrid resin composite (Esthet•X® HD) using fiber Bragg grating (FBG) sensors. The null hypothesis stated that there are no significant differences in cuspal displacement generated by the two resin composites.

2. Materials and Methods

2.1. Tooth Selection and Cavity Preparation. Twenty caries free, intact, and freshly extracted human upper third molars were collected after the patient's informed consent, as approved by the Ethical Committee of the Faculty of Medicine of Coimbra, Portugal (CE-001/2013). The teeth were cleaned and visually inspected to guarantee absence of hypoplastic defects, fractures, or cracks. Teeth were then stored at room temperature in a 10% buffered formalin solution (pH 7.0) for up to 3 months after extraction.

To standardize the dimensions of the molars, the teeth were selected based on the maximum buccal-palatal width (BPW), varying between 9.5 mm and 10.6 mm, and on the mesiodistal distance, varying between 8.1 mm and 9.7 mm, measured with a digital micrometer gauge (105–156, Mitutoyo, IL, Chicago, USA). Teeth were then randomly distributed into two groups ($n = 10$) ensuring a variance of the mean BPW between groups lower than 5%.

Each tooth was embedded in self-polymerizing acrylic resin (Orthocryl®, Dentaurum Ispringen, Germany) 2 mm apical from the cement-enamel junction (CEJ), with the long axis vertically oriented. Standardized large mesio-occluso-distal (MOD) cavities were prepared in each molar using a tungsten carbide round-ended bur in a high-speed handpiece (G848-314-031-10-ML, Diamond FG, Colténe/Whaledent AG, Switzerland) with copious water irrigation followed by a cone shaped bur of large diameter (980.040, set 4273, Komet®, Germany) mounted in a low-speed handpiece with water coolant indicated for inlay cavity preparations. The cusps were minimally frayed and all internal angles rounded. The width of the proximal boxes was approximately two-thirds the BPW (Figure 1). The occlusal isthmus was prepared approximately to half of the BPW (3.885 mm). The cavity depth was standardized to 3.5 mm from the tip of cusps at the occlusal isthmus and 1 mm above the CEJ at the cervical aspect of the proximal boxes (adapted from Palin et al. 2005 [5]).

After cavity preparation, adhesive procedures were performed using a two-step etch and rinse adhesive system (Prime&Bond®NT™, Dentsply DeTrey, Konstanz, Germany) according to manufacturer instructions followed by a 10-second light-curing exposure with a LED light-curing unit (Bluephase®, Ivoclar Vivadent, Liechtenstein) in its "low mode" program emitting 650 mW/cm² (Table 1). A second layer of adhesive was then applied and cured in similar way. Afterwards, teeth were passively surrounded by a band of polytetrafluoroethylene (Teflon®, DuPont, Wilmington, DE, USA) to prevent resin overflow during restorative procedures with either the bulk fill resin composite SDR™ or the microhybrid Esthet•X® HD (Dentsply DeTrey, Konstanz, Germany) used as control (Table 1).

2.2. Measurement of Cuspal Deflection with FBGs. A FBG is a periodic modulation of the refractive index along the core of an optical fiber. This modulation operates as a highly selective wavelength filter. When a FBG is illuminated by a broadband light source, only wavelengths that satisfy the Bragg condition are reflected, while all the others are transmitted.

The Bragg condition is given by the following:

$$\lambda_B = 2\Lambda n_{\text{eff}}, \tag{1}$$

where λ_B is the Bragg wavelength, Λ is the periodic modulation of the refractive index, and n_{eff} is the effective refractive index of the fiber core.

TABLE 1: Materials composition.

Resin composite	Manufacturer	Resin matrix	Filler	Batch #
SDR™ Microhybrid	Dentsply DeTrey	Modified UDMA EBPADMA TEGDMA	Ba-Al-F-B-Si-glass Sr-Al-F-Si-glass (68 wt%., 45 vol%)	1105141
Esthet•X® HD Microhybrid	Dentsply DeTrey	Bis-GMA adduct Bis-EMA adduct TEGDMA	Ba-F-Al-B-Si-glass Nanofiller silica (77 wt%; 60 vol%)	1006292

Adhesive	Manufacturer	Chemical composition	Instructions	Batch #
Prime&Bond®NT™ 2-step etch and rinse adhesive	Dentsply DeTrey	Di- and trimethacrylate resins PENTA Photoinitiators Stabilizers Nanofillers Acetone	Apply 36% phosphoric acid for 15 seconds; spray and rinse with water for 15 seconds; blot dry conditioned areas; apply adhesive and leave the surface wet for 20 seconds; gently dry for at least 5 seconds; polymerize for 10 seconds; apply a second layer of adhesive	1109001528

Bis-GMA (bisphenol A dimethacrylate); Bis-EMA (bisphenol A polyethylene glycol diether dimethacrylate); UDMA (urethane dimethacrylate); TEGDMA (triethylene glycol dimethacrylate); EBPADMA (ethoxylated bisphenol A dimethacrylate).

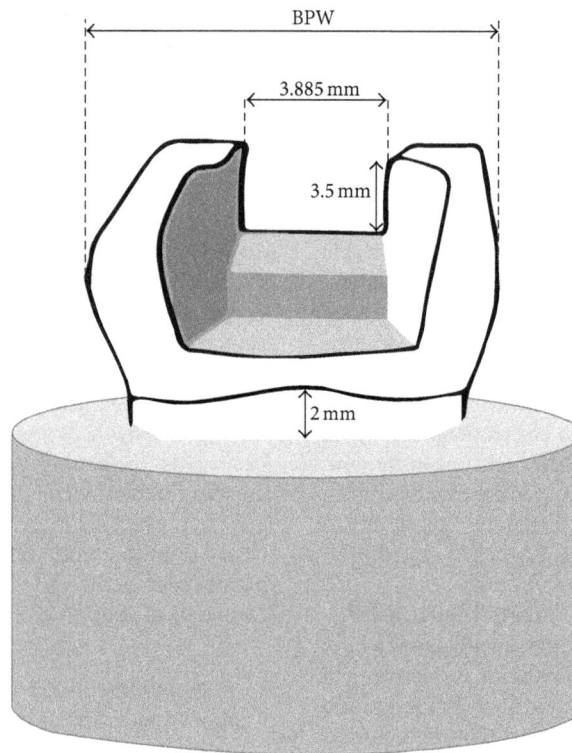

FIGURE 1: Schematic illustration of the shape and dimensions of the cavity (adapted from Palin et al. 2005 [5]).

The effective refractive index, as well as the periodic spacing between the grating planes, will be affected by changes in strain and/or temperature which will modify the center wavelength of light back reflected from the Bragg grating. Using the first equation, the shift in the Bragg grating center wavelength due to strain and temperature changes is given by the following:

$$\Delta\lambda_B = \Delta\lambda_{B,l} + \Delta\lambda_{B,T}$$
$$= 2\left(\Lambda\frac{\partial n}{\partial l} + n\frac{\partial \Lambda}{\partial l}\right)\Delta l + 2\left(\Lambda\frac{\partial n}{\partial T} + n\frac{\partial \Lambda}{\partial T}\right)\Delta T \quad (2)$$
$$= S_l\Delta l + S_T\Delta T,$$

where $\Delta\lambda_{B,l}$ is the strain induced wavelength shift and $\Delta\lambda_{B,T}$ is the thermal effect on the same parameter. S_l and S_T

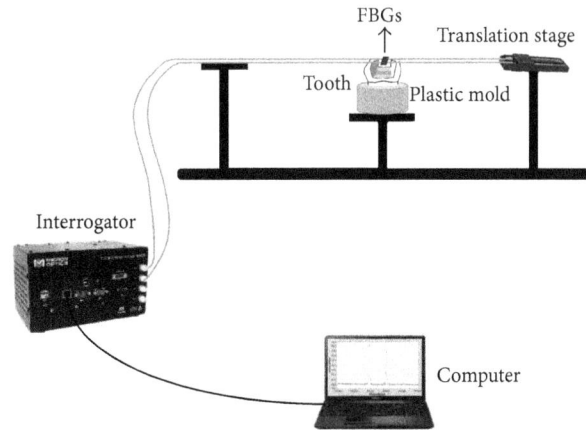

FIGURE 2: Schematic overview of the experimental setup used to measure the setting cuspal deformation and the temperature variation.

represent the strain and temperature sensitivity coefficients of the FBG sensors [23]. For this work, they were previously determined and 0.00118 nm/$\mu\varepsilon$ and 0.0089 nm/°C were obtained, respectively.

In the current study, gratings with 1 mm length were inscribed onto photosensitive optical fiber (FiberCore PS 1250/1500) with a UV light (248 nm) from a KrF excimer laser, using the phase mask technique.

One drawback of the FBG based sensors is their cross-sensitivity to both strain and temperature. For that, in this study two FBGs were used to measure the setting cuspal deformation and the temperature variation. One of them, sensitive to strain and temperature variations, was placed perpendicular to the buccal cusp. This grating was previously tensioned (about 500 $\mu\varepsilon$), allowing the FBG sensor to detect not only the increasing of the distance between the cusps but also its approximation. For this, one side of the fiber was glued (Loctite®, Henkel, Germany, and cyanoacrylate accelerator, Pekecho®, Spain) to a controllable translation stage and the other one to the cusp farthest from it. After the glue drying time (15 min), the fiber was tensioned and bonded to the other cusp (nearest of the controllable translation stage). A new waiting period of 15 min was performed. The second FBG was placed parallel to the first one, but not bonded nor pretensioned, being only sensitive to temperature variations. The gratings' wavelength was measured using a sm 125–500 interrogation system (Micron Optics Inc., Atlanta, USA) with a measurement range of 1510–1590 nm, wavelength resolution of 1 pm, and an acquisition frequency of 2 Hz. Figure 2 shows a schematic representation of the experimental apparatus.

After the setup, the cavities were gently bulk filled with resin composite according to two groups (SDR™ or Esthet•X® HD) and the same LED light-curing unit was used for polymerization in the soft-start mode, running at 650 mW/cm^2 for the first 5 seconds and at 1200 mW/cm^2 in the following period. Light-curing tip was placed 1 mm above the samples, with an incidence angle near 90°. The samples were initially irradiated for 30 seconds and after a 5-minute break a second light-curing period of 30 seconds was applied. In all tests, data were continuously acquired for 10 minutes from the beginning of the polymerization.

The experiment was repeated alternately for each resin, performing a total number of ten samples per group. Beside these tests, the thermal variation caused by the light-curing unit was further investigated using a similar experimental setup, but applying only the temperature sensitive sensor without filling the cavity with resin composite.

All the experiments were performed under controlled room temperature conditions (21°C).

2.3. Statistical Analysis. Statistical analysis was performed using SPSS 20.0® (SPSS Inc., Chicago, IL, EUA). Cuspal deflection variation between groups was determined with ANCOVA, considering temperature as a covariate. Repeated measures ANOVA considering Greenhouse correction was used to analyze cuspal deflection variation within time for each group. Significance level was set at $\alpha = 0.05$.

3. Results

Temperature rise induced by the irradiation with the LED light-curing unit is represented in Figure 3. In the soft-start mode, the first and second irradiation periods reached a temperature rise of 7.9°C and 8.3°C, respectively.

The mean temperature rise registered from all tests during light curing of the restorations is shown in Figure 4. For Esthet•X® HD, the temperature increased 36.7°C and 28.9°C at the final of the first and second irradiation period, respectively. In the case of the SDR™, a temperature variation of 38.2°C and 30.8°C was obtained for the same periods.

The average cuspal deformation, resulting from the ten tests of each resin, during the 10 minutes of monitoring is represented in Figure 5. The curves were obtained by subtracting the effect of the temperature, obtained with the grating inscribed in the fiber that was not bonded, from the measurements accomplished by the other sensor, which is sensitive to both temperature and strain variations.

TABLE 2: Descriptive statistics analysis.

Time (min)	Resin composite	Deformation ($\mu\varepsilon$) mean (Std. Deviation)	Minimum ($\mu\varepsilon$)	Maximum ($\mu\varepsilon$)	Cuspal deflection (μm) mean (Std. Deviation)
0.5 min	SDR™	−646.3 (218.5)	−457.3	−1195.5	−2.5 (0.8)
	Esthet• X® HD	−1299.5 (190.5)	−1077.9	−1582.0	−5.0 (0.7)
5.5 min	SDR™	−2074.1 (137.1)	−1910.5	−2371.7	−8.0 (0.5)
	Esthet• X® HD	−2296.9 (256.5)	−1786.7	−2608.6	−8.9 (1.0)
6 min	SDR™	−1166.4 (286.4)	−873.9	−1747.3	−4.5 (1.1)
	Esthet• X® HD	−1751.4 (286.3)	−1194.1	−2095.1	−6.8 (1.1)
10 min	SDR™	−2001.2 (179.9)	−1655.2	−2341.5	−7.8 (0.7)
	Esthet• X® HD	−2277.2 (260.6)	−1757.5	−2600.6	−8.8 (1.0)

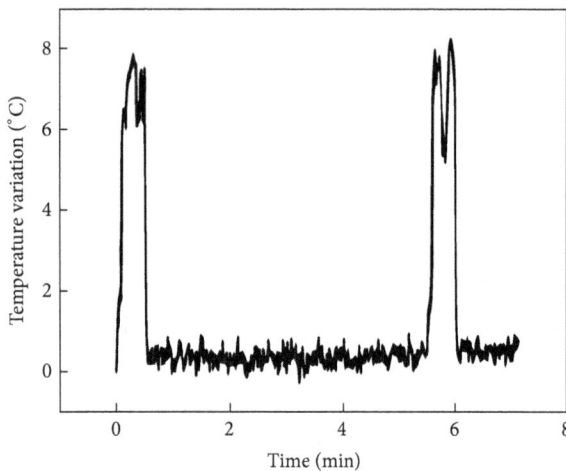

FIGURE 3: Temperature variation induced by the LED light-curing unit.

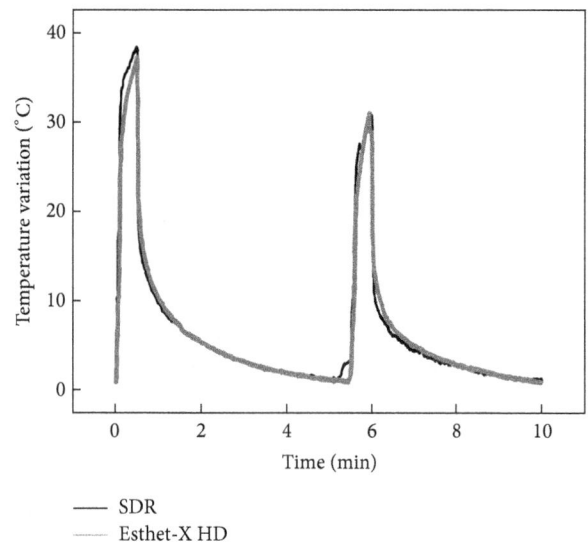

— SDR
— Esthet-X HD

FIGURE 4: Average temperature variation obtained during the light curing of the restorations with the SDR™ and Esthet•X® HD resins.

Descriptive statistics of cuspal deflection in each period is shown in Table 2. Between-subjects analysis of covariance (ANCOVA) is summarized in Table 3.

Over the time of polymerization, the two resin composites presented a significant variation in material behavior ($F(1.42, 24.19) = 245.37$, $p < 0.01$, partial $\eta^2 = 0.935$), as expressed in Figure 5. After the first polymerization period, both groups present statistically significant cuspal inwards deformation in relation to any other measurement ($p < 0.01$ for all comparisons between 0.5 minutes and 5.5, 6, and 10 minutes within each group). A similar behavior was found after the second polymerization period (6 to 10 minutes) but with lower deformation ($p < 0.01$). Only SDR presented no differences in cuspal deformation after the postpolymerization clearance periods ($p = 0.051$ for the 5.5- and 10-minute comparison).

When light curing started, cuspal deformation experienced a slight expansion curve for both materials evaluated. This peak reached the maximum value at 0.05 minutes for both resin composites. Esthet•X® achieved in average 27.1 $\mu\varepsilon$ and SDR™ 91.8 $\mu\varepsilon$. Thereafter, curves decreased (Figure 5). At 30 seconds, the mean values of cuspal deflection were −1299.5 $\mu\varepsilon$ and −646.3 $\mu\varepsilon$ for Esthet•X® HD and SDR™, respectively (Figure 5 and Table 2). ANCOVA analysis considering temperature as the covariate revealed a statistically significant difference in cuspal deflection between the two resin composite resins at 30 seconds of polymerization ($F(1, 17) = 52.69$, $p < 0.01$, partial $\eta^2 = 0.756$). t-test for independent samples equality of means revealed deformation values statistically significantly higher in Esthet•X® HD compared to SDR™ (Table 3).

In the mean time until the second polymerization (0.5 to 5.5 minutes), Esthet•X® HD curve deflection decreased continuously until 2 minutes pass. From this moment, values remained constant until the terminus of this light-curing free period, reaching a mean of −2296.9 $\mu\varepsilon$ at 5.5 minutes. The SDR™ curve declines at a slower rate. At 5.5 minutes, samples restored with SDR™ presented a cuspal deflection of −2074.1 $\mu\varepsilon$ (Figure 5 and Table 2). There was a statistically significant difference in cuspal deflection between two groups at this point of the curing protocol ($p = 0.016$) (Table 3).

TABLE 3: Mean differences of deformation for samples restored with Esthet• X® HD and SDR™, for each time period (ANCOVA).

Time (min)	Levene's test (p)	F	p	Mean difference (SDR - Esthet X) ($\mu\varepsilon$)	Std. error difference	Mean difference (SDR - Esthet X) (μm)
0.5	0.751	52.69	<0.01*	626.8	86.4	2.4
5.5	0.120	7.15	0.016*	255.6	95.6	1.0
6	0.733	21.64	<0.01*	601.4	129.3	2.3
10	0.620	7.37	0.015*	274.6	101.1	1.1

*Statistically significant differences.

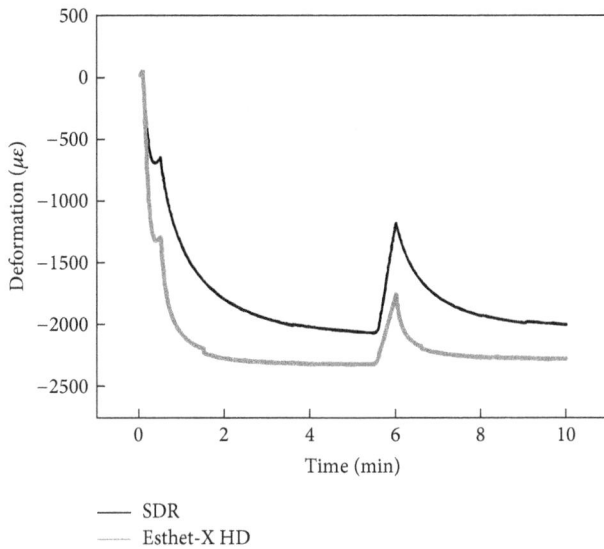

FIGURE 5: Average cuspal deformation induced by SDR™ and Esthet•X® HD resins during the 10 minutes of real-time monitoring.

During the second polymerization period, a new expansion peak occurred for both resin composites. After that, at 6 minutes, Esthet•X® HD curve presented −1751.4 $\mu\varepsilon$ of cuspal deflection and SDR™ −1166.3 $\mu\varepsilon$ (Table 2 and Figure 5). According to that, between the beginning and the ending of the second polymerization period, cusps expanded approximately 560.1 $\mu\varepsilon$ in samples restored with Esthet•X® HD and 907.8 $\mu\varepsilon$ in SDR™ samples. Mixed ANOVA considering Greenhouse correction revealed a statistically significant difference in cuspal deflection between the two resin composites at 6 minutes ($F(1, 10) = 11.81$, $p = 0.006$, partial $\eta^2 = 0.542$) (Table 3), meaning that cuspal deflection of Esthet•X® HD at this time point was significantly higher than SDR™.

In the period from 6 to 10 minutes, both deformation curves decreased and for the last evaluated point the values remained constant (Figure 5). The average results obtained with Esthet•X® HD and SDR™, at 10 minutes, were −2277.2 $\mu\varepsilon$ and −2001.2 $\mu\varepsilon$, respectively (Table 2). There was a statistically significant difference in cuspal deflection between the two materials at 10 minutes ($p = 0.015$) (Table 3).

In order to determine the shortening distance between the cusps (cuspal deflection), deformation expressed in

microstrain ($\mu\varepsilon$) was converted to micrometers (μm), according to the following equation:

Cuspal deflection (μm)

$$= \frac{\text{Deformation} (\mu\varepsilon) * \text{Initial distant between cuspids} (3885\,\mu m)}{1 * 10^6}. \quad (3)$$

The average results obtained in micrometers (μm) for the more relevant periods of 0.5 and 10 min were 5.0 μm (0.13%) and 8.8 μm (0.23%) in the case of the Esthet•X® HD and 2.5 μm (0.06%) and 7.8 μm (0.20%) for the SDR™ resin, respectively.

4. Discussion

Despite numerous studies published, the highly complex phenomenon of polymerization shrinkage that develops in polymeric dental restorative materials is not yet fully understood and remains a significant clinical concern. This phenomenon becomes even more complex when the composite is bonded into cavities of variable configurations. For resin composites bonded to enamel and dentin, polymerization shrinkage is constrained and polymerization stress development becomes more complex due to the generation of interfacial stresses usually unevenly distributed along the cavity walls and the bonded composite surfaces [10, 12].

The present study measured the tooth deformation instead of shrinkage stresses, which can be assumed as an indirect indicator associated with internal stress [3, 4]. Regrettably, the methods most commonly used for cuspal deflection monitoring depend primarily on measuring the difference between precuring and postcuring values, but they do not provide detailed data regarding how this phenomenon occurs in a real-time process [9]. In fact, the methods of measurement of cusp deflection during composite restoration have been reported to produce considerably varying results. When applying contact methods, the use of reproducible reference points on the cusps seems to be critical to avoid erroneous results between samples [5]. With FBG sensors this point is not an issue since the measurement of dimensional changes is made by the Bragg sensor, which is not directly attached to the tooth structure. Therefore, it can be expected that under identical experimental conditions, cusp deflection measurements for the same restoration protocol can result in different data according to the experimental device used and the inherent tooth compliance. Caution is needed when comparing results across studies as mean cuspal deflections of up to 50 μm were recorded using a wide range of techniques [2–4, 6–9, 11].

Human molars were used to assess cuspal deflection using FBG sensors. A relatively high variation among teeth dimensions in experimental studies may affect the outcomes, impairing the comparison between studies with the same purpose. Despite the careful attempts to achieve standardization of cavity preparation, some inevitable discrepancies may be present when natural teeth are used. In this study, molar teeth were selectively allocated in order to promote a maximum difference of 5% in BPW between the groups. Despite tooth standardization, the dispersion values (SD) of each group reported high variability among samples, which could be due to the employment of natural teeth with non-homogenous morphological and structural characteristics as previously described for other experimental models using natural teeth. In fact, it was not possible to determine the residual thickness of both dentin and enamel, which in conjunction with slight differences in the amount of the resin used to fill the tooth could compromise the compliance behavior and consequently induce different cuspal displacements. Notwithstanding this, molar teeth choice together with the large Class II MOD cavity design used in the present study has the advantage of wider surface area for fiber bonding, contrarily to premolars mostly used in other studies [2, 6, 7] while providing an in vitro simulation of some clinical situations of weakened remaining tooth structure, favorable to cuspal deflection during restorative procedures. Large deflections for MODs cavities can be explained by the loss of tooth rigidity when the marginal ridges are removed [3]. Additionally, the cusps that remain after an MOD cavity preparation were reported to act as a cantilever beam under occlusal load, which increases with cavity depth, while the prepared cavity floor acts as a fulcrum for cusp bending. Biomechanical principles refer to the fact that deflection is proportional to the cubed power of the length and to the inverse of the thickness of the cantilever cusp cubed [7, 8].

FBG sensors methodology allowed a real-time monitoring of the deformations occurring during resin composite curing as well as the thermal behavior during this procedure [17, 20]. For cavity preparations restored with resin composites under shrinkage, loadings and displacements will occur in multiple directions. While knowing that shrinking composite develops a triaxial stress state, as reported with finite element analysis [10, 12], our measurements registered only the forces developing uniaxially in the long axis of each tooth, expressed as a single value. Nevertheless, the use of a standardized tooth cavity allowed a well-balanced homogenization of compliance, cavity configuration, and composite volume, which have been found to be the most critical variables related to stress development in a clinical situation [10].

The present study compared the marketed low-shrinkage flowable RBC SDR™ with a conventional microhybrid RBC Esthet•X® HD assessing their performance in bulk fill adhesive MOD molar restorations by measuring cuspal deformation with FBG sensors during a two-step curing protocol in order to deliver a reliable energy density to the restoration. Although conventional microhybrid RBCs have restricted indications for bulk fill placement technique and SDR™ application advocates the use of a microhybrid resin composite for the final covering layer, the purpose was to equalize the restorative protocol in order to isolate and evaluate the individual biomechanical behavior of both materials in similar one-increment situations related to the potential advantage evocated for the "low-shrinkage" one. Significant differences were found between the two materials; therefore the null hypothesis was rejected.

The greatest difference was detected at the end of the first 30 seconds curing period. At this point, Esthet•X® HD induced significantly more deformation (meaning cuspal deflection) than SDR™, which could enhance faster and higher stress development at the tooth/restoration adhesive interface. At the end of the subsequent five-minute pause period both RBCs reached higher shrinkage values, although the polymerization kinetics of SDR™ seems to develop more gradually than that achieved by Esthet•X® HD. This may be related to the functionality of the polymerization modulator incorporated in SDR™ resin matrix. In theory, when this modulator interacts with the photoinitiator (camphorquinone) the polymerization kinetics can be controlled by delaying the gel point, by slowing the rate of polymerization and elastic modulus development, and by reducing polymer cross-linking and, consequently, shrinkage stress [16, 24].

Cuspal deformation curves showed three expansion peaks during the experimental curing period for both resin composites. As soon as the first light irradiation period started, a discrete expansion and transitory peak were detected. However, at the beginning of the second curing light exposure, an expressive and prolonged expansion peak could be observed during all the 30-second irradiation time. These events can be interpreted as the thermal expansion effect caused either by the heat from the curing light or by the exothermic nature of the free radical polymerization of dimethacrylate monomers, as pointed out by other authors [4, 19, 20]. When polymerization shrinkage exceeds thermal expansion in the first expansion peak, fast overall material shrinking takes place, evidenced by a sudden increase contraction strain [19, 20]. In opposite, the persistent expansion peak observed along the second irradiation period occurs when a considerable cross-linking of the monomer has already been achieved, meaning that the thermal effect has greater relative influence on the dimensional behavior of both resin composites. Additionally, the inherent temperature rise can be implied in the resin composite glass transition temperature attainment, at which the polymer goes from the glassy to the rubbery state [25, 26]. If this occurs, a significant increase in polymer chain mobility is expected, favoring additional cross-linking and stress relief [27]. This expansion was significantly more pronounced for samples restored with SDR™ than Esthet•X® HD. This can be further explained by the lower filler content exhibited by SDR™, as an inverse linear relationship between coefficient of linear thermal expansion and the filler volume fraction of the resin composite has been observed by different researchers [25]. Another intermediate and discrete expansion peak was obtained immediately few seconds before the end of the first irradiation period. Possibly, a cumulative thermal effect induced by the high power density emitted by the curing unit at this point leads to the development of a new expansion phase. These thermal expansions can be considered internal constraints that will be added to the total amplitude internal stress [1].

In the last measurement (10 minutes), the mean total cuspal deflection was 8.8 μm (0.23%) and 7.8 μm (0.20%) for the maxillary molar teeth restored with Esthet•X® HD and SDR™ resins, respectively. SDR™ presented significantly less final cuspal deflection than Esthet•X® HD. Indeed, in the few studies available concerning SDR™, polymerization stress was reported to be considerably lower than that of conventional flowable resin composites, being comparable to other marketed low shrinking resin composites [8, 16] and marginal integrity appeared as good as that obtained with a conventionally layered resin composite [8]. Previous studies showed volumetric polymerization shrinkage around 3% for Esthet•X® [28] and around 3.1% for SDR™ [15]. Differences between those findings and the results of the present work could be due to the fact that shrinkage stress development is not exclusively associated with the volumetric shrinkage behavior. Moorthy et al. [8] showed that two bulk fill flowable RBC bases (SDR™ and x-tra base) have significantly reduced cuspal deflection during light irradiation when compared to a conventional RBC (GrandioSO), reporting a total mean cuspal deflection of 4.63 μm (1.19), 4.73 μm (0.99), and 11.26 μm (2.56) for SDR™, x-tra base, and GrandioSO, respectively. Other recent studies conducted by Tauböck et al. [24] reported that SDR™ generated significantly lower shrinkage forces compared with the microhybrid Esthet•X® HD when irradiation was performed at continuous high irradiance, even though axial polymerization shrinkage of the SDR composite exceeded that of the microhybrid. Nevertheless, the authors revealed no benefit of SDR regarding shrinkage force generation when modulated curing protocols with low initial irradiance were applied, such as in the soft-start mode, arguing a low responsiveness of SDR to modulated photoactivation due to the predominant effect of the polymerization modulator on reaction kinetics and stress development. In the present study, to control shrinkage-induced tensions, both resin composites were polymerized using a soft-start curing mode, with the expectation that this approach could reduce cuspal movement and improvement of restoration interfacial integrity. One can speculate that if a continuous high level curing irradiance had been used, higher cuspal displacement could have taken place, particularly in the microhybrid Esthet•X® HD samples.

Some drawbacks can be pointed out to the methodology employed in this study concerning FBG sensors. Cross-sensitivity to both strain and temperature requires specific techniques to compensate for the thermal influence, which was dealt with as an additional Bragg grating. The information obtained is wavelength encoded and extracting real information from the wavelength shift involves the development of particular applications. Other disadvantages consist on the inherent fiber fragility, which makes the manipulation and the bonding of the fiber to the teeth difficult [17].

Several authors have developed in vitro simulation models to determine the shrinkage stress or cuspal deflection of RBC materials. However, the limitation of some of those in vitro experiments is related to the simplicity of the model designs, using frequently, parallel walls that ignore nonaligned stress development linked to more complex cavity geometries and to the differential compliance of the testing systems, which significantly influences stress development, particularly depending on C-factor and resin composite volume [10]. Also, they do not allow the acquisition of information in a continuous and real-time mode [7, 29, 30], which was overcome by the methodology used in the present study. Another relevant advantage of the FBG is the high resolution obtained in the cuspal displacement monitoring. Since the wavelength resolution of the interrogation system is 1 pm, a variation of 0.85 με can be noticed. Considering the initial distance between the cusps of 3885 μm, it corresponds to an absolute resolution of 0.003 μm. During the last few years, research work devoted to studying and exploring the potential application of fiber optic technology in biomedicine has increased significantly [20].

To the extent of the knowledge of the authors concerning the scientific published literature, this is the first known study to measure tooth cuspal deflection with fiber Bragg grating sensors. Fiber Bragg sensors can be extensively applied in future studies, comparing different protocols or clinical modified variables useful for shrinkage stress management in clinical practice.

5. Conclusion

Within the conditions of this research protocol, SDR™ polymerization kinetics induced less stress to dental structure than Esthet•X® HD, as the mean cuspal deflection values were statistically different between the two resin composites.

Despite the limitations of this in vitro study the optical FBG sensors seem to be a suitable measurement method to evaluate tooth dimensional changes related to cuspal deflection induced by resin composite polymerization shrinkage, with some advantages over other techniques, namely, continuous real-time assessment of tooth biomechanical behavior.

Competing Interests

The authors declare that they have no competing interests.

References

[1] J. L. Ferracane, "Buonocore lecture. Placing dental composites—a stressful experience," *Operative Dentistry*, vol. 33, no. 3, pp. 247–257, 2008.

[2] M.-R. Lee, B.-H. Cho, H.-H. Son, C.-M. Um, and I.-B. Lee, "Influence of cavity dimension and restoration methods on the cusp deflection of premolars in composite restoration," *Dental Materials*, vol. 23, no. 3, pp. 288–295, 2007.

[3] D. Tantbirojn, A. Versluis, M. R. Pintado, R. DeLong, and W. H. Douglas, "Tooth deformation patterns in molars after composite restoration," *Dental Materials*, vol. 20, no. 6, pp. 535–542, 2004.

[4] Y. Kwon, J. Ferracane, and I.-B. Lee, "Effect of layering methods, composite type, and flowable liner on the polymerization shrinkage stress of light cured composites," *Dental Materials*, vol. 28, no. 7, pp. 801–809, 2012.

[5] W. M. Palin, G. J. P. Fleming, H. Nathwani, F. J. T. Burke, and R. C. Randall, "In vitro cuspal deflection and microleakage of maxillary premolars restored with novel low-shrink dental composites," *Dental Materials*, vol. 21, no. 4, pp. 324–335, 2005.

[6] G. J. P. Fleming, S. Khan, O. Afzal, W. M. Palin, and F. J. T. Burke, "Investigation of polymerisation shrinkage strain, associated cuspal movement and microleakage of MOD cavities restored incrementally with resin-based composite using an LED light curing unit," *Journal of Dentistry*, vol. 35, no. 2, pp. 97–103, 2007.

[7] R. El-Helali, A. H. Dowling, E. L. McGinley, H. F. Duncan, and G. J. P. Fleming, "Influence of resin-based composite restoration technique and endodontic access on cuspal deflection and cervical microleakage scores," *Journal of Dentistry*, vol. 41, no. 3, pp. 216–222, 2013.

[8] A. Moorthy, C. H. Hogg, A. H. Dowling, B. F. Grufferty, A. R. Benetti, and G. J. P. Fleming, "Cuspal deflection and microleakage in premolar teeth restored with bulk-fill flowable resin-based composite base materials," *Journal of Dentistry*, vol. 40, no. 6, pp. 500–505, 2012.

[9] H. H. Hamama, N. M. Zaghloul, O. B. Abouelatta, and A. E. El-Embaby, "Determining the influence of flowable composite resin application on cuspal deflection using a computerized modification of the strain gauge method," *The American Journal of Esthetic Dentistry*, vol. 1, no. 1, pp. 48–59, 2011.

[10] L. C. C. Boaro, W. C. Brandt, J. B. C. Meira, F. P. Rodrigues, W. M. Palin, and R. R. Braga, "Experimental and FE displacement and polymerization stress of bonded restorations as a function of the C-Factor, volume and substrate stiffness," *Journal of Dentistry*, vol. 42, no. 2, pp. 140–148, 2014.

[11] J. Park, J. Chang, J. Ferracane, and I. B. Lee, "How should composite be layered to reduce shrinkage stress: incremental or bulk filling?" *Dental Materials*, vol. 24, no. 11, pp. 1501–1505, 2008.

[12] F. P. Rodrigues, R. G. Lima, A. Muench, D. C. Watts, and R. Y. Ballester, "A method for calculating the compliance of bonded-interfaces under shrinkage: validation for Class I cavities," *Dental Materials*, vol. 30, no. 8, pp. 936–944, 2014.

[13] N. Ilie, E. Jelen, and R. Hickel, "Is the soft-start polymerisation concept still relevant for modern curing units?" *Clinical Oral Investigations*, vol. 15, no. 1, pp. 21–29, 2011.

[14] J. G. Leprince, W. M. Palin, J. Vanacker, J. Sabbagh, J. Devaux, and G. Leloup, "Physico-mechanical characteristics of commercially available bulk-fill composites," *Journal of Dentistry*, vol. 42, no. 8, pp. 993–1000, 2014.

[15] J. Burgess and D. Cakir, "Comparative properties of low-shrinkage composite resins," *Compendium Continuing Education in Dentistry*, vol. 31, no. 2, pp. 10–15, 2010.

[16] N. Ilie and R. Hickel, "Investigations on a methacrylate-based flowable composite based on the SDR™ technology," *Dental Materials*, vol. 27, no. 4, pp. 348–355, 2011.

[17] N. Alberto, L. Carvalho, H. Lima, P. Antunes, R. Nogueira, and J. L. Pinto, "Characterization of different water/powder ratios of dental gypsum using fiber bragg grating sensors," *Dental Materials Journal*, vol. 30, no. 5, pp. 700–706, 2011.

[18] M. S. Milczewski, J. C. C. da Silva, I. Abe et al., "Determination of setting expansion of dental materials using fibre optical sensing," *Measurement Science and Technology*, vol. 17, no. 5, pp. 1152–1156, 2006.

[19] M. S. Milczewski, J. C. C. Silva, A. S. Paterno, F. Kuller, and H. J. Kalinowski, "Measurement of composite shrinkage using a fibre optic Bragg grating sensor," *Journal of Biomaterials Science, Polymer Edition*, vol. 18, no. 4, pp. 383–392, 2007.

[20] E. J. Anttila, O. H. Krintilä, T. K. Laurila, L. V. J. Lassila, P. K. Vallittu, and R. G. R. Hernberg, "Evaluation of polymerization shrinkage and hydroscopic expansion of fiber-reinforced bio-composites using optical fiber Bragg grating sensors," *Dental Materials*, vol. 24, no. 12, pp. 1720–1727, 2008.

[21] S. M. M. Quintero, A. M. B. Braga, H. I. Weber, A. C. Bruno, and J. F. D. F. Araújo, "A magnetostrictive composite-fiber bragg grating sensor," *Sensors*, vol. 10, no. 9, pp. 8119–8128, 2010.

[22] H. Li, H. Yang, E. Li, Z. Liu, and K. Wei, "Wearable sensors in intelligent clothing for measuring human body temperature based on optical fiber Bragg grating," *Optics Express*, vol. 20, no. 11, pp. 11740–11752, 2012.

[23] A. Othonos and K. Kali, *Fiber Bragg Gratings—Fundamentals and Applications in Telecommunications and Sensing*, Artech House, 1999.

[24] T. T. Tauböck, A. J. Feilzer, W. Buchalla, C. J. Kleverlaan, I. Krejci, and T. Attin, "Effect of modulated photo-activation on polymerization shrinkage behavior of dental restorative resin composites," *European Journal of Oral Sciences*, vol. 122, no. 4, pp. 293–302, 2014.

[25] I. Sideridou, D. S. Achilias, and E. Kyrikou, "Thermal expansion characteristics of light-cured dental resins and resin composites," *Biomaterials*, vol. 25, no. 15, pp. 3087–3097, 2004.

[26] R. Walter, E. J. Swift Jr., H. Sheikh, and J. L. Ferracane, "Effects of temperature on composite resin shrinkage," *Quintessence International*, vol. 40, no. 10, pp. 843–847, 2009.

[27] E. K. Viljanen, M. Skrifvars, and P. K. Vallittu, "Dendritic copolymers and particulate filler composites for dental applications: degree of conversion and thermal properties," *Dental Materials*, vol. 23, no. 11, pp. 1420–1427, 2007.

[28] W. Lien and K. S. Vandewalle, "Physical properties of a new silorane-based restorative system," *Dental Materials*, vol. 26, no. 4, pp. 337–344, 2010.

[29] F. Gonçalves, C. S. Pfeifer, J. L. Ferracane, and R. R. Braga, "Contraction stress determinants in dimethacrylate composites," *Journal of Dental Research*, vol. 87, no. 4, pp. 367–371, 2008.

[30] L. C. C. Boaro, F. Gonalves, T. C. Guimarães, J. L. Ferracane, A. Versluis, and R. R. Braga, "Polymerization stress, shrinkage and elastic modulus of current low-shrinkage restorative composites," *Dental Materials*, vol. 26, no. 12, pp. 1144–1150, 2010.

8

Developing a Suitable Model for Water Uptake for Biodegradable Polymers Using Small Training Sets

Loreto M. Valenzuela,[1] Doyle D. Knight,[2] and Joachim Kohn[3]

[1]Department of Chemical and Bioprocess Engineering, Research Center for Nanotechnology and Advanced Materials "CIEN-UC",
 Pontificia Universidad Católica de Chile, Vicuña Mackenna 2860, Macul, 7820436 Santiago, Chile
[2]Department of Mechanical and Aerospace Engineering, Rutgers, The State University of New Jersey, New Brunswick,
 NJ 08854-8087, USA
[3]New Jersey Center for Biomaterials, Rutgers, The State University of New Jersey, 145 Bevier Road, Piscataway, NJ 08854, USA

Correspondence should be addressed to Loreto M. Valenzuela; lvalenzr@ing.puc.cl

Academic Editor: Rosalind Labow

Prediction of the dynamic properties of water uptake across polymer libraries can accelerate polymer selection for a specific application. We first built semiempirical models using Artificial Neural Networks and all water uptake data, as individual input. These models give very good correlations ($R^2 > 0.78$ for test set) but very low accuracy on cross-validation sets (less than 19% of experimental points within experimental error). Instead, using consolidated parameters like equilibrium water uptake a good model is obtained ($R^2 = 0.78$ for test set), with accurate predictions for 50% of tested polymers. The semiempirical model was applied to the 56-polymer library of L-tyrosine-derived polyarylates, identifying groups of polymers that are likely to satisfy design criteria for water uptake. This research demonstrates that a surrogate modeling effort can reduce the number of polymers that must be synthesized and characterized to identify an appropriate polymer that meets certain performance criteria.

1. Introduction

Degradable materials are very important in fabricating biomedical devices. After implantation, they do not need to be removed; rather, under ideal conditions, the implant site repairs itself while the device is resorbed [1]. In comparison, nondegradable materials often need to be surgically removed after their purpose has been achieved, thus subjecting the patient to a second surgery that potentially exposes them to more complications [2]. Degradable devices can be used in a broad range of applications such as vascular stents, vascular bypass grafts, bone fixation devices, and soft tissue replacement scaffolds [3].

Degradable biomaterials have a wide range of requirements depending on the particular clinical application. Parameters such as chemical structure, composition, porosity, and device geometry determine surface and bulk properties of an implant, and thus, they are critical to the selection of the material [4].

One important characteristic of degradable biomaterials is their water uptake versus time, as it is crucial for the determination of how long a polymeric device will reside in the body before erosion leads to the ultimate removal of the device from the implant site [5]. Water uptake affects degradation, swelling, mechanical [6], and adhesive properties [7]; also it determines drug stability [8], drug release profile [9], and biological response [10].

Current methods used to measure water uptake versus time are labor intensive and time consuming. Depending on the polymer, water uptake can take days to weeks to equilibrate [11]. There are potentially very large libraries of polymeric biomaterials, which make it impractical to measure these parameters experimentally for each polymer. For example, a virtual library of about 40,000 polymethacrylates has been described by Kholodovych et al. [12]. This library would clearly be too large for each polymer to be characterized individually by experimental methods.

Computational modeling is a useful tool to minimize the number of experiments needed to characterize a polymer library [13]. Costache et al. [14], Gubskaya [15], and Le et al. [16] published reviews that include the most relevant models currently available for important parameters in biomaterials such as glass transition temperature (T_g), Young's modulus, air-water contact angle, water uptake, and degradation. Serna et al. (2008) built a model of equilibrium water uptake for 12 aromatic polyamides with very similar levels of water uptake (13.9%–19.1%). They found correlations between the amidic hydrogen charge and the water uptake [17].

Although empirical mathematical modeling has been successfully used to model water uptake for different polymers, all models require parameters that can only be obtained through experimentation. Fick's diffusion [18], anomalous Fickian diffusion [19], dual-stage Fick's diffusion [20], power law [18], Weibull equation [21], Langmuir theory [22], and concentration-dependent diffusion coefficient model [23] have been used. Modeling of hydration at the molecular level has been demonstrated using parameters such as free volume redistribution frequency [24], Radial Distribution Functions (RDFs) [25], 3D atomic density maps known as spatial distribution functions [26], and angular distribution functions [27]. Furthermore, from MD simulations, water absorption has been predicted for a single polymer system [28–30].

Prior works by Kholodovych et al. [12, 31], Smith et al. [32–35], Gubskaya et al. [36], and Ghosh et al. [37] showed that it is possible to build computational models of polymer properties for an entire library based upon experimental data for a small subset. In these studies, a polymer library is explored using a combined experimental and computational approach, looking for polymers that fulfill a series of design criteria to be suitable for specific applications. Smith et al. [33, 34] developed semiempirical models using molecular descriptors obtained from two-dimensional polymer structures (i.e., the descriptors were independent of the polymer conformation). These models were able to predict fibrinogen adsorption within experimental error in 38 out of the 45 polymers and rat lung fibroblast proliferation in 41 out of 48 polymers. Pearson correlation coefficient values for these predictions were 0.54 ± 0.12 and 0.54 ± 0.09, respectively. Gubskaya et al. [36] calculated descriptors from relaxed three-dimensional polymeric structures obtained from Molecular Dynamics (MD) simulations of tetramers in vacuum and implicit water. In this work, Decision Tree Analysis and ANNs were used to predict fibrinogen adsorption with a Pearson coefficient of 0.67 ± 0.13. The incorporation of three-dimensional descriptors led to important improvements in comparison with previous semiempirical models, increasing the average Pearson correlation coefficient from 0.54 ± 0.12 to 0.67 ± 0.13.

One of the challenges of biomaterials is the change of their interactions and properties over time [38]. However, all aforementioned models study and predict individual values for each polymer. They do not consider dynamic properties

that may change over time. Even Le et al. (2013), who built predictions of phase behavior over time, developed the model using each experimental value as a single input, without considering how the phase behavior changes over time [39]. Previously, we built ANN models to accurately predict drug release over time on a family of terpolymers [40] using molecular descriptors. In this study, we develop and compare models for water uptake over time, first using all individual data separately and then using a global parameter for this property.

Our research has two objectives: (i) the development of computational models for water uptake versus time based upon experimental data from a small subset of polymers in a library and (ii) the application of these models to predict water uptake for an entire library of polymers. The main challenge of this research is to model and predict properties that change over time with particular kinetics using a small set of experimental data. As a model system, a library of L-tyrosine-derived polyarylates was used. Kohn and collaborators used this library to discover promising lead polymers for several medical applications [41], such as bone pins [42], hernia repair devices, and an antibacterial sleeve that protects recipients of implanted cardiac assist devices from potentially life-threatening infections [43].

This library, consisting of A-B-type copolymers having an alternating sequence of a diphenol and a diacid [41], was obtained by copolymerizing 14 tyrosine-derived diphenols with 8 aliphatic diacids in all possible combinations resulting in 112 distinct polymers. Changes in polymer backbone or pendent chain length affect polymer properties such as T_g and hydrophobicity. In this study we investigate the effect of polymer backbone and pendent chain on the water uptake profiles of polymer films.

2. Materials and Methods

2.1. Materials. A subset of the L-tyrosine-derived polyarylates was synthesized as described previously by carbodiimide-mediated solution polycondensation of a diphenol and a diacid at room temperature [44].

2.1.1. Nomenclature. DTR = desaminotyrosyl-tyrosine alkyl ester: R = methyl (M), ethyl (E), iso-propyl (iP), butyl (B), iso-butyl (iB), sec-butyl (sB), hexyl (H), octyl (O), dodecyl (D), benzyl (Bn), 2-(2-ethoxyethoxy)ethyl (G).

HTR = hydroxyacetic acid-tyrosine alkyl ester: R = ethyl (E), hexyl (H), octyl (O).

2.2. Experimental Methods

2.2.1. Film Processing. Polymer films were compression molded and annealed at 5–10°C above T_g for 20 h before incubation, as described previously [11].

2.2.2. Water Uptake. Water uptake was obtained for the selected polymers from the L-tyrosine-derived polyarylates combinatorial library (Table 1) using ^{3}H-labeled water, as described previously [45]. Briefly, films 1 cm in diameter were incubated in ^{3}H-radiolabeled water (0.2 μCi/mL) at 37°C.

TABLE 1: Subset of the library of L-tyrosine-derived polyarylates used in this study.

Polymer[a]	M_w (kDa)[b,c]	T_g (°C)[d]	Polymer set for model	Predictions
Poly(DTO sebacate)	123 ± 1	16	●	
Poly(DTB adipate)	111 ± 3	42	●	
Poly(DTO succinate)	84 ± 6	43	●	
Poly(DTE adipate)	126 ± 7	59	●	
Poly(DTE glutarate)	80 ± 1	64	●	
Poly(DTB succinate)	145 ± 11	67	●	
Poly(HTH sebacate)	64 ± 5	23	●	
Poly(HTH adipate)	87 ± 2	40	●	
Poly(DTM sebacate)	126 ± 4	45	●	
Poly(DTiP adipate)	144 ± 2	55	●	
Poly(DTM adipate)	99 ± 3	67	●	
Poly(HTE succinate)	★	78	●	
Poly(DTO adipate)	132 ± 2	26	●	
Poly(DTsB* R(+)methyladipate*)	79 ± 3	45	●	
Poly(DTsB* R(+) glutarate)	86 ± 3	46	●	
Poly(DTM R(+) methyladipate*)	68 ± 1	53	●	
Poly(DTBn adipate)	69 ± 8	61	●	
Poly(HTE adipate)	37 ± 4	61	●	
Poly(DTO suberate)		21		●
Poly(DTH suberate)		24		●
Poly(HTH suberate)		27		●
Poly(DTO glutarate)		32		●
Poly(DTiB sebacate)		33		●
Poly(DTH R(+) methyladipate*)		33		●
Poly(DTH L(−) methyladipate*)		33		●
Poly(DTH adipate)		34		●
Poly(DTB R(+) methyladipate*)		35		●
Poly(DTB L(−) methyladipate*)		35		●
Poly(DTB suberate)		37		●
Poly(DTO diglycolate)		40		●
Poly(DTBn sebacate)		42		●
Poly(DTH glutarate)		43		●
Poly(DTH diglycolate)		45		●
Poly(DTsB* L(−) methyladipate*)		45		●
Poly(DTsB* L(−) glutarate)		46		●
Poly(DTsB* R(+) suberate)		46		●
Poly(DTsB* L(−) suberate)		46		●
Poly(DTsB* R(+) adipate)		50		●
Poly(DTsB* L(−) adipate)		50		●
Poly(DTB glutarate)		50		●
Poly(DTH succinate)		53		●
Poly(DTM L(−) methyladipate*)		53		●
Poly(HTE suberate)		54		●
Poly(DTiP R(+) methyladipate*)		54		●
Poly(DTiP L(−) methyladipate*)		54		●

TABLE 1: Continued.

Polymer[a]	M_w (kDa)[b,c]	T_g (°C)[d]	Polymer set for model	Predictions
Poly(DTM suberate)		55		●
Poly(DTBn R(+) methyladipate*)		55		●
Poly(DTBn L(−) methyladipate*)		55		●
Poly(DTiB adipate)		56		●
Poly(DTE R(+) methyladipate*)		63		●
Poly(DTE L(−) methyladipate*)		63		●
Poly(HTE R(+) methyladipate*)		63		●
Poly(HTE L(−) methyladipate*)		63		●
Poly(DTB diglycolate)		64		●
Poly(DTiB succinate)		75		●

[a]The "*" symbol indicates the presence of more than one chiral center in the polymer repeat unit.
[b]Molecular weight (M_w) was measured by THF-GPC (mean value of three different films ± standard deviation (SD)).
[c]The "★" symbol indicates the polymers that did not dissolve in THF and, thus, M_w could not be measured, and degradation could not be measured.
[d]Glass transition temperature (T_g) was measured by DSC for the dry polymer before pressing.

After 6 h and 12 h and 1, 2, 3, 4, 7, 14, 21, 28, 35, and 42 days, samples were removed from the vial, rinsed with distilled water, blotted dry, and dissolved with 3 mL of tetrahydrofuran (THF) (VWR) and 12 mL of liquid scintillation cocktail (LSC) (Ecolite). Radioactive counts were measured using a scintillation counter (Beckmann 6500), and water content ($M_{3_{H_2O}}$) was calculated using a calibration curve. Water uptake (WU) was calculated as the water content relative to the original dry weight (M_{sample}):

$$\text{WU (\%)} = 100 \cdot \frac{M_{3_{H_2O}}}{M_{\text{sample}}}. \qquad (1)$$

Table 2 lists the estimated values for equilibrium water uptake from the experimental measurements; both this parameter and individual water uptake experimental points were used to build surrogate models for water uptake.

2.3. Computational Methods. The data-mining package WEKA (Waikato Environment for Knowledge Analysis) [46] was used in this study. The methodology can be summarized in the following steps (Figure 1):

(i) Polymers were characterized using two-dimensional (2D) descriptors [32] and three-dimensional (3D) descriptors [36].

(ii) Descriptors to build the model were selected using correlation based feature selection (CFS), expectation-maximization (EM) cluster analysis, Decision Tree Analysis, and linear regression.

(iii) Either all water uptake experimental data points over time or equilibrium water uptake was used to build the model using ANNs, using 10% for testing and the rest for training.

TABLE 2: Equilibrium water uptake for 18 polymers of the L-tyrosine-derived polyarylate library.

Polymer[a]	Equilibrium water uptake (%)
Poly(DTB adipate)	18.2 ± 1.2
Poly(DTB succinate)	4.0 ± 0.3
Poly(DTBn adipate)	32.2 ± 7.2
Poly(DTE adipate)	36.2 ± 3.2
Poly(DTE glutarate)	29.6 ± 3.4
Poly(DTiP adipate)	27.6 ± 1.0
Poly(DTM adipate)	14.5 ± 3.5
Poly(DTM sebacate)	12.3 ± 2.7
Poly(DTO adipate)	6.1 ± 0.3
Poly(DTO sebacate)	2.7 ± 0.4
Poly(DTO succinate)	3.5 ± 0.6
Poly(HTE adipate)	7.8 ± 1.1
Poly(HTE succinate)	43.1 ± 10.6
Poly(HTH adipate)	18.0 ± 2.1
Poly(HTH sebacate)	2.3 ± 0.4
Poly(DTM R(+) methyladipate)	90.1 ± 8.8
Poly(DTsB R(+) glutarate)	97.4 ± 4.1
Poly(DTsB R(+) methyladipate)	136.5 ± 10.0

[a]Polymers are ordered by name used in the descriptor set.

2.3.1. Descriptors. The descriptors in this study include "2D" descriptors based on the chemical structure of the polymers [32] and "3D" descriptors based on the chemical structure of the polymers in implicit water or vacuum incorporating polymer conformation [36]. Two-dimensional descriptors for the entire library of 112 polymers were obtained by Smith et al. [34], using the basic molecular structure derived from the chemical formulae and both the Molecular Operating Environment (MOE, Chemical Computing Group Inc.)

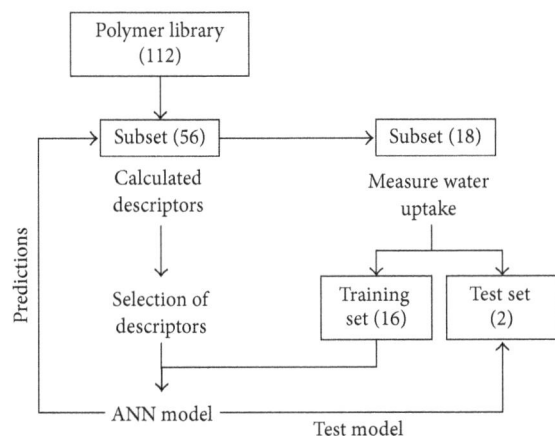

FIGURE 1: Scheme of experimental method for surrogate models of water uptake.

[47] and the Dragon (Milano Chemometrics and QSAR Research Group) [48] commercial software packages. Three-dimensional descriptors were obtained by Gubskaya et al. [36] for 56 polymers from the polyarylate library. Descriptors were obtained by the Dragon commercial software package using the 3D structures of the tetramers after structure minimization and 1 ns of MD simulations using MacroModel v.8.5 (Schrödinger) [49] commercial package with the generalized Born/surface area implicit solvent model [50] and the OPLS-all atom force field [51]. Although 3D descriptors obtained from tetramers do not capture the realistic structure of large M_w polymers, they include very important information about their structure, which allows building more accurate models, as shown previously by Gubskaya et al. [36]. Similarly, other authors had previously used monomers [52] or less than 5 repeating monomeric units [53] to obtain molecular descriptors.

2.3.2. Descriptor Selection. Starting with 2,272 descriptors taken from Gubskaya et al. [36] and Smith et al. [32], a correlation based feature selection (CFS) was used to reduce the dimensionality of the descriptors for each parameter in study. CFS is a function available in WEKA that evaluates the worth of a subset of attributes (descriptors) by considering the individual predictive ability of each feature along with the degree of redundancy between them. As a result, it selects a subset of attributes that are highly correlated with the parameter while removing irrelevant, redundant, and noisy attributes [54]. A genetic search algorithm was used in conjunction with the CFS, allowing a parallel search of the attribute space and avoiding local optima.

For each model, expectation-maximization (EM) [46] cluster analysis was employed to categorize the polymer property of study (i.e., water uptake and equilibrium water uptake) into three classes (i.e., low, medium, and high). When analyzing all data points for water uptake, both time and water uptake values were included in the cluster analysis.

The most significant descriptors were selected using a J48 Decision Tree [55], selecting descriptors that correctly partition the water uptake values and equilibrium water uptake according to the EM cluster analysis. Because Decision Tree Analysis cannot represent relationships between continuous variables, an additional descriptor was selected by linear regression, that is, the highest weight on the linear regression, for the full training set and the experimental values of water uptake. Time was also included as a descriptor for water uptake with all data points.

2.3.3. Artificial Neural Networks. Linear models are insufficient to capture the complexity of the structure-property-relationships between polymer structure and water uptake profiles. Specifically, we observed that water uptake does not yield a simple correlation with the hydrophilic factor, as defined by Todeschini et al. [56] and calculated by Smith et al. [32].

Several authors have shown that an ANN model provides more accurate predictions than a linear model [57–62]. A multilayer perceptron (MLP) was used to build ANN models for each parameter with the three descriptors selected as explained in Section 2.3.2. Two hidden layers (nodes) were used. Output nodes were unthresholded linear units [46]. Backpropagation by gradient descent was used as MLP learning method. All input variables were scaled to the unit interval while the learning rate and the momentum applied for updating the weights were 0.3 and 0.2, respectively. Training time was set on 1,000 epochs, which showed to be enough for model convergence. To perform cross-validation, 10% of data was separated as test set in each model, in all possible combinations. Randomization of the initial weights and shuffling of the training data were performed by varying the seed for the random number generator. The model obtained with each seed represents a local optimum, based on the initial weights. Thus, running enough seeds and selecting the best model among them would allow finding the global optimum. For the present models, a hundred ANN models were obtained with different seeds, from which the best model in terms of root mean squared error for the training set was selected.

3. Results and Discussion

3.1. Descriptors Selection. Table 3 summarizes the descriptors selected for both models. One 3D descriptor and five 2D descriptors were selected for the model for all time points; two 3D descriptors and one 2D were selected for the model of WU_{eq}. 2D descriptors include nCt, hydrophilic factor, SMR_VSA6, GGI3, MATS3m, and C-003. nCt is the number of tertiary carbon atoms (sp3). The hydrophilic factor is calculated from the number of hydrophilic groups (-OH, -SH, and -NH) of the molecule [63] and it was previously used to predict biological response on this polymer library [34]. SMR_VSA6 is a descriptor of subdivided surface area, based on accessible van der Waals surface area of each atom [64], and type of descriptor used before to predict fibrinogen adsorption of this polymer library [35]. GGI3 is a topological charge descriptor; similar topological descriptors have been used to predict biological response on polymethacrylate surfaces [37]. MATS3m is a Moran autocorrelation descriptor, which describes the level of correlation between molecules, and it has been used to study protein interactions [65]. C-003 is

TABLE 3: Best descriptors and their variability within the training set and within the complete set of 56 polymers.

Model	Descriptor	SD for polymers of the model	SD for the complete library
All data points	Hydrophilic factor	0.246	0.212
	SMR_VSA6	0.291	0.242
	GGI3	0.227	0.264
	MATS3m	0.256	0.273
	C-003	0.394	0.478
	G2m vacuum	0.231	0.255
WU_{eq}	nCt	0.287	0.316
	Mor25m water	0.212	0.238
	R8p+ vacuum	0.243	0.242

the number of CHR3 molecular subfragments, an atom center fragment; it gives information about structural motifs important for the molecular shape and it was used before to predict fibrinogen adsorption on polymethacrylate surfaces [37].

3D descriptors include G2m and R8p+ in vacuum and Mor25m in water. G2m is a WHIM descriptor, which captures relevant 3D information about molecular size, shape, symmetry, and atom distribution with respect to invariant reference frames [66]. WHIM descriptors were used to predict fibrinogen adsorption on polymethacrylate surfaces [37]. R8p+ is R-GETAWAY descriptor, which accounts for the local aspects of the molecule such as branching, cyclicity, and conformational changes [67].

Mor25m is a 3D-MoRSE descriptor, which provides structural information of the molecules in the space [68], and it has been suggested that this information is related to the free volume of molecules [69, 70] and, thus, responsible for the ability of the polymer to uptake water. 3D-MoRSE and GETAWAY descriptors have been also correlated with the tendency of a molecule to be solvated by water, measured by the hydrophilic index Hy [71], as defined by Todeschini and Consonni [63]. These types of descriptors encode relevant information of this polymer library that gives information of several physical and chemical processes such as water uptake and even in fibrinogen adsorption as discussed by Gubskaya et al. [36].

3.2. Model for Water Uptake. Results in Table 4 show that correlation coefficient is not the best indicator of model accuracy. Both models present high R^2 of training set (>0.92). However, the model using all data presents only 17% or less of predictions within experimental variability, for training and test sets, while the model for WU_{eq} is able to predict 67% for training and 50% for test, within experimental variability.

Results of cross-validation have to be analyzed very carefully when using all data points, because they are not independent of each other. In that case, it is likely to select for cross-validation data that belong to polymers for which there is a large data set in the training set. Thus, depending on how the cross-validation set is selected, different results will be obtained.

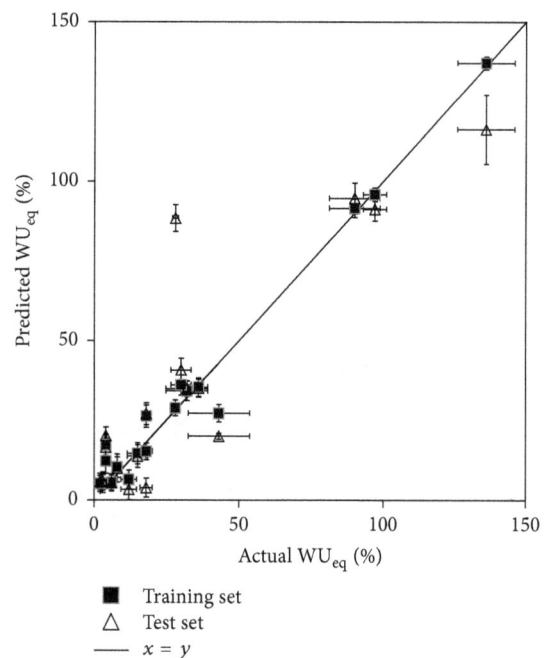

FIGURE 2: Prediction versus experimental values for WU_{eq} for polymers as part of training (■) and test (△) sets. Black line represents $x = y$. Values are presented as mean value ± SD of predictions (y-error) ± SD of experimental values (x-error).

On the other hand, the model for WU_{eq} obtains its values from several experimental measurements of each polymer after its water content is equilibrated. This gives more representative and reliable experimental data, and it captures more information than single points at the same time of incubation. With this, cross-validation that in this case includes only independent values, considering all possible combinations of leave-two-out (10%) of experimental values, gives accurate predictions in 50% of the cases from test sets, and WU_{eq} was correctly classified as high, medium, or low according to the EM cluster analysis previously done, in 83% of the cases. With this, predictions accurately represent the relative order in water uptake of the polymers studied (Figure 2 and Table 4).

TABLE 4: Summary of models for water uptake.

Model	n training set	Number of descriptors	R^2 training	Within experimental variability (training)	R^2 cross-validation (10%)	Within experimental variability (test)
All data points	189	6+ time	0.92	30/189 (16%)	0.83	3/18 (17%)
WU_{eq}	18	3	0.97	12/18 (67%)	0.78	9/18 (50%)

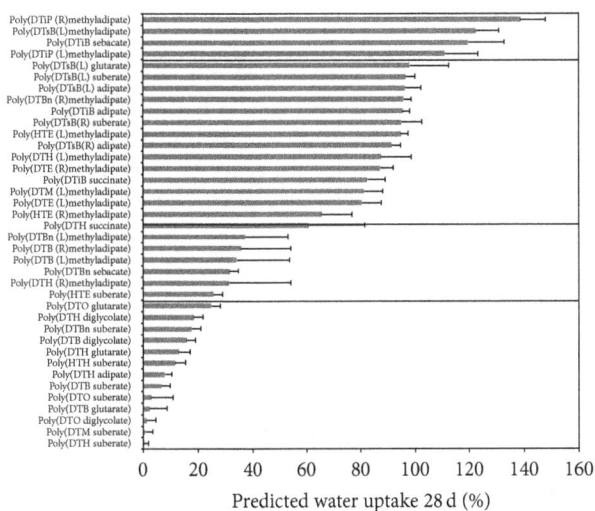

FIGURE 3: Predictions of equilibrium water uptake over the remaining 38 polymers of the polymer library. Values are presented as mean value ± SD of the predicted value for each training/test set combination. Polymers are ordered from highest to the lowest water uptake predicted values. Solid lines separate areas of very high, high, medium, and low water uptake polymers.

This result is less accurate than predictions of simple physical behaviors such as T_g [14], but it is much more accurate than predictions of more complex processes such as fibrinogen adsorption [34, 36], cell growth [34], and gene delivery efficiency [72], where the Pearson coefficient for these models was below 0.77.

3.3. Predictions of Water Uptake over Rest of the Library. For each training and test set selection, predictions of equilibrium water uptake were made for the rest of the 56-polymer library. As Figure 3 shows, the model predicts low levels of water uptake for polymers containing DTM (0%–14%), DTO (0%–25%), and HTH (5%–14%) (with the exception of methyladipates); low to intermediate levels of water uptake for polymers containing DTBn (18%–34%) and DTE (35%–37%) (with the exception of methyladipates), glutarate (0%–37%), suberate (0%–26%) (with the exception of DTsB), and sebacate (5%–32%) (with the exception of DTiB); intermediate levels of water uptake for succinate-containing polymers (13%–61%); medium to high levels of water uptake for DTiB-containing polymers (82%–120%); high levels of water uptake for DTiP methyladipates (111%–139%); and widely ranging

levels of water uptake for DTH (0%–87%), adipate (5%–96%), and methyladipate (31%–139%) polymers. It also predicts that all DTB polymers have low values of water uptake (less than 36%) and only high values of water uptake for DTsB polymers (92%–135%); it predicts low values of water uptake (10%–26%) for HTE polymers (with the exception of methyladipate) and predicts low levels of water uptake for diglycolate polymers (0%–18%).

Some of these predictions would be expected directly from the chemical structure of the polymers, but others would not be easily expected. For example, all DTO polymers would have low water uptake, which is expected from the long pendant chain (8 carbons), while the DTH polymers, with only one carbon less than the DTO, would have water uptake levels from low to medium.

3.4. Limitations of Surrogate Modeling. Limitations of this type of model include the following: (i) it needs experimental data to train the model; (ii) the descriptors give a reference of relevant parameters to the target property, but they cannot explain the mechanism; (iii) experimental measurements must be performed to validate the predictions; (iv) for new polymers outside of the sublibrary, new descriptors must be generated, which is time consuming due to the need for MD simulations. However, this last limitation is only encountered for the first property that you wish to model, for once the descriptors are generated, they can be used to build predictive models for several properties of the polymer library. The obtained models for water uptake can be improved by increasing the size of training set, by generating more meaningful descriptors, such as 3D descriptors in explicit water, by improving the descriptor selection algorithm, and by identifying other surrogate methods.

4. Conclusions

This study describes a new approach to modeling dynamic properties and demonstrates the potential value of this approach. In particular, we developed models for water uptake for a library of polymers using only a small training set and molecular descriptors for all the polymers in the library. We also demonstrate that using a consolidated parameter of water uptake, a dynamic property, gives a more accurate model than using a more conventional approach of all experimental measurement as independent values. By separating time points from one experiment, information about the slope, rate, and progression of the dynamic property is not considered in the model. And, since data points are not independent, accuracy of predictions is compromised.

A surrogate model was built to accurately predict equilibrium water uptake of a polymer sublibrary of 56 L-tyrosine-derived polyarylates using a small training set and only three descriptors selected from a large set of descriptors, calculated from either 2D or 3D structures. Those descriptors included atom counts; 3D information about electron diffraction (3D-MoRSE); and chemical properties of molecular atoms, branching, cyclicity, and conformational changes (GETAWAY). Although these descriptors can be used only for this model in this polymer library, the methodology for selecting descriptors can be applied to any polymer library and/or polymer property.

The model was able to accurately predict low, intermediate, and high levels of water uptake for up to 12 of the 18 polymers. Using this model, predictions were obtained for the rest of the sublibrary. Those predictions can be used primarily as a reference of order of magnitude and ranking of polymers in terms of water uptake.

Finally, having several semiempirical models for different polymer properties such as glass transition temperature, contact angle, fibrinogen adsorption, cell response, water uptake, and degradation for the same polymer library may be used to select a polymer for a specific application. With a known set of design criteria, a group of polymers can be selected from the mentioned models. After this selection, the actual parameters must be measured experimentally, the models must be validated, and the best polymer can be selected to begin the device development process. With this, surrogate modeling of polymer properties may accelerate the discovery and selection of rationally designed materials for a target application.

Disclosure

The content of this paper is solely the responsibility of the authors and does not necessarily represent the official views of the NIH, NIBIB, NCMHD, or CONICYT.

Competing Interests

The authors declare that they have no competing interests.

Acknowledgments

This work was supported by RESBIO (Integrated Technology Resource for Polymeric Biomaterials) funded by National Institutes of Health (NIBIB and NCMHD) under Grant P41 EB001046 and by the New Jersey Center for Biomaterials and CONICYT (FONDECYT 11121392).

References

[1] J. Kohn, W. J. Welsh, and D. Knight, "A new approach to the rationale discovery of polymeric biomaterials," *Biomaterials*, vol. 28, no. 29, pp. 4171–4177, 2007.

[2] J. M. Lantry, C. S. Roberts, and P. V. Giannoudis, "Operative treatment of scapular fractures: a systematic review," *Injury*, vol. 39, no. 3, pp. 271–283, 2008.

[3] L. G. Griffith, "Polymeric biomaterials," *Acta Biomaterialia*, vol. 48, no. 1, pp. 263–277, 2000.

[4] N. Angelova and D. Hunkeler, "Rationalizing the design of polymeric biomaterials," *Trends in Biotechnology*, vol. 17, no. 10, pp. 409–421, 1999.

[5] M. S. Shoichet, "Polymer scaffolds for biomaterials applications," *Macromolecules*, vol. 43, no. 2, pp. 581–591, 2010.

[6] H. Kranz, N. Ubrich, P. Maincent, and R. Bodmeier, "Physicomechanical properties of biodegradable poly(D,L-lactide) and poly(D,L-lactide-co-glycolide) films in the dry and wet states," *Journal of Pharmaceutical Sciences*, vol. 89, no. 12, pp. 1558–1566, 2000.

[7] C. Hopkins, P. E. McHugh, N. P. O'Dowd, Y. Rochev, and J. P. McGarry, "A combined computational and experimental methodology to determine the adhesion properties of stent polymer coatings," *Computational Materials Science*, vol. 80, pp. 104–112, 2013.

[8] C. Ahlneck and G. Zografi, "The molecular basis of moisture effects on the physical and chemical stability of drugs in the solid state," *International Journal of Pharmaceutics*, vol. 62, no. 2-3, pp. 87–95, 1990.

[9] D. Caccavo, S. Cascone, G. Lamberti, and A. A. Barba, "Modeling the drug release from hydrogel-based matrices," *Molecular Pharmaceutics*, vol. 12, no. 2, pp. 474–483, 2015.

[10] M. Tanaka, T. Hayashi, and S. Morita, "The roles of water molecules at the biointerface of medical polymers," *Polymer Journal*, vol. 45, no. 7, pp. 701–710, 2013.

[11] L. M. Valenzuela, B. Michniak, and J. Kohn, "Variability of water uptake studies of biomedical polymers," *Journal of Applied Polymer Science*, vol. 121, no. 3, pp. 1311–1320, 2011.

[12] V. Kholodovych, A. V. Gubskaya, M. Bohrer et al., "Prediction of biological response for large combinatorial libraries of biodegradable polymers: polymethacrylates as a test case," *Polymer*, vol. 49, no. 10, pp. 2435–2439, 2008.

[13] D. C. Webster and M. A. R. Meier, "Polymer libraries: preparation and applications," in *Polymer Libraries*, vol. 225 of *Advances in Polymer Science*, pp. 1–15, Springer, Berlin, Germany, 2010.

[14] A. D. Costache, J. Ghosh, D. D. Knight, and J. Kohn, "Computational methods for the development of polymeric biomaterials," *Advanced Engineering Materials*, vol. 12, no. 1-2, pp. B3–B17, 2010.

[15] A. V. Gubskaya, "Quantum-chemical descriptors in QSAR/QSPR modeling: achievements, perspectives and trends," in *Quantum Biochem*, C. F. Matta, Ed., pp. 693–721, Wiley-VCH, Weinheim, Germany, 2010.

[16] T. Le, V. C. Epa, F. R. Burden, and D. A. Winkler, "Quantitative structure-property relationship modeling of diverse materials properties," *Chemical Reviews*, vol. 112, no. 5, pp. 2889–2919, 2012.

[17] F. Serna, F. García, J. L. De La Peña, and J. M. García, "Aromatic polyisophthalamides with mononitro, dinitro and trinitroiminobenzoyl pendant groups," *High Performance Polymers*, vol. 20, no. 1, pp. 19–37, 2008.

[18] J. Crank, *The Mathematics of Diffusion*, Clarendon Press, Oxford, UK, 2nd edition, 1975.

[19] S. Roy, W. X. Xu, S. J. Park, and K. M. Liechti, "Anomalous moisture diffusion in viscoelastic polymers: modeling and testing," *Journal of Applied Mechanics*, vol. 67, no. 2, pp. 391–396, 1999.

[20] W. K. Loh, A. D. Crocombe, M. M. A. Wahab, and I. A. Ashcroft, "Modelling anomalous moisture uptake, swelling and thermal characteristics of a rubber toughened epoxy adhesive," *International Journal of Adhesion and Adhesives*, vol. 25, no. 1, pp. 1–12, 2005.

[21] W. Weibull, "A statistical distribution of wide applicability," *Journal of Applied Mechanics*, vol. 18, pp. 293–297, 1951.

[22] S. Popineau, C. Rondeau-Mouro, C. Sulpice-Gaillet, and M. E. R. Shanahan, "Free/bound water absorption in an epoxy adhesive," *Polymer*, vol. 46, no. 24, pp. 10733–10740, 2005.

[23] S. Joannès, L. Mazé, and A. R. Bunsell, "A simple method for modeling the concentration-dependent water sorption in reinforced polymeric materials," *Composites Part B: Engineering*, vol. 57, pp. 219–227, 2014.

[24] A. Noorjahan and P. Choi, "Effect of free volume redistribution on the diffusivity of water and benzene in poly(vinyl alcohol)," *Chemical Engineering Science*, vol. 121, pp. 258–267, 2015.

[25] Y. Tamai, H. Tanaka, and K. Nakanishi, "Molecular dynamics study of polymer-water interaction in hydrogels. 2. Hydrogen-bond dynamics," *Macromolecules*, vol. 29, no. 21, pp. 6761–6769, 1996.

[26] A. V. Gubskaya and P. G. Kusalik, "Molecular dynamics simulation study of ethylene glycol, ethylenediamine, and 2-aminoethanol. 1. The local Structure in pure liquids," *Journal of Physical Chemistry A*, vol. 108, no. 35, pp. 7151–7164, 2004.

[27] J. Behler, D. W. Price, and M. G. B. Drew, "Water structuring properties of carbohydrates, molecular dynamics studies on 1,5-anhydro-D-fructose," *Physical Chemistry Chemical Physics*, vol. 3, no. 4, pp. 588–601, 2001.

[28] M. Canales, D. Aradilla, and C. Alemán, "Water absorption in polyaniline emeraldine base," *Journal of Polymer Science Part B: Polymer Physics*, vol. 49, no. 18, pp. 1322–1331, 2011.

[29] T.-X. Xiang and B. D. Anderson, "Distribution and effect of water content on molecular mobility in poly(vinylpyrrolidone) glasses: a molecular dynamics simulation," *Pharmaceutical Research*, vol. 22, no. 8, pp. 1205–1214, 2005.

[30] T. X. Xiang and B. D. Anderson, "Water uptake, distribution, and mobility in amorphous poly(D,L-lactide) by molecular dynamics simulation," *Journal of Pharmaceutical Sciences*, vol. 103, no. 9, pp. 2759–2771, 2014.

[31] V. Kholodovych, J. R. Smith, D. Knight, S. Abramson, J. Kohn, and W. J. Welsh, "Accurate predictions of cellular response using QSPR: a feasibility test of rational design of polymeric biomaterials," *Polymer*, vol. 45, no. 22, pp. 7367–7379, 2004.

[32] J. R. Smith, V. Kholodovych, D. Knight, W. J. Welsh, and J. Kohn, "QSAR models for the analysis of bioresponse data from combinatorial libraries of biomaterials," *QSAR & Combinatorial Science*, vol. 24, no. 1, pp. 99–113, 2005.

[33] J. R. Smith, V. Kholodovych, D. Knight, J. Kohn, and W. J. Welsh, "Predicting fibrinogen adsorption to polymeric surfaces in silico: a combined method approach," *Polymer*, vol. 46, no. 12, pp. 4296–4306, 2005.

[34] J. R. Smith, A. Seyda, N. Weber, D. Knight, S. Abramson, and J. Kohn, "Integration of combinatorial synthesis, rapid screening, and computational modeling in biomaterials development," *Macromolecular Rapid Communications*, vol. 25, no. 1, pp. 127–140, 2004.

[35] J. R. Smith, D. Knight, J. Kohn et al., "Using surrogate modeling in the prediction of fibrinogen adsorption onto polymer surfaces," *Journal of Chemical Information and Computer Sciences*, vol. 44, no. 3, pp. 1088–1097, 2004.

[36] A. V. Gubskaya, V. Kholodovych, D. Knight, J. Kohn, and W. J. Welsh, "Prediction of fibrinogen adsorption for biodegradable polymers: integration of molecular dynamics and surrogate modeling," *Polymer*, vol. 48, no. 19, pp. 5788–5801, 2007.

[37] J. Ghosh, D. Y. Lewitus, P. Chandra et al., "Computational modeling of in vitro biological responses on polymethacrylate surfaces," *Polymer*, vol. 52, no. 12, pp. 2650–2660, 2011.

[38] S. W. Cranford, J. de Boer, C. van Blitterswijk, and M. J. Buehler, "Materiomics: an -omics approach to biomaterials research," *Advanced Materials*, vol. 25, no. 6, pp. 802–824, 2013.

[39] T. C. Le, C. E. Conn, F. R. Burden, and D. A. Winkler, "Computational modeling and prediction of the complex time-dependent phase behavior of lyotropic liquid crystals under in meso crystallization conditions," *Crystal Growth and Design*, vol. 13, no. 3, pp. 1267–1276, 2013.

[40] A. V. Gubskaya, I. J. Khan, L. M. Valenzuela, Y. V. Lisnyak, and J. Kohn, "Investigating the release of a hydrophobic peptide from matrices of biodegradable polymers: an integrated method approach," *Polymer*, vol. 54, no. 15, pp. 3806–3820, 2013.

[41] S. Brocchini, K. James, V. Tangpasuthadol, and J. Kohn, "A combinatorial approach for polymer design," *Journal of the American Chemical Society*, vol. 119, no. 19, pp. 4553–4554, 1997.

[42] K. A. Hooper, N. D. Macon, and J. Kohn, "Comparative histological evaluation of new tyrosine-derived polymers and poly (L-lactic acid) as a function of polymer degradation," *Journal of Biomedical Materials Research*, vol. 41, no. 3, pp. 443–454, 1998.

[43] H. L. Bloom, L. Constantin, D. Dan et al., "Implantation success and infection in cardiovascular implantable electronic device procedures utilizing an antibacterial envelope," *Pacing and Clinical Electrophysiology*, vol. 34, no. 2, pp. 133–142, 2011.

[44] J. Fiordeliso, S. Bron, and J. Kohn, "Design, synthesis, and preliminary characterization of tyrosine-containing polyarylates: new biomaterials for medical applications," *Journal of Biomaterials Science*, vol. 5, no. 6, pp. 496–510, 1994.

[45] L. M. Valenzuela, G. Zhang, C. R. Flach et al., "Multiscale analysis of water uptake and erosion in biodegradable polyarylates," *Polymer Degradation and Stability*, vol. 97, no. 3, pp. 410–420, 2012.

[46] I. H. Witten and E. Frank, *Data Mining: Practical Machine Learning Tools and Techniques with JAVA Implementations*, Academic Press, San Diego, Calif, USA, 1st edition, 2000.

[47] Chemical Computing Group, *MOE (The Molecular Operating Environment)*, Chemical Computing Group, Montreal, Canada, 2nd edition, 2003.

[48] R. Todeschini, V. Consonni, A. Mauri, and M. Pavan, *Dragon Web Version, 3.0*, Milano, Italy, 2003.

[49] Schrödinger, *Schrödinger Release 2005: MacroModel, V. 8.5*, Schrödinger, New York, NY, USA, 2005.

[50] W. C. Still, A. Tempczyk, R. C. Hawley, and T. Hendrickson, "Semianalytical treatment of solvation for molecular mechanics and dynamics," *Journal of the American Chemical Society*, vol. 112, no. 16, pp. 6127–6129, 1990.

[51] W. L. Jorgensen, D. S. Maxwell, and J. Tirado-Rives, "Development and testing of the OPLS all-atom force field on conformational energetics and properties of organic liquids," *Journal of the American Chemical Society*, vol. 118, no. 45, pp. 11225–11236, 1996.

[52] A. P. Toropova, A. A. Toropov, V. O. Kudyshkin, D. Leszczynska, and J. Leszczynski, "Optimal descriptors as a tool to predict the thermal decomposition of polymers," *Journal of Mathematical Chemistry*, vol. 52, no. 5, pp. 1171–1181, 2014.

[53] P. R. Duchowicz, S. E. Fioressi, D. E. Bacelo, L. M. Saavedra, A. P. Toropova, and A. A. Toropov, "QSPR studies on refractive indices of structurally heterogeneous polymers," *Chemometrics and Intelligent Laboratory Systems*, vol. 140, pp. 86–91, 2015.

[54] M. A. Hall, *Correlation-Based Feature Selection for Machine Learning*, The University of Waikato, 1999.

[55] J. R. Quinlan, *Programs for Machine Learning*, Morgan Kaufmann Publishers, San Francisco, Calif, USA, 1st edition, 1993.

[56] R. Todeschini, C. Bettiol, G. Giurin, P. Gramatica, P. Miana, and E. Argese, "Modeling and prediction by using WHIM descriptors in QSAR studies: Submitochondrial particles (SMP) as toxicity biosensors of chlorophenols," *Chemosphere*, vol. 33, no. 1, pp. 71–79, 1996.

[57] A. Afantitis, G. Melagraki, K. Makridima, A. Alexandridis, H. Sarimveis, and O. Iglessi-Markopoulou, "Prediction of high weight polymers glass transition temperature using RBF neural networks," *Journal of Molecular Structure*, vol. 716, no. 1–3, pp. 193–198, 2005.

[58] A. T. Seyhan, G. Tayfur, M. Karakurt, and M. Tanoglu, "Artificial neural network (ANN) prediction of compressive strength of VARTM processed polymer composites," *Computational Materials Science*, vol. 34, no. 1, pp. 99–105, 2005.

[59] J. W. Gao, X. Y. Wang, X. B. Li, X. Yu, and H. Wang, "Prediction of polyamide properties using quantum-chemical methods and BP artificial neural networks," *Journal of Molecular Modeling*, vol. 12, no. 4, pp. 513–520, 2006.

[60] W. Q. Liu, P. G. Yi, and Z. L. Tang, "QSPR models for various properties of polymethacrylates based on quantum chemical descriptors," *QSAR & Combinatorial Science*, vol. 25, no. 10, pp. 936–943, 2006.

[61] F. Gharagheizi, "QSPR analysis for intrinsic viscosity of polymer solutions by means of GA-MLR and RBFNN," *Computational Materials Science*, vol. 40, no. 1, pp. 159–167, 2007.

[62] J. Xu, H. Liang, B. Chen, W. Xu, X. Shen, and H. Liu, "Linear and nonlinear QSPR models to predict refractive indices of polymers from cyclic dimer structures," *Chemometrics and Intelligent Laboratory Systems*, vol. 92, no. 2, pp. 152–156, 2008.

[63] R. Todeschini and V. Consonni, *Handbook of Molecular Descriptors*, Wiley-VCH, 1st edition, 2000.

[64] S. A. Wildman and G. M. Crippen, "Prediction of physicochemical parameters by atomic contributions," *Journal of Chemical Information and Computer Sciences*, vol. 39, no. 5, pp. 868–873, 1999.

[65] J.-F. Xia, K. Han, and D.-S. Huang, "Sequence-based prediction of protein-protein interactions by means of rotation forest and autocorrelation descriptor," *Protein and Peptide Letters*, vol. 17, no. 1, pp. 137–145, 2010.

[66] R. Todeschini and P. Gramatica, "New 3D molecular descriptors: the WHIM theory and QSAR applications," *Perspectives in Drug Discovery and Design*, vol. 9, pp. 355–380, 1998.

[67] V. Consonni, R. Todeschini, and M. Pavan, "Structure/response correlations and similarity/diversity analysis by GETAWAY descriptors. 1. Theory of the novel 3D molecular descriptors," *Journal of Chemical Information and Computer Sciences*, vol. 42, no. 3, pp. 682–692, 2002.

[68] J. Gasteiger, J. Schuur, P. Selzer, L. Steinhauer, and V. Steinhauer, "Finding the 3D structure of a molecule in its IR spectrum," *Fresenius' Journal of Analytical Chemistry*, vol. 359, no. 1, pp. 50–55, 1997.

[69] W. Liu, "Prediction of glass transition temperatures of aromatic heterocyclic polyimides using an ANN model," *Polymer Engineering and Science*, vol. 50, no. 8, pp. 1547–1557, 2010.

[70] B. E. Mattioni and P. C. Jurs, "Prediction of glass transition temperatures from monomer and repeat unit structure using computational neural networks," *Journal of Chemical Information and Computer Sciences*, vol. 42, no. 2, pp. 232–240, 2002.

[71] Z. Jelcic, "Solvent molecular descriptors on poly(D, L-lactide-co-glycolide) particle size in emulsification—diffusion process," *Colloids and Surfaces A: Physicochemical and Engineering Aspects*, vol. 242, no. 1–3, pp. 159–166, 2004.

[72] A. V. Gubskaya, T. O. Bonates, V. Kholodovych et al., "Logical analysis of data in structure-activity investigation of polymeric gene delivery," *Macromolecular Theory and Simulations*, vol. 20, no. 4, pp. 275–285, 2011.

9

Improvement of Physicomechanical Properties of Pineapple Leaf Fiber Reinforced Composite

K. Z. M. Abdul Motaleb ⓘ,[1] **Md Shariful Islam,**[1] **and Mohammad B. Hoque**[2]

[1]*Department of Textile Engineering, BGMEA University of Fashion and Technology, Dhaka, Bangladesh*
[2]*Department of Textile Engineering, World University of Bangladesh, Dhaka, Bangladesh*

Correspondence should be addressed to K. Z. M. Abdul Motaleb; abdul.motaleb@buft.edu.bd

Academic Editor: Vijaya Kumar Rangari

Pineapple leaf fiber (PALF) reinforced polypropylene (PP) composites were prepared by compression molding. The fiber content varied from 25% to 45% by weight. Water uptake percentages of the composites containing various wt% of fiber were measured. All the composites demonstrated lower water uptake percentages and maximum of 1.93% for 45 wt% PALF/PP composite treated with 7(w/v)% NaOH. Tensile Strength (TS), Tensile Modulus (TM), Elongation at Break (Eb %), Bending Strength (BS), Bending Modulus (BM), and Impact Strength (IS) were evaluated for various fiber content. The 45 wt% PALF/PP composite exhibited an increase of 210% TS, 412% TM, 155% BS, 265% BM, and 140% IS compared to PP matrix. Moreover, with the increasing of fiber content, all the mechanical properties increase significantly; for example, 45 wt% fiber loading exhibited the best mechanical property. Fibers were also treated with different concentration of NaOH and the effects of alkali concentrations were observed. The composite treated with 7 (w/v)% NaOH exhibited an increase of 25.35% TS, 43.45% TM, 15.78% BS, and 52% BM but 23.11% decrease of IS compared to untreated composite. Alkali treatment improved the adhesive characteristics of fiber surface by removing natural impurities, hence improving the mechanical properties. However, over 7% NaOH concentration of the tensile strength of the composite reduced slightly due to overexposure of fibers to NaOH.

1. Introduction

Use of natural fibers in composite fabrication drew great interest from the researchers due to its biodegradability and acceptable mechanical strength. As a matrix material, polypropylene has been extensively used with natural fiber in composite preparation [1, 2]. After banana and citrus fruit, pineapple (*Ananas comosus*) is one of the most essential tropical fruits in the world [3]. Hence leaves of pineapple can be used for producing natural fibers, which necessarily are considered as waste materials. Pineapple leaf fibers (PALF) are composed of holocellulose (70–82%), lignin (5–12%), and ash (1.1%), with tremendous mechanical properties [4]. Composites, the wonder lightweight material with high strength-to-weight ratio and stiffness properties, have come a long way in replacing conventional materials like metals, woods, etc. [5].

Researchers are having great interest in finding out new sources of raw materials that possess comparable physical and mechanical properties to synthetic fibers. Various other parameters to be considered while selecting raw materials are being cheap, being ecofriendly [6], absence of health hazards, high degree of flexibility [7], lower plant's age, easy collection, and regional availability which directly influence the suitability of natural fibers [8]. Above all the natural fibers are renewable resource, thus providing a better solution of sustainable supply; for example, it has low cost, low density, least processing expenditure, no health hazards, and better mechanical and physical properties [9].

Thermoplastic matrix materials are the most important part of a composite. Polypropylene (PP) is an amorphous thermoplastic polymer and is widely used as an engineering thermoplastic, as it possesses useful properties such as transparency, dimensional stability, flame resistance, high heat distortion temperature, and high impact strength. PP is also very suitable for filling, reinforcing, and blending. PP with fibrous natural polymers of biomass origin is one of the

TABLE 1: Sample Identification.

Sl. No.	Samples	Thickness(mm)	Identification Code			
			0% NaOH	3% NaOH	5% NaOH	7% NaOH
01	PP Matrix	0.8	PM000	-	-	-
02	25 wt% PALF/PP composite	1.5	PP025	PP325	PP525	PP725
03	30 wt% PALF/PP composite	1.66	PP030	PP330	PP530	PP730
04	35 wt% PALF/PP composite	1.73	PP035	PP335	PP535	PP735
05	40 wt% PALF/PP composite	1.85	PP040	PP340	PP540	PP740
06	45 wt% PALF/PP composite	1.97	PP045	PP345	PP545	PP745

most promising routes to create natural-synthetic polymer composites [10].

However, the major limitation of natural fibers/polymers composites is the incompatibility between the hydrophobic thermoplastic matrices and the hydrophilic natural fibers [11]. Therefore it is highly important to modify the natural fiber surface to increase the fiber and matrix interaction. Chemical treatments, graft copolymerization, and use of coupling agents on natural fibers have already been explored by a number of researchers with a view to improve the fiber/matrix interaction [12]. However, mercerization of natural fibers by alkali is the most popular method nowadays to improve fiber/matrix interactions by reducing the hydrophilicity of the natural fiber. In this study, sodium hydroxide (NaOH) is used for the surface modification of the PALF.

PALF has been explored by a number of researchers in the fiber reinforced composites, and they have various alkali treatment with a view to improve the mechanical properties of the composite. However, the effect of the alkali concentration has not been explored. Therefore, the aim of this study is to manufacture the PALF/PP composite having better mechanical strength. The effects of fiber loading on the mechanical properties of PALF/PP composites are analyzed. Finally, the effects of the concentration of the alkali treatments on the mechanical properties of the composite are evaluated.

2. Materials and Methods

2.1. Materials. Granules of PP were purchased from the Cosmoplene Polyolefin Company Ltd. (401, Ayer Merbau Road, Singapore). Raw pineapple fibers were collected from Tangail, Bangladesh.

2.2. Alkali Treatment. Pineapple fibers were washed multiple times to remove the dirt and impurities attached on the fiber surface. After drying, the fibers were cut into small pieces. NaOH solutions were prepared with three different concentrations as 3, 5, and 7 (w/v) %. Then the fibers were emerged on NaOH solution and treated for 1 hour at room temperature, maintaining the fiber to solution weight ratio at 1:20 (w/v). Treatment with NaOH helps to remove the hydrogen bonding in the network structure of the fibers cellulose, thereby increasing fibers surface roughness [13]. Certain amount of wax and oils covering the external surface

of the fibers cell wall, as well as lignin, will be removed in this process. Besides, this process will expose the short length crystallites and depolymerise the native cellulose structure [14].

After 1 hour samples were washed thoroughly until the neutral pH is achieved. Then the fibers were dried by oven at 80°C for 20 hours and stored in the plastic container.

2.3. Fabrication of Composite. PP sheet was prepared from its granules (10 g) by heat press. The press was operated at 360°F. Steel plates were pressed for 5 min under a pressure of 2 tons. The plates were then cooled for 5 min in a cold press under the same pressure at room temperature. The resulting PP sheet was cut into desired size for composite fabrication. Composites were prepared by sandwiching pineapple fibers between two sheets of PP. The fiber content in the pineapple composites were 25%, 30%, 35%, 40%, and 45% by weight. The prepared composites were then packaged in polyethylene bags prior to testing.

2.4. Sample Identification. Thickness of the samples was recorded by digital slide calipers. Average results of three readings from different place along the sample have been taken for measuring each thickness. Different samples are marked as mentioned in Table 1.

2.5. Water Uptake Test. Water uptake test was carried out according to ASTM D-570. Composite samples were immersed in a beaker containing 100 ml of deionized water at room temperature (25°C) for 1 hour. Initially, weight of the samples was determined, after certain time interval; samples were taken out of the beaker and wiped using tissue papers. Their weight was taken again. In this case, it shows no uptake after 50 minutes; that is why we carried out the test up to 1 hour [19]. Water uptake test was carried out on three samples of different weight of each composite. Water uptake percentage was determined by using the following equation:

$$\text{Water up-take } (\%) = \left(\frac{W_f - W_i}{W_f} \right) \times 100\% \qquad (1)$$

where W_i is initial weight (oven dry weight) and W_f is final weight (weight after immerse in water).

2.6. Mechanical Test. The tensile strength (TS), Young's modulus (YM), and elongation at break (EB%) of the composites were measured according to the European standard (ISO/DIS 527-1:2010) by a universal tensile testing machine (H50 KS-0404) with an initial clamp separation of 20mm and a cross-head speed of 10 mm/min speed. The samples sizes were 60mm×10mm×1.6mm. The samples were conditioned at 25°C and 50% relative humidity for two days before testing and all the tests were performed under the same conditions. The samples were conditioned at 25°C and 50% relative humidity for two days before testing and all the tests were performed under the same conditions (ISO 291:2012) for consistent result.

Equations (2), (3), and (4) were used for measuring the tensile strength, elongation at break percentage, and Young's modulus, respectively.

$$\text{Tensile strength, } \sigma = \frac{Force}{Area} = \frac{F_{max}}{A} \quad (2)$$

where F_{max} is maximum load applied to the sample and A is cross-sectional area of the sample.

Percentage of elongation at break was obtained by the following relation:

$$\text{Elongation at Break, } EB\,(\%) = \left(\frac{\Delta L_b}{L_0}\right) \times 100 \quad (3)$$

where ΔL_b is extension at break point and L_0 is original length of the sample.

$$\text{Young's Modulus, } Y = \frac{d\sigma}{d\varepsilon} \quad (4)$$

where $d\sigma$ is stress at yield point and $d\varepsilon$ is strain at yield point.

Bending test was carried out for determining bending strength (BS) and bending modulus (BM) using the same universal tensile testing machine according to ISO standard (ISO 14125).

Bending test was carried out for determining bending strength (BS) and bending modulus (BM) using the same universal tensile testing machine.

Bending strength and modulus were calculated by (5) and (6), respectively.

$$\text{Bending Strength, } \sigma = \frac{3FL}{2bd^2} \quad (5)$$

where F is the load (force) at the fracture point (N), L is the length of the support span, b is width, and d is thickness.

$$\text{Bending Modulus, } E_{bend} = \frac{L^3 F}{4wh^3 d} \quad (6)$$

where w and h are the width and height of the sample, L is the distance between the two outer supports, and d is the deflection due to the load F applied at the middle of the sample.

Dynamic impact test was conducted to evaluate impact strength (IS) on unnotched mode composite specimens according to ASTM D 6110-97 standard using an Impact Tester (HT-8041B IZOD, Pendulum type, Taiwan). Mechanical property measurement for each composite was repeated for four times for accuracy, and three duplicate samples having the same dimensions were tested.

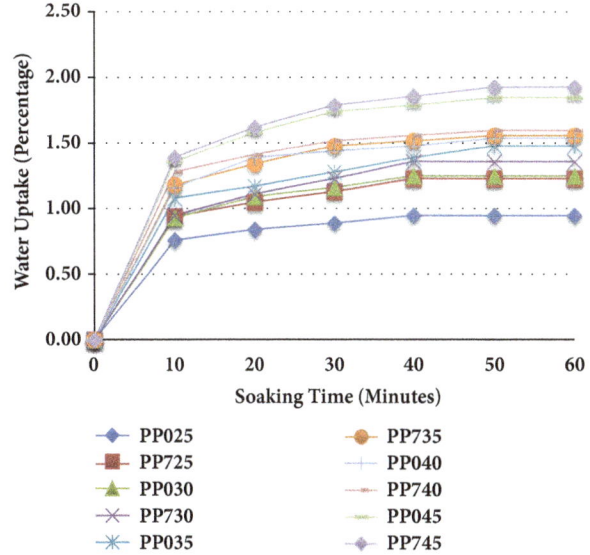

FIGURE 1: Water uptake percentage of PALF/PP composites.

3. Results and Discussion

3.1. Water Uptake. Water uptake determines the water-swelling behavior of the composites. Water uptake test was carried out on three samples of each composite and their average values were calculated. The results of water uptake are shown in Figure 1. It can be observed from Figure 1 that the water absorption percentages were increased with the increasing soaking time.

The maximum water absorption was found when the soaking time was about 50 minutes for almost all the composites and, in most of the cases, water absorption was very fast (almost 50% of maximum water uptake) in the first 10 minutes of soaking time; then it became slower gradually. It is also found that alkali-treated composite takes more water than untreated composites; for example, sample PP725 took 29% more water than sample PP025 at maximum level.

Alkali treatment removes all the noncellulosic impurities, so microgaps may be increased due to the removal of interfibrillar matrix material, such as lignin and pectin. Then the small water molecules may penetrate fiber surface easily due to having more space, resulting in more water absorption by the alkali-treated composite [19].

Besides, with the increase of fiber content (wt%) the water uptake percentage was increased; for example, sample PP045 took 94.73% more water compared to sample PP025 at maximum level. The highest water uptake percentage (1.93%) was recorded for sample PP745 at 50 minutes of soaking time. With the increased fiber loading in the composite, more absorbent sites are available, and more empty spaces resulting from the removal of the lignin and pectin are present, which leads to the higher water uptake [19].

3.2. Effect of Fiber Loading on Mechanical Properties. Tensile strength (TS), tensile modulus (TM), elongation at break (Eb%), bending strength (BS), bending modulus (BM), and

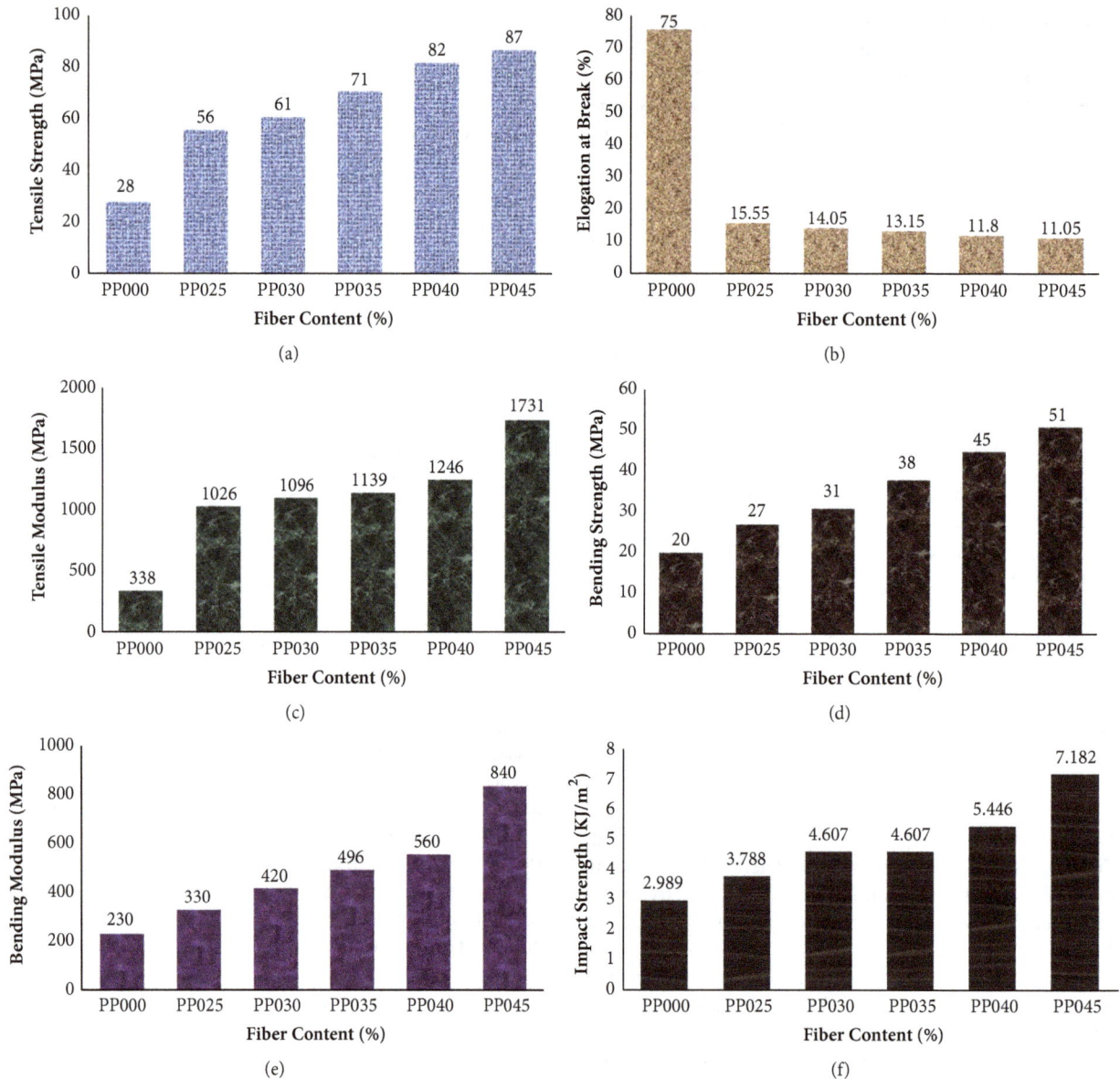

FIGURE 2: Effect of fiber content on the mechanical properties of PALF/PP composites.

impact strength (IS) of the prepared composites were studied and the data are given in Figure 2. The TS, TM, and Eb% of PP matrix were found to be 28 MPa, 338 MPa, and 75%, respectively, for sample PP025 (25 wt% fiber). Composite with 40 wt% fibers content (sample PP040) showed an increase of 192% TS and 268% TM. In all cases of fiber addition to composite, Eb% was lowered compared to PP. This was because PALF in general have low elongation percentage compared to PP [20]. So, the elasticity of polypropylene decreases with the addition of pineapple fiber and, therefore, the elongation at break also decreases which shows that composite became brittle after the increase in volume. It was noticed that composite with 45 wt% fiber content (PP045) brought out higher BS and BM values. The results revealed that, due to 45 wt% reinforcement by PALF (PP045), an increase of 155% BS and 265% BM was observed. The IS

values of PP045 were obtained 7.182 kJ/m^2. It was noticed that 140% IS was increased for PP045 compared to PP025. From Figure 1, it is clear indication that PP000 < PP025 < PP030 < PP035 < PP040 < PP045 for all the properties except elongation at break. That means that, with the increasing of fiber content, all the mechanical properties increase. PALF composites gained huge mechanical properties over the matrix material and thus indicated good fiber matrix adhesion.

The superiority of mechanical properties of pineapple leaf fiber is related to high content of alpha-cellulose content with low microfibrillar angle (14°). As reinforcing agent PALF has both qualities, that is, high content of alpha-cellulose content with low microfibrillar angle (14°), the result of PALF based polymer composites shows excellent stiffness and strength compared to other cellulose based composite materials.

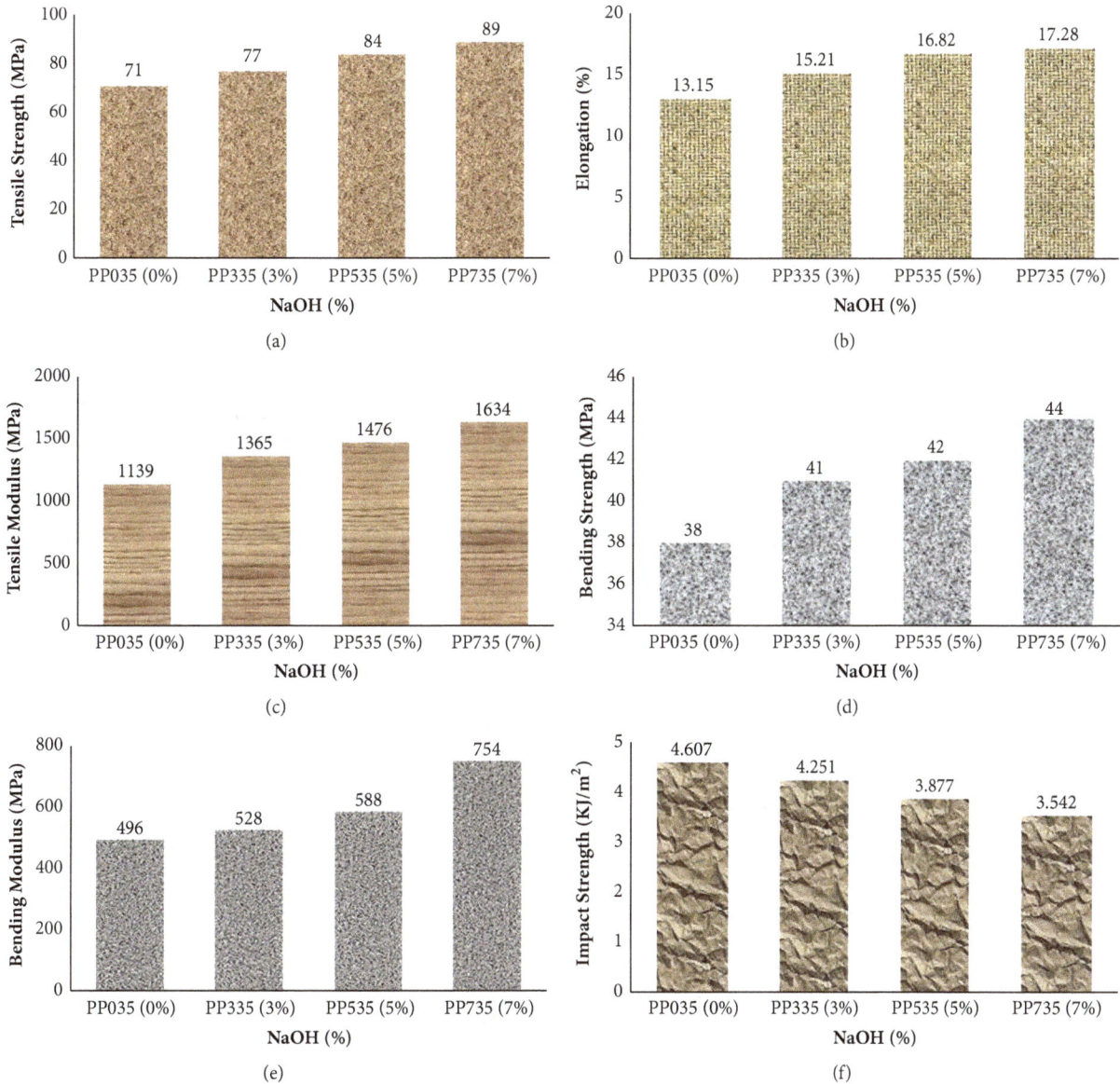

Figure 3: Effect of alkali treatment on the mechanical properties of PALF/PP composites.

3.3. Effect of Alkali Concentration on Mechanical Properties.
Figure 3 depicts the TS, EB, TM, BS, BM, and IS of alkali-treated and alkali-untreated PALF/PP composites.

Before PALF was treated with 3, 5, and 7 (w/v) % of NaOH, in all the cases fiber content was 35 wt%. Then all the mechanical properties of all the samples were measured. The samples were found as PP035 < PP335 < PP535 < PP735 for all the properties. For example, TS, EB, and TM were increased 25.35%, 31.4%, and 43.45%, respectively, for sample PP735 compared to sample PP035. It is observed that the bending strength and modulus were also increased by 15.78% and 52%, respectively, but the impact strength was decreased 23.11% for sample PP735 compared to sample PP035.

From the figure it is clear that alkali treatment improves the mechanical properties of the PALF reinforced polypropylene composites. It is also evident that the percentage of

NaOH is directly related to all the mechanical properties and composite, treated with 7 (w/v) % NaOH showing better mechanical properties. The increasing of tensile strength and modulus value of alkali-treated composites is due to the fact that the alkali treatment improved the adhesive characteristics of fiber surface by removing natural and artificial impurities.

Moreover, alkali treatment leads to fiber fibrillation, that is, breaking down of the fiber bundles into smaller fibers and results in the increasing of interface between matrix and fiber [21, 22]. This also increases in the number of reactive sites and changes in chemical compositions of the fibers [23]. The improvement of the mechanical properties may also be achieved by removing hemicellulose and lignin contents from the fiber [24]. However, when the composite was treated with 10% NaOH, it reduces the tensile strength of the composite

TABLE 2: Comparison of mechanical properties.

Sl.	Name of the composite	Tensile strength (MPa)	References
1	PALF/PP	89	This study
2	Jute-Coir hybrid fibre/PP	27	[15]
3	Jute fibre/PP	55	[16]
4	Banana fibre/Epoxy resin	45.57	[17]
5	Jute-Banana hybrid fibre/Epoxy resin	18.96	[18]
6	Palm-Coir hybrid fiber/PP	30	[11]

to 81 MPa. The reason for this may be that the higher alkali concentration weakens the fiber by excess delignification [25].

4. Comparison of the Tensile Strength of PALF/PP Composite with Other Composites

It can be seen from Table 2 that PALF/PP fiber exhibits superior tensile strength than other natural fiber reinforced composites. Therefore PALF/PP fiber can be a potential natural fiber reinforced composite for various applications.

5. Conclusions

Pineapple leaf fiber is very common in tropical regions and it is very simple to extract fibers from its leaves. The utilization of pineapple leaf fiber in composite material is a new source of materials which can be economic, ecofriendly, and recyclable. It was found that PALF/PP composite shows better tensile strength than other natural fiber reinforced composites. Besides, better mechanical properties of composite were observed with the increase of fiber loading and alkali treatment. Moreover, water uptake percentage increased with alkali treatment. Thus, PALF may be used in fabrication of ecofriendly composite products for diversified applications and, thus, synthetic fibers can easily be replaced with PALF.

Conflicts of Interest

The authors declare that there are no conflicts of interest.

References

[1] L. P. Wu, "Structures and properties of low-shrinkage polypropylene composites," *Journal of Applied Polymer Science*, p. 134, 2017.

[2] P. Uawongsuwan, Y. Yang, and H. Hamada, "Long jute fiber-reinforced polypropylene composite: Effects of jute fiber bundle and glass fiber hybridization," *Journal of Applied Polymer Science*, vol. 132, no. 15, 2015.

[3] R. M. N. Arib, S. M. Sapuan, M. A. M. M. Hamdan, M. T. Paridah, and H. M. D. K. Zaman, "A literature review of pineapple fibre reinforced polymer composites," *Polymers and Polymer Composites*, vol. 12, no. 4, pp. 341–348, 2004.

[4] S. Mishra, M. Misra, S. S. Tripathy, S. K. Nayak, and A. K. Mohanty, "Potentiality of pineapple leaf fibre as reinforcement in PALF-polyester composite: Surface modification and mechanical performance," *Journal of Reinforced Plastics and Composites*, vol. 20, no. 4, pp. 321–334, 2001.

[5] S. Nangia and S. Biswas, "Jute composite: technology & business opportunities," in *Proceedings of the international conference on advances in composites*, IISc & HAL, Bangalore, India, 2000.

[6] M. S. Sreekala, M. G. Kumaran, and S. Thomas, "Oil palm fibers: morphology, chemical composition, surface modification, and mechanical properties," *Journal of Applied Polymer Science*, vol. 66, no. 5, pp. 821–835, 1997.

[7] P. J. Herrera-Franco and A. Valadez-González, "A study of the mechanical properties of short natural-fiber reinforced composites," *Composites Part B: Engineering*, vol. 36, no. 8, pp. 597–608, 2005.

[8] M. Abdelmouleh, S. Boufi, M. N. Belgacem, and A. Dufresne, "Short natural-fibre reinforced polyethylene and natural rubber composites: effect of silane coupling agents and fibres loading," *Composites Science and Technology*, vol. 67, no. 7-8, pp. 1627–1639, 2007.

[9] L. Yan, N. Chouw, and X. Yuan, "Improving the mechanical properties of natural fibre fabric reinforced epoxy composites by alkali treatment," *Journal of Reinforced Plastics and Composites*, vol. 31, no. 6, pp. 425–437, 2012.

[10] R. A. Khan, M. A. Khan, S. Sultana, M. Nuruzzaman Khan, Q. T. H. Shubhra, and F. G. Noor, "Mechanical, degradation, and interfacial properties of synthetic degradable fiber reinforced polypropylene composites," *Journal of Reinforced Plastics and Composites*, vol. 29, no. 3, pp. 466–476, 2010.

[11] M. M. Haque, M. Hasan, M. S. Islam, and M. E. Ali, "Physico-mechanical properties of chemically treated palm and coir fiber reinforced polypropylene composites," *Bioresource Technology*, vol. 100, no. 20, pp. 4903–4906, 2009.

[12] H. Ku, H. Wang, N. Pattarachaiyakoop, and M. Trada, "A review on the tensile properties of natural fiber reinforced polymer composites," *Composites Part B: Engineering*, vol. 42, no. 4, pp. 856–873, 2011.

[13] B.-H. Lee, H.-J. Kim, and W.-R. Yu, "Fabrication of long and discontinuous natural fiber reinforced polypropylene biocomposites and their mechanical properties," *Fibers and Polymers*, vol. 10, no. 1, pp. 83–90, 2009.

[14] X. Li, S. Panigrahi, and L. G. Tabil, "A study on flax fiber-reinforced polyethylene biocomposites," *Applied Engineering in Agriculture*, vol. 25, no. 4, pp. 525–531, 2009.

[15] S. Siddika, F. Mansura, M. Hasan, and A. Hassan, "Effect of reinforcement and chemical treatment of fiber on the properties of jute-coir fiber reinforced hybrid polypropylene composites," *Fibers and Polymers*, vol. 15, no. 5, pp. 1023–1028, 2014.

[16] M. R. Rahman, M. M. Huque, M. N. Islam, and M. Hasan, "Improvement of physico-mechanical properties of jute fiber

reinforced polypropylene composites by post-treatment," *Composites Part A: Applied Science and Manufacturing*, vol. 39, no. 11, pp. 1739–1747, 2008.

[17] M. A. Maleque, F. Y. Belal, and S. M. Sapuan, "Mechanical properties study of pseudo-stem banana fiber reinforced epoxy composite," *Arabian Journal for Science and Engineering*, vol. 32, no. 2, pp. 359–364, 2007.

[18] M. Boopalan, M. Niranjanaa, and M. J. Umapathy, "Study on the mechanical properties and thermal properties of jute and banana fiber reinforced epoxy hybrid composites," *Composites Part B: Engineering*, vol. 51, pp. 54–57, 2013.

[19] J. Zhu, H. Abhyankar, and J. Njuguna, "Effect of fibre treatment on water absorption and tensile properties of flax/tannin composites," in *Proceedings of the ICMR*, 2013.

[20] R. M. N. Arib, S. M. Sapuan, M. M. H. M. Ahmad, M. T. Paridah, and H. M. D. K. Zaman, "Mechanical properties of pineapple leaf fibre reinforced polypropylene composites," *Materials and Corrosion*, vol. 27, no. 5, pp. 391–396, 2006.

[21] J. P. Siregar, S. M. Sapuan, M. Z. A. Rahman, and H. M. D. K. Zaman, "The effect of alkali treatment on the mechanical properties of short pineapple leaf fibre (PALF) reinforced high impact polystyrene (HIPS) composites," *Journal of Food, Agriculture and Environment (JFAE)*, vol. 8, no. 2, pp. 1103–1108, 2010.

[22] Y. Li, Y.-W. Mai, and L. Ye, "Sisal fibre and its composites: A review of recent developments," *Composites Science and Technology*, vol. 60, no. 11, pp. 2037–2055, 2000.

[23] H. Obasi et al., "Influence of alkali treatment and fibre content on the properties of oil palm press fibre reinforced epoxy biocomposites," *American Journal of Engineering Research*, vol. 3, no. 2, pp. 117–123, 2014.

[24] J. Jayaramudu, B. R. Guduri, and A. V. Rajulu, "Characterization of natural fabric sterculia urens," *International Journal of Polymer Analysis and Characterization*, vol. 14, no. 2, pp. 115–125, 2009.

[25] X. Li, L. G. Tabil, and S. Panigrahi, "Chemical treatments of natural fiber for use in natural fiber-reinforced composites: a review," *Journal of Polymers and the Environment*, vol. 15, no. 1, pp. 25–33, 2007.

10

Synthesis and Characterization of Chitosan Nanoaggregates from Gladius of *Uroteuthis duvauceli*

J. R. Anusha and Albin T. Fleming

Department of Advanced Zoology and Biotechnology, Loyola College, Chennai, Tamil Nadu 600 034, India

Correspondence should be addressed to J. R. Anusha; anushajr@gmail.com

Academic Editor: Rosalind Labow

We report the synthesis, characterization, and biological properties of chitosan nanoaggregates from gladius of squid, *Uroteuthis duvauceli*. β-Chitin extracted from gladius was deacetylated to chitosan and further reduced to nanosize using ionic gelation process. The morphology and occurrence of chitosan nanoaggregates (CSNA) were observed using transmission electron microscopy (TEM). The degree of deacetylation (DD%) calculated from Fourier transform infrared (FTIR) spectrum showed high value (~94 ± 1.25%) for chitosan. The CSNA depicts low molecular weight, stable positive zeta potential, and less ash and moisture content with high water and fat binding capacity. The antimicrobial activity was tested against pathogenic microorganisms, which depicted significant rate of inhibition against *Staphylococcus aureus* and *Escherichia coli* due to high cellular uptake. The antioxidant analysis for CSNA demonstrated high reducing power and scavenging activity towards superoxide radicals compared with the commercially available chitosan. Furthermore, nanoaggregates exhibited low cytotoxic behavior in biological *in vitro* tests performed using cervical cancer cell line. These results indicate that chitosan nanoaggregates synthesized from waste gladius will be highly efficient and safe candidate for biological applications as food packing film, drug carrier, and tissue engineering.

1. Introduction

In recent years, the biopolymers have been considered as potential eco-friendly substitute for the use of non-biodegradable and renewable materials. One among such naturally abundant biopolymers is chitin, a mucopolysaccharide formed of 2-acetamido-2-deoxy-β-D-glucose through β (1 → 4) linkages. Depending on the source, chitin exists in three different polymorphic forms: α, β, and γ. Rhombic α-chitin resembles chain-like structure arranged in an antiparallel direction with strong intermolecular hydrogen bonds found in shells of crabs, shrimps, and other arthropods [1, 2]. The pure form of monoclinic β-chitin is patterned in parallel direction with weak intermolecular hydrogen bonds occurring in gladius of squid fish commonly known as squid pen [3]. γ-Chitin is extracted from fungal microorganisms which is a combination of α- and β-chitin [4]. Chitin is a hard, inelastic material which is insoluble in most of the solvents due to its compact structure [5]. Hence, the acetyl groups were removed from chitin through deacetylation process, to form a deacetylated derivative called chitosan [6]. The structure of chitosan has N-atom with unshared pair of electrons that can potentially be donated, and amino groups are mostly protonated with no possibility of donating electrons. In solution, chitosan has to be acidified in order to dissolve the polymer; pK_a of chitosan is ~6.3. It has attracted considerable interest in the field of research owing to excellent biodegradability, biocompatibility, and bioactive and nontoxic nature [7, 8]. Due to its promising properties, now it is used in vast range of applications such as waste water treatment, additives for cosmetics, fibres for textiles, photographic papers, biodegradable films, biomedical devices, and microcapsule implants for controlled release in drug delivery [9–11].

With the aid of nanotechnology, nanosized biopolymers with enhanced properties have been considered as a promising option in the field of food packing, drug delivery, biosensors, and other biomedical applications [12–14].

Nanoparticles synthesized using chemical as well as mechanical method exhibit much improved properties compared with normal sized biopolymers due to high aspect ratio and surface area. Moreover, these nanoparticles with high antimicrobial profile have been devoted for the development of food packaging films in addition to improved mechanical barrier, rheological and thermal properties [15–18]. From the recent literatures, researchers have clearly demonstrated the low or nontoxic activity of chitosan nanoparticles with different chemical modifications [19–21]. In addition, chitosan scaffolds have been used in tissue engineering for skin repair and wound healing with regeneration mechanism [22]. Moreover, the mucoadhesive character facilitates the administration of poorly absorbable drugs as well as macromolecules such as nucleic acids, growth factors, and antigens across epithelial barriers [21, 23]. It was reported that the chitosan/polyguluronate nanoparticles with low cytotoxicity were used for the effective delivery of siRNA to HEK 293FT and HeLa cells [24].

For the commercial production of chitosan, chitin from shrimp and crab shells was used as raw material, which required high production cost and multiple chemical processes such as demineralization, deproteinization, and decolourization [25]. Gladius of squid is another rich source of chitosan, which is a transparent material thrown as waste from seafood processing industries. Chitosan production from gladius is cost-effective and prevents the usage of excess acids and alkaline pollutants due to its low impurities and absence of coloured compounds. In addition, it shows better reactivity, solubility, and swelling than from other sources due to much weaker molecular hydrogen bonding [7]. Most frequently, nanoparticles were synthesized according to bottom-up approach as a result of self-assembling or cross-linking process [26].

From the above-mentioned application, it is evident that the chitosan and its nanoparticles were highly effective in biomedical applications. Meanwhile, a well suitable, abundant source and easy processing steps are necessary for chitosan nanoparticle synthesis. In the present study, the chitosan nanoaggregate has been extracted from the gladius of squid, *Uroteuthis duvauceli*. The gladius was subjected to deproteinization and demineralization to form β-chitin and further deacetylated to chitosan. The nanoaggregate of chitosan was synthesized by ionic gelation process using tripolyphosphate solution. The physical and chemical properties of as-prepared chitosan nanoaggregate were analyzed and discussed in detail. In addition, the antimicrobial, antioxidant, and cytotoxic activities were evaluated to identify bioactive potential of synthesized nanostructure.

2. Experimental

2.1. Materials.
Commercially available chitosan (medium molecular weight, 190 to 300 kDa, DD, 75 to 85%) and sodium tripolyphosphate were purchased from Sigma Aldrich, Korea. The chemicals for biological assays were from Himedia (Mumbai). All other chemicals were of analytical grade and used without further purification. The squid, *Uroteuthis*

duvauceli, was from the coastal region of Kanyakumari District, Tamil Nadu, India. The gladius was separated from fish and sealed in plastic bags and stored in 4°C before further analysis.

2.2. Isolation of β-Chitin.
The gladius collected from squids (*U. duvauceli*) was cleaned to remove muscle debris and washed thoroughly with water to remove adherent proteins and soluble organic materials. The cleaned gladius was dried at room temperature and pulverised using mortar and pestle. β-Chitin was extracted from powdered gladius by sequence of processes such as deproteinization using 1 M NaOH at 50°C for 5 h and demineralization in 1 M HCl at room temperature for 2 h under constant stirring. The extract was filtered and washed with deionized (DI) water until it attains neutral pH followed by dehydration using methanol and acetone. Finally, the obtained chitin powder was dried overnight in a vacuum oven at 60°C. As-prepared chitin powder was subjected to deacetylation in 50% of 1 M NaOH solution at 60°C under nitrogen atmosphere for 6 h. The obtained white chitosan precipitate was washed several times with DI water to attain neutral pH.

2.3. Synthesis of Chitosan Nanoaggregate (CSNA).
As-prepared chitosan particles were deduced to nanosized particles by ionic gelation process using sodium tripolyphosphate. The 100 mL of chitosan solution was added to 40 mL of TPP (1.0 g/L) solution, stirred for 2 h at an ambient temperature to form optimized quantity of opalescent solution. The opalescent solution was centrifuged at 10,000 rpm and the pellets were rinsed with DI water, dried, and used for further analysis (Scheme 1).

2.4. Physicochemical Characterization.
The morphology of CSNA was recorded using transmission electron microscopy (TEM, JEOL model JEM 2011). The Fourier transform infrared (FTIR) spectra of CSNA were recorded using Perkin Elmer US/Spectrum GX spectrometer by KBr pellet technique. The degree of deacetylation (DD%) and functional groups were evaluated from FTIR absorbance peak of Std.CS and CSNA at 1655 cm^{-1} for amide I and at 3450 cm^{-1} for OH group. The DD% was calculated using the modified baseline technique [27] as shown in

$$DD\% = \left(\frac{A_{1655}}{A_{3450}} \right) \times 115, \qquad (1)$$

where A_{1655} and A_{3450} are the absorbance at 1655 cm^{-1} and 3450 cm^{-1}, respectively.

XRD patterns of CSNA were obtained with X-ray diffractometer (D/MAX-2200, Rigaku Co., Japan) using Cu Kα radiation with a wavelength of 0.154 nm. The scattering intensities were measured over an angle range from 10 to 60° (2θ) with a scanning rate 2°/min.

The average molecular weight was measured from viscosities by Mark-Houwink equation. The samples were dissolved in 0.17 M (1%, v/v) acetic acid, filtered through a sinter glass, and the viscosity of solution was measured with

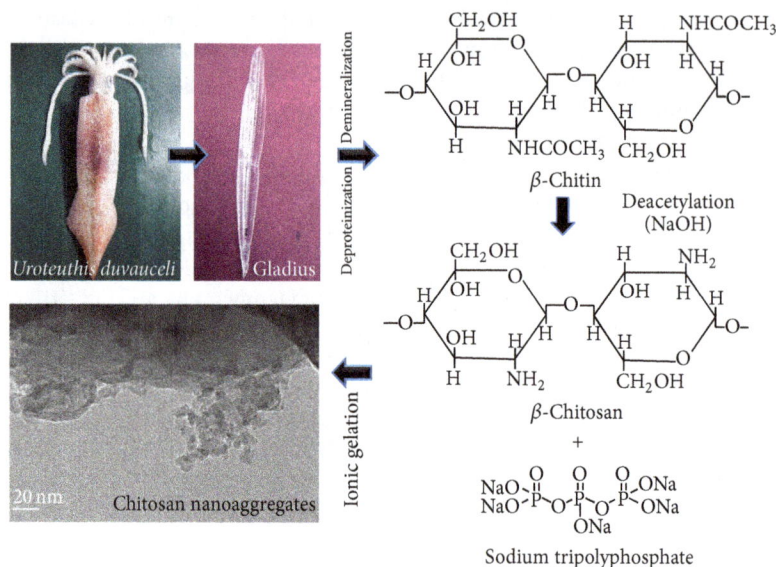

SCHEME 1: Illustrating the steps involved in the synthesis of chitosan nanoaggregates from gladius of squid, *Uroteuthis duvauceli*.

Ubbelohde-type capillary viscometer at $25 \pm 1°C$. Then, M_v was calculated using Mark-Houwink equation:

$$[\eta] = K (M_v)^{\alpha}, \qquad (2)$$

where $[\eta]$ and M_v represented intrinsic viscosity and viscosity molecular weight and K and α are viscosity constant with literature values 1.81×10^{-5} and 0.93, respectively.

In addition, zeta potential of CSNA was analyzed using Zetasizer (Malvern Instruments, UK) based on dynamic light scattering method. The ash and moisture content were estimated as described methods from the Association of Official Analytical Chemistry [28].

For the analysis of water binding capacity (WBC) and fat binding capacity (FBC), 0.5 g of sample was kept in 50 mL centrifuge tube and 10 mL of water or soybean oil was added and mixed thoroughly for 1 min on a vortex mixer. The tubes were incubated under constant shaking for 30 min at room temperature and centrifuged at 2500 rpm for 25 min. The supernatant was decanted and the tube was weighed. From the measurement, WBC and FBC were calculated as follows:

$$WBC (\%) = \frac{\text{water bound (g)}}{\text{sample weight (g)}} \times 100,$$

$$\qquad (3)$$

$$FBC (\%) = \frac{\text{fat bound (g)}}{\text{sample weight (g)}} \times 100.$$

2.5. Antimicrobial Activity. Antimicrobial assay was performed using Kirby-Bauer disk diffusion method. For antibacterial assay, the bacteria were lawn cultured on nutrient agar plates with CSNA loaded disc along with the reference standard antibiotics, ofloxacin, and incubated at 37°C for 24 h. *In vitro* antifungal assay was performed in the surface of Sabouraud dextrose agar (SDA) plates seeded with 0.1 mL of spore suspension (104 spores/mL). Disc loaded

with CSNA was placed in SDA plates and incubated at 25°C for 3–5 days. The minimal inhibitory concentrations (MICs) of the CSNA and Std.CS were determined for each antimicrobial activity against selected microorganisms. The zone of growth inhibition was measured in triplicate and the mean ± standard deviation (SD) is recorded.

2.6. Antioxidant Assay. The antioxidant capacities of CSNA were determined using reducing power and superoxide radical scavenging assay. The reducing property was quantified by assessing the abilities of chitosan samples to reduce $FeCl_3$ solution described by Oyaizu [29]. The superoxide radical scavenging ability of CSNA and Std.CS was also assessed by *in vitro* method reported by Jing and Zhao [30]. The resultant absorbance was measured at 590 nm with UV visible spectrometer (Shimadzu, UV-1800, Japan). The scavenging activity (%) was calculated using

$$\text{Scavenging activity (\%)}$$
$$= \left[1 - \frac{\text{absorbance of test}}{\text{absorbance of blank}} \right] \times 100. \qquad (4)$$

2.7. Cell Toxicity Analysis. HeLa (human cervical carcinoma) cells were used to test the *in vitro* cytotoxicity of CSNA and Std.CS. Cells were seeded in 96-well plates at a density of 50,000 cells/well and cultured in Dulbecco's Modified Eagle Medium (DMEM, Gibco), supplemented with 10% fetal bovine serum (FBS) and penicillin. The culture was incubated overnight at 37°C in 95% humidified air along with 5% CO_2 environment. After 24 h of postincubation with CSNA and Std.CS, cell proliferation was determined by adding standard 3-(4,5-dimethylthiazol-2-yl)-2,5-diphenyltetrazolium bromide (MTT) to each well. Then, the optical density was measured at 570 nm after 2 h of incubation.

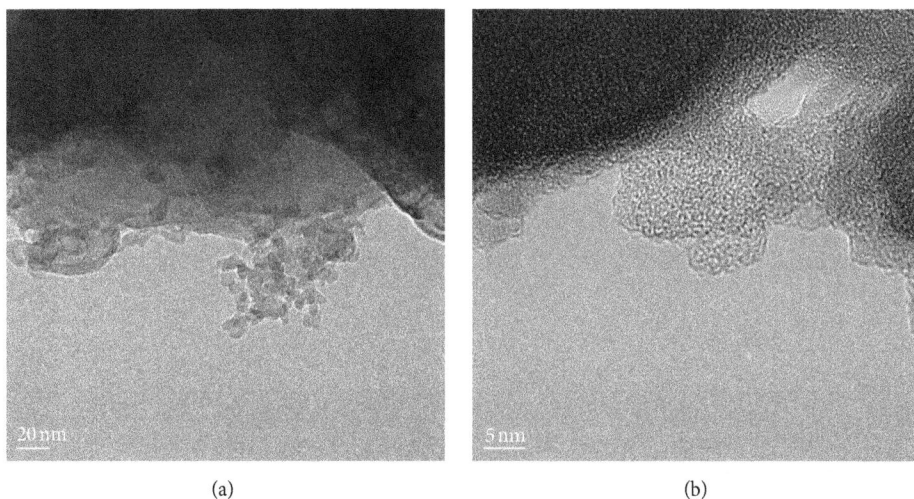

FIGURE 1: (a) TEM and (b) HRTEM images showing CSNA.

2.8. Statistical Analysis. Each experiment was performed in triplicate and the results shown were mean values ± standard deviation (SD) of data obtained. The difference between pairs of means was resolved by confidence intervals using Tukey's tests with the level of significance set at $P < 0.05$. In addition, OriginPro 8 and Microsoft Excel, 2010, programs were used for analysis.

3. Results and Discussion

3.1. Characterization of Chitosan Nanoaggregate. The yield of CS from β-chitin isolated from gladius of *U. duvauceli* was 49.28%, which was higher or even comparable with the results of other squid species [31, 32]. The obtained β-chitin was reduced to nanosized particles by ionic gelation process using tripolyphosphate. As-prepared CSNA morphology was analyzed by TEM image which showed an agglomeration of chitosan particles of size ≤ 50 nm in the form of nanoaggregates (Figure 1(a)). Figure 1(b) shows the resultant high resolution (HRTEM) image of CSNA. The image clearly displays the lattice fringes of sample which represents high degree of crystallinity. The chitosan nanoaggregates were formed by the interaction between positively charged amino groups of chitosan and negative charged tripolyphosphate ions at room temperature [26]. Moreover, the gel chitosan was reduced to opalescence while reacting with tripolyphosphate which relied on the formation of inter- and intramolecular cross-linkages medicated by the anionic molecule [33].

The functional groups of CSNA and Std.CS were analyzed from the FTIR spectra shown in Figure 2(a). The spectrum for CSNA showed a band at 3273 cm^{-1} which corresponds to NH$_2$ and OH stretching. The bands at 2871 cm^{-1} are responsible for aliphatic CH stretching whereas the band at 1561 cm^{-1} was attributed to amide stretching of C=O. Meanwhile, the C-O-C stretching vibrational modes were found at 1064 and 1029 cm^{-1}, respectively. Similar functional groups vibration was noticed in Std.CS which well agreed with the previous reports. In CSNA, the peak observed at

1204 cm^{-1} and 890 cm^{-1} for P=O stretching and P-O bending clearly represents the presence of TPP which was relevant to the reported literature [33]. Moreover, NH$_2$ absorption peak was found in 1596 cm^{-1} for Std.CS, while in the case of CSNA, the NH$_2$ stretch vibration peak slightly drifted to a low wave number, which confirmed that the phosphate group linked to amino group and formed strong intermolecular hydrogen bonds to produce nanoaggregates [26].

The degree of deacetylation depends on the number of glucosamine units of biopolymer chain with respect to the total number of units, which depicts the formation of chitosan from chitin. Moreover, DD% of chitosan could dictate its solubility, swelling behavior, crystallinity, and material degradation [27]. The DD% was calculated from FTIR and found to be 94 ± 1.25% for chitosan synthesized from *U. duvauceli* which is comparatively higher than Std.CS (~85%). The DD% was higher than the result reported on chitosan isolated from the gladius of *Sepioteuthis lessoniana* (87.93%) [34] and comparable with gladius of *Todarodes pacificus* (96.2%) [35], since the DD% values may increase with increase in NaOH concentration and deacetylation time and also depend upon the source.

X-ray diffraction was used to detect the crystallinity and crystallite sizes of CSNA and Std.CS were shown in Figure 2(b). Both the spectra show a broad diffraction peak at $20°$ was assigned to be the prominent diffraction peak of (110) chitosan [36]. The peak intensity of CSNA was found to be higher than Std.CS representing higher crystallinity of CSNA than standard sample. The relative crystallinity index (CI) of CSNA and Std.CS was calculated using the following equation:

$$\text{Crystallinity Index (CI\%)} = \frac{(I_{110} - I_{am})}{I_{110}} \times 100, \quad (5)$$

where I_{110} (arbitrary units) is the maximum intensity of the (110) peak, which is usually around $2\theta = 20°$, and I_{am} (arbitrary units), which is the amorphous diffraction at $2\theta = 16°$. From this, the crystallinity index of Std.CS is

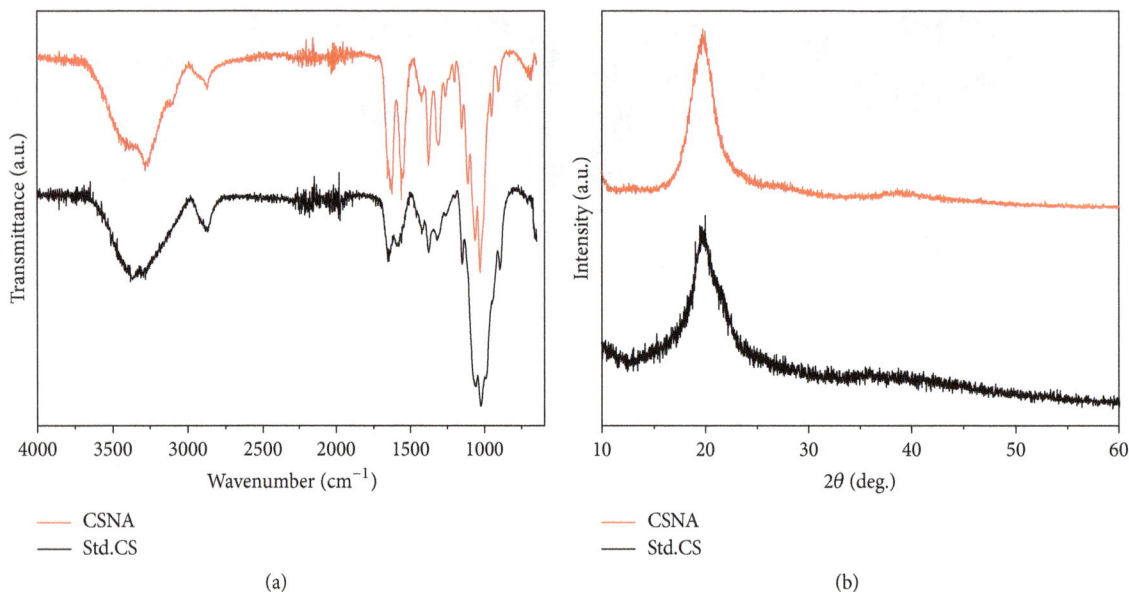

FIGURE 2: (a) Comparative FTIR spectra and (b) XRD pattern of CSNA and Std.CS.

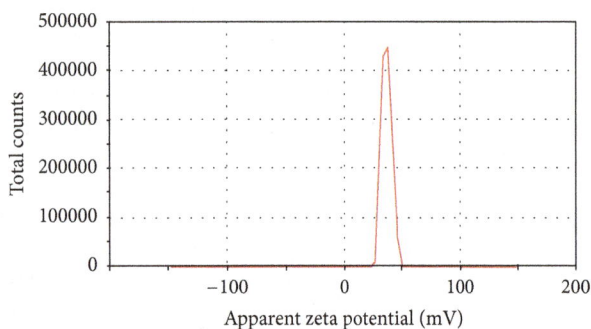

FIGURE 3: Zeta potential of CSNA synthesized from gladius of squid.

The molecular weight determined was found to be 109.35 kDa for CSNA synthesized from gladius of *U. duvauceli*. This low molecular weight may be due to the source of chitosan and also varies with other parameters such as intermolecular hydrogen bonding, dissolved oxygen concentration, chitin concentration, particle size, reaction time, concentration of alkali, and high temperature [35]. Additionally, the ash content and moisture content for CSNA analyzed were $0.3 \pm 0.05\%$ and $0.62 \pm 0.15\%$ comparable to Std.CS (0.57% ash and 0.93% moisture content). The physical properties such as water and fat binding capacity of CSNA were $569.3 \pm 1.25\%$ and $324.2 \pm 2.25\%$ whereas commercial chitosan was $509 \pm 0.25\%$ and $298 \pm 1.05\%$, respectively. The chitosan nanoaggregates showed high water absorbance which reveals the property of higher hygroscopicity nature [31, 38].

64.75%, CSNA with superior crystallinity of 85.47%. The apparent crystallite size $D_{app}(110)$ of CSNA in the direction perpendicular to the (110) crystal plane was calculated with the aid of the Scherrer equation:

$$D_{app}(110) = \frac{k\lambda}{\beta(\cos(\theta))}, \qquad (6)$$

where β is the half-width of the reflection corrected for instrumental broadening (in radians); k is constant, indicative of crystallite perfection, which was assumed to be 0.9; and λ is the wavelength of used X-ray radiation. The apparent crystallite size of CSNA was calculated to be ~2.8 nm which was in well agreement with those crystallites highlighted in the HRTEM image of CSNA (Figure 1(b)).

Zeta potential (Zp) of CSNA possesses mean positive potential of 38.5 ± 1.25 mV, ranging from +10 to +60 mV under acidic condition (Figure 3). The surface charge potential of CSNA was found to be physically stable, which determines the biological properties of biopolymer. The zeta potential was highly comparable with previous reported result [37].

3.2. Antimicrobial Activity.
To compare the antimicrobial activity of CSNA and Std.CS, the samples were tested against various pathogenic microorganisms and the results were listed in Table 1. From the results, both the chitosan samples showed antimicrobial property due to the presence of amino groups. Highest level of zone of inhibition was found in gram positive bacteria, *Staphylococcus aureus* (22 mm), followed by gram negative bacteria *Escherichia coli* (20 mm) for chitosan nanoaggregates at 1.0 mg/mL concentration. In case of fungal organisms, chitosan nanoaggregates showed high activity for *Aspergillus niger* with 19 mm, zone of inhibition. In case of Std.CS, significant result was shown against *E. coli* (18 mm) and few other microorganisms, which was evident with another report [39]. This was in agreement with the various reports on antimicrobial efficiency of chitosan with different molecular weight which showed high activity against *S. aureus* compared to *E. coli* may due to increased cellular uptake [15–17]. The MIC was tested against low concentrations of samples. The results revealed that the MIC against *S.*

TABLE 1: Antimicrobial activity of CSNA and Std.CS against various microorganisms.

Microorganisms[a]	CSNA			Std.CS		
	Concentration (mg/mL)					
	0.25	0.50	1.0	0.25	0.50	1.0
Micrococcus sp.	5 ± 1.2	7 ± 0.6	12 ± 1.0	3 ± 1.0	5 ± 0.5	10 ± 1.5
Staphylococcus aureus	11 ± 1.5	17 ± 1.2	22 ± 2.5	7 ± 1.2	11 ± 2.0	16 ± 2.0
Streptococcus sp.	6 ± 1.0	8 ± 1.5	14 ± 0.5	2 ± 0.0	3 ± 0.0	5 ± 0.5
Enterobacter sp.	2 ± 0.6	5 ± 0.5	9 ± 1.5	1 ± 0.0	1 ± 0.0	3 ± 0.75
Escherichia coli	10 ± 1.5	16 ± 1.0	20 ± 2.5	6 ± 0.8	10 ± 1.2	18 ± 1.0
Proteus vulgaris	4 ± 2.0	6 ± 2.5	11 ± 1.0	0 ± 0.0	5 ± 1.05	8 ± 2.5
Aspergillus niger	10 ± 0.5	14 ± 0.45	19 ± 1.05	4 ± 0.5	7 ± 0.25	11 ± 0.45
Rhizopus sp.	4 ± 1.0	7 ± 2.5	10 ± 2.2	2 ± 1.5	4 ± 1.4	9 ± 2.0
Candida sp.	3 ± 0.25	5 ± 1.50	7 ± 1.0	2 ± 0.0	3 ± 0.0	4 ± 0.0

[a]DMSO was used as control and showed no inhibition zone.

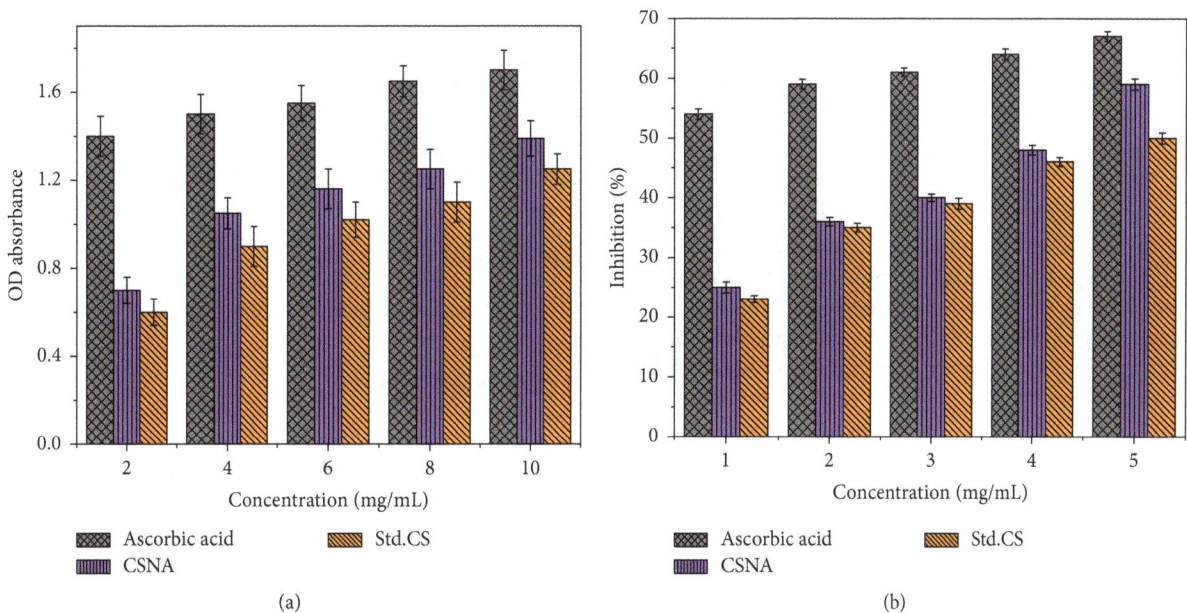

FIGURE 4: (a) Antioxidant reducing power and (b) superoxide radical scavenging ability of CSNA and Std.CS.

aureus and E. coli was lower than the tested microorganisms, whereas highest MIC was recorded against C. albicans. Moreover, the degree of antimicrobial effects of tested samples was higher with increase in concentration. Better antimicrobial activity of CSNA may be due to the interaction mechanism between nanoaggregates and the cell wall [40], in which the positively charged CSNA molecule and negative charged microbial cell wall constituents will interact and interrupt the normal cell mechanism related to DNA and thus inhibit RNA synthesis [41]. These results were consistent with the previous finding of nanosized chitosan against E. coli for potential application as food packing antimicrobial films [16]. Moreover, the antimicrobial property of chitosan showed strong bactericidal effect which may probably associated with molecular weight, degree of deacetylation, concentration of chitosan, and bacterial inoculum size [38, 42, 43].

3.3. Antioxidant Potential of CSNA. In order to study the antioxidant activity of CSNA and Std.CS, different concentrations of samples were prepared and analyzed for reducing power and scavenging property. All the samples showed highly effective reducing power as shown in Figure 4(a). The CSNA with 10 mg/mL concentration showed high reducing power of 1.39. When compared to Std.CS (1.25 for 10 mg/mL), the nanoaggregate of chitosan from gladius of squid showed high reducing power values. Reducing power of CSNA correlated with the concentration of sample; that is, increase in concentration will increase the absorbance which indicates higher reducing power and vice versa. The reducing ability of CSNA may vary with the degree of deacetylation, deacetylation times, and molecular weight [38]. Figure 4(b) shows concentration dependent scavenging activity against superoxide radicals. At 1–5 mg/mL concentration, the scavenging

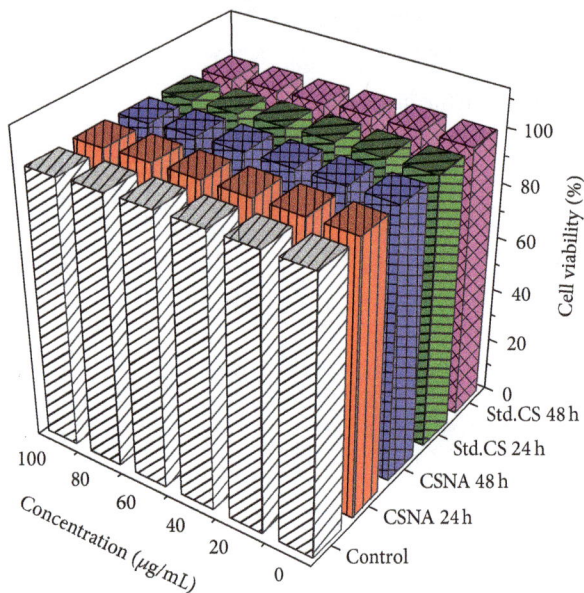

Figure 5: *In vitro* cell proliferation of HeLa against CSNA and Std.CS (24 h and 48 h) measured using MTT assay.

percentage of CSNA against superoxide radical ranged from 25 to 59% which was found to be higher than Std.CS (23–50%). These depict the concentration dependent scavenging effect of chitosan derivatives for superoxide anion radicals. The scavenging ability of Std.CS was comparable with the reported results [9], whereas the CSNA showed higher activity than its normal. These results revealed good reducing power and scavenging ability promising an excellent source of antioxidant supplement.

3.4. In Vitro Cell Toxicity Assessment. The cytotoxicity of CSNA and Std.CS was analyzed by MTT assay which is based on the conversion of MTT to formazan crystals by mitochondrial dehydrogenases (Figure 5). Surprisingly, the results did not evidence significant effect on cell viability. The cell shows a viability of over 99% during the experiments with those concentrations of nanoaggregates when compared with the negative control. The highest concentration of nanoaggregates evidenced 99.86% of cellular viability. Similarly, the commercial chitosan was also found to be less toxic (98.52%) towards the cell line. Moreover, up to 60 μg/mL concentration the chitosan nanoaggregates were found to induce cell proliferation in 24 h of incubation. No significant differences in cell viability were observed among CSNA concentrations [19]. However, the low cytotoxic CSNA against HeLa cell were consistent with the previous finding for the application of siRNA delivery [24, 44]. Chitosan was acknowledged as more suitable for drug delivery across epithelial membrane based on the favourable biological characteristics [45, 46]. Several studies have demonstrated that the nanoparticles are transported across cell membrane more rapidly than microparticles [37, 47]. CSNA can be easily prepared under mild conditions and it can easily incorporate with biological macromolecules. In addition, the less toxic

CSNA were reported as potential edible food packaging films [12]. These findings indicate that the chitosan nanoaggregates with low toxicity could be a potential edible drug delivery agent for cancer therapy.

4. Conclusions

Chitosan nanoaggregates were successfully synthesized by ionic gelation technique from the gladius of squid, *U. duvauceli*. The physiochemical properties reveal high degree of deacetylation, stable positive zeta potential, and low molecular weight along with very low ash content and high water binding capacity. CSNA also showed significant antioxidant reducing power and scavenging ability against superoxide radicals. Moreover, the excellent antimicrobial ability proposes that the CSNA can be used to control, suppress, or inhibit the growth of bacterial and fungal organisms. The nanoaggregates exhibited low toxicity in contact with HeLa cell line, which is an encouraging indicator of their cytocompatibility and safety. Therefore, the gladius thrown as waste could be utilized for the production of chitosan nanoaggregates for potential food packing film and biomedical application especially for drug delivery and in context of tissue engineering strategies.

Conflict of Interests

The authors declare that there is no conflict of interests.

Acknowledgment

The authors would like to express their gratitude to the Management of Loyola College, Chennai, India, for providing research facilities.

References

[1] I. Younes, S. Hajji, V. Frachet, M. Rinaudo, K. Jellouli, and M. Nasri, "Chitin extraction from shrimp shell using enzymatic treatment. Antitumor, antioxidant and antimicrobial activities of chitosan," *International Journal of Biological Macromolecules*, vol. 69, pp. 489–498, 2014.

[2] L. Manni, O. Ghorbel-Bellaaj, K. Jellouli, I. Younes, and M. Nasri, "Extraction and characterization of chitin, chitosan, and protein hydrolysates prepared from shrimp waste by treatment with crude protease from *Bacillus cereus* SV1," *Applied Biochemistry and Biotechnology*, vol. 162, no. 2, pp. 345–357, 2010.

[3] D. K. Youn, H. K. No, and W. Prinyawiwatkul, "Preparation and characteristics of squid pen β-chitin prepared under optimal deproteinisation and demineralisation condition," *International Journal of Food Science and Technology*, vol. 48, no. 3, pp. 571–577, 2013.

[4] A. Zamani and M. J. Taherzadeh, "Production of superabsorbents from fungal chitosan," *Iranian Polymer Journal*, vol. 21, no. 12, pp. 845–853, 2012.

[5] E. L. Mogilevskaya, T. A. Akopova, A. N. Zelenetskii, and A. N. Ozerin, "The crystal structure of chitin and chitosan," *Polymer Science—Series A*, vol. 48, no. 2, pp. 116–123, 2006.

[6] C.-H. Chen, F.-Y. Wang, and Z.-P. Ou, "Deacetylation of β-chitin. I. Influence of the deacetylation conditions," *Journal of Applied Polymer Science*, vol. 93, no. 5, pp. 2416–2422, 2004.

[7] A. Chandumpai, N. Singhpibulporn, D. Faroongsarng, and P. Sornprasit, "Preparation and physico-chemical characterization of chitin and chitosan from the pens of the squid species, *Loligo lessoniana* and *Loligo formosana*," *Carbohydrate Polymers*, vol. 58, no. 4, pp. 467–474, 2004.

[8] R. C. Mundargi, V. R. Babu, V. Rangaswamy, P. Patel, and T. M. Aminabhavi, "Nano/micro technologies for delivering macromolecular therapeutics using poly(d,l-lactide-co-glycolide) and its derivatives," *Journal of Controlled Release*, vol. 125, no. 3, pp. 193–209, 2008.

[9] N. Subhapradha, P. Ramasamy, V. Shanmugam, P. Madeswaran, A. Srinivasan, and A. Shanmugam, "Physicochemical characterisation of β-chitosan from *Sepioteuthis lessoniana* gladius," *Food Chemistry*, vol. 141, no. 2, pp. 907–913, 2013.

[10] J. R. Anusha, C. Justin Raj, B.-B. Cho, A. T. Fleming, K.-H. Yu, and B.-C. Kim, "Amperometric glucose biosensor based on glucose oxidase immobilized over chitosan nanoparticles from gladius of *Uroteuthis duvauceli*," *Sensors and Actuators B: Chemical*, vol. 215, pp. 536–543, 2015.

[11] S. A. Agnihotri, N. N. Mallikarjuna, and T. M. Aminabhavi, "Recent advances on chitosan-based micro- and nanoparticles in drug delivery," *Journal of Controlled Release*, vol. 100, no. 1, pp. 5–28, 2004.

[12] S. B. Schreiber, J. J. Bozell, D. G. Hayes, and S. Zivanovic, "Introduction of primary antioxidant activity to chitosan for application as a multifunctional food packaging material," *Food Hydrocolloids*, vol. 33, no. 2, pp. 207–214, 2013.

[13] J. R. Anusha, A. T. Fleming, H.-J. Kim, B. C. Kim, K.-H. Yu, and C. J. Raj, "Effective immobilization of glucose oxidase on chitosan submicron particles from gladius of *Todarodes pacificus* for glucose sensing," *Bioelectrochemistry*, vol. 104, pp. 44–50, 2015.

[14] V. R. Babu, P. Patel, R. C. Mundargi, V. Rangaswamy, and T. M. Aminabhavi, "Developments in polymeric devices for oral insulin delivery," *Expert Opinion on Drug Delivery*, vol. 5, no. 4, pp. 403–415, 2008.

[15] P. K. Dutta, S. Tripathi, G. K. Mehrotra, and J. Dutta, "Perspectives for chitosan based antimicrobial films in food applications," *Food Chemistry*, vol. 114, no. 4, pp. 1173–1182, 2009.

[16] A. N. B. Romainor, S. F. Chin, S. C. Pang, and L. M. Bilung, "Preparation and characterization of chitosan nanoparticles-doped cellulose films with antimicrobial property," *Journal of Nanomaterials*, vol. 2014, Article ID 710459, 10 pages, 2014.

[17] M. Sadeghi-Kiakhani, M. Arami, and K. Gharanjig, "Application of a biopolymer chitosan-poly(propylene)imine dendrimer hybrid as an antimicrobial agent on the wool fabrics," *Iranian Polymer Journal (English Edition)*, vol. 22, no. 12, pp. 931–940, 2013.

[18] S. Adibzadeh, S. Bazgir, and A. A. Katbab, "Fabrication and characterization of chitosan/poly(vinyl alcohol) electrospun nanofibrous membranes containing silver nanoparticles for antibacterial water filtration," *Iranian Polymer Journal*, vol. 23, no. 8, pp. 645–654, 2014.

[19] A. Grenha, M. E. Gomes, M. Rodrigues et al., "Development of new chitosan/carrageenan nanoparticles for drug delivery applications," *Journal of Biomedical Materials Research—Part A*, vol. 92, no. 4, pp. 1265–1272, 2010.

[20] K. Konecsni, N. H. Low, and M. T. Nickerson, "Chitosan-tripolyphosphate submicron particles as the carrier of entrapped rutin," *Food Chemistry*, vol. 134, no. 4, pp. 1775–1779, 2012.

[21] G.-F. Li, J.-C. Wang, X.-M. Feng, Z.-D. Liu, C.-Y. Jiang, and J.-D. Yang, "Preparation and testing of quaternized chitosan nanoparticles as gene delivery vehicles," *Applied Biochemistry and Biotechnology*, vol. 175, no. 7, pp. 3244–3257, 2015.

[22] P. T. Sudheesh Kumar, V.-K. Lakshmanan, T. V. Anilkumar et al., "Flexible and microporous chitosan hydrogel/nano ZnO composite bandages for wound dressing: *in vitro* and *in vivo* evaluation," *ACS Applied Materials & Interfaces*, vol. 4, no. 5, pp. 2618–2629, 2012.

[23] C. Roney, P. Kulkarni, V. Arora et al., "Targeted nanoparticles for drug delivery through the blood-brain barrier for Alzheimer's disease," *Journal of Controlled Release*, vol. 108, no. 2-3, pp. 193–214, 2005.

[24] D. W. Lee, K.-S. Yun, H.-S. Ban, W. Choe, S. K. Lee, and K. Y. Lee, "Preparation and characterization of chitosan/polyguluronate nanoparticles for siRNA delivery," *Journal of Controlled Release*, vol. 139, no. 2, pp. 146–152, 2009.

[25] M.-K. Jang, B.-G. Kong, Y.-I. Jeong, C. H. Lee, and J.-W. Nah, "Physicochemical characterization of α-chitin, β-chitin, and γ-chitin separated from natural resources," *Journal of Polymer Science Part A: Polymer Chemistry*, vol. 42, no. 14, pp. 3423–3432, 2004.

[26] K.-S. Huang, Y.-R. Sheu, and I.-C. Chao, "Preparation and properties of nanochitosan," *Polymer—Plastics Technology and Engineering*, vol. 48, no. 12, pp. 1239–1243, 2009.

[27] A. Baxter, M. Dillon, K. D. Anthony Taylor, and G. A. F. Roberts, "Improved method for i.r. determination of the degree of N-acetylation of chitosan," *International Journal of Biological Macromolecules*, vol. 14, no. 3, pp. 166–169, 1992.

[28] AOAC, *Official Methods of Analysis of the Association of Official Analytical Chemistry*, The Association of Official Analytical Chemistry, Washington, DC, USA, 14th edition, 1984.

[29] M. Oyaizu, "Studies on products of browning reactions: antioxidative activities of products of browning reaction prepared from glucosamine," *Japanese Journal of Nutrition*, vol. 44, no. 6, pp. 307–315, 1986.

[30] T. Y. Jing and X. Y. Zhao, "The improved pyrogallol method by using terminating agent for superoxide dismutase measurement," *Progress in Inorganic Biochemistry and Biophysics*, vol. 22, pp. 84–86, 1995.

[31] E. S. Abdou, K. S. A. Nagy, and M. Z. Elsabee, "Extraction and characterization of chitin and chitosan from local sources," *Bioresource Technology*, vol. 99, no. 5, pp. 1359–1367, 2008.

[32] J. Huang, W.-W. Chen, S. Hu et al., "Biochemical activities of 6-carboxy β-chitin derived from squid pens," *Carbohydrate Polymers*, vol. 91, no. 1, pp. 191–197, 2013.

[33] A. Rampino, M. Borgogna, P. Blasi, B. Bellich, and A. Cesàro, "Chitosan nanoparticles: preparation, size evolution and stability," *International Journal of Pharmaceutics*, vol. 455, no. 1-2, pp. 219–228, 2013.

[34] N. Subhapradha, R. Saravanan, P. Ramasamy, A. Srinivasan, V. Shanmugam, and A. Shanmugam, "Hepatoprotective effect of β-chitosan from gladius of *Sepioteuthis lessoniana* against carbon tetrachloride-induced oxidative stress in wistar rats," *Applied Biochemistry and Biotechnology*, vol. 172, no. 1, pp. 9–20, 2014.

[35] D. K. Youn, H. K. No, and W. Prinyawiwatkul, "Preparation and characterisation of selected physicochemical and functional properties of β-chitosans from squid pen," *International Journal*

of Food Science and Technology, vol. 48, no. 8, pp. 1661–1669, 2013.

[36] F. C. Vasconcellos, G. A. S. Goulart, and M. M. Beppu, "Production and characterization of chitosan microparticles containing papain for controlled release applications," *Powder Technology*, vol. 205, no. 1–3, pp. 65–70, 2011.

[37] H. Liu and C. Gao, "Preparation and properties of ionically cross-linked chitosan nanoparticles," *Polymers for Advanced Technologies*, vol. 20, no. 7, pp. 613–619, 2009.

[38] M. Y. Arancibia, A. Alemán, M. M. Calvo, M. E. López-Caballero, P. Montero, and M. C. Gómez-Guillén, "Antimicrobial and antioxidant chitosan solutions enriched with active shrimp (*Litopenaeus vannamei*) waste materials," *Food Hydrocolloids*, vol. 35, pp. 710–717, 2014.

[39] D. Celis, M. I. Azocar, J. Enrione, M. Paez, and S. Matiacevich, "Characterization of salmon gelatin based film on antimicrobial properties of chitosan against *E. coli*," *Procedia Food Science*, vol. 1, pp. 399–403, 2011.

[40] G. Azadi, M. Seward, M. U. Larsen, N. C. Shapley, and A. Tripathi, "Improved antimicrobial potency through synergistic action of chitosan microparticles and low electric field," *Applied Biochemistry and Biotechnology*, vol. 168, no. 3, pp. 531–541, 2012.

[41] K. Saita, S. Nagaoka, T. Shirosaki, M. Horikawa, S. Matsuda, and H. Ihara, "Preparation and characterization of dispersible chitosan particles with borate crosslinking and their antimicrobial and antifungal activity," *Carbohydrate Research*, vol. 349, pp. 52–58, 2012.

[42] C.-P. Chen, C.-T. Chen, and T. Tsai, "Chitosan nanoparticles for antimicrobial photodynamic inactivation: characterization and in vitro investigation," *Photochemistry and Photobiology*, vol. 88, no. 3, pp. 570–576, 2012.

[43] A. J. Friedman, J. Phan, D. O. Schairer et al., "Antimicrobial and anti-inflammatory activity of chitosan-alginate nanoparticles: a targeted therapy for cutaneous pathogens," *Journal of Investigative Dermatology*, vol. 133, no. 5, pp. 1231–1239, 2013.

[44] W. E. Rudzinski and T. M. Aminabhavi, "Chitosan as a carrier for targeted delivery of small interfering RNA," *International Journal of Pharmaceutics*, vol. 399, no. 1-2, pp. 1–11, 2010.

[45] S. Honary and F. Zahir, "Effect of zeta potential on the properties of nano-drug delivery systems—a review (part 1)," *Tropical Journal of Pharmaceutical Research*, vol. 12, no. 2, pp. 255–264, 2013.

[46] K. S. Soppimath, T. M. Aminabhavi, A. R. Kulkarni, and W. E. Rudzinski, "Biodegradable polymeric nanoparticles as drug delivery devices," *Journal of Controlled Release*, vol. 70, no. 1-2, pp. 1–20, 2001.

[47] S. P. Thakker, A. P. Rokhade, S. S. Abbigerimath et al., "Inter-polymer complex microspheres of chitosan and cellulose acetate phthalate for oral delivery of 5-fluorouracil," *Polymer Bulletin*, vol. 71, no. 8, pp. 2113–2131, 2014.

11

Encapsulation of Nicardipine Hydrochloride and Release from Biodegradable Poly(D,L-lactic-co-glycolic acid) Microparticles by Double Emulsion Process: Effect of Emulsion Stability and Different Parameters on Drug Entrapment

Nopparuj Soomherun,[1] Narumol Kreua-ongarjnukool,[1] Sorayouth Chumnanvej,[2] and Saowapa Thumsing[1]

[1]Department of Industrial Chemistry, Faculty of Applied Science, King Mongkut's University of Technology North Bangkok, Bangkok, Thailand
[2]Neurosurgery Unit, Surgery Department, Faculty of Medicine Ramathibodi Hospital, Mahidol University, Bangkok, Thailand

Correspondence should be addressed to Narumol Kreua-ongarjnukool; narumol.k@sci.kmutnb.ac.th

Academic Editor: Ravin Narain

Poly(D,L-lactic-co-glycolic acid) (PLGA) is an important material used in drug delivery when controlled release is required. The purpose of this research is to design and characterize PLGA microparticles (PLGA MPs) implants for the controlled release of nicardipine hydrochloride (NCH) *in vitro*. This study used the water-in-oil-in-water ($w_1/o/w_2$) double emulsion and solvent diffusion/evaporation approach to prepare PLGA MPs. Optimal processing conditions were found, such as polymer content, surfactant type, stabilizer concentration, inner and outer aqueous phase volumes, and stirring speed. The PLGA MPs for use as nicardipine hydrochloride (NCH) loading and release had spherical morphology, and the average diameter was smaller than $5.20 \pm 0.25\,\mu$m. The release kinetics were modeled to elucidate the possible mechanism of drug release. *In vitro* release studies indicated that the NCH release rate is slow and continuous. PLGA MPs are an interesting alternative drug delivery system, especially for use with NCH for biomedical applications.

1. Introduction

Currently, in general medical practice, the use of nicardipine hydrochloride (NCH) for the long-term treatment of hypertension (high blood pressure) is common. NCH prevents calcium ion (Ca^{2+}) entry into vascular smooth and cardiac muscle through calcium channels [1]. Vasoconstriction, or a blood vessel constriction, is dependent on Ca^{2+}, which increases blood pressure. The inhibition of Ca^{2+} movement can be used to treat high blood pressure, angina, and subarachnoid haemorrhage in patients with hypertension [2]. NCH is part of a class of calcium channel blockers and is a hydrophilic drug, as shown in Figure 1(a). As a result, it has a short half-life and has degradable ester linkages in

its structure [3, 4]. Therefore, a patient may need to take medication frequently, which may result in toxic effects.

To solve the above problem, researchers have attempted to find a biomaterial for controlled NCH delivery. Biopolymers are biomaterials that can reduce side effects and promote continuous therapeutic medication levels for extended periods of time [5]. Among all biopolymers, poly(D,L-lactic-co-glycolic acid) (PLGA) has shown immense potential as a good drug delivery material. Chemically, PLGA is a copolymer of poly(D,L-lactic acid) (PLA) and poly(glycolic acid) (PGA), as shown in Figure 1(b). As a result of this structure, PLGA is highly biocompatible and biodegradable [6]. PLGA degrades in the body by hydrolysis of the backbone ester linkages into oligomers and finally into nontoxic monomeric compounds.

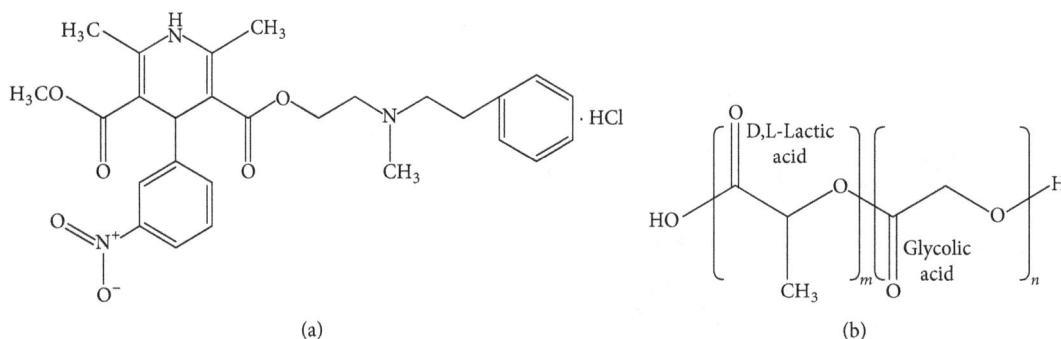

FIGURE 1: Chemical structure of nicardipine hydrochloride (a) and poly(D,L-lactic-co-glycolic acid) (b).

Above all, this material has been approved by the FDA for biomedical applications [7].

In addition, Laura C. et al. modeled the release profiles of risperidone loaded PLGA microparticles (PLGA MPs) *in vitro* and *in vivo* in rats, using varying ratios of the D,L-lactic acid and glycolic acid (50 : 50, 65 : 35, 75 : 25, and 85 : 15). The PLGAs had typical biphasic release profiles extending over 20, 40, 55, and 90 days, respectively [8]. The profiles also showed continued release with increasing D,L-lactic acid. The presence of the methyl side groups in D,L-lactic acid makes it more hydrophobic than glycolic acid; therefore, PLGA is less hydrophilic, absorbs less water, and subsequently degrades more slowly, as PLGA degrades by hydrolysis of its ester linkages in the presence of water [7, 9]. Usually, PLGA is used to prepare capsules for controlling the release of a substance.

The majority of research based on this technology has been focused on a targeted drug delivery system with polymer encapsulation. The polymer encapsulation enhances the specific activity of the main drug, leading to an improvement in pharmacokinetics, modifications to the toxicities associated with a particular drug, protection of the drug from deactivation, and preservation of its activity during transport to the target organ [10, 11]. Current encapsulation techniques include various methods, such as coacervation, spray drying, ionic gelation, and emulsion, each of which obtains a different particle size [12].

The emulsion technique is the best suited to encapsulate hydrophilic and hydrophobic drugs and to improve drug bioavailability, and the particles are highly stable and can be administered in many ways, such as via the gastrointestinal system, the eyes, the skin, or the nose or even via injection into a vein. This is particularly interesting for applications in the pharmaceutical, cosmetic, and food industries [13]. The water-in-oil-in-water ($w_1/o/w_2$) double emulsion is a technique used to prepare PLGA MPs with an encapsulated hydrophilic drug distributed in the PLGA. Drugs formulated in such polymeric devices are released by either Fickian or non-Fickian diffusion through the PLGA barrier and by the erosion of the PLGA material. In addition to their biocompatibility, drug compatibility, suitable biodegradation kinetics, and mechanical properties, PLGA MPs can be easily processed and prepared in various sizes [14]. This is particularly interesting for pharmaceutical and biomedical applications.

With the above motivations, the hypothesis in this research was that these PLGA MPs can effect controlled drug release. For the drug delivery system to be successful, the stable encapsulation of NCH is necessary. PLGA MPs were prepared with the ratio of D,L-lactic acid to glycolic acid at 50 : 50 because this research focuses on controlling NCH release over approximately 15 days [8]. The main goal of the research is to study for the first time the influence of various parameters on the PLGA MPs. This research used the $w_1/o/w_2$ double emulsion and solvent diffusion evaporation approach method to prepare PLGA MPs. PLGA MPs were prepared to find the optimum processing parameters [7], such as polymer content, surfactant type, stabilizer concentration, inner and outer aqueous phase volumes, and speed of stirring for the emulsification process. Finally, the PLGA MPs were compared in terms of size, polydispersity, morphology, encapsulation efficiency, and NCH release.

2. Materials and Methods

2.1. Materials. Acid-terminated poly(D,L-lactic-co-glycolic acid) (PLGA) with an average molecular weight of 24,000–38,000 Da and a copolymer ratio of D,L-lactide to glycolide at 50 : 50 was purchased from Sigma-Aldrich (USA) to be used as the wall material for microparticles. Sorbitan monooleate (Span 80, Fluka, Switzerland) and poly(vinyl alcohol) (PVA, Mw. ~31,000 Da, Sigma-Aldrich, USA) are often used in foods and cosmetics and were used here as stabilizers. Nicardipine hydrochloride (NCH, LR IMPERIAL, Philippines) and normal saline solution (NSS, Thai-Otsuka, Thailand) were obtained as from Medicine Ramathibodi Hospital, Mahidol University, Thailand. Dichloromethane and acetone were purchased from Lab-Scan (Thailand). All chemical agents were of analytical grade and used without further purification.

2.2. Preparation of the PLGA MPs. An appropriate amount of PLGA was dissolved in DCM (15.0 mL) and acetone [15] and mixed until a clear solution was formed. Then, the inner aqueous phase (w_1) solution, consisting of 3% v/v NCH in NSS, was added to the PLGA organic solution, which was subsequently added to the surfactant. This mixture was homogenized (EL Dorado Labtech) for the 1st emulsion thoroughly in an ice bath, forming the w_1/o single emulsion.

TABLE 1: Parameters used in different sets of experiments. The "bold" values show the parameters, which were changed in their respective recipes.

Studied parameters	$w_1/o/w_2$ double emulsion's parameter					
	PLGA (mg)	w_1 (mL)	Acetone (mL)	PVA (% w/v)	w_2 (mL)	Speed (rpm)
PLGA amount used	**10.0**	2.0	0.5	3.0	30.0	8,000
	25.0	2.0	0.5	3.0	30.0	8,000
	50.0	2.0	0.5	3.0	30.0	8,000
	100.0	2.0	0.5	3.0	30.0	8,000
w_1 volume	50.0	**1.0**	0.5	3.0	30.0	8,000
	50.0	**2.0**	0.5	3.0	30.0	8,000
	50.0	**3.0**	0.5	3.0	30.0	8,000
	50.0	**5.0**	0.5	3.0	30.0	8,000
Acetone volume	50.0	1.0	**Control**	3.0	30.0	8,000
	50.0	1.0	**0.5**	3.0	30.0	8,000
	50.0	1.0	**1.0**	3.0	30.0	8,000
	50.0	1.0	**2.0**	3.0	30.0	8,000
	50.0	1.0	**3.0**	3.0	30.0	8,000
PVA concentration	50.0	1.0	0.5	**1.0**	30.0	8,000
	50.0	1.0	0.5	**3.0**	30.0	8,000
	50.0	1.0	0.5	**5.0**	30.0	8,000
	50.0	1.0	0.5	**10.0**	30.0	8,000
w_2 volume	50.0	1.0	0.5	5.0	**15.0**	8,000
	50.0	1.0	0.5	5.0	**20.0**	8,000
	50.0	1.0	0.5	5.0	**25.0**	8,000
	50.0	1.0	0.5	5.0	**30.0**	8,000
Stirring speed for the 1st emulsion	50.0	1.0	0.5	5.0	20.0	**8,000**
	50.0	1.0	0.5	5.0	20.0	**12,000**
	50.0	1.0	0.5	5.0	20.0	**18,000**

Next, the w_1/o single emulsion was added to an outer aqueous (w_2) solution containing PVA as a stabilizer. For phase separation of PLGA, this was stirred with a magnetic stirrer for the 2nd emulsion, using a specific PLGA amount, w_1 volume, acetone volume, PVA concentration, and w_2 volume and speed for the 1st emulsion, as listed in Table 1. The system was thermally maintained in an ice bath to achieve the $w_1/o/w_2$ double emulsion. After that, DCM and acetone were allowed to evaporate. The resulting PLGA MPs were collected by centrifugation, washed three times with DI water, and finally resuspended in 10 mL of phosphate-buffered saline (PBS). The schematic of the setup in this experiment is shown in Figure 2.

2.3. Emulsion Stability Measurements

2.3.1. Preparation of Emulsion. The selection of the surfactant is crucial in the formation of emulsion and its long-term stability. First, 2.0 mL of NSS was added to the PLGA solution phase consisting of 100.0 mg of PLGA in 15.0 mL of DCM and 0.5 mL of acetone. Then, 10.0 mg/mL of nonionic surfactants was added to the PLGA solution, and this mixture was homogenized thoroughly at 8,000 rpm. The solution was

then allowed to mix in the same manner as the w_1/o single emulsion. The choice of nonionic surfactant was intended to represent a wide range of hydrophilic-lipophilic balance (HLB) values, as Span 85, Span 80, Tween 80, and PVA are known to have HLB values of 1.8, 4.3, 15.0, and 18.0, respectively [16].

2.3.2. Bottle Test. The stability of the w_1/o emulsions as a function of the 10.0 mg/mL of nonionic surfactant (Span 85, Span 80, Tween 80, and PVA) was investigated using a bottle test by observation of the phase separation of samples over time. Freshly prepared emulsions were transferred into 10.0 mL graduated glass bottles sealed with a plastic cap and stored for 480 min at room temperature. The phase separation of the emulsions was visually monitored at regular time intervals. Nonseparated phases were observed for all emulsions after 480 min. The percentage of each phase volume in relation to the total volume was calculated. Analyses were performed in triplicate ($n = 3$) [17].

2.3.3. Emulsion Stability Index (ESI). To evaluate the emulsion stability, the extent of emulsion separation by gravity was assessed by ESI. For this test, 1.0 mL of the freshly

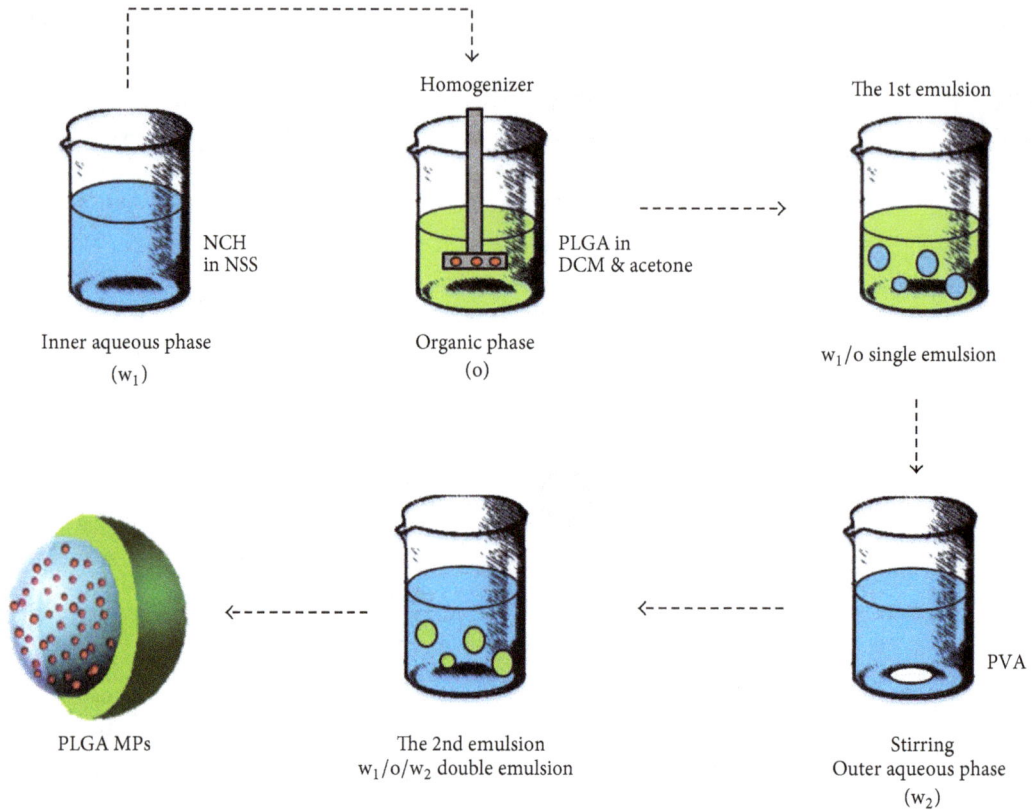

FIGURE 2: Schematic illustration of the process of forming PLGA MPs via $w_1/o/w_2$ double emulsion.

prepared, diluted emulsion was transferred to a 20.0 mL test tube and capped to prevent evaporation; tubes were stored at room temperature for 480 min. The absorbance of the diluted emulsions was measured with a UV-Visible spectrophotometer (GENESYS 20 UV-VIS, Thermo Fisher Scientific, USA) at 500 nm in 1 cm path length cuvettes. The percentage of ESI can be identified as the following equation [18]:

$$\% \text{ ESI} = \left(\frac{2.303 A \Delta t}{\Delta T} \right), \qquad (1)$$

where T is turbidity at 0 min, A is absorbance at 500 nm, ΔT is change in turbidity over a 480 min period, and Δt is time interval (480 min).

2.4. Measurement of Physicochemical Properties of PLGA MPs

2.4.1. Critical Micelle Concentration (CMC). Using nonionic surfactants that were found to result in a highly stable emulsion, this research studied the surfactant concentration. The surfactant stock solution was diluted (in the range of 1.0–100.0 mg/mL). Then, the molar conductivity of different concentrations of the surfactant stock solution was determined by electrical conductometry (EUTECH Instruments con 510) at room temperature (25 ± 1°C). Analyses were performed in triplicate ($n = 3$) [19].

2.4.2. Swelling of the PLGA MPs. The swelling behavior of the PLGA MPs that had been prepared with different amounts

of PLGA was investigated after incubation at 37°C in 10 mL of 10 mM PBS. After 24 h, the size of the swollen PLGA MPs was analyzed according to the procedure described in the previous subsection. For each sample type, at least 100 microspheres were analyzed, and their swelling behavior was quantified using the following equation [20]:

$$\text{Swelling ratio} = \left(\frac{D_{\text{swell}}}{D_{\text{dry}}} \right) \times 100, \qquad (2)$$

where D_{swell} and D_{dry} represent the size of the PLGA MPs after and before the incubation, respectively.

2.4.3. Characterization of the PLGA MPs. A scanning electron microscope (SEM, JEOL JSM 6400) was used to obtain electron microscopic images of the PLGA MPs. The PLGA MPs were mounted on double-sided conductive tape attached to the SEM specimen holders. The PLGA MPs were then sputtered with a layer of gold by spraying them with gold vapor for 20 minutes under an argon atmosphere. The PLGA MPs were observed by SEM with an accelerating voltage of 15 kV and under high vacuum. In this work, the diameter sizes for nonstable dispersion were measured from optical microscope (DM 4000 M) by the SemAfore 5.21 software ($n = 200$).

2.5. Encapsulation Efficiency (% EE) of NCH in the PLGA MPs. PLGA MPs encapsulating NCH (NCH/PLGA MPs) were prepared by the $w_1/o/w_2$ double emulsion technique. NCH was dissolved in the NSS phase (3.0% v/v). This NCH solution was added to the solution of PLGA in DCM (15.0 mL) and acetone, into which was subsequently added 30.0 mg/mL of Span 80. This mixture was homogenized thoroughly in an ice bath to yield the w_1/o single emulsion. Next, the w_1/o single emulsion was added to an aqueous PVA solution and further stirred, with a specific PLGA amount, w_1 volume, acetone volume, PVA concentration, and w_2 volume and speed, as listed in Table 1. The amount of NCH encapsulated in PLGA MPs was measured by using a UV-Visible spectrophotometer at a wavelength of 240 nm. The % EE was calculated using the following equation:

$$\% \ EE = 100 - \left(\frac{B}{A} \times 100 \right) - C, \qquad (3)$$

where A was the total amount of NCH, B was the unencapsulated NCH in PLGA MPs, and C was the NCH adsorbed in w_2.

2.6. Effect of the Stability of the NCH Entrapment in PLGA MPs. The optimum of parameters indicated in Table 1 was used to prepare the PLGA MPs. A set of experiments was performed to study the effect of two different temperatures, 4 and 37°C, in PBS solution (pH 7.4). The NCH remaining was determined using a UV-Visible spectrophotometer at a wavelength of 240 nm.

2.7. In Vitro Drug Release Profiles. The *in vitro* release kinetics of the PLGA MPs were investigated in PBS solution (pH 7.4) at 37°C. First, this research prepared a 3.0% v/v solution of NCH in 1.0 mL of NSS loaded into 50.0 mg of PLGA in DCM (15.0 mL) and acetone (0.5 mL). The Span 80 concentration was 30.0 mg/mL, and a stirring speed of 8,000 rpm was used to yield the w_1/o single emulsion. The w_1/o single emulsion was then added to a 20 mL solution of 5.0% w/v of PVA and further stirred. The system was thermally maintained in an ice bath to achieve the w_1/o/w_2 double emulsion. Finally, the preparation of PLGA MPs was suspended in a 50 mL test tube containing 20 mL of PBS solution. The sample tubes were then placed in an incubator and shaken horizontally at 70 rpm. At predetermined times following the beginning of the incubation, the PLGA MPs were retrieved from the solution by centrifugation at 1,000 rpm. The supernatant release medium was removed and replaced with fresh PBS solution each day, and the NCH concentration in the supernatant was determined using a UV-Visible spectrophotometer at a wavelength of 240 nm.

2.8. Statistical Analysis. All of results in this research are expressed as the mean ± SD. Significant differences between groups were tested using randomized complete block design (RCBD), followed by a one-way analysis of variance (ANOVA) in SPSS (SPSS, USA). The differences were considered statistically significant when $P \leq 0.05$.

3. Results and Discussion

The purpose of this work was to study preparations of NCH/PLGA MPs by the w_1/o/w_2 double emulsion method to achieve release over longer times than NCH without PLGA MPs. The experimental portion is divided into three parts. The first part was a study of parameters that affect emulsion stability. The second part was a study of the parameters affecting the w_1/o/w_2 double emulsion method used to prepare the NCH/PLGA MPs, which are shown in Table 1. After the PLGA MPs were prepared, drug encapsulation efficiency and release *in vitro* were studied.

3.1. Effects on Emulsion Stability

3.1.1. Effect of Surfactant Type. The surfactant type has an effect on emulsion stability. In this work, this research characterized Span 85, Span 80, Tween 80, and PVA. The results shown in Figure 3(a) imply that all the surfactants can be used to prepare the w_1/o single emulsion, but emulsion stability was affected by the type of surfactant. For the Span 80 stabilized emulsions, a visual observation of the emulsion indicates that phase separation occurs after 90 min. On the other hand, Span 85, Tween 80, and PVA showed phase separation of the emulsion after 20 min, which appears as the transparent phase shown in Figure 4. This process results from external forces, primarily gravity. When such forces exceed the thermal motion of the droplets (Brownian motion), a concentration gradient builds up in the system with the larger droplets moving faster to the top or the bottom of the container. Emulsion stability refers to the ability of emulsion to resist changes to its concentration gradient over time. Emulsion stability is expressed as ESI [18], shown in Figure 3(b). The stability of emulsions made from Span 80 was significantly higher than that of emulsions prepared from another surfactant type ($P \leq 0.05$).

The selection of different surfactants in the preparation of either o/w or w/o emulsions is often still made on an empirical basis. The HLB is a measure of the percentage of hydrophilic to lipophilic groups in the surfactant molecules, as described by Griffin [16]. The Span 80 (HLB = 4.3) and Span 85 (HLB = 1.8) structures contain 1 and 3 fatty acid groups, respectively. The HLB value decreases with increasing fatty acid content. Thus, surfactants with HLB < 7 tend to form w/o emulsions. Moreover, Tween 80 (HLB = 15.0) and PVA (HLB = 18.0) are hydrophilic in nature. Their advantage, with HLBs > 7.0, is their ability to support the formation of o/w emulsions [21]. In this study, we prepared NCH/PLGA MPs by the w_1/o/w_2 double emulsion method. In the first stage, this research prepared w_1/o single emulsions. Taken together, these results suggest that Span 80 was the optimal surfactant for this research.

3.1.2. Effect of Surfactant Concentration. The results showed that Span 80 conferred the highest emulsion stability in this research and this research studied the effect of different Span 80 concentrations. Then, the CMC was determined by measuring the molar conductivity of different concentrations of Span 80. The CMC is determined with a tensiometer by measuring the surface tension of a concentration series [19]. For Span 80, increasing the concentration decreased the molar conductivity. Figure 5(a) shows this relationship, and the CMC corresponds to the point on the curve at which

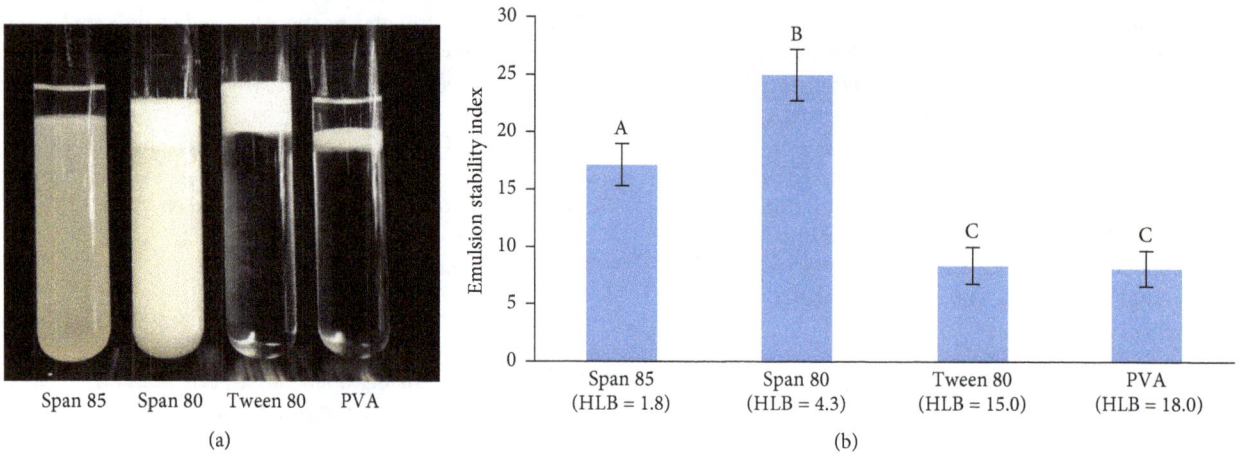

FIGURE 3: Digital photographs (a) and emulsion stability index (b) of the w_1/o single emulsion with the addition of surfactants, taken 480 min after preparation (A, B, and C are significantly different at the $P \leq 0.05$ level, with comparisons among the type of surfactants in each group).

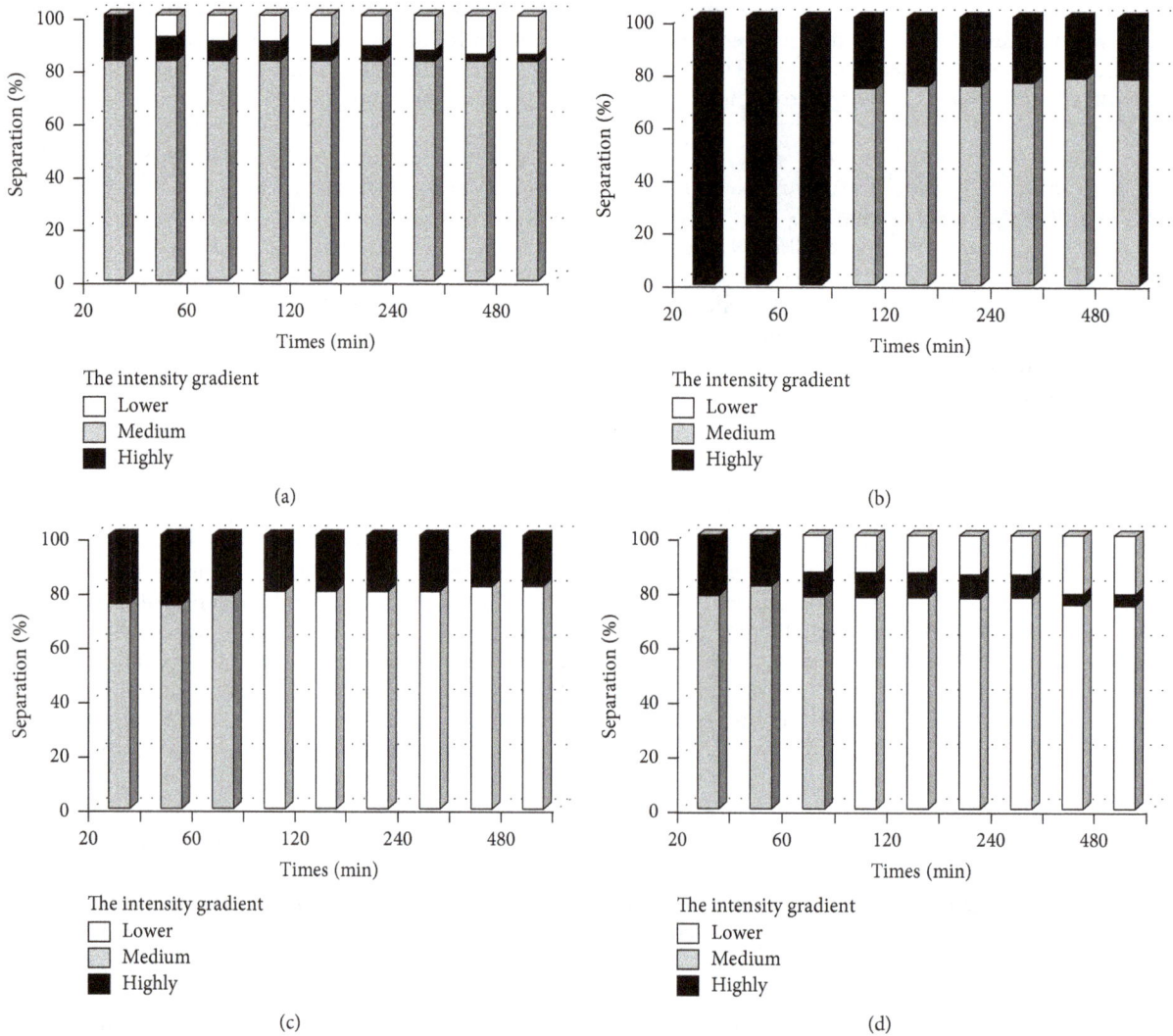

FIGURE 4: Bottle test results demonstrating phase separation of Span 85 (a), Span 80 (b), Tween 80 (c), and PVA (d).

(a)

(b)

(c)

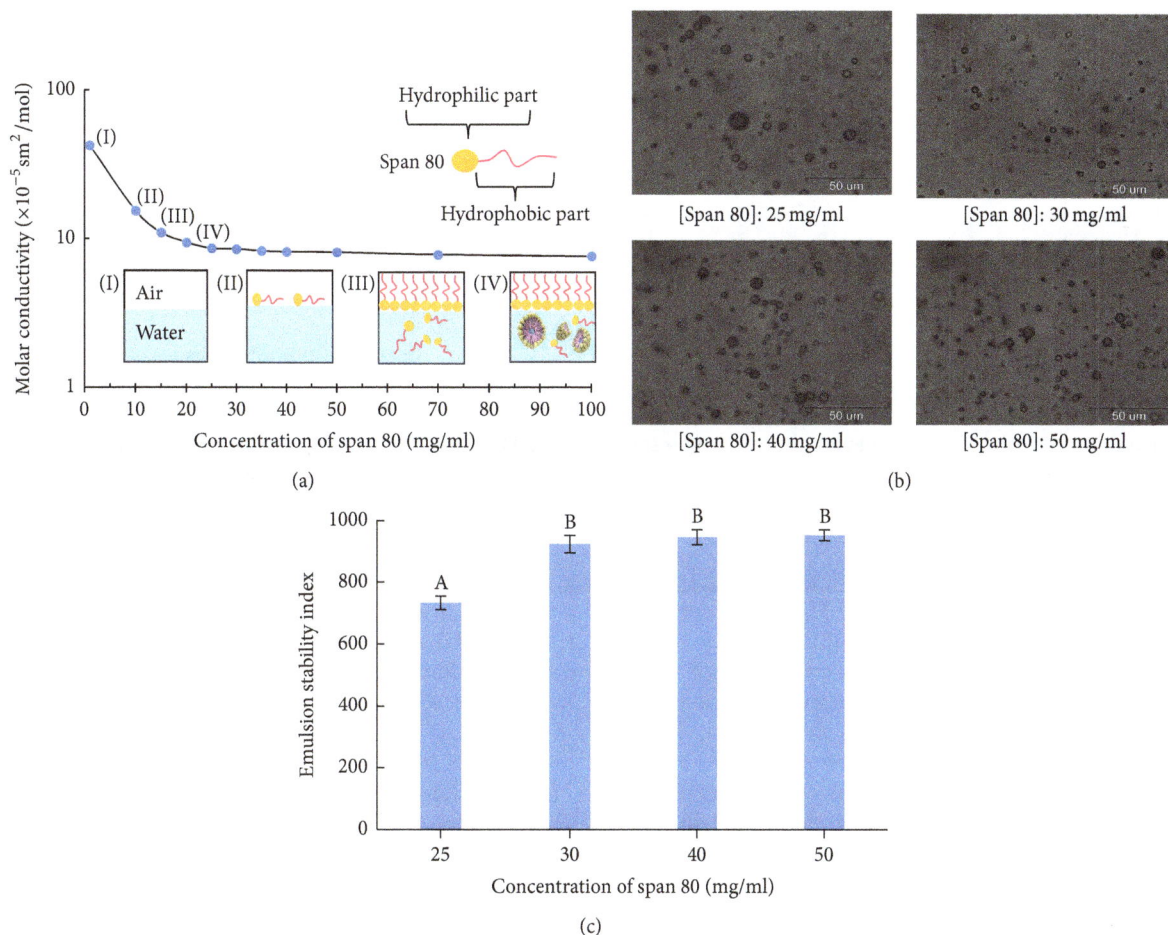

FIGURE 5: Molar conductivity of Span 80 at different concentrations (a), optical microphotographs following the preparation process of the PLGA MPs (magnification 20x) (b), and emulsion stability index (after 480 min) of the PLGA MPs stabilized by surfactants (c) (A and B are significantly different at the $P \leq 0.05$ level with comparisons among the PLGA particles in each group).

a sharp change of slope occurs (15.0 mg/mL). Below the CMC point are few or no micelles, while, beginning at the CMC point, a sharp increase in micelle concentration occurs. Concentrations above 25.0 mg/mL had constant molar conductivity. Therefore, this research studied concentrations in this range (i.e., 25.0 to 50.0 mg/mL). The results shown in Figures 5(b) and 5(c) imply that all concentrations in the selected range can form micelles, but the ESI at 25.0 mg/mL was significantly lower than that of emulsions prepared from another surfactant concentration ($P \leq 0.05$). Therefore, a Span 80 concentration of 30.0 mg/mL was selected to study the effects of the PLGA amount.

3.2. Effects of Different Parameters on Particle Size and NCH Encapsulation Efficiency

3.2.1. Effect of PLGA Amount. Microparticles of PLGA were prepared to fabricate a biodegradable polymeric carrier for NCH by the $w_1/o/w_2$ double emulsion method. The PLGA amount is an important factor influencing the properties of the particles, such as the PLGA MPs size and encapsulation efficacy. In this work, PLGA was used as the polymer, and

the effects of different amounts of PLGA were investigated by keeping all other conditions constant (Table 1). The results shown in Figure 6(a) imply that increasing the PLGA amount led to an increase in the viscosity and chain entanglement, resulting in a significant increase in the size of PLGA MPs. On the other hand, the size distribution of PLGA MPs diminished with increasing PLGA amount, as illustrated in Figure 7. SEM micrographs showed that the PLGA MPs were distributed on the PVA film under all conditions. These issues were resolved when this research studied the effect of w_2 volume.

The swelling ratio and the swelling behavior of the PLGA MPs made with different amounts of PLGA before and after swelling in PBS were studied at 37°C for 24 h [20]. The diameters of the PLGA MPs from at least 100 freeze-dried and wet samples were measured and calculated, shown in Figure 6(b). The PLGA MPs demonstrated the ability to absorb water and increase in size. The swelling ratios were measured to be 1.2–2.0. Different PLGA amounts did influence the swelling ratio; the swelling ratios decreased significantly when the swelling ratios in each group were compared.

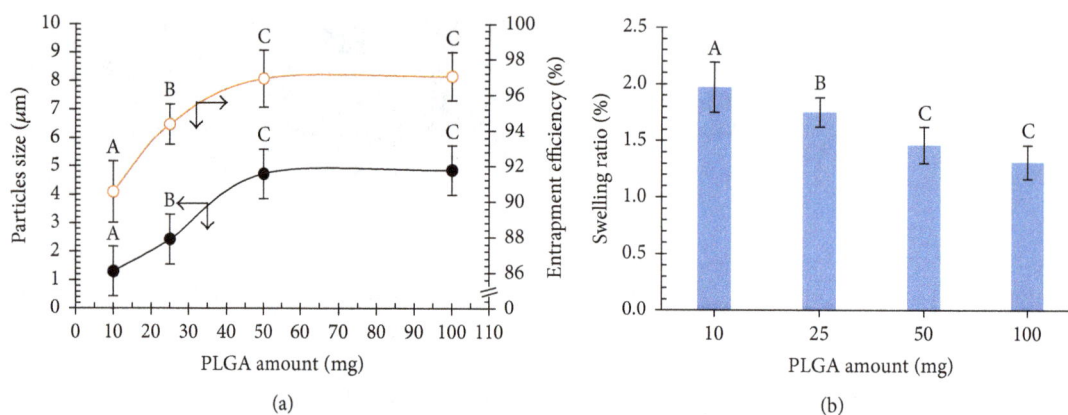

FIGURE 6: Effects of the PLGA amount on the particle size (blue solid circle) and the entrapment efficiency (orange vacant circle) of the PLGA MPs (a). The ability of PLGA MPs to absorb water and increase in size after swelling in phosphate-buffered saline was studied at 37°C and pH 7.4 for 24 h (A, B, and C are significantly different at the $P \leq 0.05$ level when comparing in each group).

Figure 6(a) illustrates the loading of NCH as a suspension into the PLGA MPs. As the amount of PLGA increased, the encapsulation efficiency increased significantly in each group. However, 50.0 mg PLGA caused the lowest swelling ratios and the highest encapsulation efficiency. An increase in the PLGA amount led to a reduction in the partitioning of the NCH into the w_2. Indeed, as stated before, PLGA is a copolymer of PLA and PGA. PLA is a more hydrophobic polymer than PGA because of its chemical structure. PGA is a highly crystalline polymer [7]. Therefore, the swelling ratios decrease and the encapsulation efficiency increased when the PLGA amount was increased. These results revealed that, by increasing the amount of PLGA, the particle size and NCH encapsulation efficiency were increased, although the swelling ratio and size were also decreased. A PLGA amount of 50.0 mg was selected as the optimal condition for this research.

3.2.2. Effect of the Inner Aqueous Phase (w_1) Volume.

In this work, this research used different volumes of w_1, including 1, 2, 3, and 5 mL of NSS, to determine the effect on the PLGA MPs. Increasing the volume of the w_1 does not affect the PLGA MPs size. High amounts of polymer in the solution cause faster coagulation, and chain entanglement occurred when stirring with a high-speed homogenizer when necessary (Table 1). Moreover, the NCH encapsulation efficiency was affected by w_1 because NCH is a hydrophilic drug whose solubility increases when the w_1 volume increases. An increase in the volume of the w_1 led to a decrease in NCH entrapment. The highest NCH encapsulation efficiency was achieved at a w_1 volume of 1.0 mL. The effect of the w_1 volume is shown in Figure 8.

3.2.3. Effect of Adding Acetone to the Organic Phase.

Acetone is a volatile organic cosolvent of PLGA. The PLGA solubility increased upon the addition of a suitable volume of acetone to the organic solution of PLGA [22]. Figure 9 shows that the addition of acetone led to a decrease in the size of the PLGA MPs compared to their size in the absence of acetone. The electrically charged surfaces of the PLGA MPs changed when

acetone was added to DCM. Acetone decreased the interfacial tension and increased the movement of the dimensions of the polymer chains in the organic phase, resulting in a decrease in the size of PLGA MPs.

The NCH encapsulation efficiency in the PLGA MPs decreased as increasing amounts of acetone were added. The PLGA solubility increased with increasing acetone volume. Therefore, the NCH diffused easily out of the PLGA MPs. The acetone enhanced the affinity between aqueous and organic phases because of its amphiphilic nature [15]. It promoted PLGA solubility in the organic phase, with similar solubility parameters (PLGA, DCM, and acetone = 23.8, 20.3, and 19.9 MPa$^{1/2}$, resp.) [23, 24]. An acetone volume of 0.5 mL did not affect the NCH/PLGA MPs, but it made their sizes smaller than those made in the absence of acetone, as shown in Figure 9. Therefore, for this research, 0.5 mL of acetone was selected for further study.

3.2.4. Effects of the PVA Concentration in the Outer Aqueous Phase (w_2).

The addition of a suitable concentration of PVA plays a key role in the formation of PLGA MPs by w_1/o/w_2 double emulsion. Normally, PVA is a biocompatible polymer which has shown a capability to improve pH consistency and lifelong temperature, so PVA is used for the highest stabilizer of PLGA MPs. The properties of PVA can promote suitable mediator for the fabrication of PLGA MPs using in medical applications [25]. The concentration of PVA as a stabilizer affects the size of the PLGA MPs and the NCH encapsulation efficiency, which is illustrated in Figure 10. In the 1.0% w/v of PVA, the encapsulation of NCH in PLGA MPs was very low, and the PLGA MP size was larger than 4.51 μm. Thus, the increasing concentrations of PVA drastically improved the NCH loading and the PLGA MP size. The largest PLGA MP size was that observed in the absence of PVA. Because the PVA is a high molecular weight polymer, the increased concentration of PVA led to an increase in the viscosity of and chain entanglements in w_2. The small size of the PLGA MPs prepared was achieved with a PVA concentration of 1.0% w/v, but 5.0% w/v of PVA prepared PLGA MPs with high NCH

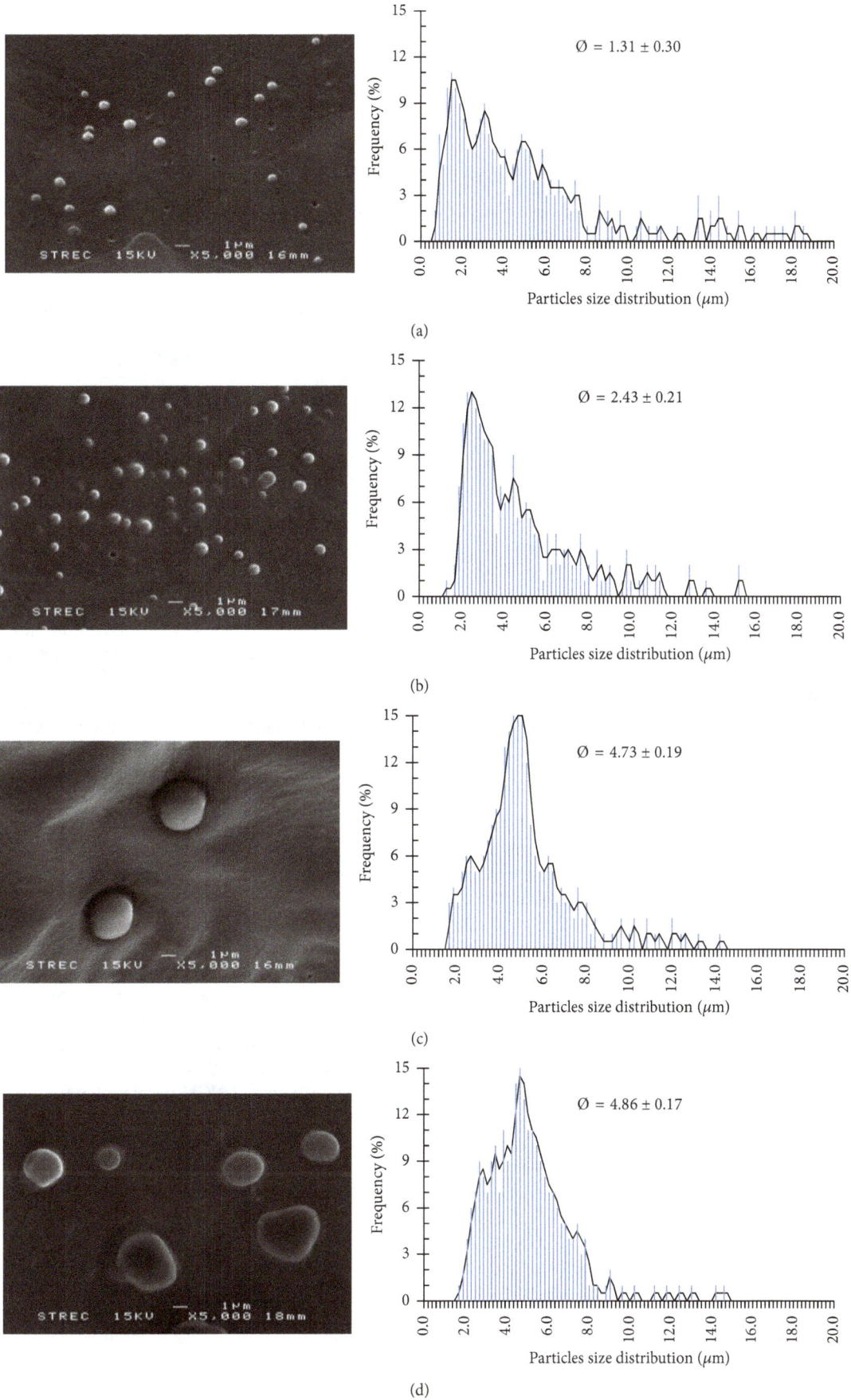

FIGURE 7: The influence of the PLGA amounts of 10.0 (a), 25.0 (b), 50.0 (c), and 100.0 (d) mg on the morphology and particle size distribution of PLGA MPs from SEM imaging and diameter sizes for nonstable dispersion distribution diagram (magnification 5,000x).

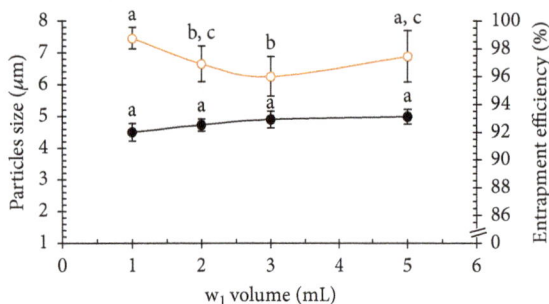

FIGURE 8: Effects of the w_1 volume on the particles size (blue solid circle) and the entrapment efficiency (orange vacant circle) of the PLGA MPs (a, b, and c are significantly different at the $P \leq 0.05$ level when comparing among the w_1 volumes in each group).

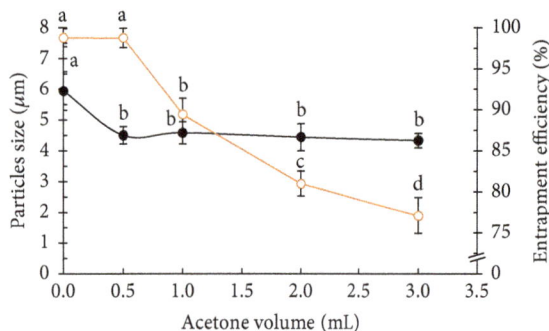

FIGURE 9: Effect of acetone volume on the particle size (blue solid circle) and the entrapment efficiency (orange vacant circle) of the PLGA MPs (a, b, c, and d are significantly different at the $P \leq 0.05$ level when comparing among the acetone volumes in each group).

loading. The focus of this research is to prepare PLGA MPs with high NCH entrapment. This research selected a PVA concentration of 5.0% w/v because it resulted in the highest NCH entrapment and a PLGA MP size smaller than the size of red blood cells (~10 μm) [26].

3.2.5. Effect of the Outer Aqueous Phase (w_2) Volume.

In this process, the $w_1/o/w_2$ double emulsions are created from a w_1/o single emulsion distributed in a w_2. The w_2 has greater volume than w_1, which allows the preparation of the $w_1/o/w_2$ double emulsion. Figure 11(a) shows four samples that were prepared with different volumes of w_2. Increasing the volume of w_2 decreased the size of the PLGA MPs. On the other hand, the NCH/PLGA MPs increased as an effect of this, but the volumes used in this study, starting from 20.0 mL, did not affect the NCH entrapment. This may be due to a decrease in the viscosity of the emulsion in higher volumes of w_2, causing more efficient shearing forces that reduced the PLGA MP size. The effect of this parameter was studied by comparing the SEM photographs of two samples, of which the first used a w_2 volume of 15.0 mL and the second a volume of 20.0 mL. From the SEM micrographs (Figure 11(b)), it was found that the PLGA MPs prepared by using 15.0 mL of w_2 did not have smooth surface morphology. On the other hand, the particles prepared with 20.0 mL of w_2 had regular rounded shapes and

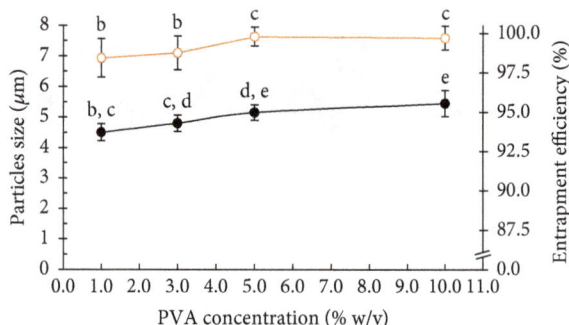

FIGURE 10: Effect of PVA concentration on the particle size (blue solid circle) and the entrapment efficiency (orange vacant circle) of the PLGA MPs (a, b, c, and d are significantly different at the $P \leq 0.05$ level when comparing among the PVA concentrations in each group).

a narrow size distribution, resulting in an increase in the NCH entrapment. The issue of the distribution of PLGA MPs on the PVA film, as shown in Figure 7, was addressed by washing the PLGA MPs three times with DI water. Therefore, 20.0 mL of the w_2 volume was selected as the optimal condition for this research.

3.2.6. Effect of Stirring Speed for the 1st Emulsion.

An emulsion is a mixture of two or more immiscible liquids. High energy is applied to the system. In this work, the necessary energy was supplied to the system via high stirring using a homogenizer as shown in Table 1. It is obvious from Figure 12 that changes in the stirring speed for the 1st emulsion during emulsion preparation had no influence on the size of PLGA MPs and the NCH encapsulation efficiency [27]. The effect on both was a slight change with extreme changes in stirring speed. The system agglomerated fast because of the chain entanglement of the PLGA solution.

3.3. Effects on the Stability of the Entrapment in PLGA MPs.

In water, NCH degrades by hydrolysis of the ester linkages in the structure [3, 4]. Due to the potential for hydrolysis of NCH, this research believed that PLGA MPs in this research would afford more stability than the absence of PLGA MPs [28]. This investigation was carried out as a demonstration at two different temperatures, that is, 4 and 37°C in PBS solution (pH 7.4). The remnants of NCH decreased at both temperatures, but the NCH/PLGA MPs retained much more NCH alone. The PLGA MPs can preserve the stability of NCH, as shown in Figure 13.

At 4 and 37°C, NCH/PLGA MPs had 99.68% and 97.90% of the remaining NCH, respectively. Therefore, the PLGA MPs are colloids. In general, the disperse phase of a colloid is thermodynamically unstable with respect to the bulk [27]. The increase in temperature had a tendency to cause agglomeration of the PLGA MPs. This phenomenon was responsible for the NCH degradation.

3.4. In Vitro Drug Release Profiles.

The aim of this study was to prepare the NCH entrapped in a polymeric system for

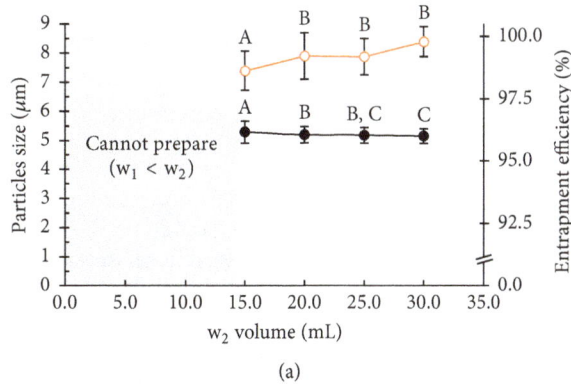

(a)

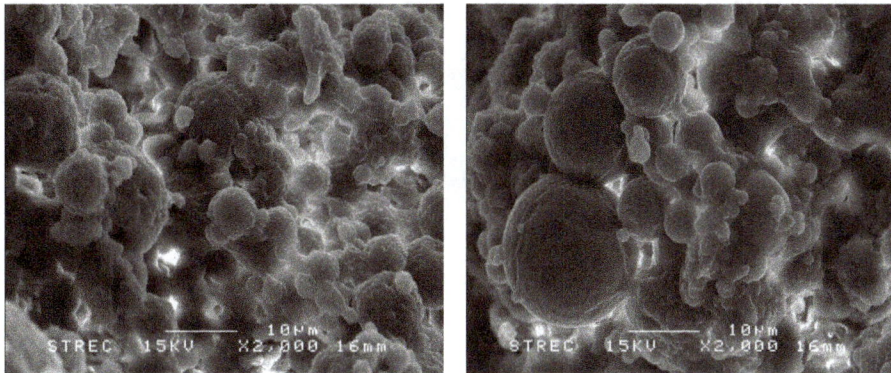

(b)

FIGURE 11: Effects of w_2 volume on the particle size (blue solid circle) and the entrapment efficiency (orange vacant circle) of the PLGA MPs (a) and SEM image of PLGA MPs fabricated with 15.0 and 20.0 mL of w_2 volume (b); the scale bar represents 10 μm. (A, B, and C are significantly different at the $P \leq 0.05$ level when comparing among the w_2 volumes in each group).

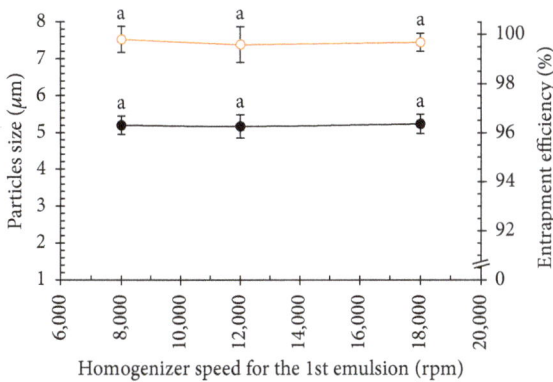

FIGURE 12: Effects of homogenizer speed for the 1st emulsion on the particle size (blue solid circle) and the entrapment efficiency (orange vacant circle) of the PLGA MPs (a is significantly different at the $P \leq 0.05$ level when comparing among the homogenizer speeds in each group).

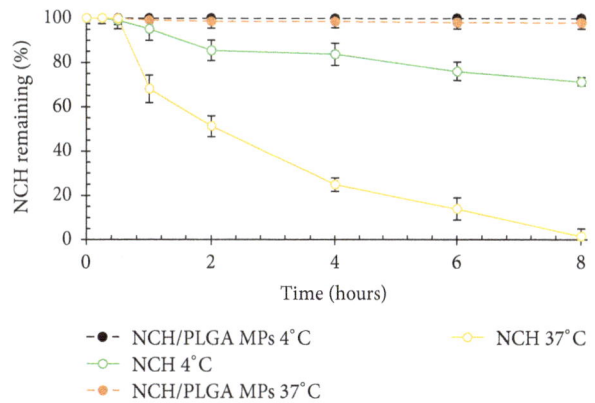

FIGURE 13: NCH remaining when entrapped in PLGA MPs and in the absence of PLGA MPs at 4 and 37°C in PBS solution (pH 7.4).

controlled release as a drug delivery system for vasodilatation agent. This agent is very important for neurosurgeons to prevent cerebral vasospasm after intracranial aneurysmal clipping. The samples were immersed in PBS (pH 7.4) and incubated in a water bath at 37°C with constant shaking at 70 rpm. Whereas *in vitro* release studies on the NCH/PLGA MPs indicated that the NCH is released continuously, NCH

release leveled off after the first 33 days. *In vitro* release profiles demonstrated a biphasic modulation. The first phase was characterized by a relatively rapid initial release and followed by a second slower phase. As observed, the *in vitro* release profile of the NCH/PLGA MPs was stopped at 50% (1.5 mg) which is sufficient to sustain an effective drug [29]. Hence, this research has demonstrated the *in vitro* release profile of NCH released from PLGA MPs at destined rate to prolong drug concentration. On the other hand, the NCH *in vitro*

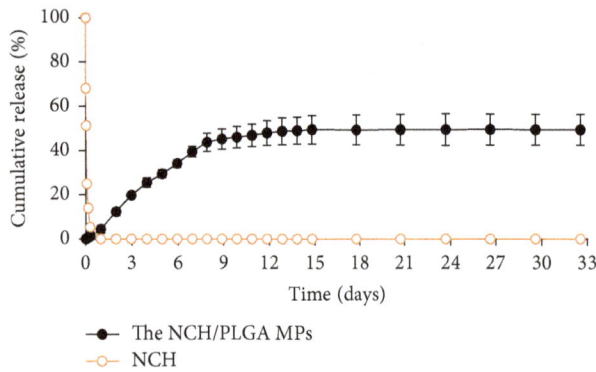

FIGURE 14: The *in vitro* release profiles of NCH from PLGA MPs incubated in testing medium with 20.0 mL PBS at pH 7.4 and 37°C to mimic the body fluids in normal tissues (*n* = 3).

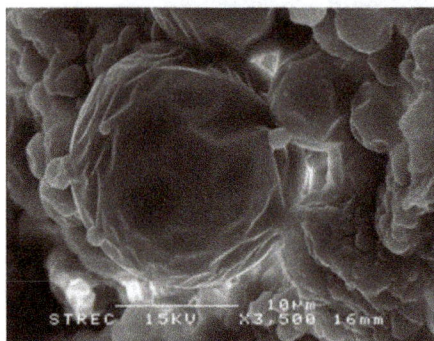

(a)

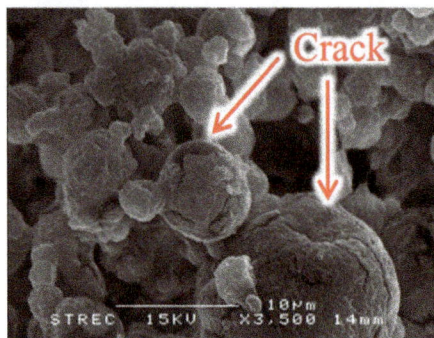

(b)

FIGURE 15: Scanning electron microphotographs illustrating morphology of the PLGA MPs before *in vitro* release (a) and after *in vitro* release (b); (magnification 3,500x).

release in the absence of PLGA MPs demonstrated NCH degradation over a period of 1 day. NCH has a short half-life [4], as shown in Figure 14; hence NCH was entrapped in PLGA MPs. PLGA MPs have been shown to be excellent delivery systems for controlling the administration of NCH due to their biocompatibility and biodegradability [30]. In Figure 15, the morphology of the NCH/PLGA MPs before and after releasing in PBS solution is studied. In our experiment, the result from Figure 15(a) illustrated that the NCH/PLGA MPs before releasing process provided a sphere shape with smooth surface. Figure 15(b) showed that it was impossible

for drug release for 33 days to see a cracked NCH/PLGA MPs due to PLGA degradation from hydrolysis reaction.

4. Conclusions

The optimal processing conditions for the $w_1/o/w_2$ double emulsion method of preparing PLGA MPs include 3.0% v/v of NCH in 1.0 mL of NSS, and this NCH solution is added to organic phase consisting of 50.0 mg of PLGA in DCM (15.0 mL) and acetone (0.5 mL). The surfactant is 30.0 mg/mL of Span 80 with a stirring speed of 8,000 rpm to yield the w_1/o single emulsion. Next, the w_1/o single emulsion is added to 20 mL of a 5.0% w/v solution of PVA and further stirred. The system is thermally maintained in an ice bath to form the NCH/PLGA MPs. The average size of the NCH/PLGA MPs was approximately at 5.20 μm, and percentage encapsulation efficiency of NCH in the NCH/PLGA MPs was approximately 99%. The NCH release from PLGA MPs was constant. As a result, this drug delivery system will be an option to demonstrate the choice of treatment for intracranial vasospasm problem.

Conflicts of Interest

The authors declare that they have no conflicts of interest.

Acknowledgments

The authors gratefully acknowledge the Neurosurgery Unit, Surgery Department, Faculty of Medicine Ramathibodi Hospital, Mahidol University, Bangkok, Thailand, for supplying the nicardipine hydrochloride. Special thanks are given for partial financial support from the Industrial Chemistry Department, Faculty of Applied Science, King Mongkut's University of Technology North Bangkok, Thailand (in house Grant no. 5841103).

References

[1] M. Lin, O. Aladejebi, and G. H. Hockerman, "Distinct properties of amlodipine and nicardipine block of the voltage-dependent Ca 2+ channels Ca v1.2 and Ca v2.1 and the mutant channels Ca v1.2/Dihydropyridine insensitive and Ca v2.1/Dihydropyridine sensitive," *European Journal of Pharmacology*, vol. 670, no. 1, pp. 105–113, 2011.

[2] L. Michalewicz and F. H. Messerli, "Cardiac effects of calcium antagonists in systemic hypertension," *American Journal of Cardiology*, vol. 79, no. 10, pp. 39–46, 1997.

[3] C. M. Fernandes, P. Ramos, A. C. Falcão, and F. J. B. Veiga, "Hydrophilic and hydrophobic cyclodextrins in a new sustained release oral formulation of nicardipine: in vitro evaluation and bioavailability studies in rabbits," *Journal of Controlled Release*, vol. 88, no. 1, pp. 127–134, 2003.

[4] M. P. Curran, D. M. Robinson, and G. M. Keating, "Intravenous nicardipine: its use in the short-term treatment of hypertension and various other indications," *Drugs*, vol. 66, no. 13, pp. 1755–1782, 2006.

[5] N. Rescignano, E. Fortunati, S. Montesano et al., "PVA bio-nanocomposites: a new take-off using cellulose nanocrystals

and PLGA nanoparticles," *Carbohydrate Polymers*, vol. 99, pp. 47–58, 2014.

[6] I. Banerjee, D. Mishra, and T. K. Maiti, "PLGA microspheres incorporated gelatin scaffold: microspheres modulate scaffold properties," *International Journal of Biomaterials*, vol. 2009, Article ID 143659, 9 pages, 2009.

[7] H. K. Makadia and S. J. Siegel, "Poly Lactic-*co*-Glycolic Acid (PLGA) as biodegradable controlled drug delivery carrier," *Polymer*, vol. 3, no. 3, pp. 1377–1397, 2011.

[8] L. C. Amann, M. J. Gandal, R. Lin, Y. Liang, and S. J. Siegel, "In vitro-in vivo correlations of scalable PLGA-Risperidone implants for the treatment of schizophrenia," *Pharmaceutical Research*, vol. 27, no. 8, pp. 1730–1737, 2010.

[9] M. Ramchandani and D. Robinson, "In vitro and in vivo release of ciprofloxacin from PLGA 50:50 implants," *Journal of Controlled Release*, vol. 54, no. 2, pp. 167–175, 1998.

[10] N. S. Barakat and A. A. E. Ahmad, "Diclofenac sodium loaded-cellulose acetate butyrate: effect of processing variables on microparticles properties, drug release kinetics and ulcerogenic activity," *Journal of Microencapsulation*, vol. 25, no. 1, pp. 31–45, 2008.

[11] D. C. Drummond, C. O. Noble, Z. Guo et al., "Improved pharmacokinetics and efficacy of a highly stable nanoliposomal vinorelbine," *The Journal of Pharmacology and Experimental Therapeutics*, vol. 328, no. 1, pp. 321–330, 2009.

[12] M. Vemmer and A. V. Patel, "Review of encapsulation methods suitable for microbial biological control agents," *Biological Control*, vol. 67, no. 3, pp. 380–389, 2013.

[13] U. S. Schmidt, R. Bernewitz, G. Guthausen, and H. P. Schuch-mann, "Investigation and application of measurement techniques for the determination of the encapsulation efficiency of O/W/O multiple emulsions stabilized by hydrocolloid gelation," *Colloids and Surfaces A: Physicochemical and Engineering Aspects*, vol. 475, no. 1, pp. 55–61, 2015.

[14] D. Suvakanta, N. M. Padala, N. Lilakanta, and C. Prasanta, "Kinetic modeling on drug release from controlled drug delivery systems," *Acta Poloniae Pharmaceutica. Drug Research*, vol. 67, no. 3, pp. 217–223, 2010.

[15] T. Niwa, H. Takeuchi, T. Hino, M. Nohara, and Y. Kawashima, "Biodegradable submicron carriers for peptide drugs: Preparation of dl-lactide/glycolide copolymer (PLGA) nanospheres with nafarelin acetate by a novel emulsion-phase separation method in an oil system," *International Journal of Pharmaceutics*, vol. 121, no. 1, pp. 45–54, 1995.

[16] W. C. Griffin, "Calculation of HLB values of non-ionic surfactants," *Journal of the Society of Cosmetic Chemists*, vol. 5, pp. 249–256, 1954.

[17] A. Nesterenko, A. Drelich, H. Lu, D. Clausse, and I. Pezron, "Influence of a mixed particle/surfactant emulsifier system on water-in-oil emulsion stability," *Colloids and Surfaces A: Physicochemical and Engineering Aspects*, vol. 457, no. 1, pp. 49–57, 2014.

[18] M. B. Sabolović, S. R. Brnčić, and V. Lelas, "Emulsifying properties of tribomechanically treated whey proteins," *Mljekarstvo*, vol. 63, no. 2, pp. 64–71, 2013.

[19] A. Zdziennicka, K. Szymczyk, J. Krawczyk, and B. Jańczuk, "Critical micelle concentration of some surfactants and thermodynamic parameters of their micellization," *Fluid Phase Equilibria*, vol. 322-323, pp. 126–134, 2012.

[20] S. Thumsing, N. Israsena, C. Boonkrai, and P. Supaphol, "Preparation of bioactive glycosylated glial cell-line derived neurotrophic factor-loaded microspheres for medical applications," *Journal of Applied Polymer Science*, vol. 131, no. 8, Article ID 40168, 2014.

[21] T. F. Tadros, *Emulsion Formation and Stability*, Wiley-VCH Verlag GmbH & Co. KGaA, Weinheim, Germany, 2013.

[22] T. Niwa, H. Takeuchi, T. Hino, N. Kunou, and Y. Kawashima, "Preparations of biodegradable nanospheres of water-soluble and insoluble drugs with D,L-lactide/glycolide copolymer by a novel spontaneous emulsification solvent diffusion method, and the drug release behavior," *Journal of Controlled Release*, vol. 25, no. 1-2, pp. 89–98, 1993.

[23] S. Schenderlein, M. Lück, and B. W. Müller, "Partial solubility parameters of poly(D,L-lactide-co-glycolide)," *International Journal of Pharmaceutics*, vol. 286, no. 1-2, pp. 19–26, 2004.

[24] Z. M. Wu, X. D. Guo, L. J. Zhang et al., "Solvent mediated microstructures and release behavior of insulin from pH-sensitive nanoparticles," *Colloids and Surfaces B: Biointerfaces*, vol. 94, pp. 206–212, 2012.

[25] M. Iqbal, J.-P. Valour, H. Fessi, and A. Elaissari, "Preparation of biodegradable PCL particles via double emulsion evaporation method using ultrasound technique," *Colloid and Polymer Science*, vol. 293, no. 3, pp. 861–873, 2015.

[26] A. Z. Wilczewska, K. Niemirowicz, K. H. Markiewicz, and H. Car, "Nanoparticles as drug delivery systems," *Pharmacological Reports*, vol. 64, no. 5, pp. 1020–1037, 2012.

[27] E. R. Leite and C. Ribeiro, *Crystallization and Growth of Colloidal Nanocrystals*, Springer Science & Business Media, Berlin, Germany, 2011.

[28] H. Katas, M. A. G. Raja, and K. L. Lam, "Development of chitosan nanoparticles as a stable drug delivery system for protein/siRNA," *International Journal of Biomaterials*, vol. 2013, Article ID 146320, 9 pages, 2013.

[29] Y. Kishi, F. Okumura, and H. Furuya, "Haemodynamic effects of nicardipine hydrochloride. Studies during its use to control acute hypertension in anaesthetized patients," *British Journal of Anaesthesia*, vol. 56, no. 9, pp. 1003–1007, 1984.

[30] S. D'Souza, J. A. Faraj, S. Giovagnoli, and P. P. DeLuca, "IVIVC from long acting olanzapine microspheres," *International Journal of Biomaterials*, vol. 2014, Article ID 407065, 11 pages, 2014.

12

The use of Pulsed Electromagnetic Fields to Promote Bone Responses to Biomaterials *In Vitro* and *In Vivo*

Carlo Galli ⓘD,[1] Giuseppe Pedrazzi,[1] Monica Mattioli-Belmonte ⓘD,[2] and Stefano Guizzardi[1]

[1]*Dep. of Medicine and Surgery, University of Parma, Italy*
[2]*DISCLIMO, Department of Clinical and Molecular Sciences, Polytechnic University of Marche, Ancona, Italy*

Correspondence should be addressed to Carlo Galli; carlo.galli@unipr.it

Academic Editor: Esmaiel Jabbari

Implantable biomaterials are extensively used to promote bone regeneration or support endosseous prosthesis in orthopedics and dentistry. Their use, however, would benefit from additional strategies to improve bone responses. Pulsed Electromagnetic Fields (PEMFs) have long been known to act on osteoblasts and bone, affecting their metabolism, in spite of our poor understanding of the underlying mechanisms. Hence, we have the hypothesis that PEMFs may also ameliorate cell responses to biomaterials, improving their growth, differentiation, and the expression of a mature phenotype and therefore increasing the tissue integration of the implanted devices and their clinical success. A broad range of settings used for PEMFs stimulation still represents a hurdle to better define treatment protocols and extensive research is needed to overcome this issue. The present review includes studies that investigated the effects of PEMFs on the response of bone cells to different classes of biomaterials and the reports that focused on in vivo investigations of biomaterials implanted in bone.

1. Biomaterials and Bone Regeneration

Biomaterials play an important role in bone regenerative strategies [1] in both orthopedics and dentistry as scaffolds [2] or as a support for prosthesis, e.g., hip or dental implants [3]. In all these clinical situations the challenge biomaterials must face is to integrate in the host and promote bone healing along its surfaces [4], albeit with noticeable differences. Most scaffolds are made of resorbable materials, because common opinion dictates that scaffolds should progressively be replaced by native tissue [5], whereas prostheses are mostly permanent implants and their purpose is to last and function as long as possible in patients, usually while withstanding relevant mechanical forces in the process [6]. Thus, most scaffolds currently used in bone are made of bioceramics, predominantly calcium phosphates, because of their chemical similarity to the inorganic matrix of bone [7], which makes them osteoconductive [8, 9]. Furthermore, bioceramics are rigid and their mechanical properties have been shown to positively affect cell differentiation along the osteoblastic lineage [10, 11]. Last but not least, this class of biomaterials

is usually very biocompatible and resorbable within a time span that appears to quite closely meet the requirements for implantation into natural bone [8]. Although bioceramics can be loaded with biologically active ions [12] or biomolecules [13] to improve bone formation, they are not as versatile and customizable as polymers, whose structure can be modified almost *ad libitum*, enabling researchers to add functional groups and control their polymerization, their chemical behavior, their mechanical properties, and resorbability [14–16]. Polymers have opened up hitherto unexplored possibilities, such as injection of photopolymerizable compounds [17] or easy 3D printing [18].

In contrast, implantable prostheses are still mostly made of titanium and its alloys, although novel and highly resistant ceramics, i.e., zirconia, could represent a viable alternative [19, 20]. Titanium is a very biocompatible metal, which has been shown to represent an efficient material for orthopedic and dental implants [21]. A lot of effort has gone into investigating optimal surface treatments to optimize bone response and speed up tissue healing after surgery [4, 22]. What bioceramics, most polymers, and metals still lack is, however, specific biochemical cues that can control cell behavior toward desired clinical goals, beside generic stimuli, such as calcium release from resorbable bioceramics or stiffness-related mechanical stimulation of cell differentiation, unless of course these materials are loaded with bioactive compounds [13, 23]. Most of these materials still offer the organism just a viable framework within which to heal or regenerate, supporting the process but fundamentally relying on the drive to healing that is intrinsic to many tissues, especially epithelia and bone. This means that those numerous clinical situations where the tissue regenerative potential has been compromised due to age or pathology are still a serious challenge and adjunctive or ancillary therapies are still an issue of interest and hot debate. This is where additional, physical therapies such as electromagnetic fields could play an important, if not vital, role.

2. Electromagnetic Fields and Bone

Electromagnetic fields (EMF) are created by the interaction of electrically charged objects and permeate our whole reality [24]. Our world is flooded with artificial EMFs created by electrical and electronic devices [25] and although these have become a source of potential health concerns [26–31], research has long sought a way to harness their therapeutic potential [32]. To this purpose, different sources of low frequency EMFs have been actively investigated. These can be further divided into Pulsed EMFs (PEMFs), where the EMF signal is delivered in pulses of different shape interspersed with gaps and sinusoidal EMFs (SEMFs), where the superposition of the EMF signal continuously and gradually varies along a sine waveform [33].

It is known that the effects of electromagnetic fields on living beings are complex. Organisms are composed of cells, which possess an electrically charged membrane and tightly regulate the concentration of ions, electrically charged particles, e.g., Ca^{2+} or Na^+, which they use as potent signal mediators [34]. It is therefore likely that most of the effects of EMFs in cells occur or are triggered at the membrane level. There is abundant evidence suggesting that EMFs can act on Ca^{2+} concentration [35–37] and Ca-dependent pathways [38], and more recently Vincenzi et al. have convincingly shown a regulation of Adenosine receptors by PEMFs [39]. Actually the recent evidence by Yan et al. [40] and Xie et al. [41] of a role of primary cilia in transducing EMF effects in cells could be a part of a broader activity on membrane trafficking, including receptor trafficking. Further mechanisms are likely to be involved as PEMFs have been shown to modulate defenses against Reactive Oxygen Species [42] and the production of bioactive factors [40, 43–45] and to activate intracellular pathways such as the sAC–cAMP–PKA–CREB signaling pathway [46].

Most life science and biomedical research has been focused on the biological effects of PEMFs of different waveform, frequency, and intensity on different tissues and in different clinical situations. Bone has long been recognized as a suitable target for EMF treatment [47].

Indeed EMFs have been investigated as a tool to promote bone healing in several preclinical studies of bone defect healing in rodents, encompassing diverse defect models, e.g., limb or facial defects [48–56], bone loss due to (a) hyperparathyroidism [57], (b) glucocorticoids or ovariectomy [58–66], (c) disuse [67–69], or (d) diabetes [70], or even osteoporotic fractures [71] or osteoarthritis [72]. Different animal models, e.g., horses, were used as well for PEMF testing [73, 74], with positive results.

EMFs have also a long clinical story as an aid to reduce bone loss in osteoporosis [75–77], to improve osteotomies or nonunions [78–93], and different research groups have investigated frequencies, intensities, durations of exposure, pulses [94–97], or waveforms [98].

Actually EMFs can be administered in a vast range of modalities. Stimuli can be delivered as single pulses, or discrete pulses, or even complex arrays of pulse bursts, also known as Pulsed Radio Frequencies (PRF), similarly to FM radio receivers. In this case the single pulses that constitute the carrier frequency reach the kHz range, but these are modulated into sets or trains of pulses that cycle at slower frequency, often 15 Hz. Using high carrier frequency increases the penetration of EMFs throughout the body, which then is able to demodulate the signal and perceive the modulating frequency, which exerts the biological effect [99]. Intensities range across a wide spectrum as well, from μT to a few mTesla. However, a fundamental lack of understanding of the mechanisms of actions of EMFs on cells and tissues has been presented to reach a consensus on a set of clinical parameters to maximize the effects of EMFs [47].

To further compound this problem, it must be remembered that different biomaterials may require different stimulations to optimize the outcome and this has also hindered proving their clinical effectiveness, in spite of promising results [100–103].

Therefore, the present study will review the available literature on the effects of EMF treatment on osteoblasts and bone in vitro and in preclinical animal models in vivo.

3. The Effects of PEMFs on Osteoblasts

Several parameters have been shown to affect cell responses, e.g., PEMF waveform, its frequency, its intensity, or the duration of exposure. A study by J. Zhou et al. investigated the effects of EMF waveform on primary rat calvaria cells [98]. When comparing 50 Hz, 1.8 mT sinusoidal, triangular, square, or serrated EMFs on primary osteoblasts, the authors observed that only square waves significantly increased cell proliferation and that sinusoidal waves decreased it. Interestingly, only triangular and sinusoidal waves, however, significantly increased cell differentiation, as assessed by Alkaline Phosphatase activity or mineralization assays. Although the group by Zhang et al. reported similar findings [33], other studies report conflicting evidence.

Martino et al. [104] exposed human osteosarcoma SaOS-2 cells to 0.9 mT, 15 Hz PRF PEMF quasi square bursts of 4 kHz square pulses for 4 hours/day, and they observed an increase in ALP activity and the deposition of mineralized nodules although no effect on cell proliferation was reported. Their results were confirmed by Hannay et al., who applied a similar stimulation (15 Hz PRF bursts of trapezoidal pulses)

TABLE 1: The table summarizes the in vitro and in vivo studies on the effects of PEMF stimulation on osteoblastic primary cells and cell lines on calcium phosphate biomaterials. Studies are listed in chronological order.

Experimental model	Biomaterial	PEMF	Field intensity (mT unless otherwise specified)	PEMF waveform	Exposure	PEMF Generator	Reference
Defects in proximal tibia of rabbits	Porous hydroxyapatite (HA) or tricalcium phosphate (TCP) nails	1.5 Hz, 26 ms-long PEMF bursts of 3.8 kHz pulses	0.18	Quasi square	8 hours/day for up to 6 weeks	American Medical Electronics (Dallas, TX, U.S.A.)	(Shimizu et al., 1988)
Defects in rabbit tibia	Natural or synthetic hydroxyapatite granules	50 Hz	8	Triangular	30 min/12 hours for up to 4 weeks	In-house built generator	(Ottani et al., 2002)
Defects in rabbit femur (condyles)	Synthetic HA rods obtained by granule sintering	1.3 ms-long, 75 Hz	1.6	Trapezoidal	6 hours/day for 3 weeks	BIOSTIM, Igea, Carpi, Italy	(Milena Fini et al., 2002)
Defects in rabbit femurs (cortical bone, mid-diaphysis)	Synthetic HA rods obtained by granule sintering	1.3 ms-long, 75 Hz	1.6	Trapezoidal	6 hours/day for 3 weeks	BIOSTIM, Igea, Carpi, Italy	(M. Fini, Giavaresi, Giardino, Cavani, & Cadossi, 2006)
Commercially available human mesenchymal stem cells	Commercially available calcium phosphate discs	4.5 ms-long, 15 Hz bursts of 4.4 kHz, 225 μs-long pulses	1.6	Quasi-square (with trapezoidal pulses)	8 hours/day	Electro-Biology Inc., Parsippany, NJ	(Z. Schwartz et al., 2008)
Commercially available mesenchymal stem cells, normal human osteoblasts, MG-63 or Saos-2	Commercially available calcium phosphate discs	4.5 ms-long, 15 Hz bursts of 4.4 kHz, 225 μs-long pulses	1.6	Quasi-square (with trapezoidal pulses)	8 hours/day	Electro-Biology Inc., Parsippany, NJ	(Zvi Schwartz, Fisher, Lohmann, Simon, & Boyan, 2009)
Human osteosarcoma Saos-2 cells	Commercially available discs of porous bovine natural apatite	1.3 ms pulses at 75 Hz	2	Trapezoidal	24 hours/day for 22 days	BIOSTIM, Igea, Carpi, Italy	(Lorenzo Fassina et al., 2010)

with a 1.6 mT intensity to Saos-2 and observed significant increase in ALP activity [105]. Other cell models, such as human osteosarcoma MG-63 [43, 106–108], mouse calvaria osteoblastic cell line MC3T3-E1 [36, 95, 109–114], rat primary calvaria cells [37, 40, 41, 45, 115, 116], primary human osteoblasts [42, 117–119], adipocyte-derived mesenchymal stem cells [118, 120–122], or bone marrow stromal cells [120, 123–133] were tested as well. As anticipated, most studies on osteoblast-related cell models rely on the 50-75 Hz range of stimulation [40, 41, 107, 108, 134–137] or, alternatively, on the use of 15 Hz PRF burst system [43–45, 105, 111, 112, 132, 138, 139]. The spectrum of intensities used is quite broad but, taken together, most works focus on the 0.6-2 mT [40, 41, 110, 137].

When osteoblastic cells grow on biomaterials however, a further layer of complexity is added. For the sake of simplicity, these studied were divided according to the nature of the biomaterial used.

4. PEMFs and Calcium Phosphate Scaffolds

All the studies on EMFs and calcium phosphate scaffolds included in the present review are listed in Table 1. One of the first studies to investigate the effects of PEMFs on bone response to bioceramics was performed by Shimizu et al. who implanted porous hydroxyapatite (HA) or tricalcium phosphate (TCP) cylinders in the proximal tibia of rabbits, which were then exposed to 1.5 Hz, 26 ms-long PFR PEMF bursts at 0.18 mT intensity for 8 hours/day. They were able to demonstrate a beneficial effect of PEMF stimulation on bone ingrowth into HA samples, with a higher amount of newly formed bone in and around HA, in both the cortical and medullary area, up to 4 weeks after surgery, but not around TCP implants [140]. A morphological evaluation of bone ingrowth into natural or synthetic hydroxyapatite granules implanted into rabbit tibia defects was conducted by Ottani et al. using 50 Hz triangular-shaped PEMF pulses at an intensity of 8 mT for 30'-long sessions twice a day. The sacrifice and subsequent TEM and SEM observation with electron backscattering at 2 and 4 weeks after surgery showed that PEMF treatment promoted a more advanced bone formation around the granules, which appeared cemented into the healing defect [141]. In the same year a study by Fini et al. was published, which investigated the effects of PEMFs on the integration of synthetic HA rods obtained by granule sintering in bone defects created in rabbit femoral condyles. The group used 1.35 ms-long trapezoidal PEMF pulses, repeated at a 75 Hz frequency, with an intensity of 1.6 mT for 6 hours/day for 3 weeks. Although histomorphometry did not reveal any increase in bone architectural parameters after PEMF stimulation at either 3 or 6 weeks after surgery, the bone-to-implant contact (BIC) was increased in the PEMF-treated group at both time points. The same happened with the mechanical properties of the treated bones, as assessed by hardness to microindentation [142]. The same research group adopted this stimulation model again to evaluate the integration of synthetic HA rods in the cortical bone of rabbit femurs and observed that PEMFs were able to significantly increase bone-to-implant contact, Mineral Apposition Rate (MAR), and Bone Formation Rate (BFR) at both time points. They also confirmed that the mechanical properties of treated bones were increased by PEMFs, using both indentation and push-out tests [143]. The cellular effects of PEMFs on the response of human Saos-2 osteosarcoma cells to discs of porous bovine natural apatite were investigated by Fassina et al., who exposed cells to 1.3 ms trapezoidal pulses at 75 Hz, 2 mT in bioreactors for 24 hours/day for 22 days [144]. In response to PEMFs the authors observed an increase in cell proliferation and the deposition of components of the extracellular matrix.

The group by Schwartz et al. investigated the effects of electromagnetic fields on human mesenchymal stem cells, using an established stimulation model of 4.5 ms PEMF bursts at 15 Hz frequency, with each burst composed of 225 μs-long pulses. Cells were grown on commercially available calcium phosphate discs and were exposed to PEMFs for 8 hours/day. Although, in their model, they did not observe significant effects of PEMFs on cell number or differentiation markers, the group found that electromagnetic fields synergistically stimulated cell responses to BMP-2 and promoted Alkaline Phosphatase (ALP) activity, Osteocalcin expression, and the release of TGFβ1 [145].

Interestingly, BMPs have been shown to be involved in the responses of rat calvaria osteoblasts to PEMFs in a study by Bodamyali et al. [45] and by Yan et al. [40]. Selvamu-rugan et al. demonstrated that PEMFs and BMP-2 may act synergistically in rat osteoblasts and this could be indicative of similar or overlapping signaling pathways in bone cells [115]. The group by Schwartz et al. also investigated the response of mesenchymal stem cells, commercially available normal human osteoblasts, or osteoblastic cells from two well established cell lines (MG-63 and Saos-2 cells) to 8-hour long exposures to 4.5 ms-long pulse bursts repeated at 15 Hz [146]. Their study showed that PEMFs were able to increase OPG expression in cell lines when cultured on calcium phosphate discs and synergistically increase OPG when administered together with BMP-2 in mesenchymal stem cells, while not affecting RANKL. Given the relevance of the OPG-RANKL system in bone, the effects of PEMFs on these molecular effects have been extensively studied in several osteoblastic models, also in the absence of biomaterials, and most studies agree with the results from Schwartz's groups in observing an increase in OPG following PEMF exposure. This is of obvious interest to bone researcher, because of the role of OPG and RANKL for tissue metabolism [147–150]. Schwartz's results were confirmed in cell cultures on plastic by Borsje et al. and similarly by Jansen et al. using BMMSCs [129] and even in human marrow macrophages cultures [132]. The group by Chang et al. showed that 7.5 Hz 0.3 ms long PEMF pulses increased OPG secretion [151] in mouse bone marrow cells [151]. They also observed that PEMFs enhanced OPG and hampered RANKL expression in mouse primary calvaria cells [152].

5. PEMFs and Titanium Surfaces or Implantable Devices

The effects of PEMFs on metal devices have been investigated in several studies. Though stainless steel implants in rabbit tibia and femurs were investigated by Spadaro et al., who observed an increase in the amount of formed bone in the medullary canal of femurs around moveable steel wires after 15 Hz PRF PEMF stimulation [153], most of the subsequent research focused on titanium and titanium alloy-based biomaterials. Saos-2 cells were used as a model of osteoblastic cells on titanium fiber-mesh scaffolds and continuously stimulated with 1.3 ms trapezoidal pulses at 75 Hz, 2 mT in bioreactors for 22 days. It was shown that PEMFs increased the expression of TGF-β and upregulated the deposition of matrix on the scaffolds, by increasing the expression of Decorin, Osteopontin, and Type I collagen [154]. The same group investigated the effects of PEMFs using the same cell and stimulation model on sintered titanium grids [155], observing similar findings. Wang et al. stimulated primary rat calvaria cells with 15 Hz, 5 ms long bursts of 4.5 kHz pulses, 0.9 mT, on polished, sand-blasted/acid-etched or anodized nanotubular titanium surfaces [156]. Interestingly, PEMF stimulation increased protein adsorption and cell adhesion on all titanium surfaces, cell proliferation up to 7 days, and cell mineralization on all surfaces. PEMF also affected cell morphology and induced more pseudopodia and cytoskeletal reorganization that aligned cells along their main axis. Interestingly, PEMFs also increased BMP-2 expression, beside differentiation markers. Bloise et al. [157] recently stimulated human BMMSCs nanostructured TiO$_2$ surfaces obtained through cluster-assembly by a pulsed microplasma cluster source [158, 159] with 1.3 ms long, 75 Hz PEMFs at 2 mT intensity for 10 min/day. The authors observed an increase in osteogenic differentiation in PEMF-stimulated cells, an increase in the intracellular levels of Ca^{2+}, and an increase in the extracellular Ca^{2+} deposition.

Using TiZr or titanium discs with different topography, Atalay et al. showed that the proliferative response of primary calvaria cells to 100 Hz PEMFs was clearly dependent on the microgeometry and physicochemical properties of the substrate [160].

The group of Jing et al. used 15 Hz, 5 ms long PEMF bursts with 2 mT intensity to stimulate MC3T3-E1 cells on porous titanium scaffolds (70% porosity, 750 μm pore size) for 2 hours/day for 3 days [161]. Besides observing an increase in cell proliferation and expression of differentiation markers Runx2 and Osterix, two important transcription factors activated in osteoblasts, the group reported that PEMF treatment increased β-catenin, Lrp6, and Wnt1 expression, important components of the canonical Wnt pathway, at the mRNA and protein levels. Remarkably, these findings were confirmed in vivo after implanting porous titanium scaffolds in cylindrical defects in the femur of rabbits, which were then treated for up to 12 weeks with PEMFs. MicroCT analysis of the defects showed that PEMF treatment significantly improved bone architectural parameters, e.g., BV/TV, Trabecular Number (Tb.N), and spacing (Tb.Sp), and dynamic histomorphometry demonstrated that MAR, Mineralizing Surface, and BFR were significantly higher in rabbits treated with PEMFs than control animals. Moreover, real time PCR indicated an increase in the expression of BMP-2, consistently with Lohmann [145, 146], but also Wnt1, Lrp6 and β-catenin as observed in vitro.

These results are in agreement with Single Pulsed EMF (sPEMF) exposure of MC3T3 cells on plastic culture substrates [114]. The authors exposed this cell line to 0.2 Hz, 5 ms long, 1 T PEMF pulses for up to 20 days and observed an increase in the expression of Wnt1, Wnt3a, Wnt10b, and Wnt receptor frizzled 9 and an increase, albeit not significant, of the Wnt coreceptor Lrp6. Similarly, Zhai et al. [110] observed that 2 mT, 15 Hz bursts of 4.5 kHz PEMF pulses for 2h/day for 3 days increased the expression of Wnt1, Lrp6, and β-catenin in MC3T3-E1 cells.

Buzzà et al. used 85 μs long pulses at 20 MHz for 30′/day for up to 42 days to stimulate titanium implants in rabbit tibias but failed to observe any significant increase in removal torque [162]. A slightly lower PEMF frequency (1 MHz, 25 μs long pulses, 0.8 mT) was used by do Nascimento et al. for 20′/day for 2 weeks to stimulate postextractive dental implants in dog mandibles. The authors observed a slight increase in bone tissue formed around the implants, although no quantification was provided [163]. Matsumoto et al. investigated the effects of 100 Hz, 25 μs PEMFs at 0.2, 0.3, or 0.8 mT for 4 or 8 hours/day on the integration of Ti-6Al-4V dental implants with anodized surface into rabbit femurs and reported that BIC was higher after exposure to 0.2 or 0.3 mT PEMFs for 4 or 8 hours [164]. This stimulation model was also used with Ti-6Al-4V dental implants inserted in

rabbit mandibles. The animals were stimulated with PEMFs for 2 weeks and sacrificed right after 2 weeks or 6 more weeks (without PEMF application). Remarkably, although no differences were observed at 2 weeks and 6 weeks after PEMF stimulation a dramatic increase in labial and lingual bone was observed in treated animals, together with higher osteoblast counts, indicating that PEMF could promote a long-acting bone formation [165]. A similar PEMF stimulation model was used by Akca et al. to investigate the effects of PEMFs on the integration of cylindrical titanium implants in tibias of ovariectomized rats. The animals were stimulated for 4 hours/day for 14 days and PEMF stimulation increased Bone Volume and trabecular number in the peri-implant bone, as determined by microCT [166]. A study by Grana et al. investigated the effects of 60 ms, 1.9 Hz PEMF bursts of 50 Hz sinusoidal trains at an intensity of 72 mT administered for 30′/twice a day on bone healing around titanium mini implants in rat tibias and found a significant increase in the amount of newly formed bone around implants at 10 and 20 days after surgeries [167]. Ten Hz, 0.4 mT PEMFs were investigated as a tool to improve the bone integration of commercially available titanium dental implants inserted in rabbit tibias in a more recent study [168]. Most noticeably, PEMFs were generated by a portable device which was installed on the implant, via a screw-retained connection, not unlike common prosthetic components. The device generated a magnetic field that was concentrated around the coronal area of the implant and steeply decreased in the surrounding areas. When considering the coronal area alone, where the signal was stronger, Bone Volume/Total Volume around test implants was 56% and 68% significantly higher than control implants at 2 and 4 weeks of healing, respectively, with corresponding increased Tb.N and smaller Tb.Sp. Moreover, by 2 weeks BIC was 15% higher around stimulated implants [168]. The idea of installing intraoral devices to stimulate implants with PEMFs was explored in several papers, as devices generating 10 Hz PRF PEMF bursts at 2 mT were proposed [169] (or even neodymium-iron-bore magnets placed in the implants and generating static magnetic fields [170]). Twenty-five μs PEMFs at 10 Hz and 0.2 mT were also investigated as a tool to promote the integration of porous titanium implants in the diaphysis of rabbit humerus bones for 5 or 10 hours/day and shown to increase bone ingrowth by a 14-day stimulation [171]. Cai et al. showed that 15 Hz, 5 ms PEMF bursts of 4.5 kHz pulses 2 hours/day for 8 weeks improved bone turnover serum markers and bone architecture parameters in rabbits with alloxan-induced type 1 diabetes mellitus (T1DM). More importantly for our current review, when cylindrical sintered Ti2448 implants were inserted into the lateral condyle of these rabbits, the 8-week treatment improved bone ingrowth into the scaffold and MAR around and inside the implants, which caused an increase in the mechanical properties of the trabecular bone around the implants [172]. For a list of the studies on EMFs and titanium biomaterials included in the present review, please see Table 2.

TABLE 2: The table summarizes the in vitro and in vivo studies on the effects of PEMF stimulation on osteoblastic primary cells and cell lines on titanium-based biomaterials. Studies are listed in chronological order.

Experimental model	Biomaterial	PEMF	Field intensity (mT)	PEMF waveform	Exposure	PEMF Generator	Reference
Placement in the medullary canal of femur and tibia in rabbits	Implants of 316 L stainless steel wire	5 ms, 15 Hz PEMF bursts of 4 kHz pulses	n/a	Quasi-square (trapezoidal pulses)	4 hours/day for 2 weeks	American Medical Electronics (Dallas, TX, U.S.A.)	(Spadaro et al., 1990)
Diaphysis of rabbit humerus	Bead-covered titanium implants	25 μs PEMF pulses at 10 Hz	0.2	n/a	5-10 hours/day for 2 weeks	n/a	(Ijiri et al., 1996)
Placement in rabbit femurs	Commercially available Ti-6Al-4V dental implants with anodized surface	100 Hz, 25 μs PEMFs	0.2, 0.3, 0.8	n/a	4 or 8 hours/day for up to 4 weeks	Riken Electromagnetic Field Pulse Generator, Institute of Physical and Chemical Research, Saitama, Japan	(Matsumoto et al., 2000)
Placement in rabbit tibias	Commercially available titanium dental implants	85 μs-long pulses at 20 MHz	1 W	n/a	30 minutes/day for 21 or 42 days	Healtec-Celular, Healtec Eletromedicina Ltd., Brazil	(Buzzá et al., 2003)
Placement in rabbit mandibles	Custom Ti-6Al-4V dental implants	100 Hz, 25 μs PEMFs	0.2	n/a	4 hours/day for 14 days	In-house built	(Özen et al., 2004)
Placement in tibias of ovariectomized rats	Cylindrical titanium implants	100 Hz, 25 μs PEMFs	0.2	n/a	4 hours/day for 14 days	In-house built	(Akca et al., 2007)
Human osteosarcoma Saos-2 cells	Titanium fiber-mesh sheets	1.3 ms pulses at 75 Hz	2	Trapezoidal	24 hours/day for 22 days	BIOSTIM, Igea, Carpi, Italy	(Fassina et al., 2008b)
Human osteosarcoma Saos-2 cells	Sintered titanium grids	1.3 ms pulses at 75 Hz	2	Trapezoidal	24 hours/day for 22 days	BIOSTIM, Igea, Carpi, Italy	(Fassina et al., 2008a)
Placement in rat tibias	Custom cylindrical threaded titanium implants	60 ms, 1.9 Hz PEMF bursts of 50 Hz trains	72	Quasi-square (with sinusoidal pulses)	30 minutes/twice a day	Magnetherp (Meditea Electromédica, Buenos Aires, Argentina)	(Grana et al., 2008)
Dog mandibles, immediate post-extraction placement	Commercially available titanium dental implants	1 MHz, 25 μs-long pulses	0.8	n/a	20 minutes/day for 2 weeks	n/a	(do Nascimento et al., 2012)
Primary rat calvaria cells	Commercially pure titanium or TiZr discs	100 Hz, 25 μs PEMFs	0.2	n/a	2 hours/day for up to 72 hours	In-house built	(Atalay et al., 2013)

TABLE 2: Continued.

Experimental model	Biomaterial	PEMF	Field intensity (mT)	PEMF waveform	Exposure	PEMF Generator	Reference
Primary rat calvaria cells	Polished, sand-blasted/acid-etched or anodized nanotubular titanium surfaces	15 Hz, 5 ms-long bursts of 4.5 kHz pulses	0.96	Quasi-square (with square pulses)	Up to 7 days	GHY-III, FMMU, Xi'an, China	(Wang et al., 2014)
Placement in rabbit tibias	Commercially available titanium dental implants	10 Hz	0.4-0.2	n/a	24 hours/day for 2 or 4 weeks	n/a	(Barak et al., 2016)
Murine MC3T3-E1 osteoblastic cells	Porous titanium scaffolds by electron beam melting system	15 Hz, 5 ms-long bursts of 4.5 kHz pulses	2	Quasi-square (with square pulses)	2 hours/day for 3 days	GHY-III, FMMU, Xi'an, China	(Jing et al., 2016)
Defects in rabbit femurs (condyles)	Porous titanium scaffolds by electron beam melting system	15 Hz, 5 ms-long bursts of 4.5 kHz pulses	2	Quasi-square (with square pulses)	2 hours/day for 6 or 12 weeks	GHY-III, FMMU, Xi'an, China	(Jing et al., 2016)
Placement in rabbit femurs (condyles)	Cylindrical sintered Ti2448 implants	5 Hz, 5 ms PEMF bursts of 4.5 kHz pulses	2	Quasi-square (with square pulses)	2 hours/day for 8 weeks	GHY-III, FMMU, Xi'an, China	(Cai et al., 2018)
Human BMMSCs	Nano-TiO2 surfaces	1.3 ms-long, 75 Hz	2	Trapezoidal	10 min/day	BIOSTIM, Igea, Carpi, Italy	(Bloise et al., 2018)

6. PEMFs and Polymers

Table 3 summarizes all the studies on polymer scaffolds and EMFs that were included in the present review. Polymer scaffolds were tested for cell responses to PEMFs as well. Electrospun poly(caprolactone) nanofibrous scaffolds were used as substrate to culture adipose tissue-derived stem cells, which were then stimulated with 50 Hz, 1 mT PEMFs for 6 hours/day in normal or osteogenic medium [173]. PEMFs increased cell proliferation, mineralization, and the expression of differentiation markers, such as Runx2, Osteocalcin, Osteonectin, and ALP activity. The group of Tsai et al. cultured rat calvaria osteoblasts on highly porous poly(DL-lactic-co-glycolic acid) (PLGA) scaffolds in bioreactors and stimulated them for 2 or 8 hours/day with 300 μs long rectangular pulses at 7.5 Hz. The magnetic field they used had an intensity of 0.13, 0.24, or 0.32 mT. Interestingly, stimulation with 0.13 mT PEMFs was able to significantly increase cell number on the scaffolds up to day 12 of culture, while more intense 0.32 mT PEMFs significantly decreased cell number compared to the control group up to day 18 of culture. However, not surprisingly, the highest intensity

was also most effective in increasing ALP activity and thus cell differentiation [174]. Lin et al. used an *in vitro* inflammation model to study the effects of 75 Hz, 1.5 mT PEMFs, using previously well described instrumentation [108] in 7F2 murine osteoblasts cultured on 3D chitosan scaffolds exposed to 9 hours of treatment [135]. The osteoblastic cells were cocultured with LPS-activated RAW 264.7 macrophages. The investigators detected higher Nitric Oxide levels after PEMF treatment, consistently with the previous literature [112, 175, 176], but increased osteoblast viability and collagen expression, although reduced differentiation, as measured by ALP activity and Osteocalcin levels. In agreement with their observations, Ehnert et al. [42] exposed primary human osteoblasts to 16 Hz 0.28 mT PEMF bursts for 7 minutes/day and demonstrated an increase in defenses against reactive oxygen species after PEMF stimulation [119], which actually appears necessary for PEMF effect [42].

The response of human osteosarcoma MG-63 cells to trapezoidal 1.3 ms long, 75 Hz, 2.3. mT PEMF pulses [134] when cultured on poly-methylmethacrylate (PMMA) scaffolds or PMMA-alpha Tricalcium Phosphate (α-TCP) composite scaffolds was investigated by Torricelli et al. [177].

TABLE 3: The table summarizes the in vitro and in vivo studies on the effects of PEMF stimulation on osteoblastic primary cells and cell lines on polymer-based biomaterials. Studies are listed in chronological order.

Experimental model	Biomaterial	PEMF	Field Intensity (mT)	PEMF waveform	Exposure	PEMF Generator	Reference
Human osteosarcoma MG-63 cells	poly-methyl methacrylate (PMMA) scaffolds or PMMA-alpha Tricalcium Phosphate (α-TCP) composite scaffolds	1.3 ms-long, 75 Hz	2.3	Trapezoidal	12 hours/day for 3 days	Igea, Carpi, Italy	(Torricelli et al., 2003)
Primary rat calvaria osteoblasts	Porous poly(DL-lactic-co-glycolic acid) (PLGA) scaffolds	300 μs-long pulses at 7.5 Hz	0.13, 0.24 or 0.32	Rectangular	2 or 8 hours/day	PIC/16C54 series, Microchip Technology Inc., AZ	(Tsai et al., 2007)
7F2+ RAW 264.7	3D chitosan scaffolds	1.3 ms-long, 75 Hz	1.5	Trapezoidal	9 hours	BIOSTIM, Igea, Carpi, Italy	(Lin and Lin, 2011)
Human osteosarcoma Saos-2 cells	Methacrylamide-modified gelatin type B scaffolds	1.3 ms pulses at 75 Hz	2	Trapezoidal	24 hours/day for 22 days	BIOSTIM, Igea, Carpi, Italy	(Fassina et al., 2012)
Osteochondral defects in rabbit medial femoral condyles.	Commercially available equine collagen scaffolds with or w/o BMC	1.3 ms-long, 75 Hz	1.5	Trapezoidal	4 hours/day for 40 days	I-ONE, Igea, Carpi, Italy	(Veronesi et al., 2015)
Rat calvaria defects	Commercially available collagen sponges loaded with 2.5-10 μg rhBMP-2	12 μs pulses, 60 Hz	1	n/a	8 hours/day for 5 days	In-house built	(Yang et al., 2015)
Human adipose tissue-derived stem cells	Electrospun poly(caprolactone) nanofibrous scaffolds	50 Hz	1	n/a	6 hours/day for up to 21 days	n/a	(Arjmand et al., 2018)

Cells were stimulated for 12 hours/day for 3 days, and PEMFs were able to increase the expression of Osteocalcin, C-terminal procollagen type 1, and TGFβ1 in cells on composite scaffolds, while decreasing IL-6 expression by 6 days of culture. An involvement of TGF-β in PEMF stimulation was highlighted by several researches in MG63 cells [43], in serum-starved MC3T3 cells [111], and in human BMMSCs, where PEMFs increased Smad-2 and miRNA21, a microRNA targeting Smad-7, a TGF-β signaling inhibitor [131].

Veronesi et al. showed that 75 Hz, 1.5 mT PEMF stimulation for 4 hours/day improved 40-day healing in osteochondral defects in rabbit knees, when used together with collagen scaffolds [178]. Collagen sponges loaded with increasing doses of recombinant human BMP-2 were also implanted in calvaria defects in rats and treated with 1 mT, 60 Hz PEMF stimulation for 8 hours/day for 5 days [179]. Computer microtomography 4 weeks after surgery revealed that PEMF stimulation increased Bone Volume and Bone Mineral Density in the absence or in the presence of rhBMP-2 but not with the highest, 10 μg, dose, where no additional effect was observed. In the samples implanted with 2.5 micrograms as well PEMF stimulation significantly increased also Tb.N. and decreased Tb.Sp. Similarly, histology showed that PEMFs were able to increase bone regeneration in the central area of the defect without the addition of rhBMP-2.

Hydrogels were also explored together with PEMF exposure. Fassina et al. [180] cultured Saos-2 cells in bioreactors on methacrylamide-modified gelatin type B using the same exposure model as previously described [144, 181] and observed an increase in the deposition of Extracellular Matrix. Some research groups are also creating EMF-responsive hydrogels, which can release their bioactive load under EMD stimulation, e.g., methacrylated chondroitin sulfate (MA-CS) hydrogels coated with iron-based magnetic nanoparticles for PDGF release [182] and Ca^{2+}-crosslinked Alginate/Xanthan gum hydrogels with magnetite particles for dopamine delivery [183, 184], although these studies were not included in the present review as EMFs were used only as a release-triggering stimulus and not to elicit biological effects.

7. Conclusions

The world of biomaterials is as diverse as the clinical applications that rely on them; therefore it stands to reason that there is no easy solution to improve their performance and the responses of the organisms to implanted material and devices. We nevertheless attempted at simplifying the wealth of available materials by dividing them into three main categories, which are however broad as well. A few conclusions can be drawn.

PEMFs have been repeatedly shown to possess the potential to affect osteoblast behavior on different biomaterials and thus represent a potential tool to improve the clinical outcome of several regenerative and prosthetic therapies in orthopedics and dentistry and should be more thoroughly investigated by proper clinical trials.

The response of cells and tissues to PEMF in the presence of titanium devices, for orthopedic or dental use, has been investigated using a vast range of PEMF approaches and settings but besides a few attempts in the early 2000s with 100 Hz PEMF pulses with very light intensities, around 0.2 mT (following the seminal work by Matsumoto et al. [164]), most recent studies are narrowing down their focus to 15 Hz PRF PEMF stimulation or 75 Hz trapezoidal stimuli, with higher intensity, around 1-2 mT. Similar conclusions can be achieved considering the biological responses to bioceramic and polymer scaffolds. However broader screening studies testing cell or tissue responses across a spectrum of frequencies are still missing, though they would be sorely needed to better understand and possibly overcome the differences that exist among schools, with the purpose of establishing better and more reliable clinical protocols for this powerful technology.

Conflicts of Interest

The authors have no conflicts of interest to disclose.

References

[1] J. J. Li, M. Ebied, J. Xu, and H. Zreiqat, "Current approaches to bone tissue engineering: the interface between biology and engineering," *Advanced Healthcare Materials*, vol. 7, no. 6, Article ID 1701061, 2018.

[2] W. Wang and K. W. K. Yeung, "Bone grafts and biomaterials substitutes for bone defect repair: a review," *Bioactive Materials*, vol. 2, no. 4, pp. 224–247, 2017.

[3] S. Shanbhag, N. Pandis, K. Mustafa, J. R. Nyengaard, and A. Stavropoulos, "Bone tissue engineering in oral peri-implant defects in preclinical in vivo research: A systematic review and meta-analysis," *Journal of Tissue Engineering and Regenerative Medicine*, vol. 12, no. 1, pp. e336–e349, 2018.

[4] F. Rupp, L. Liang, J. Geis-Gerstorfer, L. Scheideler, and F. Hüttig, "Surface characteristics of dental implants: A review," *Dental Materials*, vol. 34, no. 1, pp. 40–57, 2018.

[5] Pearlin, S. Nayak, G. Manivasagam, and D. Sen, "Progress of regenerative therapy in orthopedics," *Current Osteoporosis Reports*, vol. 16, no. 2, pp. 169–181, 2018.

[6] Z. Li, R. Müller, and D. Ruffoni, "Bone remodeling and mechanobiology around implants: Insights from small animal imaging," *Journal of Orthopaedic Research*, vol. 36, no. 2, pp. 584–593, 2017.

[7] S. Kuttappan, D. Mathew, and M. B. Nair, "Biomimetic composite scaffolds containing bioceramics and collagen/gelatin for bone tissue engineering - A mini review," *International Journal of Biological Macromolecules*, vol. 93, pp. 1390–1401, 2016.

[8] S. V. Dorozhkin, "Calcium orthophosphate-based bioceramics," *Materials* , vol. 6, no. 9, pp. 3840–3942, 2013.

[9] N. Eliaz and N. Metoki, "Calcium Phosphate Bioceramics: A Review of Their History, Structure, Properties, Coating Technologies and Biomedical Applications," *Materials* , vol. 10, no. 4, p. 334, 2017.

[10] W. L. Murphy, T. C. McDevitt, and A. J. Engler, "Materials as stem cell regulators," *Nature Materials*, vol. 13, no. 6, pp. 547–557, 2014.

[11] A. Kumar, J. K. Placone, and A. J. Engler, "Understanding the extracellular forces that determine cell fate and maintenance," *Development*, vol. 144, no. 23, pp. 4261–4270, 2017.

[12] M. Nabiyouni, T. Brückner, H. Zhou, U. Gbureck, and S. B. Bhaduri, "Magnesium-based bioceramics in orthopedic applications," *Acta Biomaterialia*, vol. 66, pp. 23–43, 2018.

[13] H. Begam, S. K. Nandi, B. Kundu, and A. Chanda, "Strategies for delivering bone morphogenetic protein for bone healing," *Materials Science and Engineering C: Materials for Biological Applications*, vol. 70, pp. 856–869, 2017.

[14] C. Ribeiro, V. Sencadas, D. M. Correia, and S. Lanceros-Méndez, "Piezoelectric polymers as biomaterials for tissue engineering applications," *Colloids and Surfaces B: Biointerfaces*, vol. 136, pp. 46–55, 2015.

[15] K. Jahan and M. Tabrizian, "Composite biopolymers for bone regeneration enhancement in bony defects," *Biomaterials Science*, vol. 4, no. 1, pp. 25–39, 2016.

[16] D. M. R. Gibbs, C. R. M. Black, J. I. Dawson, and R. O. C. Oreffo, "A review of hydrogel use in fracture healing and bone regeneration," *Journal of Tissue Engineering and Regenerative Medicine*, vol. 10, no. 3, pp. 187–198, 2016.

[17] C. Arakawa, R. Ng, S. Tan, S. Kim, B. Wu, and M. Lee, "Photopolymerizable chitosan–collagen hydrogels for bone tissue engineering," *Journal of Tissue Engineering and Regenerative Medicine*, vol. 11, no. 1, pp. 164–174, 2017.

[18] G. Turnbull, J. Clarke, F. Picard et al., "3D bioactive composite scaffolds for bone tissue engineering," *Bioactive Materials*, vol. 3, no. 3, pp. 278–314, 2018.

[19] R. Martins, T. M. Cestari, R. V. N. Arantes et al., "Osseointegration of zirconia and titanium implants in a rabbit tibiae model evaluated by microtomography, histomorphometry and fluorochrome labeling analyses," *Journal of Periodontal Research*, vol. 53, no. 2, pp. 210–221, 2018.

[20] K. Sivaraman, A. Chopra, A. I. Narayan, and D. Balakrishnan, "Is zirconia a viable alternative to titanium for oral implant? A critical review," *Journal of Prosthodontic Research*, vol. 62, no. 2, pp. 121–133, 2018.

[21] D. D. Bosshardt, V. Chappuis, and D. Buser, "Osseointegration of titanium, titanium alloy and zirconia dental implants: current knowledge and open questions," *Periodontology 2000*, vol. 73, no. 1, pp. 22–40, 2017.

[22] S. V. Kellesarian, V. R. Malignaggi, T. V. Kellesarian, H. Bashir Ahmed, and F. Javed, "Does incorporating collagen and chondroitin sulfate matrix in implant surfaces enhance osseointegration? A systematic review and meta-analysis," *International Journal of Oral and Maxillofacial Surgery*, vol. 47, no. 2, pp. 241–251, 2018.

[23] S. H. Rao, B. Harini, R. P. K. Shadamarshan, K. Balagangadharan, and N. Selvamurugan, "Natural and synthetic polymers/bioceramics/bioactive compounds-mediated cell signalling in bone tissue engineering," *International Journal of Biological Macromolecules*, vol. 110, pp. 88–96, 2018.

[24] M. F. Iskander, *Electromagnetic Fields and Waves*, Waveland Press, Long Grove, IL, USA, 2013, http://cds.cern.ch/record/1529891.

[25] L. E. Birks, B. Struchen, and M. Eeftens, "Spatial and temporal variability of personal environmental exposure to radio frequency electromagnetic fields in children in Europe," *Environment International*, vol. 117, pp. 204–214, 2018.

[26] N. Wertheimer and E. Leeper, "Electrical wiring configurations and childhood cancer," *American Journal of Epidemiology*, vol. 109, no. 3, pp. 273–284, 1979.

[27] N. Wertheimer and E. Leeper, "Adult cancer related to electrical wires near the home," *International Journal of Epidemiology*, vol. 11, no. 4, pp. 345–355, 1982.

[28] C. D. Robinette, C. Silverman, and S. Jablon, "Effects upon health of occupational exposure to microwave radiation (radar)," *American Journal of Epidemiology*, vol. 112, no. 1, pp. 39–53, 1980.

[29] P. A. Valberg, R. Kavet, and C. N. Rafferty, "Can low-level 50/60 hz electric and magnetic fields cause biological effects?" *Journal of Radiation Research*, vol. 148, no. 1, pp. 2–21, 1997.

[30] M. L. Pall, "Microwave frequency electromagnetic fields (EMFs) produce widespread neuropsychiatric effects including depression," *Journal of Chemical Neuroanatomy*, vol. 75, pp. 43–51, 2016.

[31] M. L. Pall, "Wi-Fi is an important threat to human health," *Environmental Research*, vol. 164, pp. 405–416, 2018.

[32] K. Hug and M. Röösli, "Therapeutic effects of whole-body devices applying pulsed electromagnetic fields (PEMF): A systematic literature review," *Bioelectromagnetics*, vol. 33, no. 2, pp. 95–105, 2012.

[33] X. Zhang, J. Zhang, X. Qu, and J. Wen, "Effects of Different Extremely Low-Frequency Electromagnetic Fields on Osteoblasts," *Electromagnetic Biology and Medicine*, vol. 26, no. 3, pp. 167–177, 2007.

[34] E. Pchelintseva and M. B. A. Djamgoz, "Mesenchymal stem cell differentiation: Control by calcium-activated potassium channels," *Journal of Cellular Physiology*, vol. 233, no. 5, pp. 3755–3768, 2017.

[35] X. Zhang, X. Liu, L. Pan, and I. Lee, "Magnetic fields at extremely low-frequency (50 Hz, 0.8 mT) can induce the uptake of intracellular calcium levels in osteoblasts," *Biochemical and Biophysical Research Communications*, vol. 396, no. 3, pp. 662–666, 2010.

[36] J. Tong, L. Sun, B. Zhu et al., "Pulsed electromagnetic fields promote the proliferation and differentiation of osteoblasts by reinforcing intracellular calcium transients," *Bioelectromagnetics*, vol. 38, no. 7, pp. 541–549, 2017.

[37] J. Kuan-Jung Li, J. Cheng-An Lin, H. Liu et al., "Comparison of ultrasound and electromagnetic field effects on osteoblast growth," *Ultrasound in Medicine & Biology*, vol. 32, no. 5, pp. 769–775, 2006.

[38] S. Wu, Q. Yu, A. Lai, and J. Tian, "Pulsed electromagnetic field induces Ca2+ -dependent osteoblastogenesis in C3H10T1/2 mesenchymal cells through the Wnt-Ca 2+ /Wnt-β-catenin signaling pathway," *Biochemical and Biophysical Research Communications*, vol. 503, no. 2, pp. 715–721, 2018.

[39] K. Varani, F. Vincenzi, A. Ravani et al., "Adenosine receptors as a biological pathway for the anti-inflammatory and beneficial effects of low frequency low energy pulsed electromagnetic fields," *Mediators of Inflammation*, vol. 2017, Article ID 2740963, 11 pages, 2017.

[40] J. Yan, J. Zhou, H. Ma et al., "Pulsed electromagnetic fields promote osteoblast mineralization and maturation needing the existence of primary cilia," *Molecular and Cellular Endocrinology*, vol. 404, pp. 132–140, 2015.

[41] Y.-F. Xie, W.-G. Shi, J. Zhou et al., "Pulsed electromagnetic fields stimulate osteogenic differentiation and maturation of osteoblasts by upregulating the expression of BMPRII localized at the base of primary cilium," *Bone*, vol. 93, pp. 22–32, 2016.

[42] S. Ehnert, A. Fentz, A. Schreiner et al., "Extremely low frequency pulsed electromagnetic fields cause antioxidative defense mechanisms in human osteoblasts via induction of •O2– and H2O2," *Scientific Reports*, vol. 7, no. 1, Article ID 14544, 2017.

[43] C. H. Lohmann, Z. Schwartz, Y. Liu et al., "Pulsed electromagnetic field stimulation of MG63 osteoblast-like cells affects differentiation and local factor production," *Journal of Orthopaedic Research*, vol. 18, no. 4, pp. 637–646, 2000.

[44] Y. Sakai, T. E. Patterson, M. O. Ibiwoye et al., "Exposure of mouse preosteoblasts to pulsed electromagnetic fields reduces the amount of mature, type I collagen in the extracellular matrix," *Journal of Orthopaedic Research*, vol. 24, no. 2, pp. 242–253, 2006.

[45] T. Bodamyali, B. Bhatt, F. J. Hughes et al., "Pulsed electromagnetic fields simultaneously induce osteogenesis and upregulate transcription of bone morphogenetic proteins 2 and 4 in rat osteoblasts in vitro," *Biochemical and Biophysical Research Communications*, vol. 250, no. 2, pp. 458–461, 1998.

[46] Y. Wang, X. Pu, W. Shi et al., "Pulsed electromagnetic fields promote bone formation by activating the sAC-cAMP-PKA-CREB signaling pathway," *Journal of Cellular Physiology*, 2018.

[47] C. Daish, R. Blanchard, K. Fox, P. Pivonka, and E. Pirogova, "The Application of Pulsed Electromagnetic Fields (PEMFs) for Bone Fracture Repair: Past and Perspective Findings," *Annals of Biomedical Engineering*, vol. 46, no. 4, pp. 525–542, 2018.

[48] J. Huegel, D. S. Choi, C. A. Nuss et al., "Effects of pulsed electromagnetic field therapy at different frequencies and durations on rotator cuff tendon-to-bone healing in a rat model," *Journal of Shoulder and Elbow Surgery*, vol. 27, no. 3, pp. 553–560, 2018.

[49] H. M. Bilgin, F. Çelik, M. Gem et al., "Effects of local vibration and pulsed electromagnetic field on bone fracture: A comparative study," *Bioelectromagnetics*, vol. 38, no. 5, pp. 339–348, 2017.

[50] A. B. Sarker, A. N. Nashimuddin, and K. M. Islam, "Effect of PEMF on fresh fracture-healing in rat tibia," *Bangladesh Medical Research Council Bulletin*, vol. 19, no. 3, pp. 103–112, 1993.

[51] K. F. Taylor, N. Inoue, B. Rafiee, J. E. Tis, K. A. McHale, and E. Y. S. Chao, "Effect of pulsed electromagnetic fields on maturation of regenerate bone in a rabbit limb lengthening model," *Journal of Orthopaedic Research*, vol. 24, no. 1, pp. 2–10, 2006.

[52] D. C. Fredericks, J. V. Nepola, J. T. Baker, J. Abbott, and B. Simon, "Effects of pulsed electromagnetic fields on bone healing in a rabbit tibial osteotomy model," *Journal of Orthopaedic Trauma*, vol. 14, no. 2, pp. 93–100, 2000.

[53] R. J. Midura, M. O. Ibiwoye, K. A. Powell et al., "Pulsed electromagnetic field treatments enhance the healing of fibular osteotomies," *Journal of Orthopaedic Research*, vol. 23, no. 5, pp. 1035–1046, 2005.

[54] P. S. Landry, K. K. Sadasivan, A. A. Marino, and J. A. Albright, "Electromagnetic Fields Can Affect Osteogenesis by Increasing the Rate of Differentiation," *Clinical Orthopaedics and Related Research*, vol. 338, pp. 262–270, 1997.

[55] T. Takano-Yamamoto, M. Kawakami, and M. Sakuda, "Effect of a pulsing electromagnetic field on demineralized bone-matrix-induced bone formation in a bony defect in the premaxilla of rats," *Journal of Dental Research*, vol. 71, no. 12, pp. 1920–1925, 1992.

[56] E. Kapi, M. Bozkurt, C. T. Selcuk et al., "Comparison of effects of pulsed electromagnetic field stimulation on platelet-rich plasma and bone marrow stromal stem cell using rat zygomatic bone defect model," *Annals of Plastic Surgery*, vol. 75, no. 5, pp. 565–571, 2015.

[57] C. Liu, Y. Zhang, T. Fu et al., "Effects of electromagnetic fields on bone loss in hyperthyroidism rat model," *Bioelectromagnetics*, vol. 38, no. 2, pp. 137–150, 2017.

[58] D. Jing, G. Shen, J. Huang et al., "Circadian rhythm affects the preventive role of pulsed electromagnetic fields on ovariectomy-induced osteoporosis in rats," *Bone*, vol. 46, no. 2, pp. 487–495, 2010.

[59] Y. Jiang, H. Gou, S. Wang, J. Zhu, S. Tian, and L. Yu, "Effect of pulsed electromagnetic field on bone formation and lipid metabolism of glucocorticoid-induced osteoporosis rats through canonical wnt signaling pathway," *Evidence-Based Complementary and Alternative Medicine*, vol. 2016, Article ID 4927035, 13 pages, 2016.

[60] J. Zhou, H. He, L. Yang et al., "Effects of pulsed electromagnetic fields on bone mass and Wnt/β-catenin signaling pathway in ovariectomized rats," *Archives of Medical Research*, vol. 43, no. 4, pp. 274–282, 2012.

[61] J. Zhou, S. Chen, H. Guo et al., "Pulsed electromagnetic field stimulates osteoprotegerin and reduces RANKL expression in ovariectomized rats," *Rheumatology International*, vol. 33, no. 5, pp. 1135–1141, 2013.

[62] J. Zhou, Y. Liao, Y. Zeng, H. Xie, C. Fu, and N. Li, "Effect of intervention initiation timing of pulsed electromagnetic field on ovariectomy-induced osteoporosis in rats," *Bioelectromagnetics*, vol. 38, no. 6, pp. 456–465, 2017.

[63] J. Zhou, Y. Liao, H. Xie et al., "Effects of combined treatment with ibandronate and pulsed electromagnetic field on ovariectomy-induced osteoporosis in rats," *Bioelectromagnetics*, vol. 38, no. 1, pp. 31–40, 2017.

[64] D. Jing, F. Li, M. Jiang et al., "Pulsed electromagnetic fields improve bone microstructure and strength in ovariectomized rats through a Wnt/Lrp5/β-catenin signaling-associated mechanism," *PLoS ONE*, vol. 8, no. 11, Article ID e79377, 2013.

[65] T. Lei, Z. Liang, F. Li et al., "Pulsed electromagnetic fields (PEMF) attenuate changes in vertebral bone mass, architecture and strength in ovariectomized mice," *Bone*, vol. 108, pp. 10–19, 2018.

[66] K. Chang and W. H.-S. Chang, "Pulsed electromagnetic fields prevent osteoporosis in an ovariectomized female rat model: a prostaglandin E2-associated process," *Bioelectromagnetics*, vol. 24, no. 3, pp. 189–198, 2003.

[67] D. Jing, J. Cai, Y. Wu et al., "Pulsed electromagnetic fields partially preserve bone mass, microarchitecture, and strength by promoting bone formation in hindlimb-suspended rats," *Journal of Bone and Mineral Research*, vol. 29, no. 10, pp. 2250–2261, 2014.

[68] B. Li, J. Bi, W. Li et al., "Effects of pulsed electromagnetic fields on histomorphometry and osteocalcin in disuse osteoporosis rats," *Technology and Health Care*, vol. 25, no. S1, pp. 13–20, 2017.

[69] W.-W. Shen and J.-H. Zhao, "Pulsed electromagnetic fields stimulation affects BMD and local factor production of rats with disuse osteoporosis," *Bioelectromagnetics*, vol. 31, no. 2, pp. 113–119, 2010.

[70] J. Li, Z. Zeng, Y. Zhao et al., "Effects of low-intensity pulsed electromagnetic fields on bone microarchitecture, mechanical strength and bone turnover in type 2 diabetic db/db mice," *Scientific Reports*, vol. 7, no. 1, Article ID 10834, 2017.

[71] C. Androjna, B. Fort, M. Zborowski, and R. J. Midura, "Pulsed electromagnetic field treatment enhances healing callus biomechanical properties in an animal model of osteoporotic fracture," *Bioelectromagnetics*, vol. 35, no. 6, pp. 396–405, 2014.

[72] X. Yang, H. He, Y. Zhou et al., "Pulsed electromagnetic field at different stages of knee osteoarthritis in rats induced by low-dose monosodium iodoacetate: Effect on subchondral trabecular bone microarchitecture and cartilage degradation," *Bioelectromagnetics*, vol. 38, no. 3, pp. 227–238, 2017.

[73] V. Canè, P. Botti, D. Farneti, and S. Soana, "Electromagnetic stimulation of bone repair: A histomorphometric study," *Journal of Orthopaedic Research*, vol. 9, no. 6, pp. 908–917, 1991.

[74] V. Cane, P. Botti, and S. Soana, "Pulsed magnetic fields improve osteoblast activity during the repair of an experimental osseous defect," *Journal of Orthopaedic Research*, vol. 11, no. 5, pp. 664–670, 1993.

[75] D. E. Garland, R. H. Adkins, N. N. Matsuno, and C. A. Stewart, "The effect of pulsed electromagnetic fields on osteoporosis at the knee in individuals with spinal cord injury," *The Journal of Spinal Cord Medicine*, vol. 22, no. 4, pp. 239–245, 1999.

[76] F. L. Tabrah, P. Ross, M. Hoffmeier, and F. Gilbert Jr., "Clinical Report on Long-Term Bone Density after Short-Term EMF Application," *Bioelectromagnetics*, vol. 19, no. 2, pp. 75–78, 1998.

[77] F. Tabrah, M. Hoffmeier, F. Gilbert, S. Batkin, and C. A. L. Bassett, "Bone density changes in osteoporosis-prone women exposed to pulsed electromagnetic fields (PEMFs)," *Journal of Bone and Mineral Research*, vol. 5, no. 5, pp. 437–442, 1990.

[78] C. A. Bassett, A. A. Pilla, and R. J. Pawluk, "A non-operative salvage of surgically-resistant pseudarthroses and non-unions by pulsing electromagnetic fields," *Clinical Orthopaedics and Related Research*, no. 124, pp. 128–143, 1977.

[79] C. A. Bassett, S. N. Mitchell, and S. R. Gaston, "Treatment of ununited tibial diaphyseal fractures with pulsing electromagnetic fields.," *The Journal of Bone & Joint Surgery*, vol. 63, no. 4, pp. 511–523, 1981.

[80] R. B. Simonis, E. J. Parnell, P. S. Ray, and J. L. Peacock, "Electrical treatment of tibial non-union: A prospective, randomised, double-blind trial," *Injury*, vol. 34, no. 5, pp. 357–362, 2003.

[81] M. Lazovic, M. Kocic, L. Dimitrijevic, I. Stankovic, M. Spalevic, and T. Ciric, "Pulsed electromagnetic field during cast immobilization in postmenopausal women with Colles' fracture," *Srpski Arhiv za Celokupno Lekarstvo*, vol. 140, no. 9-10, pp. 619–624, 2012.

[82] G. L. Y. Cheing, J. W. H. Wan, and S. Kai Lo, "Ice and pulsed electromagnetic field to reduce pain and swelling after distal radius fractures," *Journal of Rehabilitation Medicine*, vol. 37, no. 6, pp. 372–377, 2005.

[83] P. F. Hannemann, B. A. Essers, J. P. Schots, K. Dullaert, M. Poeze, and P. R. Brink, "Functional outcome and cost-effectiveness of pulsed electromagnetic fields in the treatment of acute scaphoid fractures: a cost-utility analysis," *BMC Musculoskeletal Disorders*, vol. 16, no. 1, p. 84, 2015.

[84] C. Faldini, M. Cadossi, D. Luciani, E. Betti, E. Chiarello, and S. Giannini, "Electromagnetic bone growth stimulation in patients with femoral neck fractures treated with screws: Prospective randomized double-blind study," *Current Orthopaedic Practice*, vol. 21, no. 3, pp. 282–287, 2010.

[85] S. Adie, I. A. Harris, J. M. Naylor et al., "Pulsed Electromagnetic Field Stimulation for Acute Tibial Shaft Fractures," *The Journal of Bone and Joint Surgery-American Volume*, vol. 93, no. 17, pp. 1569–1576, 2011.

[86] A. Assiotis, N. P. Sachinis, and B. E. Chalidis, "Pulsed electromagnetic fields for the treatment of tibial delayed unions and nonunions. A prospective clinical study and review of the literature," *Journal of Orthopaedic Surgery and Research*, vol. 7, no. 1, p. 24, 2012.

[87] H. Shi, J. Xiong, Y. Chen et al., "Early application of pulsed electromagnetic field in the treatment of postoperative delayed union of long-bone fractures: a prospective randomized controlled study," *BMC Musculoskeletal Disorders*, vol. 14, no. 1, p. 35, 2013.

[88] A. Streit, B. C. Watson, J. D. Granata et al., "Effect on Clinical Outcome and Growth Factor Synthesis With Adjunctive Use of Pulsed Electromagnetic Fields for Fifth Metatarsal Nonunion Fracture," *Foot & Ankle International*, vol. 37, no. 9, pp. 919–923, 2016.

[89] H. Refai, D. Radwan, and N. Hassanien, "Radiodensitometric Assessment of the Effect of Pulsed Electromagnetic Field Stimulation Versus Low Intensity Laser Irradiation on Mandibular Fracture Repair: A Preliminary Clinical Trial," *Journal of Maxillofacial and Oral Surgery*, vol. 13, no. 4, pp. 451–457, 2014.

[90] A. Abdelrahim, H. R. Hassanein, and M. Dahaba, "Effect of pulsed electromagnetic field on healing of mandibular fracture: a preliminary clinical study," *Journal of Oral and Maxillofacial Surgery*, vol. 69, no. 6, pp. 1708–1717, 2011.

[91] A. T. Barker, R. A. Dixon, W. J. W. Sharrard, and M. L. Sutcliffe, "Pulsed Magnetic Field Therapy for Tibial Non-Union. Interim Results of a Double-Blind Trial," *The Lancet*, vol. 1, no. 8384, pp. 994–996, 1984.

[92] G. Scott and J. B. King, "A prospective, double-blind trial of electrical capacitive coupling in the treatment of non-union of long bones," *The Journal of Bone & Joint Surgery*, vol. 76, no. 6, pp. 820–826, 1994.

[93] W. Sharrard, "A double-blind trial of pulsed electromagnetic fields for delayed union of tibial fractures," *The Journal of Bone & Joint Surgery*, vol. 72, no. 3, pp. 347–355, 1990.

[94] C. Liu, J. Yu, Y. Yang et al., "Effect of 1 mT sinusoidal electromagnetic fields on proliferation and osteogenic differentiation of rat bone marrow mesenchymal stromal cells," *Bioelectromagnetics*, vol. 34, no. 6, pp. 453–464, 2013.

[95] K. Li, S. Ma, Y. Li et al., "Effects of PEMF exposure at different pulses on osteogenesis of MC3T3-E1 cells," *Archives of Oral Biology*, vol. 59, no. 9, pp. 921–927, 2014.

[96] M. S. Markov, "Magnetic Field Therapy: A Review," *Electromagnetic Biology and Medicine*, vol. 26, no. 1, pp. 1–23, 2007.

[97] T. Lei, F. Li, Z. Liang et al., "Effects of four kinds of electromagnetic fields (EMF) with different frequency spectrum bands on ovariectomized osteoporosis in mice," *Scientific Reports*, vol. 7, no. 1, p. 553, 2017.

[98] J. Zhou, J.-Q. Wang, B.-F. Ge et al., "Different electromagnetic field waveforms have different effects on proliferation, differentiation and mineralization of osteoblasts in vitro," *Bioelectromagnetics*, vol. 35, no. 1, pp. 30–38, 2014.

[99] D. K. Hubbard and R. Dennis, "Pain relief and tissue healing using pemf therapy: a review of stimulation waveform effects," *Asia Health Care Journal*, vol. 1, no. 2, pp. 26–35, 2012.

[100] X. L. Griffin, M. L. Costa, N. Parsons, and N. Smith, "Electromagnetic field stimulation for treating delayed union or nonunion of long bone fractures in adults," *Cochrane Database of Systematic Reviews*, 2011.

[101] H. H. Handoll and J. Elliott, "Rehabilitation for distal radial fractures in adults," *Cochrane Database of Systematic Reviews*, 2015.

[102] P. F. W. Hannemann, E. H. H. Mommers, J. P. M. Schots, P. R. G. Brink, and M. Poeze, "The effects of low-intensity pulsed ultrasound and pulsed electromagnetic fields bone growth stimulation in acute fractures: a systematic review

and meta-analysis of randomized controlled trials," *Archives of Orthopaedic and Trauma Surgery*, vol. 134, no. 8, pp. 1093–1106, 2014.

[103] L. Massari, G. Caruso, and V. Sollazzo, "Pulsed electromagnetic fields and low intensity pulsed ultrasound in bone tissue," *Clinical Cases in Mineral Bone Metabolism*, vol. 6, no. 2, pp. 149–154, 2009.

[104] C. F. Martino, D. Belchenko, V. Ferguson, S. Nielsen-Preiss, and H. J. Qi, "The effects of pulsed electromagnetic fields on the cellular activity of SaOS-2 cells," *Bioelectromagnetics*, vol. 29, no. 2, pp. 125–132, 2008.

[105] G. Hannay, D. Leavesley, and M. Pearcy, "Timing of pulsed electromagnetic field stimulation does not affect the promotion of bone cell development," *Bioelectromagnetics*, vol. 26, no. 8, pp. 670–676, 2005.

[106] B. Noriega-Luna, M. Sabanero, M. Sosa, and M. Avila-Rodriguez, "Influence of pulsed magnetic fields on the morphology of bone cells in early stages of growth," *Micron*, vol. 42, no. 6, pp. 600–607, 2011.

[107] V. Sollazzo, A. Palmieri, F. Pezzetti, L. Massari, and F. Carinci, "Effects of pulsed electromagnetic fields on human osteoblast-like cells (MG-63): a pilot study," *Clinical Orthopaedics and Related Research*, vol. 468, no. 8, pp. 2260–2277, 2010.

[108] M. De Mattei, N. Gagliano, C. Moscheni et al., "Changes in polyamines, c-myc and c-fos gene expression in osteoblast-like cells exposed to pulsed electromagnetic fields," *Bioelectromagnetics*, vol. 26, no. 3, pp. 207–214, 2005.

[109] A. Soda, T. Ikehara, Y. Kinouchi, and K. Yoshizaki, "Effect of exposure to an extremely low frequency-electromagnetic field on the cellular collagen with respect to signaling pathways in osteoblast-like cells," *The Journal of Medical Investigation*, vol. 55, no. 2, pp. 267–278, 2008.

[110] M. Zhai, D. Jing, S. Tong et al., "Pulsed electromagnetic fields promote in vitro osteoblastogenesis through a Wnt/β-catenin signaling-associated mechanism," *Bioelectromagnetics*, vol. 37, no. 3, pp. 152–162, 2016.

[111] T. E. Patterson, Y. Sakai, M. D. Grabiner et al., "Exposure of murine cells to pulsed electromagnetic fields rapidly activates the mTOR signaling pathway," *Bioelectromagnetics*, vol. 27, no. 7, pp. 535–544, 2006.

[112] P. Diniz, K. Soejima, and G. Ito, "Nitric oxide mediates the effects of pulsed electromagnetic field stimulation on the osteoblast proliferation and differentiation," *Nitric Oxide: Biology and Chemistry*, vol. 7, no. 1, pp. 18–23, 2002.

[113] P. Diniz, K. Shomura, K. Soejima, and G. Ito, "Effects of Pulsed Electromagnetic Field (PEMF) Stimulation on Bone Tissue Like Formation Are Dependent on the Maturation Stages of the Osteoblasts," *Bioelectromagnetics*, vol. 23, no. 5, pp. 398–405, 2002.

[114] C.-C. Lin, R.-W. Lin, C.-W. Chang, G.-J. Wang, and K.-A. Lai, "Single-pulsed electromagnetic field therapy increases osteogenic differentiation through Wnt signaling pathway and sclerostin downregulation," *Bioelectromagnetics*, vol. 36, no. 7, pp. 494–505, 2015.

[115] N. Selvamurugan, S. Kwok, A. Vasilov, S. C. Jefcoat, and N. C. Partridge, "Effects of BMP-2 and pulsed electromagnetic field (PEMF) on rat primary osteoblastic cell proliferation and gene expression," *Journal of Orthopaedic Research*, vol. 25, no. 9, pp. 1213–1220, 2007.

[116] R. A. Hopper, J. P. Verhalen, O. T. Tepper et al., "Osteoblasts stimulated with pulsed electromagnetic fields increase HUVEC

proliferation via a VEGF-A independent mechanism," *Bioelectromagnetics*, vol. 30, no. 3, pp. 189–197, 2009.

[117] S. Barnaba, R. Papalia, L. Ruzzini, A. Sgambato, N. Maffulli, and V. Denaro, "Effect of pulsed electromagnetic fields on human osteoblast cultures," *Physiotherapy Research International*, vol. 18, no. 2, pp. 109–114, 2013.

[118] S. Ehnert, M. van Griensven, M. Unger et al., "Co-Culture with Human Osteoblasts and Exposure to Extremely Low Frequency Pulsed Electromagnetic Fields Improve Osteogenic Differentiation of Human Adipose-Derived Mesenchymal Stem Cells," *International Journal of Molecular Sciences*, vol. 19, no. 4, p. 994, 2018.

[119] S. Ehnert, K. Falldorf, A.-K. Fentz et al., "Primary human osteoblasts with reduced alkaline phosphatase and matrix mineralization baseline capacity are responsive to extremely low frequency pulsed electromagnetic field exposure - Clinical implication possible," *Bone Reports*, vol. 3, pp. 48–56, 2015.

[120] G. Ceccarelli, N. Bloise, M. Mantelli et al., "A comparative analysis of the in vitro effects of pulsed electromagnetic field treatment on osteogenic differentiation of two different mesenchymal cell lineages," *BioResearch Open Access*, vol. 2, no. 4, pp. 283–294, 2013.

[121] L. Ferroni, I. Tocco, A. De Pieri et al., "Pulsed magnetic therapy increases osteogenic differentiation of mesenchymal stem cells only if they are pre-committed," *Life Sciences*, vol. 152, pp. 44–51, 2016.

[122] Y. Yin, P. Chen, Q. Yu, Y. Peng, Z. Zhu, and J. Tian, "The Effects of a Pulsed Electromagnetic Field on the Proliferation and Osteogenic Differentiation of Human Adipose-Derived Stem Cells," *Medical Science Monitor*, vol. 24, pp. 3274–3282, 2018.

[123] Y.-C. Fu, C.-C. Lin, J.-K. Chang et al., "A novel single pulsed electromagnetic field stimulates osteogenesis of bone marrow mesenchymal stem cells and bone repair," *PLoS ONE*, vol. 9, no. 3, Article ID e91581, 2014.

[124] L. Petecchia, F. Sbrana, R. Utzeri et al., "Electro-magnetic field promotes osteogenic differentiation of BM-hMSCs through a selective action on Ca^{2+}-related mechanisms," *Scientific Reports*, vol. 5, 2015.

[125] M. Jazayeri, M. A. Shokrgozar, N. Haghighipour, B. Bolouri, F. Mirahmadi, and M. Farokhi, "Effects of electromagnetic stimulation on gene expression of mesenchymal stem cells and repair of bone lesions," *Cell*, vol. 19, no. 1, pp. 34–44, 2017.

[126] M.-T. Tsai, W.-J. Li, R. S. Tuan, and W. H. Chang, "Modulation of osteogenesis in human mesenchymal stem cells by specific pulsed electromagnetic field stimulation," *Journal of Orthopaedic Research*, vol. 27, no. 9, pp. 1169–1174, 2009.

[127] L.-Y. Sun, D.-K. Hsieh, P.-C. Lin, H.-T. Chiu, and T.-W. Chiou, "Pulsed electromagnetic fields accelerate proliferation and osteogenic gene expression in human bone marrow mesenchymal stem cells during osteogenic differentiation," *Bioelectromagnetics*, vol. 219, no. 75, pp. 209–219, 2009.

[128] L.-Y. Sun, D.-K. Hsieh, T.-C. Yu et al., "Effect of pulsed electromagnetic field on the proliferation and differentiation potential of human bone marrow mesenchymal stem cells," *Bioelectromagnetics*, vol. 30, no. 4, pp. 251–260, 2009.

[129] J. H. W. Jansen, O. P. van der Jagt, B. J. Punt et al., "Stimulation of osteogenic differentiation in human osteoprogenitor cells by pulsed electromagnetic fields: an in vitro study," *BMC Musculoskeletal Disorders*, vol. 11, no. 1, p. 188, 2010.

[130] E. Kaivosoja, V. Sariola, Y. Chen, and Y. T. Konttinen, "The effect of pulsed electromagnetic fields and dehydroepiandrosterone

on viability and osteo-induction of human mesenchymal stem cells," *Journal of Tissue Engineering and Regenerative Medicine*, vol. 9, no. 1, pp. 31–40, 2015.

[131] N. Selvamurugan, Z. He, D. Rifkin, B. Dabovic, and N. C. Partridge, "Pulsed Electromagnetic Field Regulates MicroRNA 21 Expression to Activate TGF- β Signaling in Human Bone Marrow Stromal Cells to Enhance Osteoblast Differentiation," *Stem Cells International*, vol. 2017, Article ID 2450327, 17 pages, 2017.

[132] Z. He, N. Selvamurugan, J. Warshaw, and N. C. Partridge, "Pulsed electromagnetic fields inhibit human osteoclast formation and gene expression via osteoblasts," *Bone*, vol. 106, pp. 194–203, 2018.

[133] M. Esposito, A. Lucariello, I. Riccio, V. Riccio, V. Esposito, and G. Riccardi, "Differentiation of human osteoprogenitor cells increases after treatment with pulsed electromagnetic fields," *In Vivo (Brooklyn)*, vol. 26, no. 2, pp. 299–304, 2012.

[134] M. De Mattei, A. Caruso, G. C. Traina, F. Pezzetti, T. Baroni, and V. Sollazzo, "Correlation between pulsed electromagnetic fields exposure time and cell proliferation increase in human osteosarcoma cell lines and human normal osteoblast cells in vitro," *Bioelectromagnetics*, vol. 20, no. 3, pp. 177–182, 1999.

[135] H.-Y. Lin and Y.-J. Lin, "In vitro effects of low frequency electromagnetic fields on osteoblast proliferation and maturation in an inflammatory environment," *Bioelectromagnetics*, vol. 32, no. 7, pp. 552–560, 2011.

[136] J. Wang, N. Tang, Q. Xiao et al., "Pulsed electromagnetic field may accelerate in vitro endochondral ossification," *Bioelectromagnetics*, vol. 36, no. 1, pp. 35–44, 2015.

[137] L. Bagheri, A. Pellati, P. Rizzo et al., "Notch pathway is active during osteogenic differentiation of human bone marrow mesenchymal stem cells induced by pulsed electromagnetic fields," *Journal of Tissue Engineering and Regenerative Medicine*, vol. 12, no. 2, pp. 304–315, 2017.

[138] R. A. Luben, C. D. Cain, M. C.-Y. Chen, D. M. Rosen, and W. R. Adey, "Effects of electromagnetic stimuli on bone and bone cells in vitro: Inhibition of responses to parathyroid hormone by low-energy low-frequency fields," *Proceedings of the National Acadamy of Sciences of the United States of America*, vol. 79, no. 13, pp. 4180–4184, 1982.

[139] C. H. Lohmann, Z. Schwartz, Y. Liu et al., "Pulsed electromagnetic fields affect phenotype and connexin 43 protein expression in MLO-Y4 osteocyte-like cells and ROS 17/2.8 osteoblast-like cells," *Journal of Orthopaedic Research*, vol. 21, no. 2, pp. 326–334, 2003.

[140] T. Shimizu, J. E. Zerwekh, T. Videman et al., "Bone ingrowth into porous calcium phosphate ceramics: Influence of pulsing electromagnetic field," *Journal of Orthopaedic Research*, vol. 6, no. 2, pp. 248–258, 1988.

[141] V. Ottani, M. Raspanti, D. Martini et al., "Electromagnetic stimulation on the bone growth using backscattered electron imaging," *Micron*, vol. 33, no. 2, pp. 121–125, 2002.

[142] M. Fini and R. Cadossi, "The effect of pulsed electromagnetic fields on the osteointegration of hydroxyapatite implants in cancellous bone: a morphologic and microstructural in vivo study," *Journal of Orthopaedic Research*, vol. 20, pp. 756–763, 2002.

[143] M. Fini, G. Giavaresi, R. Giardino, F. Cavani, and R. Cadossi, "Histomorphometric and mechanical analysis of the hydroxyapatite-bone interface after electromagnetic stimulation," *The Journal of Bone & Joint Surgery (British Volume)*, vol. 88-B, no. 1, pp. 123–128, 2006.

[144] L. Fassina, E. Saino, M. S. Sbarra et al., "In vitro electromagnetically stimulated SAOS-2 osteoblasts inside porous hydroxyapatite," *Journal of Biomedical Materials Research Part A*, vol. 93, no. 4, pp. 1272–1279, 2010.

[145] Z. Schwartz, B. J. Simon, M. A. Duran, G. Barabino, R. Chaudhri, and B. D. Boyan, "Pulsed electromagnetic fields enhance BMP-2 dependent osteoblastic differentiation of human mesenchymal stem cells," *Journal of Orthopaedic Research*, vol. 26, no. 9, pp. 1250–1255, 2008.

[146] Z. Schwartz, M. Fisher, C. H. Lohmann, B. J. Simon, and B. D. Boyan, "Osteoprotegerin (OPG) production by cells in the osteoblast lineage is regulated by pulsed electromagnetic fields in cultures grown on calcium phosphate substrates," *Annals of Biomedical Engineering*, vol. 37, no. 3, pp. 437–444, 2009.

[147] T. J. Martin and N. A. Sims, "RANKL/OPG; Critical role in bone physiology," *Reviews in Endocrine and Metabolic Disorders*, vol. 16, no. 2, pp. 131–139, 2015.

[148] W. S. Simonet, D. L. Lacey, C. R. Dunstan et al., "Osteoprotegerin: a novel secreted protein involved in the regulation of bone density," *Cell*, vol. 89, no. 2, pp. 309–319, 1997.

[149] H. Yasuda, N. Shima, N. Nakagawa et al., "Osteoclast differentiation factor is a ligand for osteoprotegerin/osteoclastogenesis-inhibitory factor and is identical to TRANCE/RANKL," in *Proceedings of the National Academy of Sciences of the USA*, vol. 95, pp. 3597–3602, 1998.

[150] E. Tsuda, M. Goto, S.-I. Mochizuki et al., "Isolation of a novel cytokine from human fibroblasts that specifically inhibits osteoclastogenesis," *Biochemical and Biophysical Research Communications*, vol. 234, no. 1, pp. 137–142, 1997.

[151] K. Chang, W. H.-S. Chang, S. Huang, S. Huang, and C. Shih, "Pulsed electromagnetic fields stimulation affects osteoclast formation by modulation of osteoprotegerin, RANK ligand and macrophage colony-stimulating factor," *Journal of Orthopaedic Research*, vol. 23, no. 6, pp. 1308–1314, 2005.

[152] W. H.-S. Chang, L.-T. Chen, J.-S. Sun, and F.-H. Lin, "Effect of pulse-burst electromagnetic field stimulation on osteoblast cell activities," *Bioelectromagnetics*, vol. 25, no. 6, pp. 457–465, 2004.

[153] J. A. Spadaro, S. A. Albanese, and S. E. Chase, "Electromagnetic effects on bone formation at implants in the medullary canal in rabbits," *Journal of Orthopaedic Research*, vol. 8, no. 5, pp. 685–693, 1990.

[154] L. Fassina, E. Saino, L. Visai et al., "Electromagnetic enhancement of a culture of human SAOS-2 osteoblasts seeded onto titanium fiber-mesh scaffolds," *Journal of Biomedical Materials Research Part A*, vol. 87, no. 3, pp. 750–759, 2008.

[155] L. Fassina, E. Saino, L. Visai, and G. Magenes, "Electromagnetically enhanced coating of a sintered titanium grid with human SAOS-2 osteoblasts and extracellular matrix," in *Proceedings of the 30th Annual International Conference of the IEEE Engineering in Medicine and Biology Society, EMBS'08*, pp. 3582–3585, IEEE, August 2008.

[156] J. Wang, Y. An, F. Li et al., "The effects of pulsed electromagnetic field on the functions of osteoblasts on implant surfaces with different topographies," *Acta Biomaterialia*, vol. 10, no. 2, pp. 975–985, 2014.

[157] N. Bloise, L. Petecchia, G. Ceccarelli et al., "The effect of pulsed electromagnetic field exposure on osteoinduction of human mesenchymal stem cells cultured on nano-TiO2 surfaces," *PLoS ONE*, vol. 13, no. 6, Article ID e0199046, 2018.

[158] R. Carbone, I. Marangi, and A. Zanardi, "Biocompatibility of cluster-assembled nanostructured TiO$_2$ with primary and cancer cells," *Biomaterials*, vol. 27, no. 17, pp. 3221–3229, 2006.

[159] M. Vercellino, G. Ceccarelli, and F. Cristofaro, "Nanostructured TiO$_2$ surfaces promote human bone marrow mesenchymal stem cells differentiation to osteoblasts," *Nanomaterials*, vol. 6, no. 7, 2016.

[160] B. Atalay, B. Aybar, M. Ergüven et al., "The Effects of Pulsed Electromagnetic Field (PEMF) on Osteoblast-Like Cells Cultured on Titanium and Titanium-Zirconium Surfaces," *The Journal of Craniofacial Surgery*, vol. 24, no. 6, pp. 2127–2134, 2013.

[161] D. Jing, M. Zhai, S. Tong et al., "Pulsed electromagnetic fields promote osteogenesis and osseointegration of porous titanium implants in bone defect repair through a Wnt/β-catenin signaling-associated mechanism," *Scientific Reports*, vol. 6, Article ID 32045, pp. 1–13, 2016.

[162] E. P. Buzzá, J. A. Shibli, R. H. Barbeiro, and J. R. D. A. Barbosa, "Effects of electromagnetic field on bone healing around commercially pure titanium surface: Histologic and mechanical study in rabbits," *Implant Dentistry*, vol. 12, no. 2, pp. 182–187, 2003.

[163] C. Do Nascimento, J. P. M. Issa, A. S. Da Silva Mello, and R. F. De Albuquerque Junior, "Effect of electromagnetic field on bone regeneration around dental implants after immediate placement in the dog mandible: A pilot study," *Gerodontology*, vol. 29, no. 2, pp. 1249–1251, 2012.

[164] H. Matsumoto, M. Ochi, Y. Abiko, Y. Hirose, T. Kaku, and K. Sakaguchi, "Pulsed electromagnetic fields promote bone formation around dental implants inserted into the femur of rabbits," *Clinical Oral Implants Research*, vol. 11, no. 4, pp. 354–360, 2000.

[165] J. Özen, A. Atay, S. Orucß, M. Dalkiz, B. Beydemir, and S. Develi, "Evaluation of pulsed electromagnetic fields on bone healing after implant placement in the rabbit mandibular model," *Turkish Journal of Medical Sciences*, vol. 34, no. 2, pp. 91–95, 2004.

[166] K. Akca, E. Sarac, U. Baysal, M. Fanuscu, T. Chang, and M. Cehreli, "Micro-morphologic changes around biophysically-stimulated titanium implants in ovariectomized rats," *Head & Face Medicine*, vol. 3, no. 28, 2007.

[167] D. R. Grana, H. J. A. Marcos, and G. A. Kokubu, "Pulsed electromagnetic fields as adjuvant therapy in bone healing and peri-implant bone formation: an experimental study in rats," *Acta Odontologica Latinoamericana*, vol. 21, no. 1, pp. 77–83, 2018.

[168] S. Barak, M. Neuman, G. Iezzi, A. Piattelli, V. Perrotti, and Y. Gabet, "A new device for improving dental implants anchorage: A histological and micro-computed tomography study in the rabbit," *Clinical Oral Implants Research*, vol. 27, no. 8, pp. 935–942, 2016.

[169] A. Y. Chan, "Development of an intra-oral bone growth stimulator for titanium dental implants," in *Proceedings of the Canadian Medical and Biological Engineering Society*, vol. 30, 2007.

[170] F. Bambini, A. Santarelli, and A. Putignano, "Use of supercharged cover screw as static magnetic field generator for bone healing, 2nd part: in vivo enhancement of bone regeneration in rabbits," *Journal of Biologucal Regulators and Homeostatic Agents*, vol. 31, no. 2, pp. 481–485, 2017.

[171] K. Ijiri, S. Matsunaga, and K. Fukuyama, "The effect of pulsing electromagnetic field on bone ingrowth into a porous coated implant," *Anticancer Research*, vol. 16, no. 5A, pp. 2853–2856, 1996.

[172] J. Cai, W. Li, T. Sun, X. Li, E. Luo, and D. Jing, "Pulsed electromagnetic fields preserve bone architecture and mechanical properties and stimulate porous implant osseointegration by promoting bone anabolism in type 1 diabetic rabbits," *Osteoporosis International*, vol. 29, no. 5, pp. 1177–1191, 2018.

[173] M. Arjmand, A. Ardeshirylajimi, H. Maghsoudi, and E. Azadian, "Osteogenic differentiation potential of mesenchymal stem cells cultured on nanofibrous scaffold improved in the presence of pulsed electromagnetic field," *Journal of Cellular Physiology*, vol. 233, no. 2, pp. 1061–1070, 2018.

[174] M.-T. Tsai, W. H.-S. Chang, K. Chang, R.-J. Hou, and T.-W. Wu, "Pulsed electromagnetic fields affect osteoblast proliferation and differentiation in bone tissue engineering," *Bioelectromagnetics*, vol. 28, no. 7, pp. 519–528, 2007.

[175] M. Schnoke and R. J. Midura, "Pulsed electromagnetic fields rapidly modulate intracellular signaling events in osteoblastic cells: Comparison to parathyroid hormone and insulin," *Journal of Orthopaedic Research*, vol. 25, no. 7, pp. 933–940, 2007.

[176] A. Patruno, P. Amerio, M. Pesce et al., "Extremely low frequency electromagnetic fields modulate expression of inducible nitric oxide synthase, endothelial nitric oxide synthase and cyclooxygenase-2 in the human keratinocyte cell line HaCat: Potential therapeutic effects in wound healing," *British Journal of Dermatology*, vol. 162, no. 2, pp. 258–266, 2010.

[177] P. Torricelli, M. Fini, G. Giavaresi, R. Botter, D. Beruto, and R. Giardino, "Biomimetic PMMA-based bone substitutes: A comparativein vitro evaluation of the effects of pulsed electromagnetic field exposure," *Journal of Biomedical Materials Research Part B: Applied Biomaterials*, vol. 64A, no. 1, pp. 182–188, 2003.

[178] F. Veronesi, M. Cadossi, G. Giavaresi et al., "Pulsed electromagnetic fields combined with a collagenous scaffold and bone marrow concentrate enhance osteochondral regeneration: an in vivo study," *BMC Musculoskeletal Disorders*, vol. 16, no. 233, 2015.

[179] H. J. Yang, R. Y. Kim, and S. J. Hwang, "Pulsed electromagnetic fields enhance bone morphogenetic protein-2 dependent-bone regeneration," *Tissue Engineering Part: A*, vol. 21, no. 19, pp. 2629–2637, 2015.

[180] L. Fassina, E. Saino, L. Visai et al., "Electromagnetic Stimulation to Optimize the Bone Regeneration Capacity of Gelatin-Based Cryogels," *International Journal of Immunopathology and Pharmacology*, vol. 25, no. 1, pp. 165–174, 2012.

[181] L. Fassina, E. Saino, L. Visai, and G. Magenes, "Electromagnetically enhanced coating of a sintered titanium grid with human SAOS-2 osteoblasts and extracellular matrix," in *Proceedings of the 30th Annual International Conference of the IEEE Engineering in Medicine and Biology Society, EMBS'08*, pp. 3582–3585, August 2008.

[182] E. D. Silva, P. S. Babo, R. Costa-Almeida et al., "Multifunctional magnetic-responsive hydrogels to engineer tendon-to-bone interface," *Nanomedicine: Nanotechnology, Biology and Medicine*, 2017.

[183] S. Kondaveeti, A. T. Semeano, D. R. Cornejo, H. Ulrich, and D. F. Petri, "Magnetic hydrogels for levodopa release and cell stimulation triggered by external magnetic field," *Colloids and Surfaces B: Biointerfaces*, vol. 167, pp. 415–424, 2018.

[184] S. Kondaveeti, D. R. Cornejo, and D. F. S. Petri, "Alginate/magnetite hybrid beads for magnetically stimulated release of dopamine," *Colloids and Surfaces B: Biointerfaces*, vol. 138, pp. 94–101, 2016.

13

Fabrication of Polycaprolactone/Polyurethane Loading Conjugated Linoleic Acid and its Antiplatelet Adhesion

Ho Hieu Minh,[1] **Nguyen Thi Hiep,**[1] **Nguyen Dai Hai,**[2,3] **and Vo Van Toi**[1]

[1]*Tissue Engineering and Regenerative Medicine Laboratory, Department of Biomedical Engineering,*
International University of Vietnam National Universities, Ho Chi Minh City 700000, Vietnam
[2]*Institute of Applied Materials Science, Vietnam Academy of Science and Technology, 01 Mac Dinh Chi, District 1,*
Ho Chi Minh City, Vietnam
[3]*Graduate University of Science and Technology, Vietnam Academy of Science and Technology, Hanoi, Vietnam*

Correspondence should be addressed to Nguyen Thi Hiep; nthiep1981@gmail.com

Academic Editor: Kheng-Lim Goh

Polycaprolactone/polyurethane (PCL/PU) fibrous scaffold was loaded with conjugated linoleic acid (CLA) by electrospinning method to improve the hemocompatibility of the polymeric surface. Fourier Transform Infrared Spectroscopy (FT-IR) analysis and Scanning Electron Microscopy (SEM) observation were employed to characterize the chemical structure and the changing morphology of electrospun PCL/PU and PCL/PU loaded with CLA (PCL/PU-CLA) scaffolds. Platelet adhesion and whole blood clot formation tests were used to evaluate the effect of CLA on antithrombotic property of PCL/PU-CLA scaffold. Endothelial cells (EC) were also seeded on the scaffold to examine the difference in the morphology of EC layer and platelet attachment with and without the presence of CLA. SEM results showed that CLA supported the spreading and proliferation of EC and PCL/PU-CLA surface induced lower platelet adhesion as well as attachment of other blood cells compared to the PCL/PU one. These results suggest that electrospinning method can successfully combine the antiplatelet effects of CLA to improve hemocompatibility of PCL/PU scaffolds for applications in artificial blood vessels.

1. Introduction

Various approaches have been investigated to improve blood compatibility of polymeric surfaces for artificial blood vessels (ABVs) application including chemical and biological modification [1, 2]. However, polymeric materials for small ABVs are still undergoing investigation and most of them induce thrombosis [3], infection, and calcination [4]. Among these, the main cause of failure in ABVs is thrombosis [5]. When implanted inside human body, polymeric materials exposed to physiological fluids can initiate a complex cascade of surface induced thrombotic events [6] as follows: adhesion and activation of platelets result in liberation of different agents such as adenosine diphosphate (ADP), arachidonic acid (AA), thromboxane A_2 (TXA_2), and thrombin, which help circulating fibrinogen bind to platelet scaffold and connect platelets together [7]. The activities of these aggregating agents act as signal for further platelets' activation and aggregation to form a loose plug at wounded area. Then thrombin converts fibrinogen into fibrin to form a mesh-like stable plug, which is called a thrombus.

As explained above, fibrinogen is a key factor in thrombus formation. Fibrous coagulant in the blood significantly increases the risk of cardiovascular diseases, one of the leading causes of death and disabilities including hypertension, ischemia, myocardial infarction, stroke, and limb loss. Platelets also have a very important role in hemostasis and thrombogenesis. Platelet is a catalyzing coagulation reaction agent, which leads to the formation of fibrin, but does not react negatively with other blood cells [8]. Therefore, one of the requirements of polymeric materials for ABVs application is antithrombus property, including antiplatelet adhesion, and reduces blood protein (fibrinogen) attraction.

To enhance antithrombotic property, ABVs polymeric surfaces can be modified to become very hydrophobic (e.g., PDMS graft) [9] or hydrophilic (e.g., PEG graft) [10, 11]

or using anticoagulation agents such as heparin [12] and albumin [13]. There is another antiplatelet agent, CLA, which can improve the hemocompatibility of biomaterials. In 2006, CLA grafted to polyacrylonitrile (PAN-CLA) to improve their hemocompatibility was investigated by Kung et al. The result showed that PNA-CLA offered a new composite for hemodialysis [14, 15]. CLA is both a cis and trans unsaturated fatty acid and has been known as famous drug for antithrombus [16]. CLA was found as inhibitor of platelet aggregation induced by different aggregating agents such as collagen, ADP, AA, and thrombin [17]. CLA was also found to decrease activity of cyclooxygenase-1 (COX_1), an enzyme which converts AA into TXA_2, resulting in opposition to platelet's aggregation [18].

In our previous study, a hybrid electrospun PU/PCL scaffold satisfying the requirements of blood vessel prosthesis with suitable mechanical properties, sufficient pore size (ranging from 5 to 150 μm) for nutrient diffusion, and high biocompatibility was successfully fabricated [19]. The result showed that the PCL/PU scaffold had good biocompatibility and mechanical properties. Blend PCL/PU composite has high potential for ABV scaffold because PCL provides favorable EC attachment and proliferation [20] combined with high tensile strength, pressure strength, and blood compatibility (blood-contacting devices) of PU [21]. In continuation for this research, the electrospun PCL/PU scaffold was loaded with CLA due to its inhibitory effects on platelet function which is expected to offer a new hemocompatibility component for ABV applications. Electrospinning is a technique that can produce polymeric fibers from polymer solution by using electric force and heat to drive the spinning process [22]. Electrospinning is one of the simplest among all methods for preparation of fibrous mat used widely for a lot of biomedical applications [23]. Electrospun fibers not only mimic extra cellular matrix but also play another role as carrier of anticoagulant agent like in this investigation and were found to support mechanical and burst strength for ABVs in our previous paper. Several methods can be used to introduce CLA to PCL/PU scaffold such as blending, coating, and grafting [24]. However, blending was chosen due to its simplicity and not requiring complicated apparatus.

The ultimate goal of this investigation was to employ electrospinning method to fabricate electrospun PCL/PU loading CLA scaffolds for ABV applications. The direct addition of CLA to the PCL/PU blend is attempted to enhance blood compatibility of polymeric materials implanted inside human body. Hemocompatibility of electrospun PCL/PU and PCL/PU-CLA scaffolds was evaluated with blood clotting and platelet adhesion tests using fresh human blood.

2. Experimental Procedure

2.1. Materials.
As starting materials, polyurethane (PU, Sigma), polycaprolactone (PCL, Mn 80,000, Sigma), conjugated linoleic acid (CLA, Sigma), tetrahydrofuran (THF, minimum 99%, Sigma), dimethylformamide (DMF, 99%, Sigma), ethanol (EtOH, 99%, Merck), and phosphate buffer saline (PBS, GIBCO) were used.

2.2. Preparation of PU/PCL and PU/PCL-CLA Electrospun Fibrous Scaffolds.
PCL/PU and PCL/PU-CLA scaffolds were fabricated as mentioned in our previous study [19]. In brief, electrospinning of PCL/PU and PCL/PU-CLA scaffolds was fabricated by 12 wt% of blend PCL : PU (1:1) in DMF : THF (1:1) solution. Then, 0.2 ml CLA was added to 10 ml polymer solution. The solution was stirred for 12 hours in room temperature until azeotrope solution. The solutions were electrospun directly with 27 kV power supply (NNC-30 kV-2 mA portable type, Korea) using vertical electrospinning setup. A grounded steel cylinder, 15 cm away from the tip of the syringe needle (ID = 0.25 mm), was used for collection of the nanofiber mats. Flow rates of the PCL/PU solutions (0.5 ml/hour) were controlled by syringe pump (luer-lock type, Korea).

2.3. Structure Characterization

2.3.1. Morphology Analysis.
Morphology of electrospun PCL/PU and PCL/PU-CLA fibrous mats was observed by SEM (JEOL JSM-IT100, Japan) with gold sputter coating (JEOL Smart Coater, Japan).

2.3.2. Fourier Transform Infrared Spectroscopy (FT-IR).
Electrospun PCL/PU and PCL/PU-CLA mats were characterized by attenuated reflectance Fourier transform spectroscopy (Spectrum GX, PerkinElmer, USA). The infrared spectra of the samples were measured over a wavelength range of 4,000–400 cm^{-1}. All spectra were taken in the spectral range by the accumulation of 64 scans with a resolution of 4 cm^{-1}.

2.4. Hemocompatibility Testing.
For this test, specimens of sample were fixed (circle shape, $R = 1$ cm).

2.4.1. Platelet Adhesion Test.
Platelet adhesion test was done on four types of scaffolds: EC seeded (PCL/PU/EC and PCL/PU-CLA/EC) and non-EC seeded (PCL/PU and PCL/PU-CLA) before addition of platelets. The purpose of seeding EC on these scaffolds is based on the fact that when ABVs are implanted in human body, migration, adhesion, and proliferation will occur creating a single layer of EC on ABVs surface. Therefore the difference in antithrombotic property of scaffolds with and without presence of EC must be investigated. For preparation of cell-seeded scaffolds, healthy EC were prepared carefully by changing media regularly every 2 days and keeping them below 90% cell confluence. Then EC were washed by PBS and detached by trypsin-EDTA. Addition of fresh media was done to make a suspension of cells and the pellet was obtained after centrifuging and was resuspended. The final suspension was used for seeding cells. 10^5 of EC were seeded on 1 cm^2 electrospun scaffolds which had been precultured for 14 days before using in platelet adhesion testing. The cultured media was changed every 3 days to obtain healthy sheet of EC on electrospun scaffolds.

For preparation of platelets, a volume of 180 ml fresh human blood (approved and supplied by Hospital Ethics Committee of Blood Transfusion Hematology Hospital, Ho

(a) (b)

FIGURE 1: SEM morphology of electrospun PCL/PU (a) and PCL/PU-CLA (b) scaffolds.

Chi Minh City, Vietnam) containing 20 ml of 3.8 v/v% sodium citrate in PBS solution as anticoagulant (sodium citrate: blood ratio, 1:9) was centrifuged at 4°C, 700 g, for 20 min [25]. Electrospun PCL/PU and PCL/PU-CLA scaffolds (with and without EC seeding) were equilibrated with PBS overnight before immersing in platelets at 37°C with shaking (100 rpm) in an incubator. After 8 h of incubation, the samples were taken out, rinsed five times with PBS, and fixed by immersing in 2% v/v glutaraldehyde in PBS solution at room temperature for 2 h. After fixation, the samples were dehydrated with a series of ethanol solutions (50, 60, 70, 80, 90, and 100 v/v%) for 15 min per each step. Then they were dried in atmosphere overnight. The dried samples were coated with evaporated gold, and the adherent platelets were observed with SEM (JEOL, Japan).

2.4.2. Whole Blood Clotting Formation. After washing thoroughly with PBS, PCL/PU and PCL/PU-CLA scaffolds were immersed twice in citrated human whole blood. Then the scaffolds were washed again with PBS before being incubated in recalcified whole blood (citrated human whole blood which contained 0.010 M $CaCl_2$) for 20 min at room temperature. Prior to this testing, these surfaces were washed extensively with PBS until a prothrombin time for a dilution of the PBS wash into plasma was observed to be normal compared to controls [26] and then examined for the presence of thrombus by camera (Nikon P90). For SEM observation, the adsorbed whole blood on the electrospun scaffolds was rinsed five times with PBS and fixed by immersing the sample in 2 v/v% glutaraldehyde in PBS solution at room temperature for 2 h. After that fixation, the samples were dehydrated as foresaid. Then the dried samples were observed with SEM (JEOL, Japan).

3. Results and Discussion

Although CLA has been known to possess antithrombotic properties both in vitro and in vivo for many years, very little research have developed CLA as an additive to increase hemocompatibility of ABVs. Furthermore, the number of methods used to immobilize CLA onto ABVs' surface is still very limited, including esterification [27] and grafting [14]. In our previous study, electrospinning method was

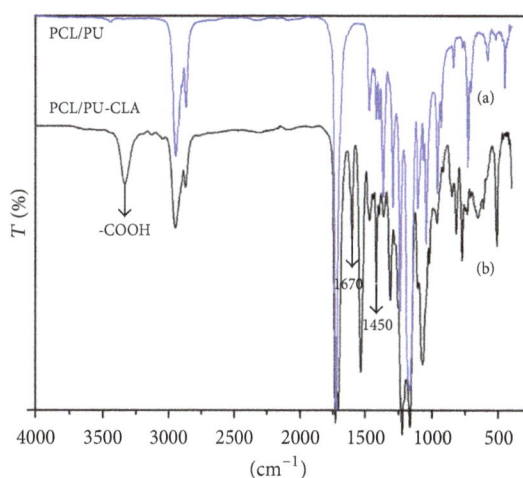

FIGURE 2: FT-IR spectrum of electrospun PCL/PU (a) and PCL/PU-CLA (b) scaffolds.

chosen to fabricate PCL/PU scaffold and in our current study CLA loading onto the membrane is carried out due to its advantage in controlling chemical ratios and examining the variation for different types of scaffolds. Figure 1 shows the difference of SEM morphology between electrospun PCL/PU (Figure 1(a)) and PCL/PU-CLA (Figure 1(b)) fiber mats. The fiber's diameters in both types of scaffolds are approximately 1 μm; however the number of fibers with such diameter size is greater for PCL/PU-CLA scaffold than the PCL/PU one. In addition, comparing the two morphologies, PCL/PU-CLA fibers were distributed with linear pattern while PCL/PU fibers convolute on top with diameter size varying in a wide range. Addition of CLA to PU/PCL solution electrolytes increases the electrical conductivity of polymer solution. That explains why average diameter size for PCL/PU-CLA fibers is smaller (1 μm) while the number of fibers at such diameter size for PCL/PU scaffold is less. This agrees with what was found in Angammana and Jayaram report: average fiber diameter decreases with the increase in electrical conductivity of the solution [28].

The existence of CLA loaded on PCL/PU scaffold was confirmed by employing FT-IR analysis. Typical FT-IR spectra of electrospun PCL/PU and PCL/PU-CLA mats are shown in Figure 2. From electrospun PCL/PU mat (Figure 2(a)),

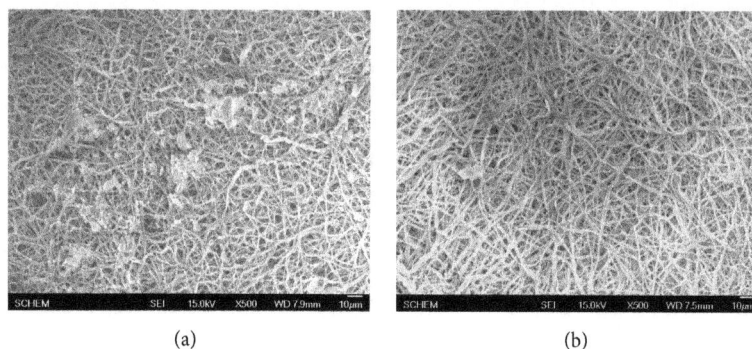

(a)

(b)

FIGURE 3: SEM morphology of platelets adhered on electrospun PCL/PU (a) and PCL/PU-CLA (b) scaffolds.

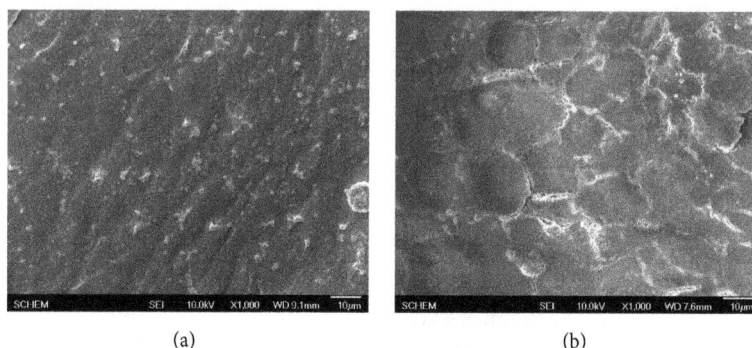

(a)

(b)

FIGURE 4: SEM morphology of platelets adhered on electrospun PCL/PU (a) and PCL/PU-CLA (b) scaffolds seeded with endothelial cells for 14 days.

peaks appearing for PU hard segments are observed at $3320 \, cm^{-1}$ (urethane N-H, stretch), $1735 \, cm^{-1}$ (urethane C=O, free from hydrogen bonding), $1710 \, cm^{-1}$ (urethane C=O, hydrogen bonded), and $1535 \, cm^{-1}$ (C-N-H, bending) [29]. The characteristic absorbance of ester in PCL is shown at $1734 \, cm^{-1}$ [30]. However, peaks $1670 \, cm^{-1}$ and $1450 \, cm^{-1}$ are detected due to the C=C and C=O groups of CLA [14] in the FT-IR spectrum of the PCL/PU-CLA mat (Figure 2(b)), apart from peaks of PU and PCL as described above. In addition, an obvious peak is also found at $3300 \, cm^{-1}$ representing the O-H stretch in carboxyl group of CLA. Thus it proves that PCL/PU scaffold loaded with CLA was fabricated successfully by using electrospinning method.

Platelet adhesion on ABVs' surface is an essential test in evaluation of hemocompatibility of biomaterials as platelets play the main role in the formation of blood clot. In this study, we performed various simple tests which used both platelet rich plasma to examine the effect of PCL/PU-CLA scaffold on platelet function and human fresh whole blood to evaluate the adhesion of other blood cells. Interaction of platelets with electrospun PCL/PU and PCL/PU-CLA surfaces was examined using platelet rich plasma (PRP) prepared from human whole blood (Figures 3 and 4). In Figure 3, SEM morphology images show that platelets deposited and aggregated onto the fibrous PCL/PU scaffold (Figure 3(a)) while electrospun PCL/PU-CLA scaffold exhibits little platelet adhesion (Figure 3(b)). Based on the SEM images, electrospun PCL/PU-CLA displayed high antiplatelet activity with

nonplatelet adhesion. In addition, to mimic the conditions of human body, endothelial cells were seeded on the scaffold and the difference in the effect of CLA with and without EC was also examined. To examine antiplatelet ability after the formation and proliferation of EC, PCL/PU and PCL/PU-CLA scaffolds were seeded with EC and the results of platelet adhesive behavior of EC seeded scaffolds are shown in Figure 4. To ensure that EC covered the scaffold surface, electrospun PCL/PU and PCL/PU-CLA were incubated for 14 days before testing. The results show that electrospun PCL/PU scaffold was good for EC spreading and elongation, the sheet of EC covered with dense surface. However, electrospun PCL/PU/EC also attracted platelet adhesion as shown in Figure 4(a) with distribution of platelets on EC layer. In contrast, although electrospun PCL/PU-CLA scaffold did not create an excellent sheet of EC as compared to the PCL/PU, platelet adhesion was not found on electrospun PCL/PU-CLA/EC scaffold (Figure 4(b)).

The initial step of platelet adhesion in thrombogenesis is very important and determines mural thrombosis that occurs assisted by other blood cells in later steps [8]. To make sure CLA has the ability of antithrombus, electrospun PCL/PU and PCL/PU-CLA scaffolds after incubation in whole fresh blood were washed thoroughly in PBS for detection of thrombus. The result shows that thrombus formation is reduced at test times of 15, 30, and 60 mins on PCL/PU-CLA scaffold as compared to the PCL/PU one (Figure 5). This proves that CLA were loaded successfully and supported electrospun

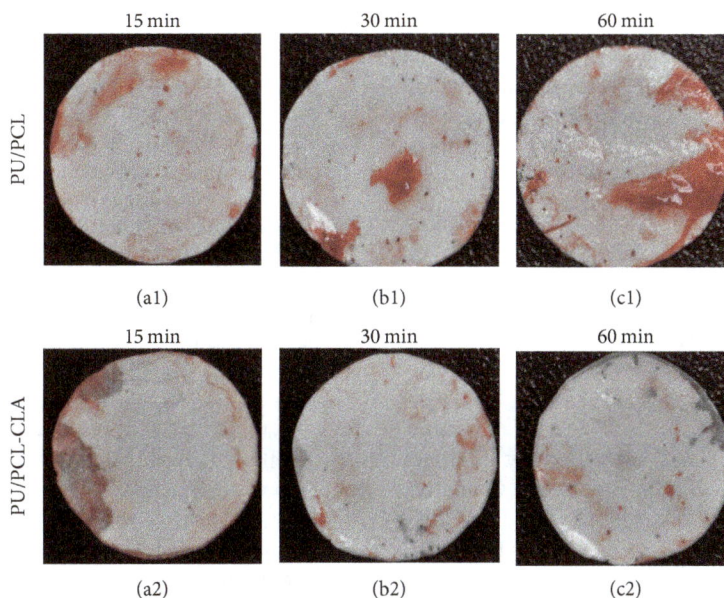

FIGURE 5: Photographs of whole blood clotting on electrospun PCL/PU (a1, b1, and c1) and PCL/PU-CLA (a2, b2, and c2) scaffolds for 15, 30, and 60 minutes.

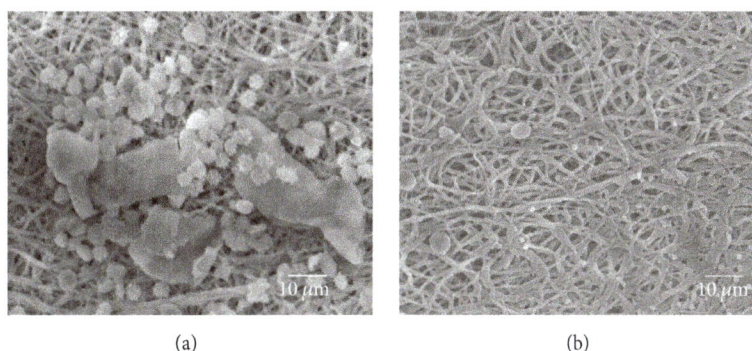

FIGURE 6: SEM morphology of whole fresh blood adhered on electrospun PCL/PU (a) and PCL/PU-CLA (b) scaffolds.

PCL/PU scaffold improving its antithrombogenic activity. Under the same conditions, red blood cells may comprise a large proportion of total thrombus mass and contribute to chemical factors that influence platelet reactivity [8]. Figure 6(a) shows that electrospun PCL/PU scaffold induced platelet adhesion resulting in higher level of deposition and aggregation of red blood cells and white blood cells on the scaffold. Meanwhile, electrospun PCL/PU-CLA scaffold did not attract platelets and only a few blood cells were observed (Figure 6(b)).

The results show that electrospun PCL/PU-CLA scaffold significantly reduces the adhesion of platelets and almost no red or white blood cells as well as no thrombus are found on the scaffold. In addition, although CLA was shown to have less support for the proliferation of EC, its antiplatelet property was retained with the presence of EC on the scaffold. These findings confirm again the antiplatelet and antithrombotic properties of CLA, suggesting that CLA can be used as a new additive agent to increase hemocompatibility of ABVs and therefore decrease the amount of anticoagulant injected to human body after the implantation. From this study, electrospinning is also proved to be effective in PCL/PU scaffold fabrication and loading CLA as anticoagulant agent, which offers a new direction for different research on blood-contacting polymeric biomaterials.

4. Conclusion

In this study, PCL/PU scaffold loaded with CLA was electrospun as substrate for ABV applications. FT-IR results demonstrated that CLA was successfully immobilized in PCL/PU blend. Electrospun PCL/PU-CLA fibers were also found to be smoother and bind to each other as compared to those without CLA. Hemocompatibility tests showed that PCL/PU-CLA scaffold significantly decreased platelet adhesion and thrombus formation with no attachment of red and white blood cells. SEM morphology of endothelial cell layer on both PCL/PU and PCL/PU-CLA scaffolds demonstrated that CLA had less support of EC spreading and elongation but the antiplatelet property of CLA was retained with the presence of EC. These results show that CLA enhances hemocompatibility of PCL/PU scaffold in terms of platelet adhesion and thrombus formation. Further research which focuses on investigating CLA concentration to optimize the anticoagulant effect of CLA in PCL/PU scaffold needs to be conducted before being used in human body.

Conflicts of Interest

The authors declare that there are no conflicts of interest regarding the publication of this paper.

Acknowledgments

The facility of this research was supported by Vietnam National University, Ho Chi Minh City, under Grant no. 1161/QĐ-ĐHQG-KHCN. The labor was supported by Office of Navy Research (ONR) under Grant no. N62909-14-1-N011-P00001.

References

[1] S. Ravi, Z. Qu, and E. L. Chaikof, "Polymeric materials for tissue engineering of arterial substitutes," *Vascular*, vol. 17, supplement 1, pp. S45-S54, 2009.

[2] V. A. Kumar, L. P. Brewster, J. M. Caves, and E. L. Chaikof, "Tissue engineering of blood vessels: functional requirements, progress, and future challenges," *Cardiovascular Engineering and Technology*, vol. 2, no. 3, pp. 137-148, 2011.

[3] H. Yamanaka, P. Soman, W. J. Weiss, and C. A. Siedlecki, "In-vitro evaluation of blood compatibility of polyurethane biomaterials," *ASAIO Journal*, vol. 52, no. 2, p. 20A, 2006.

[4] L. L. Demer and Y. Tintut, "Vascular calcification: pathobiology of a multifaceted disease," *Circulation*, vol. 117, no. 22, pp. 2938-2948, 2008.

[5] D. G. Castner and B. D. Ratner, "Biomedical surface science: foundations to frontiers," *Surface Science*, vol. 500, no. 1-3, pp. 28-60, 2002.

[6] J. Hong, A. Larsson, K. N. Ekdahl, G. Elgue, R. Larsson, and B. Nilsson, "Contact between a polymer and whole blood: sequence of events leading to thrombin generation," *Journal of Laboratory and Clinical Medicine*, vol. 138, no. 2, pp. 139-145, 2001.

[7] S. Palta, R. Saroa, and A. Palta, "Overview of the coagulation system," *Indian Journal of Anaesthesia*, vol. 58, no. 5, pp. 515-523, 2014.

[8] J. H. Kim and S. C. Kim, "PEO-grafting on PU/PS IPNs for enhanced blood compatibility—effect of pendant length and grafting density," *Biomaterials*, vol. 23, no. 9, pp. 2015-2025, 2002.

[9] S. Pinto, P. Alves, C. M. Matos, A. C. Santos, L. R. Rodrigues, J. A. Teixeira et al., "Poly(dimethyl siloxane) surface modification by low pressure plasma to improve its characteristics towards biomedical applications," *Colloids and Surfaces B: Biointerfaces*, vol. 81, no. 1, pp. 20-26, 2010.

[10] G. A. Abraham, A. A. A. D. Queiroz, and J. S. Román, "Immobilization of a nonsteroidal antiinflammatory drug onto commercial segmented polyurethane surface to improve haemocompatibility properties," *Biomaterials*, vol. 23, no. 7, pp. 1625-1638, 2002.

[11] M. T. Khorasani and H. Mirzadeh, "*In vitro* blood compatibility of modified PDMS surfaces as superhydrophobic and superhydrophilic materials," *Journal of Applied Polymer Science*, vol. 91, no. 3, pp. 2042-2047, 2004.

[12] Z. Yang, J. Wang, R. Luo et al., "The covalent immobilization of heparin to pulsed-plasma polymeric allylamine films on 316L stainless steel and the resulting effects on hemocompatibility," *Biomaterials*, vol. 31, no. 8, pp. 2072-2083, 2010.

[13] S. Guha Thakurta and A. Subramanian, "Evaluation of in situ albumin binding surfaces: a study of protein adsorption and platelet adhesion," *Journal of Materials Science: Materials in Medicine*, vol. 22, no. 1, pp. 137-149, 2011.

[14] F.-C. Kung and M.-C. Yang, "Effect of conjugated linoleic acid grafting on the hemocompatibility of polyacrylonitrile membrane," *Polymers for Advanced Technologies*, vol. 17, no. 6, pp. 419-425, 2006.

[15] A. P. Torres-Duarte and J. Y. Vanderhoek, "Conjugated linoleic acid exhibits stimulatory and inhibitory effects on prostanoid production in human endothelial cells and platelets," *Biochimica et Biophysica Acta: Molecular Cell Research*, vol. 1640, no. 1, pp. 69-76, 2003.

[16] L. Yu, D. Adams, and M. Gabel, "Conjugated linoleic acid isomers differ in their free radical scavenging properties," *Journal of Agricultural and Food Chemistry*, vol. 50, no. 14, pp. 4135-4140, 2002.

[17] P. Benito, G. J. Nelson, D. S. Kelley, G. Bartolini, P. C. Schmidt, and V. Simon, "The effect of conjugated linoleic acid on platelet function, platelet fatty acid composition, and blood coagulation in humans," *Lipids*, vol. 36, no. 3, pp. 221-227, 2001.

[18] G. Li, D. Butz, B. Dong, Y. Park, M. W. Pariza, and M. E. Cook, "Selective conjugated fatty acids inhibit guinea pig platelet aggregation," *European Journal of Pharmacology*, vol. 545, no. 2-3, pp. 93-99, 2006.

[19] T.-H. Nguyen, A. R. Padalhin, H. S. Seo, and B.-T. Lee, "A hybrid electrospun PU/PCL scaffold satisfied the requirements of blood vessel prosthesis in terms of mechanical properties, pore size, and biocompatibility," *Journal of Biomaterials Science, Polymer Edition*, vol. 24, no. 14, pp. 1692-1706, 2013.

[20] M. R. Williamson and A. G. A. Coombes, "Gravity spinning of polycaprolactone fibres for applications in tissue engineering," *Biomaterials*, vol. 25, no. 3, pp. 459-465, 2004.

[21] H. R. Lim, H. S. Baek, M. H. Lee et al., "Surface modification for enhancing behaviors of vascular endothelial cells onto polyurethane films by microwave-induced argon plasma," *Surface and Coatings Technology*, vol. 202, no. 22-23, pp. 5768-5772, 2008.

[22] X. Li, Y. Su, S. Liu, L. Tan, X. Mo, and S. Ramakrishna, "Encapsulation of proteins in poly(L-lactide-co-caprolactone) fibers by emulsion electrospinning," *Colloids and Surfaces B: Biointerfaces*, vol. 75, no. 2, pp. 418-424, 2010.

[23] N. T. Hiep and B.-T. Lee, "Electro-spinning of PLGA/PCL blends for tissue engineering and their biocompatibility," *Journal of Materials Science: Materials in Medicine*, vol. 21, no. 6, pp. 1969-1978, 2010.

[24] E. Salimi, A. Ghaee, A. F. Ismail, M. H. D. Othman, and G. P. Sean, "Current approaches in improving hemocompatibility of polymeric membranes for biomedical application," *Macromolecular Materials and Engineering*, vol. 301, no. 7, pp. 771-800, 2016.

[25] R. Dhurat and M. S. Sukesh, "Principles and methods of preparation of platelet-rich plasma: a review and author's

perspective," *Journal of Cutaneous and Aesthetic Surgery*, vol. 7, no. 4, pp. 189–197, 2014.

[26] P. P. Vicario, Z. Lu, Z. Wang, K. Merritt, D. Buongiovanni, and P. Chen, "Antithrombogenicity of hydromer's polymeric formula F202™ immobilized on polyurethane and electropolished stainless steel," *Journal of Biomedical Materials Research - Part B Applied Biomaterials*, vol. 86, no. 1, pp. 136–144, 2008.

[27] F.-C. Kung and M.-C. Yang, "Effect of conjugated linoleic acid immobilization on the hemocompatibility of cellulose acetate membrane," *Colloids and Surfaces B: Biointerfaces*, vol. 47, no. 1, pp. 36–42, 2006.

[28] C. J. Angammana and S. H. Jayaram, "Analysis of the effects of solution conductivity on electrospinning process and fiber morphology," *IEEE Transactions on Industry Applications*, vol. 47, no. 3, pp. 1109–1117, 2011.

[29] V. Chiono, P. Mozetic, M. Boffito et al., "Polyurethane-based scaffolds for myocardial tissue engineering," *Interface Focus*, vol. 4, no. 1, 2014.

[30] B. Abderrahim, E. Abderrahman, A. Mohamed, T. Fatima, T. Abdesselam, and O. Krim, "Kinetic thermal degradation of cellulose, polybutylene succinate and a green composite: comparative study," *World Journal of Environmental Engineering*, vol. 3, no. 4, 95 pages, 2015.

14

Simplified Surface Treatments for Ceramic Cementation: Use of Universal Adhesive and Self-Etching Ceramic Primer

Heloísa A. B. Guimarães, Paula C. Cardoso, Rafael A. Decurcio, Lúcio J. E. Monteiro, Letícia N. de Almeida ⓘ, Wellington F. Martins, and Ana Paula R. Magalhães ⓘ

Restorative Dentistry, Brazilian Dental Association, Goiânia 74325-110, Brazil

Correspondence should be addressed to Ana Paula R. Magalhães; anapaulardm@gmail.com

Academic Editor: Wen-Cheng Chen

The aim of this study was to evaluate the shear bond strength of resin cement and lithium disilicate ceramic after various surface treatments of the ceramic. Sixty blocks of ceramic (IPS e.max Press, Ivoclar Vivadent) were obtained. After cleaning, they were placed in polyvinyl chloride tubes with acrylic resin. The blocks were divided into six groups (n=10) depending on surface treatment: H/S/A - 10% Hydrofluoric Acid + Silane + Adhesive, H/S -10% Hydrofluoric Acid + Silane, H/S/UA - 10% Hydrofluoric Acid + Silane + Universal Adhesive, H/UA- 10% Hydrofluoric Acid + Universal Adhesive, MBEP/A - Monobond Etch & Prime + Adhesive, and MBEP - Monobond Etch & Prime. The light-cured resin cement (Variolink Esthetic LC, Ivoclar Vivadent) was inserted in a mold placed over the treated area of the ceramics and photocured with an LED for 20 s to produce cylinders (3 mm x 3 mm). The samples were subjected to a shear bond strength test in a universal test machine (Instron 5965) by 0.5 mm/min. ANOVA and Tukey tests showed a statistically significant difference between groups ($p<0.05$). The results of the shear strength test were H/S/A $(9.61\pm2.50)^A$, H/S $(10.22\pm3.28)^A$, H/S/UA $(7.39\pm2.02)^{ABC}$, H/UA $(4.28\pm1.32)^C$, MBEP/A $(9.01\pm1.97)^{AB}$, and MBEP $(6.18\pm2.75)^{BC}$. The H/S group showed cohesive failures, and the H/UA group was the only one that presented adhesive failures. The conventional treatment with hydrofluoric acid and silane showed the best bond strength. The use of a new ceramic primer associated with adhesive bonding obtained similar results to conventional surface treatment, being a satisfactory alternative to replace the use of hydrofluoric acid.

1. Introduction

Currently, several techniques and materials, such as composite resin and porcelain, have been used to correct aesthetic problems. The increasing popularity of the use of ceramic restorations for esthetic treatments is attributed to their superior optical properties, translucency, high mechanical properties, and improved esthetics [1]. Several ceramic systems are available, and glass ceramics reinforced by lithium disilicate have shown excellent clinical outcomes with great optical/mechanical properties and high survival rates over time [2].

The bond established between the ceramic material and the tooth structure is extremely important for success and longevity of ceramic restorations. To achieve a strong and durable bond, it is important to understand the ceramic's internal structure to select the best surface treatment, resin cement, and adhesive system [3]. For veneer cementation, the light-cured resin cement is preferable due to the number of colors available and long-term color stability [4].

For ceramic surface treatment, it is important to create a micromechanical interlock between the ceramic and the resin cement [5, 6]. Although the surface treatment with HF and silane is widely used and accepted for lithium disilicate ceramics [7–12], other alternatives have been proposed to enhance the bond strength between ceramic and resin cement.

The introduction of universal adhesives presents a new simplified approach for this procedure. They contain silane and a monomer called 10-methacryloyloxydecyl dihydrogen phosphate (MDP) that helps bond the ceramic to the resin cement chemically, simplifying the bonding procedure, providing the versatility of a single-bottle product, and reducing the procedure time [13]. Although recent studies [14, 15] have

TABLE 1: Materials used in this study and respective manufactures, compositions, and batch numbers.

Material	Manufacture	Composition	#Batch number
Condac	FGM, Joinville, Brazil	10% hydrofluoric acid	060917
Monobond N	Ivoclar Vivadent, Shaan, Liechtenstein	Ethanol, 3-trimethoxysilylpropyl methacrylate, 10-MDP, disulfide acrylate	V43819
AdheSE Bonding Agent	Ivoclar Vivadent, Shaan, Liechtenstein	Dimethacrylates, Hydroxyethyl methacrylate, Highly dispersed silicon dioxide, Initiators and stabilizers	U54846
Single Bond Universal	3M ESPE, Saint Paul, USA	Organophosphate monomer (MDP), Bis-GMA, HEMA, Vitrebond copolymer, ethanol, water, initiators, silane	507329
Monobond Etch & Prime, self etching glass ceramic primer	Ivoclar Vivadent, Shaan, Liechtenstein	Tetrabutyl ammonium dihydrogen trifluoride, methacrylated phosphoric acid ester, trimethoxysilylpropyl methacrylate, alcohol, water	V09353
Variolink Esthetic LC	Ivoclar Vivadent, Shaan, Liechtenstein	Bis-GMA, UDMA, TEGDMA, ytterbium trifluoride, boroaluminofluorosilicate glass, spheroidal mixed oxide, benzoylperoxide, stabilizers, pigments	V37749

shown that the silane incorporated in a universal adhesive does not seem to produce the same adhesive strength as a silane agent applied separately, more studies are necessary to evaluate new strategies of cementation with these adhesives.

Even though it is highly used, the HF is a caustic and dangerous substance and presents a risk when contacting unprotected skin [16]. A self-etching ceramic primer (Monobond Etch & Prime, Ivoclar Vivadent) has been introduced as a single-component alternative to HF etching/silane routine surface treatment. The novel material aims to eliminate the risks associated with the HF acid as well as reduce the time required and the technique sensitivity of ceramics etching [17, 18]. Until now, few studies were available in the literature on the bonding efficiency of lithium disilicate ceramics to luting resin cements with this surface treatment. Some preliminary findings [18–20] showed that this self-etching ceramic primer presents a performance similar to that of the conventional surface treatment, but other authors showed that conventional treatment resulted in higher bond strengths than a self-etching ceramic primer [21, 22]. However, the use of this new ceramic primer should also be tested with various protocols.

Bonding effectiveness may directly influence the clinical success of ceramic restorations. It is important to identify the most reliable and effective surface treatment for ceramic before cementation. Therefore, the aim of this study was to investigate the influence of simplified ceramic surface treatments on shear bond strength of resin-luting cement and lithium disilicate ceramic. The null hypothesis tested was that various surface treatments and adhesive protocols will have no significant influence on the shear bond strength between resin cement and lithium disilicate ceramic.

2. Materials and Methods

The materials used and their respective compositions and batch numbers are displayed in Table 1.

2.1. Specimen Preparation. Sixty blocks of lithium disilicate-based ceramic (IPS e.max Press, Ivoclar Vivadent) were

TABLE 2: Group codes and surface treatments of ceramic.

Groups	Surface treatment
H/S/A	10% Hydrofluoric Acid + Silane + Adhesive
H/S	10% Hydrofluoric Acid + Silane
H/S/UA	10% Hydrofluoric Acid + Silane + Universal Adhesive
H/UA	10% Hydrofluoric Acid + Universal Adhesive
MBEP/A	Monobond Etch & Prime + Adhesive
MBEP	Monobond Etch & Prime

produced according to manufacturer instructions. The blocks were 8 mm tall, 8 mm wide, and 1 mm thick. To standardize the ceramic blocks, a wax pattern (VKS Gray Wax, Yeti Dental Produkte, Engen, Germany) was made in the dimensions of future blocks for ceramic injection. Dimensions were checked with a digital caliper (Mitutoyo Corporation, Tokyo, Japan).

Ceramic specimens were sandblasted with 50 micrometers of aluminum oxide particles for 15 s, then cleaned in an ultrasonic bath, and immersed first in distilled water and then in 92.8% ethanol, for 10 minutes each. Then they were placed in polyvinyl chloride (PVC) tubes (15 mm thick and 20 mm in diameter) with acrylic resin (Jet, Lapa, Rio de Janeiro, Brazil).

2.2. Ceramic Surface Treatments. The luting protocols for the ceramic surface treatment were performed according to the groups to which the specimens belonged, described in Table 2. Sixty specimens were divided into 6 groups (n=10).

In the first group (H/S/A), the ceramic surface was etched for 20 s with 10% HF (Condac, FGM), washed with an air/water spray for 30 s, and then dried with an air spray. The silane (Monobond N, Ivoclar Vivadent) was applied with a microbrush and allowed to react for 60 s. Subsequently, the excess was dispersed with a strong stream of air to ensure the solvent's evaporation. Finally, the adhesive agent (AdheSE Bonding Agent, Ivoclar Vivadent) was applied with a microbrush in a thin layer and polymerized using an LED curing unit (Bluephase, Ivoclar Vivadent) for 20 s. For the H/S group, the same protocol was followed; however, no adhesive was applied, just HF and silane. For the third group (H/S/UA), after etching with HF and silane application, a universal adhesive (SingleBond Universal, 3M ESPE) was applied in a thin layer with a microbrush and polymerized using an LED light-curing unit (Bluephase, Ivoclar Vivadent). For the MBEP/A group, a new ceramic primer (Monobond Etch & Prime, Ivoclar Vivadent) was applied without the use of HF or silane. Initially, the primer was applied with a microbrush with friction for 20 s and then it was allowed to sit for 40 s, and the surface was washed abundantly with an air/water spray followed by drying with an air spray for 10 s. Afterward, the adhesive agent (AdheSE Bonding Agent, Ivoclar Vivadent) was applied with a microbrush in a thin layer and polymerized using an LED light-curing unit (Bluephase, Ivoclar Vivadent) for 20 s. For the MBEP group, only the ceramic primer was applied (Monobond Etch & Prime, Ivoclar Vivadent) according to the method for the last group; however, no adhesive was applied.

2.3. Resin Cement Cylinders Production.

A special metal device was used to fix a Teflon mold, with a cylindrical cavity 3 mm wide and 3 mm deep, to the pretreated ceramic surface. The light-cured resin cement (Variolink Esthetic LC, Ivoclar Vivadent), color Neutral (translucent), was injected into the mold. The excess cement was removed using a microbrush, and the luting resin cement was photocured using an LED curing unit (Bluephase, Ivoclar Vivadent) operating at 1200 mW/cm^2 in high-power mode for 20 s. The mold was disassembled and resultant rods were examined for any composite flashes, which were removed with a sharp blade.

2.4. Shear Bond Strength Test.

The samples were stored in distilled water at 37°C for 24 h. In this study, the same device was used as in a previous study [23], with a metal strip around the cement cylinder to minimize the flexural stresses. The machine's semicircular metal attachment applied shear forces at the cement-ceramic interface, running at a crosshead speed of 0.5 mm/min, until complete failure. The maximum load to failure (in Newtons) was recorded, and the shear bond strength (in MPa) was calculated by dividing the failure load by the bonding area (mm^2), which was calculated by measuring the cement cylinder's diameter at two points with a digital caliper. The same operator carried out all procedures to avoid interoperator variability. All manufactured specimens were tested for shear bond strength, as no pretest failures were observed.

The debonded specimens were examined under an optic microscope (Discovery V8 Stereo, Carl Zeiss Microimaging GmbH, Jena, Germany) to determine the failure mode. They were classified as adhesive failure, between resin cement and ceramic (A), mixed failure (M), and cohesive in resin cement (CR) or cohesive in ceramic (CC).

2.5. Scanning Electron Microscope (SEM).

SEM images of the lithium disilicate-based ceramic (IPS e.max Press) surface were captured at various magnifications to evaluate the etching pattern/micromorphology produced by each treatment

TABLE 3: Means (MPa), standard deviations, and confidence intervals of shear bond strength for each group.

Groups	Shear bond strength (mean)	Standard deviation	Confidence interval
H/S/A	9.60^A	2.50	7.81-11.39
H/S	10.22^A	3.28	7.89-12.57
H/S/UA	7.39^{A,B,C}	2.02	5.98-8.84
H/UA	4.28^C	1.32	3.33-5.23
MBEP/A	9.00^{A,B}	1.97	7.59-10.41
MBEP	6.18^{B,C}	2.75	4.21-8.15

Values followed by different letters present statistical difference ($p<0.05$).

(no treatment, HF 10% or Monobond Etch & Prime) used according to manufacturers' instructions.

2.6. Statistical Analysis.

Data obtained on shear bond strength was analyzed in Stat Plus (Mac v.6.2.21 (Analysoft, Inc, Atlanta, USA). Initially, the data were analyzed for homogeneity (Levene's test) and normality (Kolmogorov-Smirnov test). Due to its parametric and homogeneous distribution, the ANOVA test was used with multiple comparisons with the post hoc Tukey test ($\alpha = 0.05$).

3. Results

The mean and standard deviation values of each group's shear bond strength are summarized in Table 3. Significant statistical differences were observed in shear bond strength for the surface treatments ($p<0.05$). The surface treatment with hydrofluoric acid and silane (H/S group) showed the highest values of shear bond strength; however, it did not differ statistically from H/S/A, MBEP/A and H/S/UA. The use of only Monobond Etch & Prime (MBEP group) as a surface treatment led to significantly lower values of bond strength than in the MBEP/A group and was statistically similar to the H/UA group, which obtained the lowest values of bond strength. The use of silane prior to application of the universal adhesive (H/S/UA group) promoted higher values than in the group in which the universal adhesive was used without silane (H/UA).

Failure mode was also influenced by surface treatment, according to Table 4. Cohesive failures in ceramic were not observed. The H/S/A, H/S/UA, and MBEP/A groups showed the most mixed failures and a small number of cohesive failures in resin cement. These mixed failures usually presented resin cement in the border areas of the specimen and debonding in the center of the specimen. The H/S group only showed cohesive failures in resin cement, and H/UA was the only group that showed adhesive failures.

SEM analysis (Figure 1) showed the difference between the ceramic with no treatment and the etching pattern produced by HF and by Monobond Etch & Prime surface treatment. After surface treatment with HF, it is possible to observe a deeper etching pattern with glassy dissolution and exposition of crystals. When the self-etching primer was used, e etching pattern was more superficial, showing less

TABLE 4: Distribution of failure modes in percentage (%) and absolute numbers (n) after shear bond strength test for all tested groups.

Groups	Adhesive Failure - % (n)	Cohesive Failure - % (n)	Mixed Failure - % (n)	Pre-test failures % (n)
H/S/A	0 (0)	30 (3)	70 (7)	0 (0)
H/S	0 (0)	100 (10)	0 (0)	0 (0)
H/S/UA	0 (0)	20 (2)	80 (8)	0 (0)
H/UA	20 (2)	0 (0)	80 (8)	0 (0)
MBEP/A	0 (0)	20 (2)	80 (8)	0 (0)
MBEP	0 (0)	90 (9)	10 (1)	0 (0)

micromechanical retention with smaller glassy dissolution and without crystal exposition.

4. Discussion

The clinical success of a ceramic restoration depends on the quality and durability of the bond between ceramic and resin cement [24]. The protocol established for lithium disilicate-based ceramic cementation is the etching with HF and the application of a silane agent [25]. In the present study, various surface treatments were used, simplified or not, and the results showed that multiple surface treatments and adhesive protocols promoted significant changes in shear bond strength, disproving the null hypothesis.

In the cementation of lithium disilicate-based ceramics, the surface treatment with HF is extremely important to promote irregularities and create a surface with micropores by partially dissolving the glass phase, leaving behind an active surface rich in silica [3, 7]. The silane coupling agent establishes adhesion between the inorganic phase of the ceramic and the organic phase of the resin cement, forming a siloxane bond [11, 26]. The use of silane after etching with HF is indispensable; however, the use of the adhesive is still controversial. The groups that had application of the silane (H/S/A, H/S, and H/S/UA) did not present significant statistical differences among themselves; therefore, the use of the adhesive appears dispensable; this finding corroborates with those of Garboza et al. [27]. The use of HF and silane seems to be the ideal protocol because it requires fewer steps and reduces the risk of failure.

When the failure modes (Table 4) were observed, only the H/S group showed only cohesive failures in resin cement. According to Chen et al. [28], this may suggest that the adhesive interface was very strong, and the application of adhesive after silane probably weakened the interface. On the other hand, Scherrer, Cesar, and Swain [29] affirm that, when cohesive and mixed cohesive/adhesive failures occur, the bond strength obtained is not representative of the interface adhesion but reflects the strength of the materials being tested. As only one resin cement was evaluated, the differences in bond strength obtained may represent the different surface treatments performed. According to DeHoff, Anusavice, and Wang [30], due to the known variation in bond strength with specimen preparation and design, data on the same systems may show great variability in mean and large standard deviations. Thus, these tests should be used, as in this study, as a tool to compare materials or surface

treatments to determine the effect of changing some variable for the same system and not to determine the real bond strength between resin cement and ceramic [30].

The H/S/UA and H/UA groups were statistically similar; however, when only universal adhesive was used after HF, shear bond strength values decreased. Although the universal adhesive used contains silane and 10-methacryloxydecyl dihydrogen phosphate (10-MDP), the additional salinization step enhances chemical bonding to the exposed hydroxyl groups and surface wettability with resin impregnation, which has been shown in other studies, even in the long term [31–33]. Moreover, only the H/UA group exhibited adhesive failures, which corroborates the fact that only the universal adhesive after HF is ineffective in preparing the ceramic surface because the adhesive interface proved to be flawed and fragile. When HF, silane, and adhesive were used, the adhesive bonding (H/S/A) was more effective than the universal adhesive (H/S/UA). This finding is in accordance with those of Garboza et al. [27] and can be explained by the fact that the hydrophilic part of the universal adhesive might negatively affect the bond strength. Moreover, the silane contained in universal adhesive may have increased the hydrophilicity, thereby predisposing the adhesive layer to hydrolytic degradation.

The self-etching ceramic primer (Monobond Etch & Prime) contains ammonium polyfluoride and silane in a single step. This new material aims to eliminate the toxic potential of HF and minimize the technical sensitivity of the cementation process [27]. However, the ammonium polyfluoride promotes a weaker etching pattern in the ceramic surface than HF [9, 27]. Previous studies [27, 34] showed that this primer was efficient in conditioning vitreous ceramics, presenting bond strength comparable to that of the conventional treatment. However, the conventional treatment still showed superior results, remaining a gold standard for ceramic surface treatment. A recent study [22] showed that HF/silane resulted in higher mean microshear bond strength than Monobond Etch & Prime for lithium disilicate and feldspathic ceramics; however, Monobond Etch & Prime had a more stable bond after aging.

In the present study, the MBEP group was statistically inferior to the conventional treatment (H/S), corroborating with previous studies that included shear bond evaluation [17, 21, 34]. This result is probably due to the etching pattern promoted by HF. In SEM images (Figures 1(c) and 1(d)), after surface treatment of HF 10%, it was possible to observe a porous surface with exposure of lithium disilicate crystals on

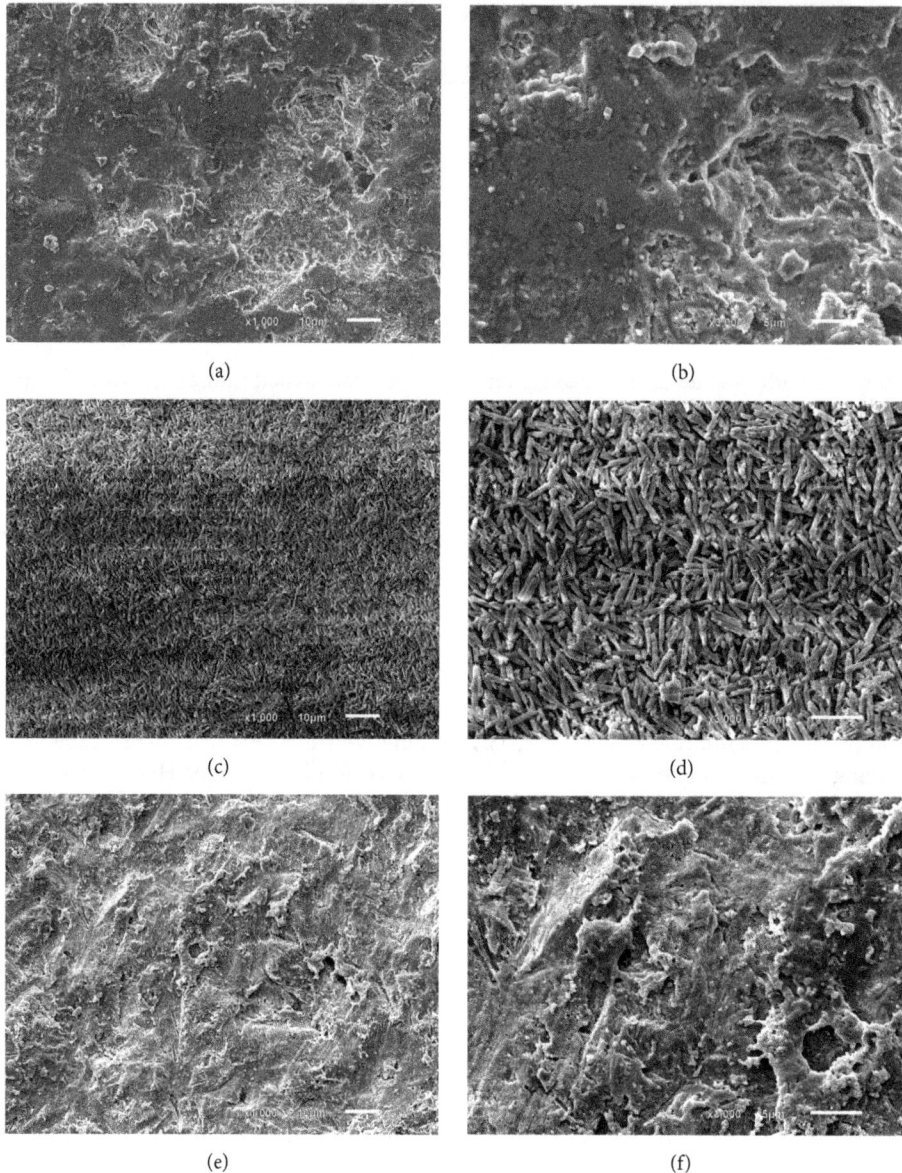

FIGURE 1: SEM images of nonetched and etched IPS e.max ceramic surfaces after different conditioning. (a) x1000 magnification, ceramic surface before etching. (b) x3000 magnification, ceramic surface before etching. (c) x1000 magnification, etching with HF 10% for 20 seconds. (d) x3000 magnification, etching with HF 10% for 20 seconds. (e) x1000 magnification, etching with Monobond Etch & Prime according to the manufacture. (f) x3000 magnification, etching with Monobond Etch & Prime according to the manufacture.

the ceramic surface, resulting in more surface area for resin bonding and promoting better chemical bonding via a silane coupling agent [35]. In Figures 1(e) and 1(f), it is possible to observe that MBEP showed almost no etching depth power, probably because of its weaker etching agent (ammonium polyfluoride), resulting in less micromechanical retention of resin cement. This finding corroborates with those of Lopes et al. [21], who evaluated the etching pattern of lithium disilicate ceramics under a field emission scanning electron microscope and showed that use of Monobond Etch&Prime resulted in the least pronounced etching pattern.

Despite these findings, when a conventional adhesive was used after ceramic primer (MBEP/A group), the resulting

bond strength was similar to that of conventional surface treatment (H/S). Although the mechanism of action and adhesion of this self-etching ceramic primer is not very clear, the use of bonding adhesive promotes better interaction between ceramic and resin cement, presenting even more mixed failures than the MBEP group, probably due to the better chemical bond established. The unfilled adhesive probably promotes the formation of a more compatible and stronger interaction between the pretreated ceramic and the resin cement. Resin adhesives are usually made up of hydrophobic dimethacrylates, which may covalently bond to silane and cement materials by means of ester bonds. Consequently, a strong intermolecular chemical interaction between ceramic and cement could be achieved, leading to the formation of a homogeneous tertiary monoblock [36, 37]. These results are still favorable because, even though the application of the adhesive agent is required, the use of the primer eliminates the use of HF and contributes to a safer procedure.

A restoration in the oral cavity is challenged in many ways: it is subjected to complex occlusal forces, immersed in saliva and exposed to food and beverages with various pH, chemistries, and temperatures. Numerous laboratory tests have attempted to simulate oral conditions in order to predict clinical bonding performance. However, no single laboratory test is able to adequately predict the clinical performance of resin-ceramic bonding [5].

The most common tests to evaluate resin-ceramic bonding measurements are shear and tensile bond strength tests [5]. As in other studies [10, 19, 26] the shear bond strength test was used in this work, even though it is known that "macro" bonding tests, due to the bigger adhesion area, tend to result in lower bond strength values [38]. This method is commonly used for ceramics bond strength evaluation, not only because it is a quick and repeatable testing option but also because it is difficult to section the ceramic for microtensile testing [5]. The use of the stainless steel tape, instead of a knife edge or a looped orthodontic wire system, for the test is justified by the possibility of reducing the stress-concentration magnitude adjacent to the interface, and the tensile and compressive stresses produced in the interface are smaller than those obtained from the other systems [23, 30, 39].

One of this study's limitations was the use of one type of light-cured resin-luting cement. Tests with multiple types of resin cements, including self-adhesive and dual-cured cements, can be interesting and could be a point for further research. Moreover, studies with long-term water storage and thermocycling are necessary to evaluate mainly the new materials. Finally, clinical studies are needed to evaluate this material's clinical performance. The use of a new self-etch ceramic primer associated with adhesive bonding is an effective alternative to simplify the clinical procedures presenting a performance similar to that of conventional surface treatment for lithium disilicate ceramics.

5. Conclusions

Even with the present study's limitations, it is possible to conclude that the surface treatment with HF and silane is an effective and simple alternative to luting lithium disilicate ceramics; the use of universal adhesive did not exempt the application of a silane, and the new ceramic self-etching primer is an effective alternative for simplified ceramic surface treatment when an adhesive agent is applied after it.

Conflicts of Interest

The authors declare that they have no conflicts of interest.

Acknowledgments

The authors thank the School of Dentistry of the Federal University of Goiás for the use of its universal testing machine.

References

[1] U. S. Beier, I. Kapferer, D. Burtscher, and H. Dumfahrt, "Clinical performance of porcelain laminate veneers for up to 20 years," *The International Journal of Prosthodontics*, vol. 25, no. 1, pp. 79–85, 2012.

[2] S. Pieger, A. Salman, and A. S. Bidra, "Clinical outcomes of lithium disilicate single crowns and partial fixed dental prostheses: a systematic review," *The Journal of Prosthetic Dentistry*, vol. 112, no. 1, pp. 22–30, 2014.

[3] M. N. Aboushelib and D. Sleem, "Microtensile bond strength of lithium disilicate ceramics to resin adhesives," *The Journal of Adhesive Dentistry*, vol. 16, no. 6, pp. 547–552, 2014.

[4] J. R. Almeida, G. U. Schmitt, M. R. Kaizer, N. Boscato, and R. R. Moraes, "Resin-based luting agents and color stability of bonded ceramic veneers," *Journal of Prosthetic Dentistry*, vol. 114, no. 2, pp. 272–277, 2015.

[5] T. Tian, J. K.-H. Tsoi, J. P. Matinlinna, and M. F. Burrow, "Aspects of bonding between resin luting cements and glass ceramic materials," *Dental Materials*, vol. 30, no. 7, pp. e147–e162, 2014.

[6] G. B. Guarda, A. B. Correr, L. S. Gonçalves et al., "Effects of surface treatments, thermocycling, and cyclic loading on the bond strength of a resin cement bonded to a lithium disilicate glass ceramic," *Operative Dentistry*, vol. 38, no. 2, pp. 208–217, 2013.

[7] A. Della Bona, K. J. Anusavice, and J. A. A. Hood, "Effect of ceramic surface treatment on tensile bond strength to a resin cement," *International Journal of Prosthodontics*, vol. 15, no. 3, pp. 248–253, 2002.

[8] R. C. R. Colares, J. R. Neri, A. M. B. de Souza, K. M. D. F. Pontes, J. S. Mendonça, and S. L. Santiago, "Effect of surface pretreatments on the microtensile bond strength of lithium-disilicate ceramic repaired with composite resin," *Brazilian Dental Journal*, vol. 24, no. 4, pp. 349–352, 2013.

[9] G. M. Iorizzo, F. Prete, B. Mazzanti, G. Timellini, R. Scotti, and P. Baldissara, "Effects of hydrofluoric acid etching on lithium disilicate," *Dental Materials*, vol. 30, p. e24, 2014.

[10] P. Kursoglu, P. F. K. Motro, and H. Yurdaguven, "Shear bond strength of resin cement to an acid etched and a laser irradiated ceramic surface," *The Journal of Advanced Prosthodontics*, vol. 5, no. 2, pp. 98–103, 2013.

[11] J. P. Matinlinna, C. Y. K. Lung, and J. K. H. Tsoi, "Silane adhesion mechanism in dental applications and surface treatments: a review," *Dental Materials*, vol. 34, no. 1, pp. 13–28, 2018.

[12] D. P. Lise, J. Perdigão, A. Van Ende, O. Zidan, and G. C. Lopes, "Microshear bond strength of resin cements to lithium disilicate substrates as a function of surface preparation," *Operative Dentistry*, vol. 40, no. 5, pp. 524–532, 2015.

[13] N. Scotti, G. Cavalli, M. Gagliani, and L. Breschi, "New adhesives and bonding techniques. Why and when?" *The international journal of esthetic dentistry*, vol. 12, no. 4, pp. 524–535, 2017.

[14] H.-Y. Lee, G.-J. Han, J. Chang, and H.-H. Son, "Bonding of the silane containing multi-mode universal adhesive for lithium disilicate ceramics," *Restorative Dentistry & Endodontics*, vol. 42, no. 2, pp. 95–104, 2017.

[15] V. K. Kalavacharla, N. C. Lawson, L. C. Ramp, and J. O. Burgess, "Influence of etching protocol and silane treatment with a universal adhesive on lithium disilicate bond strength," *Operative Dentistry*, vol. 40, no. 4, pp. 372–378, 2015.

[16] M. Özcan, A. Allahbeickaraghi, and M. Dündar, "Possible hazardous effects of hydrofluoric acid and recommendations for treatment approach: A review," *Clinical Oral Investigations*, vol. 16, no. 1, pp. 15–23, 2012.

[17] H. M. El-Damanhoury and M. D. Gaintantzopoulou, "Self-etching ceramic primer versus hydrofluoric acid etching: Etching efficacy and bonding performance," *Journal of Prosthodontic Research*, vol. 62, no. 1, pp. 75–83, 2018.

[18] F. S. Siqueira, R. S. Alessi, A. F. Cardenas, C. Kose, S. C. Souza Pinto, and M. C. Bandeca, "New single-bottle ceramic primer: 6-month case report and laboratory performance," *The Journal of Contemporary Dental Practice*, vol. 17, no. 12, pp. 1033–1039, 2016.

[19] J. L. Román-Rodríguez, J. A. Perez-Barquero, E. Gonzalez-Angulo, A. Fons-Font, and J. L. Bustos-Salvador, "Bonding to silicate ceramics: Conventional technique compared with a simplified technique," *Journal of Clinical and Experimental Dentistry*, vol. 9, no. 3, pp. e384–e386, 2017.

[20] J. Tribst, L. Anami, M. Özcan, M. Bottino, R. Melo, and G. Saavedra, "Self-etching Primers vs Acid Conditioning: Impact on Bond Strength Between Ceramics and Resin Cement," *Operative Dentistry*, vol. 43, no. 4, pp. 372–379, 2018.

[21] G. Lopes, J. Perdigão, D. Baptista, and A. Ballarin, "Does a Self-Etching Ceramic Primer Improve Bonding to Lithium Disilicate Ceramics? Bond Strengths and FESEM Analyses," *Operative Dentistry*, 2018.

[22] M. Prado, C. Prochnow, A. M. E. Marchionatti, P. Baldissara, L. F. Valandro, and V. F. Wandsher, "Ceramic surface treatment with a single-component primer: Resin adhesion to glass ceramics," *The Journal of Adhesive Dentistry*, vol. 20, no. 2, pp. 99–105, 2018.

[23] C. M. Ramos, P. F. Cesar, R. F. Lia Mondelli, A. S. Tabata, J. De Souza Santos, and A. F. Sanches Borges, "Bond strength and Raman analysis of the zirconia-feldspathic porcelain interface," *Journal of Prosthetic Dentistry*, vol. 112, no. 4, pp. 886–894, 2014.

[24] A. Attia and M. Kern, "Influence of cyclic loading and luting agents on the fracture load of two all-ceramic crown systems," *Journal of Prosthetic Dentistry*, vol. 92, no. 6, pp. 551–556, 2004.

[25] C. P. Gré, R. C. de Ré Silveira, S. Shibata, C. T. R. Lago, and L. C. C. Vieira, "Effect of silanization on microtensile bond strength of different resin cements to a lithium disilicate glass ceramic," *Journal of Contemporary Dental Practice*, vol. 17, no. 2, pp. 149–153, 2016.

[26] T. Yavuz and O. Eraslan, "The effect of silane applied to glass ceramics on surface structure and bonding strength at different temperatures," *The Journal of Advanced Prosthodontics*, vol. 8, no. 2, pp. 75–84, 2016.

[27] C. S. Garboza, S. B. Berger, R. D. Guiraldo et al., "Influence of surface treatments and adhesive systems on lithium disilicate microshear bond strength," *Brazilian Dental Journal*, vol. 27, no. 4, pp. 452–457, 2016.

[28] C. Chen, H. Xie, X. Song, M. F. Burrow, G. Chen, and F. Zhang, "Evaluation of a commercial primer for bonding of zirconia to two different resin composite cements," *The Journal of Adhesive Dentistry*, vol. 16, no. 2, pp. 169–176, 2014.

[29] S. S. Scherrer, P. F. Cesar, and M. V. Swain, "Direct comparison of the bond strength results of the different test methods: a critical literature review," *Dental Materials*, vol. 26, no. 2, pp. e78–e93, 2010.

[30] P. H. DeHoff, K. J. Anusavice, and Z. Wang, "Three-dimensional finite element analysis of the shear bond test," *Dental Materials*, vol. 11, no. 2, pp. 126–131, 1995.

[31] A. M. Cardenas, F. Siqueira, V. Hass et al., "Effect of MDP-containing Silane and Adhesive Used Alone or in Combination on the Long-term Bond Strength and Chemical Interaction with Lithium Disilicate Ceramics," *The Journal of Adhesive Dentistry*, vol. 19, no. 3, pp. 203–212, 2017.

[32] A. F. Moro, A. B. Ramos, G. M. Rocha, and C. d. Perez, "Effect of prior silane application on the bond strength of a universal adhesive to a lithium disilicate ceramic," *The Journal of Prosthetic Dentistry*, vol. 118, pp. 666–671, 2017.

[33] F. Murillo-Gómez, F. A. Rueggeberg, and M. F. De Goes, "Short- and long-term bond strength between resin cement and glass-ceramic using a silane-containing universal adhesive," *Operative Dentistry*, vol. 42, no. 5, pp. 514–525, 2017.

[34] S. K. Lyann, K. Takagaki, T. Nikaido, M. Uo, M. Ikeda, and A. Sadr, "Effect of different surface treatmens on the tensile bond strength to lithium disilicate ceramics," *The Journal of Adhesive Dentistry*, vol. 20, no. 3, pp. 261–268, 2018.

[35] M. Ozcan and C. A. Volpato, "Surface conditioning protocol for the adhesion of resin-based materials to glassy matrix ceramics: How to condition and why?" *The Journal of Adhesive Dentistry*, vol. 17, pp. 292–293, 2015.

[36] J. L. Ferracane, "Resin composite—state of the art," *Dental Materials*, vol. 27, no. 1, pp. 29–38, 2011.

[37] F. W. Machado, M. Bossardi, T. D. S. Ramos, L. L. Valente, E. A. Münchow, and E. Piva, "Application of resin adhesive on the surface of a silanized glass fiber-reinforced post and its effect on the retention to root dentin," *Journal of Endodontics*, vol. 41, no. 1, pp. 106–110, 2015.

[38] R. R. Braga, J. B. C. Meira, L. C. C. Boaro, and T. A. Xavier, "Adhesion to tooth structure: a critical review of 'macro' test methods," *Dental Materials*, vol. 26, no. 2, pp. e38–e49, 2010.

[39] M. A. C. Sinhoreti, S. Consani, M. F. De Goes, L. C. Sobrinho, and J. C. Knowles, "Influence of loading types on the shear strength of the dentin-resin interface bonding," *Journal of Materials Science: Materials in Medicine*, vol. 12, no. 1, pp. 39–44, 2001.

15

Evaluation of Sterilisation Techniques for Regenerative Medicine Scaffolds Fabricated with Polyurethane Nonbiodegradable and Bioabsorbable Nanocomposite Materials

Michelle Griffin [ORCID], [1,2,3] **Naghmeh Naderi,** [1,3,4,5] **Deepak M. Kalaskar,** [1]
Edward Malins, [6] **Remzi Becer,** [6] **Catherine A. Thornton,** [4] **Iain S. Whitaker,** [4,5]
Ash Mosahebi, [3] **Peter E. M. Butler,** [1,2,3] **and Alexander M. Seifalian** [ORCID] [7]

[1] *UCL Centre for Nanotechnology & Regenerative Medicine, University College London, Royal Free London NHS Foundation Trust, Pond Street, London NW3 2QG, UK*

[2] *The Charles Wolfson Center for Reconstructive Surgery, Royal Free London NHS Foundation Trust Hospital, London, UK*

[3] *Department of Plastic Surgery, Royal Free London NHS Foundation Trust, Pond Street, London NW3 2QG, UK*

[4] *Reconstructive Surgery & Regenerative Medicine Group, Institute of Life Science, Swansea University Medical School, Singleton Park, Swansea SA2 8PP, UK*

[5] *Welsh Centre for Burns & Plastic Surgery, ABMU Health Board, Heol Maes Egwlys, Swansea SA6 6NL, UK*

[6] *Polymer Chemistry Laboratory, School of Engineering and Materials Science, Queen Mary University of London, Mile End Road, London E1 4NS, UK*

[7] *Director/Professor Nanotechnology & Regenerative Medicine, NanoRegMed Ltd., The London BioScience Innovation Centre, London NW1 0NH, UK*

Correspondence should be addressed to Michelle Griffin; 12michellegriffin@gmail.com

Academic Editor: Rosalind Labow

An effective sterilisation technique that maintains structure integrity, mechanical properties, and biocompatibility is essential for the translation of new biomaterials to the clinical setting. We aimed to establish an effective sterilisation technique for a biodegradable (POSS-PCL) and nonbiodegradable (POSS-PCU) nanocomposite scaffold that maintains stem cell biocompatibility. Scaffolds were sterilised using 70% ethanol, ultraviolet radiation, bleach, antibiotic/antimycotic, ethylene oxide, gamma irradiation, argon plasma, or autoclaving. Samples were immersed in tryptone soya broth and thioglycollate medium and inspected for signs of microbial growth. Scaffold surface and mechanical and molecular weight properties were investigated. AlamarBlue viability assay of adipose derived stem cells (ADSC) seeded on scaffolds was performed to investigate metabolic activity. Confocal imaging of rhodamine phalloidin and DAPI stained ADSCs was performed to evaluate morphology. Ethylene oxide, gamma irradiation, argon plasma, autoclaving, 70% ethanol, and bleach were effective in sterilising the scaffolds. Autoclaving, gamma irradiation, and ethylene oxide led to a significant change in the molecular weight distribution of POSS-PCL and gamma irradiation and ethylene oxide to that of POSS-PCU ($p<0.05$). UV, ethanol, gamma irradiation, and ethylene oxide caused significant changes in the mechanical properties of POSS-PCL ($p<0.05$). Argon was associated with significantly higher surface wettability and ADSC metabolic activity ($p<0.05$). In this study, argon plasma was an effective sterilisation technique for both nonbiodegradable and biodegradable nanocomposite scaffolds. Argon plasma should be further investigated as a potential sterilisation technique for medical devices.

1. Introduction

Synthetic biomaterials are being used to replace the extracellular matrix to restore damaged and failing tissues and organs [1]. Amongst biomaterials, polymeric scaffolds have gained significant popularity due to their ease of fabrication and versatility [1]. Polymeric scaffolds for tissue engineering are either manufactured aseptically or sterilised after processing

[2, 3]. For economical and practical reasons, the latter strategy has been employed with polymeric scaffolds intended for in vivo use and is considered a more realistic approach to achieve sterile implantable scaffolds [1, 2]. Nevertheless, the challenge remains to determine an efficient and nondestructive sterilisation procedure for polymer scaffolds that preserves their structure and surface properties [3]. Sterilisation techniques may influence a material's structural, chemical, and biological properties; thus it is important to ensure the modality implemented does not affect biocompatibility [3]. Sterilisation of biomaterials accepted by the FDA for medical devices includes ethylene oxide, autoclaving, and gamma sterilisation [4]. The success of an implant for sterilisation is dependent not only on the implant remaining sterile, but also on achieving sterility without adversely affecting the material's properties. Different sterilisation agents have shown that they can attack polymers causing hydrolysis, melting, or depolymerisation [5, 6].

Our group have developed and patented two families of nanocomposite polymers for the development of organs and tissues [7–9]. The nonbiodegradable polymer incorporates POSS nanoparticles into polycarbonate-based urea-urethane (POSS-PCU, UCL-Nano). Its biodegradable counterpart modifies poly(caprolactone urea-urethane) POSS-PCL. Understanding an appropriate sterilisation technique for POSS-PCU and POSS-PCL is crucial for translation to clinical practice. A brief previous study compared three techniques of sterilising POSS-PCU and POSS-PCL scaffolds including autoclaving, gamma irradiation, and ethanol [7]. Autoclaving was found to be effective in maintaining sterilisation of the scaffolds without degrading the material. The first authors of this paper have also demonstrated that bleach may be useful for sterilising POSS-PCL scaffolds compared to ethanol and autoclaving sterilisation [8]. Adipose derived stem cells (ADSCs) were shown to adhere to the POSS-PCL scaffolds following ethanol and bleach sterilisation [8]. Lastly, a study comparing the effects of autoclave, microwave, antibiotics, and 70% ethanol sterilisation on POSS-PCL scaffolds found ethanol to be a suitable sterilisation technique with maintained fibroblast attachment [9]. The aim of this study was to compare all available sterilisation techniques for POSS-PCU and POSS-PCL in a single study including a new method using argon plasma sterilisation, building on previous studies, to understand the optimal sterilisation technique for nanocomposite scaffolds.

An increasingly popular method of modifying the surface functionality to enhance cell behaviour on scaffolds is plasma modification [10, 11]. Plasma consists of electrons, ions, energy rich neutrals, molecules, fragments, atoms, and photons. It can be under low, atmospheric, or high pressure. The distinct behaviour of gases led to the suggestion that plasma is the "fourth state of matter" [12]. Plasma modification (PM) is an easy, reliable, clean way of creating reactive functional groups on the surfaces of biomaterials and for creating anchoring sites for further chemical reactions [12].

This study compared ethylene oxide, argon plasma, bleach, antibiotic/antimycotic, ethanol, ultraviolet radiation, autoclaving, and gamma irradiation sterilisation procedures

for morphological alteration, chemical damage, effects on polymer degradation, and biocompatibility. The viability of ADSCs was assessed following the sterilisation of POSS-PCU and POSS-PCL scaffolds to assess biocompatibility of the sterilisation techniques.

2. Materials and Methods

2.1. Polymer Synthesis

2.1.1. POSS-PCU. The nanocomposite scaffolds were manufactured as previously described [7, 13]. In brief, polycarbonate polyol, 2000 mwt, and *trans*-cyclohexanechloroydrinisobutyl-silsesquioxane (Hybrid Plastics Inc.) was heated to 135°C and then cooled to 70°C. Flakes of 4,4′-methylenebis (phenyl isocyanate) (MDI) were then added to the mixture, at 75–85°C for 90 minutes to form the prepolymer. Then *N,N*-dimethylacetamide (DMAc) was to form a solution. Further chain extension was completed by the drop-wise addition of ethylenediamine and diethylamine in DMAc. This then created the POSS-modified polycarbonate urea-urethane in DMAc. All chemicals and reagents were purchased from Aldrich Limited, Gillingham, UK.

2.1.2. POSS-PCL. POSS-PCL nanocomposites solution was manufactured as described previously [7]. In brief, polycaprolactone diol (2000 g/mol) and *trans*-cyclohexanechlorohydrinisobutyl-polyhedral oligomeric silsesquioxane (POSS) were mixed and heated to 135°C. Then 9.4 g of 4,4′-methylenebis(cyclohexylisocyanate) was added to form a prepolymer. Following this, 100 g of DMAC was added to the prepolymer. Chain extension was performed by dropwise addition of 1 g of ethylenediamine in 80 g of dry DMAC. Following this, 2 g of 1-butanol in 5 g of DMAC was added to form the nanocomposite. All chemicals and reagents were purchased from Aldrich Limited, Gillingham, UK.

2.1.3. Sample Preparation. Polymers were fabricated as 3D scaffolds using the phase separation/particulate-leaching technique as described previously [13]. Firstly, NaCL (200-250 μm) was dissolved in POSS-PCL and POSS-PCU in DMAc containing Tween-20 surfactant (1:1 weight ratio). The solution was dispersed and degassed in a Thinky AER 250 mixer (Intertonics, Kidlington, UK). The polymer mixture was spread onto steel moulds. The moulds were then washed in deionised water to dissolve the solvent and DMAC for 7 days. Following washing, polymer sheets with 700-800 μm thickness were manufactured. For cell culture analysis, 16 mm polymer disks were cut from the sheets using a steel manual shape cutter.

2.2. Sterilisation

2.2.1. Gamma Irradiation. Scaffolds were irradiated with a dose of 25 kGy at room temperature, using a ^{60}Co gamma-ray source (Synergy Health, Swindon, UK). Scaffolds were exposed on a continuous path for 10 hours as described previously [7].

2.2.2. Autoclaving.
The scaffolds were exposed to steam at 121°C for 15 minutes at pressures of 115 kPa as previously described [7].

2.2.3. Ethanol.
Polymer scaffolds were submerged in 70% (v/v) ethanol on a roller mixer for 30 minutes as previously described [8]. Following alcohol sterilisation, scaffolds are washed in sterile deionised water on a roller for 15 minutes, which is then repeated five times.

2.2.4. Plasma.
Scaffolds are placed in 24-well plates for treatment by LF (radio frequency) argon plasma generator operating at 40 KHz at 100 W. Scaffolds then enter the chamber of the electrode-less, glow discharge apparatus, which is purged 3 times with argon gas (99.99% purity, BOC, UK) for 2 minutes. The chamber is then evacuated to 1.0 Torr. Plasma is then ignited by a radio frequency excitation source and was maintained at 100 W for 5 minutes. The scaffolds were treated with plasma and immediately seeded with cells for *in vitro* analysis to prevent hydrophobic recovery of the scaffolds.

2.2.5. Ethylene Oxide.
The ethylene oxide sterilisation is initiated with a preconditioning of the samples which is carried out at 41°C for 13 hours at 42% humidity. The sterilisation step is then performed by 100% ethylene oxide atmosphere at 49°C for 2 1/2 hours. Scaffold are then in air for 9 hours at 43°C.

2.2.6. Ultraviolet Irradiation.
Scaffolds were UV irradiated by placement in a UV decontamination device (40 watt wavelength 254 nm, mean density 15 kJ/cm^2) for 3 hours as previously described [8].

2.2.7. Antibiotic Antimycotic Treatment.
Scaffolds were placed in a 1% (v/v) antibiotic antimycotic solution (10 000 U/mL penicillin G, 10 mg/mL streptomycin sulphate, and 25 mg/mL amphotericin B diluted in sterile phosphate buffered saline (PBS)) for 24 hours at 4°C. Following sterilisation, the scaffolds were washed five times with sterile deionised water on a roller for 15 minutes each time.

2.2.8. Bleach (SDIC).
The scaffolds were submersed in 1000 ppm slow chlorine releasing compound sodium dichloroisocyanurate dihydrate (SDIC) at room temperature for 20 minutes as described previously [8]. The scaffolds are then washed in sterile deionised water daily for 7 days. The removal of remaining SDIC was confirmed by pH testing.

2.3. Material Characterisation

2.3.1. Tensiometry.
Tensiometry to evaluate the mechanical properties of the scaffolds following sterilisation was performed as described previously [13]. In brief, dumbbell-shaped scaffolds (dimensions of 10 × 2 mm) were tensile loaded at a loading speed of 100 mm/min ($n = 6$) using an Instron 5565 (High Wycombe, Bucks, UK). Young's modulus of elasticity at the 0-25% portion of the curve, maximum tensile strength, and elongation at break were calculated. Statistical differences in the tensile properties between sterilisation techniques were evaluated using two-way ANOVA with post hoc Turkey test.

2.3.2. Attenuated Total Reflectance Fourier Transform Infrared Spectroscopy (ATR-FTIR).
FTIR spectrophotometer was used to analyse changes in the surface chemistry of the scaffolds treated with the different sterilisation techniques as previously described [8]. Chemical groups were detected using attenuated total reflectance (ATR)-FTIR mode (Jasco FT/IR 4200 Spectrometer (JASCO Inc., USA)) (n=6). FTIR testing parameters were recorded at 20 scans at a 4 cm-1 resolution with a wavenumber range of 600 cm-1 to 4000 cm-1.

2.3.3. Gel Permeation Chromatography (GPC).
Molecular weight averages and polymer dispersity were determined by GPC as previously described [8] (n=6). In brief, samples were prepared to a 1 mg/mL concentration and passed through a 0.22 μm nylon filter. GPC analysis was conducted on the Agilent 1260 infinity system using 2 PLgel 5 μm mixed-D columns (300 × 7.5 mm), a PLgel 5 mm guard column (50 × 7.5 mm), a differential refractive index (DRI), and variable wavelength detector (VWD). Statistical differences in the GPC analysis between sterilisation techniques were evaluated using two-way ANOVA with post hoc Turkey test.

2.3.4. Water Contact Angle Measurements.
The static water contact angle of the scaffolds was performed as described previously [13]. In brief, the water contact angle was analysed using the sessile drop method (DSA 100 instrument (KRUSS, Germany)). A 5 μl volume of deionised water was used in all experiments. Measurement of a single drop was performed on six independent scaffolds (n=6). The average contact angle was calculated using the KRUSS drop shape software (version 1.90.0.14).

2.3.5. Scanning Electron Microscopy (SEM).
The surface of the scaffolds treated with the different sterilisation techniques was analysed using SEM as described previously [8] (n=6). Scaffolds were dehydrated in acetone prior to drying overnight. The scaffolds were then mounted on aluminium pin stubs using sticky carbon tape. After coating the scaffolds with a thin layer of Au/Pd (approximately 2 nm thick) using a Gatan ion beam coater the surface of the scaffolds was imaged with a Carl Zeiss LS15 Evo HD SEM.

2.4. Cytotoxicity

2.4.1. ADSC Isolation and Seeding.
ADSCs were isolated from adipose tissue according to the method described by Zuk et al. with modifications as previously described [8, 14, 15]. In brief, following removal of fibrous tissue adipose tissue was cut into small pieces of < 3 mm^3. The tissue was then further digested in Dulbecco's Modified Eagle's Medium/Nutrient Mixture F-12 Ham (DMEM/F12) containing 300 U/mL crude collagenase I (Invitrogen, Life Technologies Ltd., Paisley, UK)

for 30 minutes in an incubator (37°C, 5% CO_2). Following filtration through 70 μm Cell Strainers (BD Biosciences, Oxford, UK) the samples underwent centrifugation (290 × G, 5 min). Then the ADSC-rich cell preparation formed a pellet at the bottom of the tube. The ADSC cells were cultured for up to 2 passages. When the ADSCs reached approximately 80% confluence, subculture was performed through trypsinisation. For cell cultures analysis, 1.5 X 10^4 ADSCs at passage 2 were seeded on each polymer disk following sterilisation. Written consent was taken from all patient donors in the study and was approved by the North Scotland ethical review board, reference number 10/S0802/20.

2.4.2. AlamarBlue.
AlamarBlue, viability assay, was performed as described previously [8, 13]. Briefly following incubation of scaffolds with complete medium for 24 hours, scaffolds were seeded with 1.5 X 10^4 ADSC per well. At 1, 3, 7, and 14 days medium was removed and 10% AlamarBlue prepared in fresh media was added for 3 hours. Following incubation, AlamarBlue fluorescence was quantified at the respective excitation and emission wavelength of 540 and 595 nm. The mean fluorescent units for the six replicate cultures of three individual experiments were calculated for each exposure treatment and the mean blank value was subtracted from these.

2.4.3. Immunofluorescence: Rhodamine Phalloidin and DAPI.
To study adhesion and morphology of the ADSC onto the scaffolds, immunocytochemistry morphology staining was performed as described previously [8, 13]. At 24 hours, the media was removed and the cells were washed with PBS three times. Following this, cells were fixed with 4% (w/v) paraformaldehyde for 15 minutes. The scaffolds were then washed thrice in PBS/0.1% Tween-20 and washed in 0.1% tritonX100 to improve permeability for 5 minutes. The scaffolds were then stained with rhodamine phalloidin dye (Molecular Probes®, Life Technologies, Paisly, UK) in the ratio 1:40 (dissolved in 1 mL of methanol) in PBS for 40 minutes. Following washing the nuclei was stained with DAPI (Molecular Probes®, Life Technologies, Paisley, UK). The ADSCs on the scaffolds were visualised using confocal microscopy Zeiss LSM 710 (Zeiss, Jena, Germany). Image J (National Institute of Health, NIH) software was used to determine circularity of the ADSCs on the scaffolds.

2.5. Sterility Testing.
All samples were tested for the effectiveness of sterilisation as previously described [8]. In brief, scaffolds were immersed in tryptone soya broth (TSB) and fluid thioglycollate medium (THY) for cultivation of microorganisms (Wickham Laboratories, Hampshire) for 7 days. Sterile broth was considered the negative control and unsterilised samples as the positive control. Both broths were macroscopically observed every 1–3 days for clouding as an indicative of contamination and ineffective sterilisation. A clear broth was considered to have no infection and an effective sterilisation of the samples ($n = 9$).

2.6. Statistical Analysis.
All statistical analyses were performed using Prism software (GraphPad Inc., La Jolla, USA).

Means and standard deviations were calculated from numerical data. In figures, bar graphs represent means, whereas error bars represent 1 standard deviation (SD). A p value of ≤ 0.05 was defined as significant. The exact statistical analysis performed for each dataset is described in the figure legend.

3. Results

3.1. Material Characterisation

3.1.1. Visual Inspection after Sterilisation. All samples withstood treatment with UV, antibiotic/antimycotic, bleach, and plasma treatment. Although the POSS-PCU samples were unaffected by the autoclaving process, the POSS-PCL samples were destroyed; therefore, it was not possible to examine the autoclaved POSS-PCL samples further. Both the POSS-PCU and POSS-PCL samples held up well against gamma irradiation with slight discolouring/yellowing of the POSS-PCU samples being observed. Ethylene oxide gas caused slight yellow discolouring and ethanol visible deformation of POSS-PCL samples.

3.1.2. Tensiometry. Quantitative values of mechanical properties of POSS-PCL and POSS-PCU samples subjected to ethanol, bleach (SDIC), plasma, ethylene oxide gas, UV radiation, antibiotic/antibiotic treatment, gamma irradiation, autoclaving (only POSS-PCU), and unsterilised controls for each sterilisation method are presented in Table 1. No significant difference in elongation at break, Young's modulus, or maximum stress was seen between any of the POSS-PCU samples. The ultimate tensile stress for POSS-PCL increased from 0.56±0.08 MPa in control samples to 1.71±0.19 MPa after ethylene oxide gas treatment, to 1.45±0.04 after UV radiation, and to 1.17±0.15 after gamma irradiation ($p < 0.05$). This increase in tensile stress of ethylene oxide treated POSS-PCL translated itself to Young's modulus, which was also significantly increased compared to control (0.40±0.11 versus 0.18±0.0, $p < 0.05$). Compared to control POSS-PCL, UV radiated POSS-PCL and ethanol had a significantly shorter elongation at break ($p < 0.05$).

3.1.3. Water Contact Angle Measurements. The difference in surface hydrophilicity of POSS-PCL and POSS-PCU samples after each sterilisation technique was assessed by measuring the water contact angle (Figure 1). Plasma treated POSS-PCL and POSS-PCU samples had statistically significant lower contact angles when compared to untreated controls and other sterilisation methods ($p < 0.001$). Ethanol treatment of POSS-PCL was associated with a significant decrease in contact angles when compared to controls ($p < 0.05$). Amongst the treated POSS-PCU samples, UV radiation was associated with a statistically significant decrease in contact angle measurements compared to controls ($p < 0.05$).

3.1.4. Attenuated Total Reflectance Fourier Transform Infrared Spectroscopy (ATR-FTIR). Figure 2 shows the peak assignment in the FTIR spectra of unsterilised control POSS-PCL and POSS-PCU and POSS-PCL and POSS-PCU samples

TABLE 1: **Mechanical Properties of POSS-PCU and POSS-PCL after sterilisation processes.** Young's modulus, maximum stress, elongation at break, and thickness of control POSS-PCL and POSS-PCU and samples subjected to ethanol, bleach (SDIC), plasma, ethylene oxide, UV radiation, antibiotic/antimycotic treatment, and gamma irradiation.

Sterilisation Method		Control	Bleach (SDIC)	Ethanol	Gamma	Ethylene Oxide	UV	Antibiotic/Antimycotic	Plasma	Autoclaving
Young's Modulus (MPa)	POSS-PCL	0.18 ± 0.0	0.29 ± 0.02	0.20 ± 0.02	0.32 ± 0.04	0.40 ± 0.11	0.35 ± 0.02	0.25 ± 0.03	0.30 ± 0.05	N/A
	POSS-PCU	0.55 ± 0.04	0.57 ± 0.02	0.56 ± 0.02	0.56 ± 0.04	0.55 ± 0.02	0.55 ± 0.02	0.53 ± 0.01	0.55 ± 0.03	0.54 ± 0.02
Maximum Stress (MPa)	POSS-PCL	0.56 ± 0.08	0.61 ± 0.10	0.31 ± 0.11	1.17 ± 0.15	1.71 ± 0.19	1.45 ± 0.04	1.10 ± 0.14	0.80 ± 0.26	N/A
	POSS-PCU	0.83 ± 0.03	0.82 ± 0.06	0.85 ± 0.03	0.83 ± 0.01	0.84 ± 0.01	0.83 ± 0.01	0.81 ± 0.12	0.83 ± 0.02	0.82 ± 0.04
Elongation at break (%)	POSS-PCL	589.6 ± 26.2	549.7 ± 68	293.0 ± 61	462.2 ± 16	471.6 ± 39	387.5 ± 22	469.5 ± 11	500.1 ± 30	N/A
	POSS-PCU	283.3 ± 9.57	273.6 ± 10.01	279.2 ± 1.74	288.0 ± 3.05	288.6 ± 1.46	269.3 ± 8.55	273.8 ± 9.01	281.1 ± 4.12	280.5 ± 7.88
Thickness (mm)	POSS-PCL	2.0 ± 0	1.43 ± 0.15	1.525 ± 0.13	1.6 ± 0.22	1.2 ± 0.25	1.38 ± 0.05	1.55 ± 0.17	2.433 ± 0.12	N/A
	POSS-PCU	0.75 ± 0.04	0.77 ± 0.02	0.84 ± 0.02	0.77 ± 0.05	0.76 ± 0.02	0.77 ± 0.02	0.75 ± 0.03	0.76 ± 0.02	0.74 ± 0.06

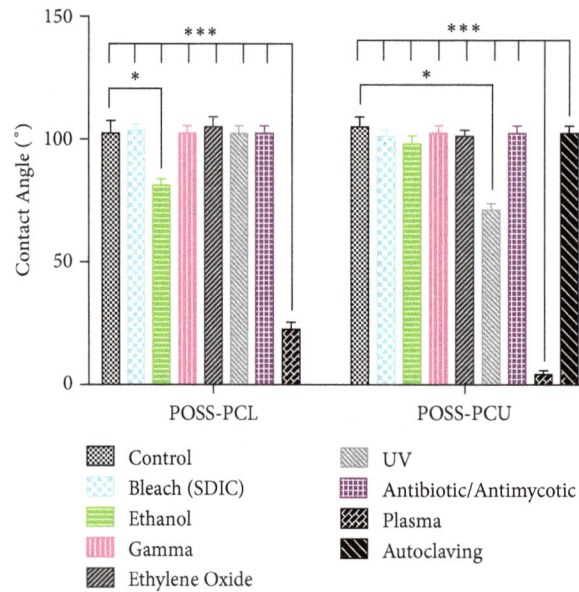

FIGURE 1: **Contact angles of POSS-PCL and POSS-PCU samples measured after different sterilisation techniques.** One-way ANOVA and Turkey's multiple comparison test was used to show statistical significance. ∗ indicates statistically significant differences (∗p < 0.05 and ∗ ∗ ∗p < 0.001).

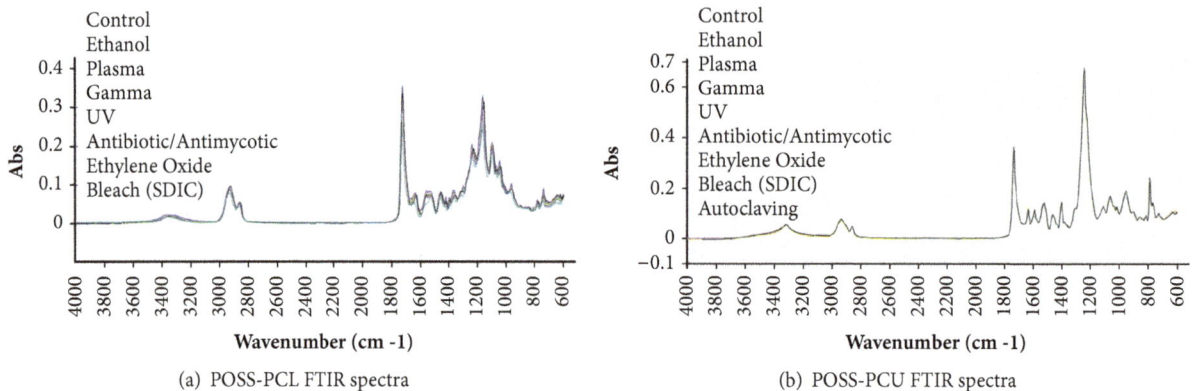

(a) POSS-PCL FTIR spectra

(b) POSS-PCU FTIR spectra

FIGURE 2: **POSS-PCL and POSS-PCU FTIR spectra after different sterilisation techniques.** [a] SDIC treatment led to a slight decrease in the peaks at 1634 cm^{-1} and 1557 cm^{-1} and an increase in the peak at 1520 cm^{-1} compared to untreated control POSS-PCL. [b] No major changes were noted in the FTIR spectra of POSS-PCU samples after different sterilisation methods.

after different sterilisation methods. Bleach (SDIC) treatment led to a slight decrease in the peaks at 1634 cm^{-1} and 1557 cm^{-1} and an increase in the peak at 1520 cm^{-1}. POSS-PCL FTIR spectra were not significantly affected by the other sterilisation techniques. FTIR spectra of POSS-PCU samples were unaffected by different sterilisation techniques.

3.1.5. Gel Permeation Chromatography (GPC). Gel permeation chromatography (GPC) results are summarised in Table 2. The untreated POSS-PCU was found to have a weight average molecular weight (M_w) of 91200 g/mol and a number average molecular weight (M_n) of 47700 g/mol, whereas POSS-PCL had an M_w of 361100 and M_n of 141000 g/mol. After ethanol, bleach, UV radiation, and antibiotic treatments

there was a negligible impact on M_w, for any of the samples. However, solely amongst POSS-PCL samples, ethanol caused a decrease of 21% in M_n, whilst UV increased M_n by 10% in comparison to unsterilised POSS-PCL. No major changes in molecular weight distributions were detected in autoclaved samples of POSS-PCU. However, autoclaved POSS-PCL samples showed a 52% decrease in M_w and 38% decrease in M_n. Exposure to gamma irradiation had a significant impact on all of the samples. M_n of POSS-PCU decreased significantly by 16%, whereas M_w increased by 48%. Meanwhile, gamma irradiation decreased both M_n and M_w of POSS-PCL by 68% and 58%, respectively. Ethylene oxide increased M_w of POSS-PCU by 23% and decreased M_n of POSS-PCU by 28%. Concurrently, it decreased M_w and M_n of POSS-PCL by 23% and 31%, respectively. Plasma had no significant impact on

TABLE 2: **Summary of gel permeation chromatography (GPC) results.** Molecular weight (M_w) and molecular number (M_n) values of POSS-PCL and POSS-PCU after different sterilisation techniques.

Sterilisation Method		Control	Bleach (SDIC)	Ethanol	Ethylene Oxide	Gamma	Plasma	Antibiotic/Antimycotic	Autoclaving	UV
Mass-average Molecular Weight (Mw) (g/mol)	POSS-PCL	361100	350900	356100	276300	151200	358600	391600	174600	365400
	POSS-PCU	91200	92000	89400	112200	135000	90800	92700	88300	93800
Number-average Molecular Weight (Mn) (g/mol)	POSS-PCL	141000	137600	111300	97500	44700	113300	160800	88100	155300
	POSS-PCU	47700	46300	47200	34200	40000	46700	46200	45500	46200

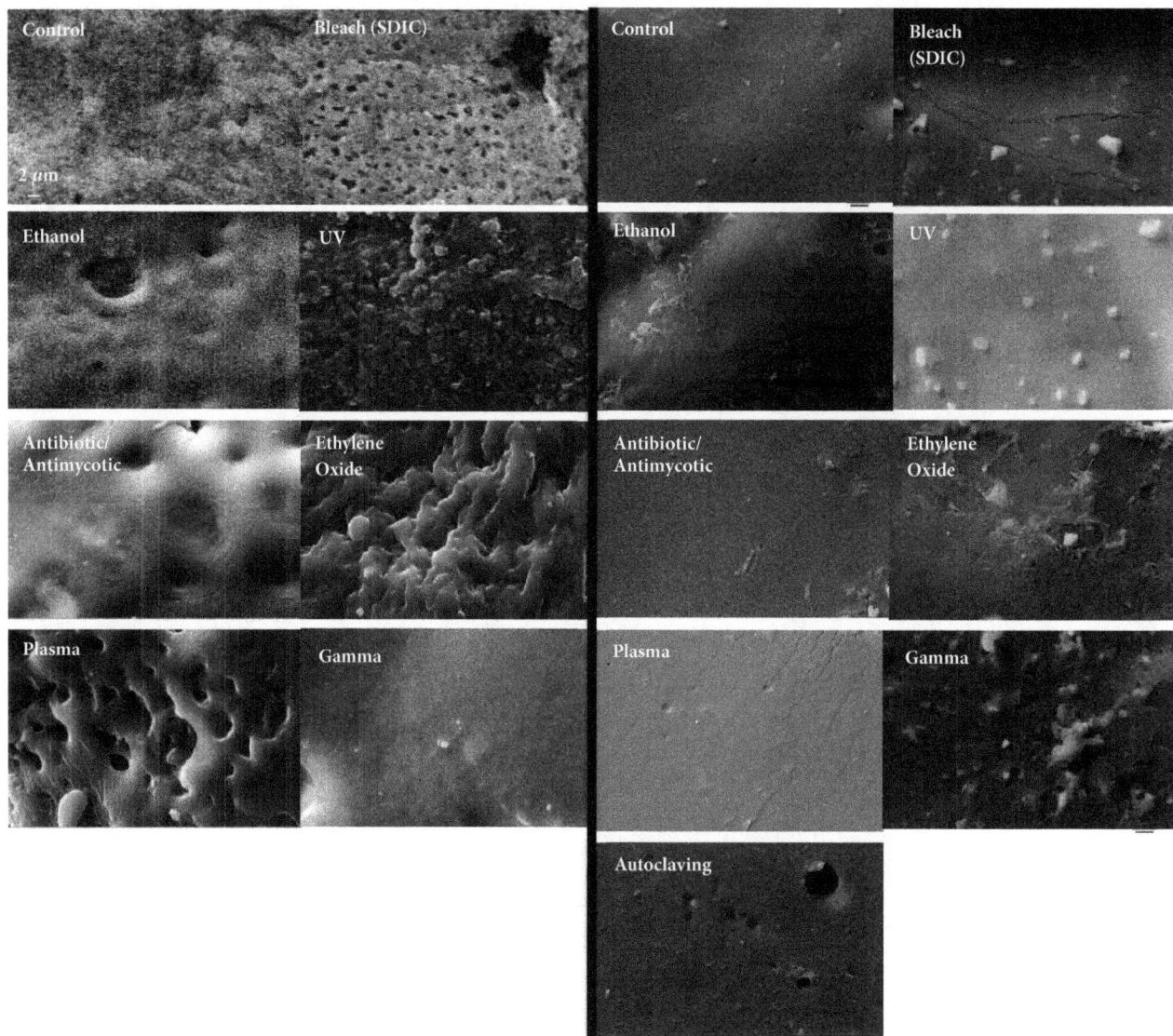

FIGURE 3: Scanning electron microscopy (SEM) images of POSS-PCL (left) and POSS-PCU (right) surfaces after different sterilisation techniques.

the molecular weight distribution of POSS-PCU or POSS-PCL polymers.

3.1.6. Scanning Electron Microscopy (SEM). SEM images of POSS-PCL and POSS-PCU after different sterilisation techniques are shown in Figure 3. Unsterilised POSS-PCL sample surface exhibits tufts and pits on the surface. Such tufts were lost after using ethanol, antibiotic/antimycotic, and gamma sterilisation with polymer melting and irregular reformation into larger and flatter ridges. SDIC treatment was associated with a marked increase in pits, whereas UV irradiation was associated with a pronounced increase in the number of tufts. Ethylene oxide was associated with larger tufts, whereas plasma treatment caused larger pits and tufts on the POSS-PCL surface (Figure 3). POSS-PCU samples, generally, showed little surface alterations after sterilisation.

Ethanol, SDIC, UV, EO, and gamma were associated with increased tufts. The antibiotic/antibiotic, autoclaving, and plasma sterilisation caused minimal changes on the POSS-PCU surface.

3.2. Cell Biocompatibility

3.2.1. AlamarBlue. The results of the alamarBlue viability assay after 1, 3, 7, 10, and 14 days of incubation are presented in Figure 4. ADSC cultured on POSS-PCU and POSS-PCL samples showed similar metabolic activity. At Days 7 and 10, plasma treated POSS-PCL was associated with the highest ADSC metabolic activity compared to any other sterilisation technique ($p < 0.05$). Ethanol and bleach (SDIC) sterilised POSS-PCL had a statistically significant higher ADSC metabolic activity compared to ADSC on gamma, ethylene

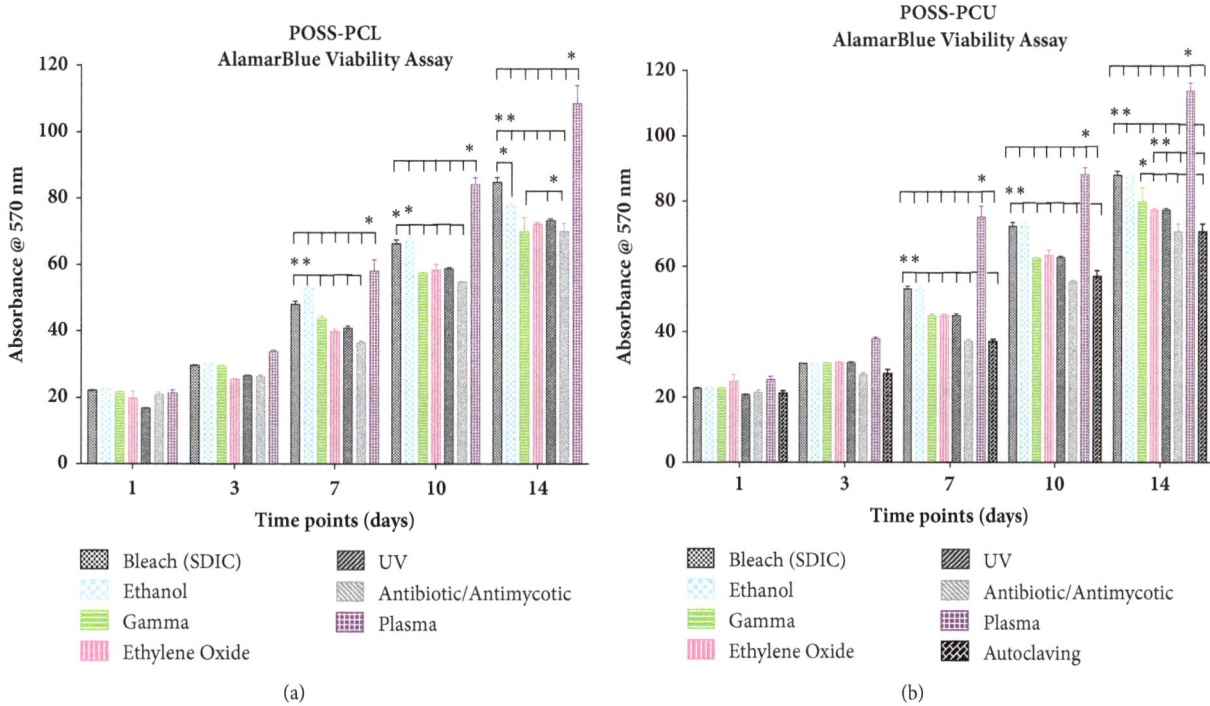

FIGURE 4: **AlamarBlue viability assay of adipose derived stem cells (ADSCs) after 1, 3, 7, 10, and 14 days of incubation on (a) POSS-PCL and (b) POSS-PCU samples**. Statistical significance was shown using two-way ANOVA and Turkey's multiple comparisons test. ∗ indicates statistically significant differences (∗p < 0.05).

oxide, UV radiation, antibiotic/antimycotic, and autoclaving sterilised POSS-PCL. At Day 14, the same observations were made. In addition, UV radiation sterilised POSS-PCL had a statistically significant higher ADSC metabolic activity compared to ADSC on antibiotic/antibiotic treated and gamma irradiated POSS-PCL (p < 0.05). Bleach (SDIC) sterilised POSS-PCL had a statistically significantly higher ADSC metabolic activity compared to ADSC on ethanol sterilised samples (p < 0.05). Amongst POSS-PCU samples, similar observations were made. At Days 7 and 10, plasma treated POSS-PCU was associated with the highest ADSC metabolic activity compared to any other sterilisation technique (p < 0.05). Ethanol and bleach sterilised POSS-PCU had a statistically significant higher ADSC metabolic activity compared to ADSC on gamma, ethylene oxide, UV radiation, antibiotic/antimycotic, and autoclaving sterilised POSS-PCU (p < 0.05). In addition, gamma, ethylene oxide, and UV radiation treated samples had a statistically significant higher ADSC metabolic activity compared to antibiotic/antimycotic and autoclaving (p < 0.05). At Day 14, the same observations were made. In addition, gamma irradiation was associated with significantly higher ADSC metabolic activity compared to UV and ethylene oxide (p < 0.05).

3.2.2. Immunofluorescence: Rhodamine Phalloidin and DAPI.
Confocal laser scanning microscopy images indicated that ADSC developed different morphologies when grown on differently sterilised surfaces (Figure 5). Cells grown on bleach sterilised POSS-PCL exhibited a more spread-out

phenotype compared to cells grown on surfaces exposed to other sterilisation techniques which demonstrated a more round character. In general, ADSC on POSS-PCL had a more round morphology compared to ADSC on POSS-PCU surfaces. Circularity measurements with Image J software showed that ADSC grown on bleach sterilised POSS-PCL had a significantly less circular morphology compared to ADSC on ethanol, ethylene oxide, gamma, antibiotic/antibiotic, and plasma (p < 0.05) but not on UV sterilised POSS-PCL (Figure 5(a)). There was no statistically significant difference between the morphology of the ADSCs on the POSS-PCU samples (Figure 5(b)).

3.3. Sterility Testing

3.3.1. Visual Inspection. The polymeric materials were incubated in TSB and THY for 7 days to test the efficiency of sterilisation with the resultant level of bacterial growth reported in Table 3. THY is a viscous growth medium with reduced oxygen levels, which tests the growth of anaerobic bacteria and other organisms capable of growing in reduced oxygen tension. No evidence of bacterial growth was observed on any of the materials tested after incubation in THY. TSB, however, is a general growth media for aerobic microorganisms and is designed for the growth of aerobic bacteria and yeasts and moulds. Amidst POSS-PCU sterilised samples, there was no sign of infection. Only the unsterilised control samples showed signs of infection. Amongst POSS-PCL samples, all

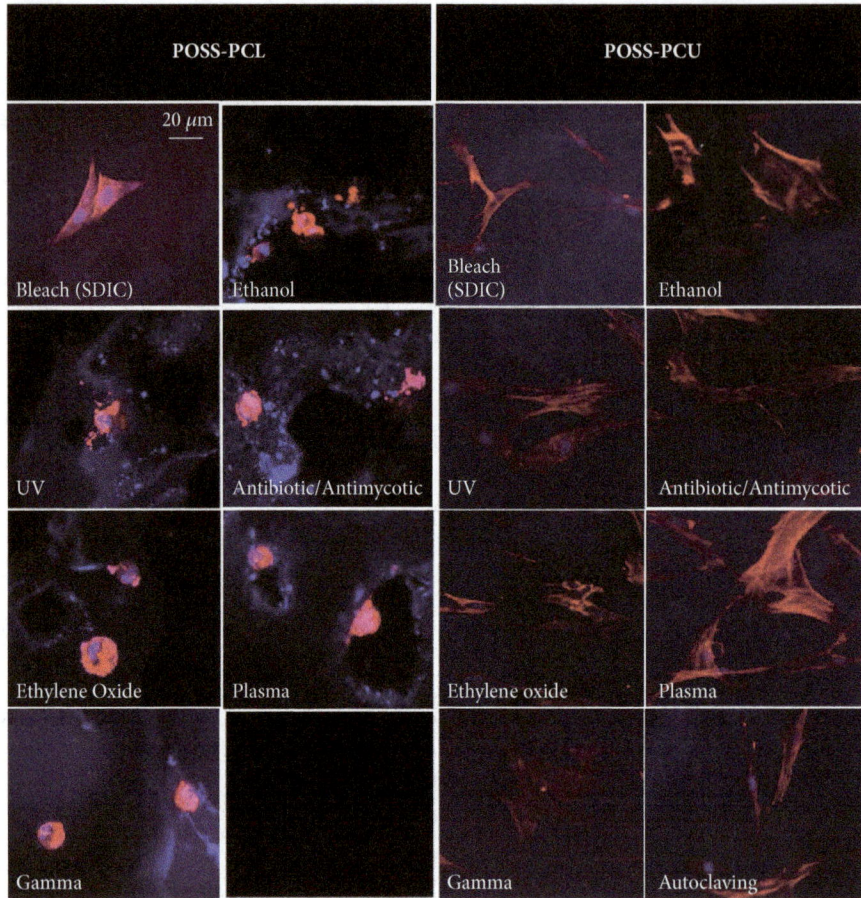

(a) Confocal microscope images of rhodamine phalloidin and DAPI stained adipose derived stem cells (ADSCs) cultured on POSS-PCL and POSS-PCU surfaces for 24 hours after different sterilisation techniques

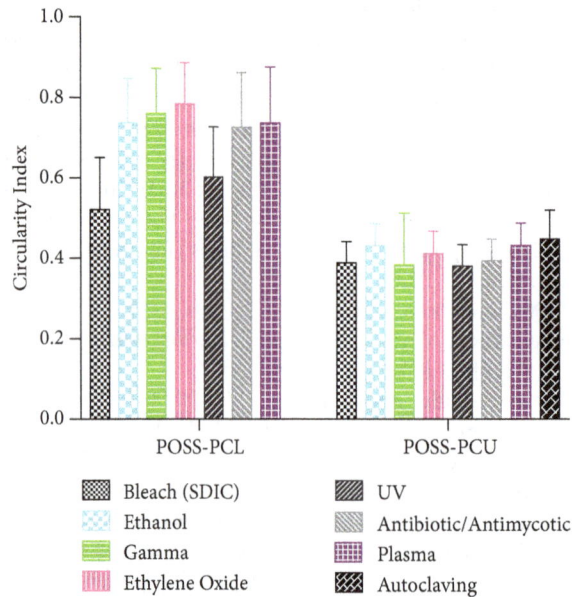

(b) The circularity quantification of the cultured adipose derived stem cells (ADSCs) seeded on POSS-PCL and POSS-PCU surfaces for 24 hours after different sterilisation techniques

FIGURE 5

TABLE 3: **Summary of POSS-PCL and POSS-PCU sterilisation efficacy for bleach (SDIC), ethanol, ethylene oxide, gamma, plasma, antibiotic/antimycotic, UV, and autoclaving sterilisation techniques.** All control samples (3) of POSS-PCU and POSS-PCL and 1 of 9 POSS-PCL samples sterilised using UV and antibiotic/antimycotic were infected.

Sterilisation Method		Control	Bleach (SDIC)	Ethanol	Ethylene Oxide	Gamma	Plasma	Antibiotic/Antimycotic	UV	Autoclaving
POSS-PCL	TSB	9/9	0/9	0/9	0/9	0/9	0/9	1/9	1/9	N/A
	THY	9/9	0/9	0/9	0/9	0/9	0/9	1/9	1/9	N/A
POSS-PCU	TSB	9/9	0/9	0/9	0/9	0/9	0/9	0/9	0/9	0/9
	THY	9/9	0/9	0/9	0/9	0/9	0/9	0/9	0/9	0/9

unsterilised control samples and one of nine samples sterilised using UV radiation and antibiotic/antimycotic showed signs of infection. Although there was minimal evidence of bacterial growth in the sterility studies of scaffolds with UV and antibiotic/antimycotic treatment, no evidence was seen in the viability testing or the morphology assessment to invalidate the assays.

4. Discussion

Polymer degradation after sterilisation techniques can be assessed using (i) macroscopic characterisation methods, providing information on "bulk" properties such as mechanical performance, (ii) microscopic characterisation, looking at molecular weight and its dispersity, (iii) characterisation of the molecular structure and composition, such as FTIR analysis, and (iv) surface characterisation, i.e., scanning electron microscopy and surface wettability [16]. According to these criteria, in this study, we performed a thorough investigation of the bulk, surface, and molecular properties of POSS-PCL and POSS-PCU scaffolds after several sterilisation techniques, including plasma, gamma irradiation, ethylene oxide, UV radiation, antibiotic/antimycotic, 70% ethanol, and bleach treatments. As the biodegradable counterpart of POSS-PCU, POSS-PCL was considerably more susceptive to changes in the molecular weight distribution, mechanical properties, and surface morphology and chemistry after several sterilisation techniques.

The three leading sterilisation methodologies with FDA approved for medical devices are gamma irradiation, autoclaving, and ethylene oxide. Autoclaving has been the earliest method for the sterilisation of biomaterials. Sterilisation with moist heat in an autoclave is usually performed at temperatures equal to or higher than 121°C; dry heat sterilisation requires considerably higher temperatures to effectively inactivate bacterial spores. The suitability of steam sterilisation has been questioned for polyurethanes, as the high temperature may soften the polymer and deform the material [17]. In this study POSS-PCU material's properties were unaffected as shown by mechanical properties, surface chemistry as shown by contact angle and FTIR, and surface topography as shown by SEM. Polymer degradation was determined by measuring changes in molecular weight and mass immediately after the sterilisation process using GPC analysis. Polymeric biomaterials of low molecular weight present less chemical and mechanical resistance in relation to the same material of higher molecular weight [18]. Autoclaving was an optimal sterilisation technique for POSS-PCU,

with no evidence of changes in molecular number and weight. However, POSS-PCL was unsuited to autoclaving as shown by visual changes after sterilisation and changes in molecular weight and number, demonstrating the polymer underwent a degree of hydrolysis. This finding is in accordance with the literature, where biodegradable polymers break down due to the high temperatures of autoclaving [19].

Gamma sterilisation involves ionising radiation from either cobalt 60 isotope or accelerated electrons. Gamma irradiation can generate free radicals in the polymer, causing surface oxidation and subsequent degradation due to chain scission and cross-linking with increasing dosages of radiation [20]. However, gamma irradiation has definite advantages in that it is penetrating and free of residues. Also, material temperatures are only moderately elevated during sterilisation, which is an advantageous feature for the sterilisation of bioresorbable implants, where temperature and dose conditions need close consideration. Microbiological validation experiments according to ISO 11137 [21] have shown gamma irradiation in dry ice at doses of 16 kGy or more effectively inactivates microorganisms. In this study, gamma irradiation was associated with effective sterilisation of POSS-PCL and POSS-PCU scaffolds. Polyurethanes have also shown some degradation after gamma radiation. POSS-PCL was found to have a significant decrease in the molecular weight and number, indicating degradation may have occurred. Several biodegradable polymers have shown to be susceptible to gamma irradiation [19]. Holy et al. found that polylactide-co-glycolide scaffolds had a 50% loss in their molecular weight after gamma irradiation [22].

Gamma irradiation has been shown to cause the release of free radicals in polymers, causing surface oxidation and subsequent degradation of the polymer. Surface oxidation of the polyurethanes is indicated by a strong yellowing of samples, which was also found in this study with some POSS-PCL scaffolds. Despite the degradation and potential surface oxidation of gamma on PCL scaffolds, the water contact of POSS-PCL did not change after gamma sterilisation compared to control. There are reports in the literature that show no change in water contact angle of biomaterials following gamma sterilisation [23, 24]. This could be due to the penetrating action of the gamma sterilisation affecting bulk properties rather than surface properties [23, 24].

Gamma irradiation showed no changes in material properties including mechanical, surface chemistry, or molecular number for POSS-PCU scaffolds. However, SEM did show some surface changes after gamma sterilisation on POSS-PCU. Although suitable for POSS-PCU, POSS-PCL was unsuitable for gamma sterilisation with significant changes in molecular weight and number.

EtO is the final sterilisation technique, which has been approved for medical implants. EtO is a liquid below 11° so in contrast to steam sterilisation it is a low temperature method [17]. In addition to being toxic and explosive causing a significant occupational health and safety hazard, there are concerns on removing all traces of the EtO from the implant after sterilisation [15]. In this study, SEM showed some surface material changes after EtO without changing the bulk properties including tensile strength or molecular weight for POSS-PCU. For POSS-PCL EtO caused significant loss in molecular weight and number and changes to the surface morphology by SEM. EtO has also been shown in the literature to be an inappropriate sterilisation technique for biodegradable scaffolds due to changes in the structural and biochemical properties [17, 22]. For example, Hooper et al. showed that EtO of polycarbonate materials caused a reduction in yield strength and faster degradation rates compared to nonsterilised controls [25].

Although not considered sterilisation techniques for medical devices, we also evaluated the effect of different disinfectant methods. Techniques such as bleach, using antibiotics, and UV radiation are used in the laboratory setting to sterilise scaffolds for *in vitro* and in vivo examination. The use of SDIC prevented contamination of the samples and was not associated with any significant changes in mechanical properties or any visual deformation of both nanocomposite polymers. Although longstanding polymer exposure to bleach has been associated with increased hydrophobicity [26], we did not observe any significant changes in surface wettability of POSS-PCL scaffolds. Bleach has been shown to increase roughness and porosity of polymeric scaffolds [26]. In this study, we observed similar submicron changes to POSS-PCL scaffold structure, mainly an increase in the number of pits. SDIC was the sole sterilisation method, which showed slight changes in the surface chemistry of the scaffolds on the FTIR spectra, likely due to hydrolysis.

Ethanol and UV are all useful sterilisation techniques for laboratory analysis and provided sterility of the POSS-PCU and POSS-PCL nanocomposite scaffolds. For POSS-PCU scaffolds, UV and ethanol did cause some changes to the surface as shown by SEM although bulk mechanical properties were not affected. UV decreased the contact angle of POSS-PCU, which is consistent with other reports of UV creating hydrophilic surfaces [27–29]. UV sterilisation may have oxidized the surface and caused a drop in the water contact angle [27–29].

POSS-PCL scaffolds also maintained their bulk mechanical properties after ethanol and UV modification. However, reports in the literature show that whilst Gram-positive, Gram-negative, acid-fast bacteria, and lipophilic viruses show high susceptibility to concentrations of ethanol in water ranging from 60 to 80%, hydrophilic viruses and bacterial spores are resistant to the microbial effects of ethanol [30]. Antibiotic solution had minimal effect on the surface or bulk properties of POSS-PCU or POSS-PCL. Furthermore, reports have shown that different microorganisms have varying sensitivity to UV [19]. For example UV easily destroys vegetative

bacteria but bacterial spores and prions are more resistant. Some viruses are considered inactivated whilst others are resistant [19]. Antibiotic solutions inactivate bacteria by interfering with DNA replication [19]. However, such solutions are only effective against vegetative bacteria and spores whilst fungi, moulds, and viruses are not affected [19]. Therefore, ethanol, UV, and antibiotics are considered to be chemical disinfectants instead of sterilising techniques and not used in the sterilisation of biomedical devices [19].

Plasma technology is defined as neutral ionised gas and includes photons, electrons, positive, and negative ions, atoms, free radicals, and nonexcited molecules [31]. Several types of plasma techniques exist but in this study radiofrequency plasma was utilised. Argon plasma demonstrated no changes in the material characteristics of both POSS-PCU including mechanical, molecular number and weight, surface chemistry by FTIR, or surface topography by SEM. However, argon plasma did cause some surface topographical changes of POSS-PCL due to potential etching effect of plasma treatment. The surface wettability significantly decreased after argon plasma for both POSS-PCU and POSS-PCL. The sterilisation process itself may affect the surface and bulk properties to negatively influence its cytocompatibility and biocompatibility. The *in vitro* compatibility of the sterilised scaffolds was performed to evaluate the release of cytotoxic molecular weight products from the sterilisation. Viability data demonstrated that significantly more cells were able to adhere to POSS-PCU and POSS-PCL after argon plasma surface modification compared to other sterilisation techniques by Day 7 (Figure 4). Several studies have found that decreasing the water contact angle of scaffolds below 90° due to plasma treatment induces a hydrophilic surface causing a greater number of cells adhere to the biomaterial surface [32, 33]. Both POSS-PCU and POSS-PCL became more hydrophilic after argon plasma surface modification, without causing bulk mechanical properties (mechanical and molecular number/weight). With optimal sterilisation efficacy, maintenance of mechanical properties, and improvement in cell biocompatibility, plasma demonstrated the optimal sterilisation process in this study.

Menashi et al. first reported the use of argon for sterilisation in 1968 after sterilising the surface of vials contaminated by bacterial spores [34]. Since then the gas type and the power of discharge have shown to influence the efficacy of the treatment. Analysis of RF plasmas has suggested that the chemical species created in the discharge are responsible for the destruction of the microorganisms and the UV or thermal effects are secondary effects [35, 36]. Argon plasma has shown to sterilise titanium implant surfaces [37]. Other types of plasma have been investigated for the sterilisation of materials. The comparison of the ethanol, dry oven, autoclave, UV radiation, and hydrogen peroxide plasma as sterilisation techniques for electrospun PLA scaffolds [38] showed that UV irradiation and hydrogen peroxide were the most effective without affecting the chemical and morphological features. The authors demonstrated the hydrogen peroxide produced destructive hydroxyl free radicals, which can attack membrane lipids, DNA, and other essential cell components to deactivate the microorganisms [38].

There were limitations to this work. During the plasma treatment the scaffolds were exposed to the unsterile environment whilst moving them from the plasma chamber. To further optimise the plasma treatment the apparatus will be placed in a sterile laminar flow cabinet in the future.

In summary, argon plasma maintained the properties of the nanocomposite scaffolds in addition to improving cell adhesion (Figure 4). Plasma sterilisation is easily transferrable to clinical practice for sterilisation of materials as it is simple to operate and the lack of toxic residuals makes it safe for the operator.

5. Conclusion

In conclusion, the FDA approved sterilisation technique of autoclaving maintained POSS-PCU characteristics but was unsuitable for biodegradable POSS-PCL scaffolds. Taking all results into account, we argue that plasma surface modification using argon gas may effectively sterilise polyurethane biomaterials as well as improve cytocompatibility of tissue engineering scaffolds by modification in surface wettability and topography. Argon plasma should be explored and optimised for the sterilisation of further biomaterials.

Disclosure

Michelle Griffin and Naghmeh Naderi are equal first authors.

Conflicts of Interest

The authors declare that they have no conflicts of interest.

Acknowledgments

We would like to thank the funding from Medical Research Council and Action Medical Research, which provided Michelle Griffin a clinical fellowship to conduct this work, GN 2339.

References

[1] M. T. Lam and J. C. Wu, "Biomaterial applications in cardiovascular tissue repair and regeneration," *Expert Review of Cardiovascular Therapy*, vol. 10, no. 8, pp. 1039–1049, 2012.

[2] J. P. Guggenbichler, "Incidence and clinical implication of nosocomial infections associated with implantable biomaterials - catheters, ventilator-associated pneumonia, urinary tract infections," *GMS Krankenhhyg Interdiszip*, vol. 6, 2011.

[3] L. Montanari, M. Costantini, E. C. Signoretti et al., "Gamma irradiation effects on poly(DL-lactictide-co-glycolide) microspheres," *Journal of Controlled Release*, vol. 56, no. 1–3, pp. 219–229, 1998.

[4] FDA standards, http://www.fda.gov/MedicalDevices/DeviceRegulationandGuidance/Standards/default.htm.

[5] N. Hirata, K.-I. Matsumoto, T. Inishita, Y. Takenaka, Y. Suma, and H. Shintani, "Gamma-ray irradiation, autoclave and ethylene oxide sterilization to thermosetting polyurethane: Sterilization to polyurethane," *Radiation Physics and Chemistry*, vol. 46, no. 3, pp. 377–381, 1995.

[6] A. Simmons, J. Hyvarinen, and L. Poole-Warren, "The effect of sterilisation on a poly(dimethylsiloxane)/poly(hexamethylene oxide) mixed macrodiol-based polyurethane elastomer," *Biomaterials*, vol. 27, no. 25, pp. 4484–4497, 2006.

[7] M. Ahmed, G. Punshon, A. Darbyshire, and A. M. Seifalian, "Effects of sterilization treatments on bulk and surface properties of nanocomposite biomaterials," *Journal of Biomedical Materials Research Part B: Applied Biomaterials*, vol. 101, no. 7, pp. 1182–1190, 2013.

[8] N. Naderi, M. Griffin, E. Malins et al., "Slow chlorine releasing compounds: A viable sterilisation method for bioabsorbable nanocomposite biomaterials," *Journal of Biomaterials Applications*, vol. 30, no. 7, pp. 1114–1124, 2016.

[9] L. Yildirimer and A. M. Seifalian, "Sterilization-induced changes in surface topography of biodegradable POSS-PCLU and the cellular response of human dermal fibroblasts," *Tissue Engineering - Part C: Methods*, vol. 21, no. 6, pp. 614–630, 2015.

[10] T. Jacobs, H. Declercq, N. De Geyter et al., "Enhanced cell-material interactions on medium-pressure plasma-treated polyhydroxybutyrate/polyhydroxyvalerate," *Journal of Biomedical Materials Research Part A*, vol. 101, no. 6, pp. 1778–1786, 2013.

[11] Y. Wan, X. Qu, J. Lu et al., "Characterization of surface property of poly(lactide-co-glycolide) after oxygen plasma treatment," *Biomaterials*, vol. 25, no. 19, pp. 4777–4783, 2004.

[12] S. G. Wise, A. Waterhouse, A. Kondyurin, M. M. Bilek, and A. S. Weiss, "Plasma-based biofunctionalization of vascular implants," *Nanomedicine*, vol. 7, no. 12, pp. 1907–1916, 2012.

[13] M. F. Griffin, R. G. Palgrave, A. M. Seifalian, P. E. Butler, and D. M. Kalaskar, "Enhancing tissue integration and angiogenesis of a novel nanocomposite polymer using plasma surface polymerisation, an in vitro and in vivo study," *Biomaterials Science*, vol. 4, no. 1, pp. 145–158, 2016.

[14] P. A. Zuk, M. Zhu, H. Mizuno et al., "Multilineage cells from human adipose tissue: implications for cell-based therapies," *Tissue Engineering Part A*, vol. 7, no. 2, pp. 211–228, 2001.

[15] M. F. Griffin, A. Ibrahim, A. M. Seifalian, P. E. M. Butler, D. M. Kalaskar, and P. Ferretti, "Chemical group-dependent plasma polymerisation preferentially directs adipose stem cell differentiation towards osteogenic or chondrogenic lineages," *Acta Biomaterialia*, vol. 50, pp. 450–461, 2017.

[16] P. Vermette, P. S. Lévesque, and H. J. Griesser, "Biomedical degradation of polyurethanes," in *Biomedical Applications of Polyurethanes*, P. Vermette, Ed., pp. 97–159, Landes Bioscience, 2001.

[17] M. K. Lamba, K. A. Woodhouse, and S. L. Cooper, *Polyurethanes in Biomedical Applications*, CRC Press, 1997.

[18] E. F. Lucas, B. G. Soares, and E. E. C. Monteiro, *Caracterização de polímeros: determinação de peso molecular e análise térmica*, E. E-papers, Ed., Rio de Janeiro, Brazil, 1st edition, 2001.

[19] Z. Dai, J. Ronholm, Y. Tian, B. Sethi, and X. Cao, "Sterilization techniques for biodegradable scaffolds in tissue engineering applications," *Journal of Tissue Engineering*, vol. 7, 2016.

[20] D. Mohr, M. Wolff, and T. Kissel, "Gamma irradiation for terminal sterilization of 17β-estradiol loaded poly-(D,L-lactide-co-glycolide) microparticles," *Journal of Controlled Release*, vol. 61, no. 1-2, pp. 203–217, 1999.

[21] ISO 11137, "Sterilisation of Health Care Products, Requirements for Validation and Routine Control, Radiation Sterilisation," 2011.

[22] C. E. Holy, C. Cheng, J. E. Davies, and M. S. Shoichet, "Optimizing the sterilization of PLGA scaffolds for use in tissue engineering," *Biomaterials*, vol. 22, no. 1, pp. 25–31, 2001.

[23] T. J. Kinnari, J. Esteban, N. Zamora et al., "Effect of surface roughness and sterilization on bacterial adherence to ultra-high molecular weight polyethylene," *Clinical Microbiology and Infection*, vol. 16, no. 7, pp. 1036–1041, 2010.

[24] R. Colaço, E. Pires, T. Costa, and A. P. Serro, "Influence of sterilization with γ-irradiation in the degradation of plasma sprayed hydroxyapatite coatings," *Materials Science Forum*, vol. 514–516, no. 2, pp. 1054–1058, 2006.

[25] K. A. Hooper, J. D. Cox, and J. Kohn, "Comparison of the effect of ethylene oxide and γ-irradiation on selected tyrosine-derived polycarbonates and poly(L-lactic acid)," *Journal of Applied Polymer Science*, vol. 63, no. 11, pp. 1499–1510, 1997.

[26] B. Madsen, "Effect of sterilisation techniques on the physicochemical properties of polysulfone hollow fibers," *Journal of Applied Polymer Science*, pp. 3429–3436, 2011.

[27] T. Tuna, M. Wein, M. Swain, J. Fischer, and W. Att, "Influence of ultraviolet photofunctionalization on the surface characteristics of zirconia-based dental implant materials," *Dental Materials*, vol. 31, no. 2, pp. e14–e24, 2015.

[28] T. Zubkov, D. Stahl, T. L. Thompson, D. Panayotov, O. Diwald, and J. T. Yates Jr., "Ultraviolet light-induced hydrophilicity effect on TiO2(110) (1×1). Dominant role of the photooxidation of adsorbed hydrocarbons causing wetting by water droplets," *The Journal of Physical Chemistry B*, vol. 109, no. 32, pp. 15454–15462, 2005.

[29] H. Zhang, S. Komasa, C. Mashimo, T. Sekino, and J. Okazaki, "Effect of ultraviolet treatment on bacterial attachment and osteogenic activity to alkali-treated titanium with nanonetwork structures," *International Journal of Nanomedicine*, vol. 12, pp. 4633–4646, 2017.

[30] J. F. Gardner and M. M. Peel, *Introduction to Sterilisation, Disinfection and Infection Control*, Churchill Livingstone, New York, NY, USA, 1991.

[31] S. G. Wise, A. Waterhouse, A. Kondyurin, M. M. Bilek, and A. S. Weiss, "Plasma-based biofunctionalization of vascular implants," *Nanomedicine*, vol. 7, no. 12, pp. 1907–1916, 2012.

[32] T. Hoshino, I. Saito, R. Kometani et al., "Improvement of neuronal cell adhesiveness on parylene with oxygen plasma treatment," *Journal of Bioscience and Bioengineering*, vol. 113, no. 3, pp. 395–398, 2012.

[33] L. Canullo, T. Genova, H.-L. Wang, S. Carossa, and F. Mussano, "Plasma of argon increases cell attachment and bacterial decontamination on different implant surfaces," *The International Journal of Oral & Maxillofacial Implants*, vol. 32, no. 6, pp. 1315–1323, 2017.

[34] S. S. Block, *Disinfection, Sterilization and Preservation*, Lippincott Williams and Wilkins, 2001.

[35] M. Laroussi and F. Leipold, "Evaluation of the roles of reactive species, heat, and UV radiation in the inactivation of bacterial cells by air plasmas at atmospheric pressure," *International Journal of Mass Spectrometry*, vol. 233, no. 1–3, pp. 81–86, 2004.

[36] M. Moreau, N. Orange, and M. G. J. Feuilloley, "Non-thermal plasma technologies: New tools for bio-decontamination," *Biotechnology Advances*, vol. 26, no. 6, pp. 610–617, 2008.

[37] M. Annunziata, L. Canullo, G. Donnarumma, P. Caputo, L. Nastri, and L. Guida, "Bacterial inactivation/sterilization by argon plasma treatment on contaminated titanium implant surfaces: In vitro study," *Medicina Oral Patología Oral y Cirugía Bucal*, vol. 21, no. 1, pp. e118–e121, 2016.

[38] A. Rainer, M. Centola, C. Spadaccio et al., "Comparative study of different techniques for the sterilization of poly-L-lactide electrospun microfibers: Effectiveness versus material degradation," *The International Journal of Artificial Organs*, vol. 33, no. 2, pp. 76–85, 2010.

16

Anticancer Activity of Chitosan, Chitosan Derivatives, and their Mechanism of Action

Hari Sharan Adhikari ⑩[1] **and Paras Nath Yadav** ⑩[2]

[1]*Department of Chemistry, Western Region Campus, Institute of Engineering, Tribhuvan University, Pokhara, Nepal*
[2]*Central Department of Chemistry, Tribhuvan University, Kathmandu, Nepal*

Correspondence should be addressed to Paras Nath Yadav; pnyadav219@gmail.com

Academic Editor: Alexander Seifalian

Tailoring of chitosan through the involvement of its amino, acetamido, and hydroxy groups can give derivatives of enhanced solubility and remarkable anticancer activity. The general mechanism of such activity is associated with the disturbances in normal functioning of cell cycle, interference to the central dogma of biological system from DNA to RNA to protein or enzymatic synthesis, and the disruption of hormonal path to biosynthesis to inhibit the growth of cancer cells. Both chitosan and its various derivatives have been reported to selectively permeate through the cancer cell membranes and show anticancer activity through the cellular enzymatic, antiangiogenic, immunoenhancing, antioxidant defense mechanism, and apoptotic pathways. They get sequestered from noncancer cells and provide their enhanced bioavailability in cancer cells in a sustained release manner. This review presents the putative mechanisms of anticancer activity of chitosan and mechanistic approaches of structure activity relation upon the modification of chitosan through functionalization, complex formation, and graft copolymerization to give different derivatives.

1. Introduction

The source of chitosan (Figure 1) is chitin $(C_8H_{13}O_5N)_n$, a natural biopolymer (Figure 2) most abundant in exoskeletons of crustaceans and insect cuticles, cell walls of fungi, shells of mollusks, etc. Chitin consists of 2-acetamido-2-deoxy- β -D-glucose monomers (N-acetyl glucosamine units) linked through β (1⟶4) linkages [1] and chitosan is a polymer of deacetyl α-(1, 4) glucosamine $(C_6H_{11}O_4N)_n$ units that can typically be obtained by deacetylation of chitin with NaOH [2, 3] (Figure 3) after demineralization and deproteinization of the crustacean shells or exoskeletons (Scheme 1).

The degree of deacetylation (DDA) of chitin ranges from 60 to 100 % and molecular weight of commercially obtained chitosan ranges from 3800 to 20,000 Daltons. [4] It behaves as a pharmaceutical excipient [5], permeation enhancer [6], and a hemostatic agent [7] utilized as nonwoven sheet in wound healing and dressing [8] and targeted drug delivery with more efficiency and less side effects [9]. It is a multipurpose material [4] due to its nontoxicity, biocompatibility, biodegradability, and adsorptive behavior [10–12]. It has been

found to exert anticancer activity with minimal toxicity on noncancer cells [13] and such activity against different cancer cell lines significantly depends upon molecular weight and DDA [10] affected by the distribution pattern of β-(1,4)-linked N-acetylglucosamine and D-glucosamine units along the oligomeric chain [14, 15]. The uptake of chitosan nanoparticles by cultured fibroblasts was found to increase with the increase in DDA [16]. Y. Xu et al. showed antiangiogenic activity of chitosan nanoparticles [17]. Soluble form of chitosan oligosaccharide with low molecular weight has been reported to show remarkable biological activities and suppression of tumor growth [10, 18–22].

2. Chitosan and Its Derivatives as Anticancer Agents

Several derivatives of improved solubility and wide applications can be synthesized as a result of chemical modification of chitosan [23–27, 27–38]. Such derivatization of chitosan due to amino group and acetamido residue has been shown to give the compounds of enhanced solubility and biological

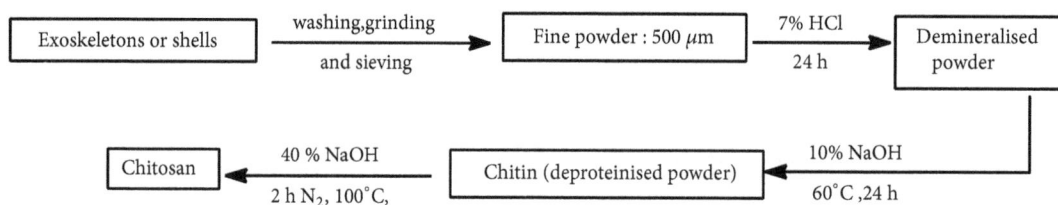

SCHEME 1: Steps involved in the preparation of chitosan.

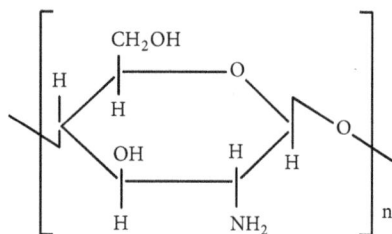

FIGURE 1: Structure of chitosan.

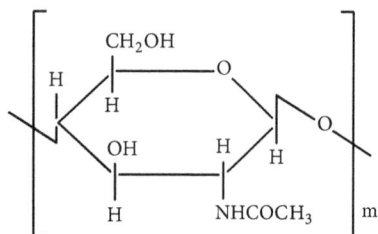

FIGURE 2: Structure of chitin.

activity [4]. Cell toxicity of 2-phenylhydrazine (or hydrazine) thiosemicarbazone chitosan is associated with its antioxidant behavior due to scavenging of cancer-causing free radicals [39], and the oxidative stress arising from imbalance between antioxidant defense and free radicals production may favor the etiological condition of cancer [40, 41]. Antitumor activity of chitosan-metal complexes is due to their interaction with deoxyribonucleic acid (DNA) [42] and free radicals scavenging behavior [42–44]. Antitumor property of the derivatives carboxymethyl chitosan (CMCS) [45], chitosan thymine conjugate [46], sulfated chitosan (SCS) and sulfated benzaldehyde chitosan (SBCS) [47], glycol-chitosan (GChi) and N-succinyl chitosan (Suc-Chi) conjugates [48], furanoallocolchicinoid chitosan conjugate [49], and polypyrrole chitosan [50, 51] from different cellular apoptotic pathways has been reported in literatures.

3. Synthetic Routes of Anticancer Derivatives of Chitosan

3.1. 2-Phenylhydrazine (or Hydrazine) Thiosemicarbazone Chitosan. Synthesis of 2-phenylhydrazine (or hydrazine) thiosemicarbazone chitosan (Zhong Zhimei et al.) (Figure 4) [39] involves stirring of the reaction mixture of phenylhydrazine (or hydrazine) dithiocarboxylate intermediate with chitosan in dimethyl sulfoxide (DMSO) at 100°C for 8 h and

cooling in acetone at 4°C for 10 h. Then its yellow precipitate is soxhlet extracted with dichloromethane for 24 h [39]. Antitumor activity of thiosemicarbazones is associated with lowering of cellular oxidative damage due to scavenging of cancer-causing free radicals [39].

The antioxidant activities of chitosan and thiosemicarbazone-chitosan derivatives measured by superoxide anion scavenging assay revealed more scavenging effect of thiosemicarbazone-chitosan than chitosan [39]. The scavenging effect of high molecular weight chitosan (HMWC) (M_w =200 k Da), water soluble chitosan (Mw=8 k Da), 2-phenylhydrazine thiosemicarbazone-chitosan (higher M_w), and hydrazine thiosemicarbazone chitosan (lower M_w) was found 0.4, 12.67, 35.23, and 43.12, respectively [39]. The data showed greater scavenging effect of thiosemicarbazone chitosan than chitosan and also more scavenging with decrease in M_w of chitosan.

3.2. Chitosan–Metal Complex. Polyfunctional nature of chitosan makes it a cationic polymer of complexing behavior with several metal ions [52]. A suitable ratio of metal ion to chitosan is essential for antitumor activity of a complex [42] and such ratio can be established by breaking the chain at weak points caused by coordinating bonds. The breakages at weak points to get complexes of uniform molecular weight can be carried out by oxidative hydrolysis with oxidants such as H_2O_2, O_3, and CH_3COOOH by controlling the coordinating conditions such as speed of stirring and rate of addition [44]. Study of degradation of chitosan by hydrogen peroxide has also shown the decrease in Mw with increase in temperature, time, and hydrogen peroxide concentration. Decrease in M_w of chitosan from 51 k Da to 1.2 k Da was found to be accompanied by structural changes, associated with 2.86 mmol/g formation of carboxyl group and deamination and with 40% loss of amino groups of the products [53]. The rate of H_2O_2 oxidative degradation of chitosan was found to increase with pH owing to degradation enhancing effect of hydroxy radicals. So, the degradation could be controlled by controlling the pH of solution [53].

Square planar chitosan copper (II) complex of potential antitumor activity is prepared by the reaction of 0.5 g of chitosan in 50 ml of 1% acetic acid containing copper sulfate in 1:0.4 molar ratio of chitosan to $CuSO_4.5H_2O$. The solution is neutralized by dilute ammonia solution, stirred for three hours at 80°C, and cooled down to room temperature and the green precipitate of the complex is obtained by the addition of ethanol [42] (Figure 5).

Tumor cell lines 293 and HeLa and normal lung fibroblast cell line HLF plated in a 100 μL/well at a density of

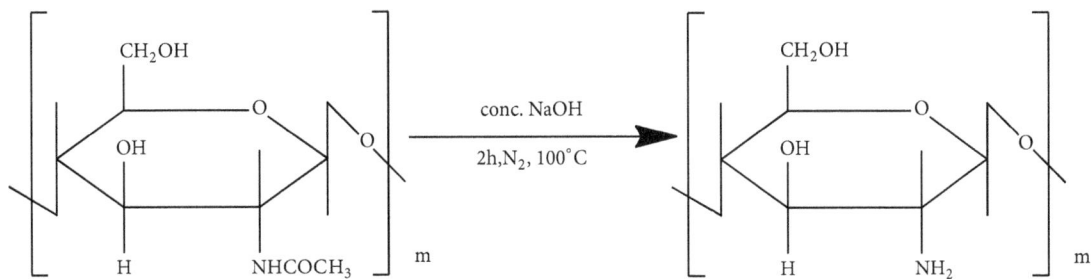

FIGURE 3: Deacetylation of chitin.

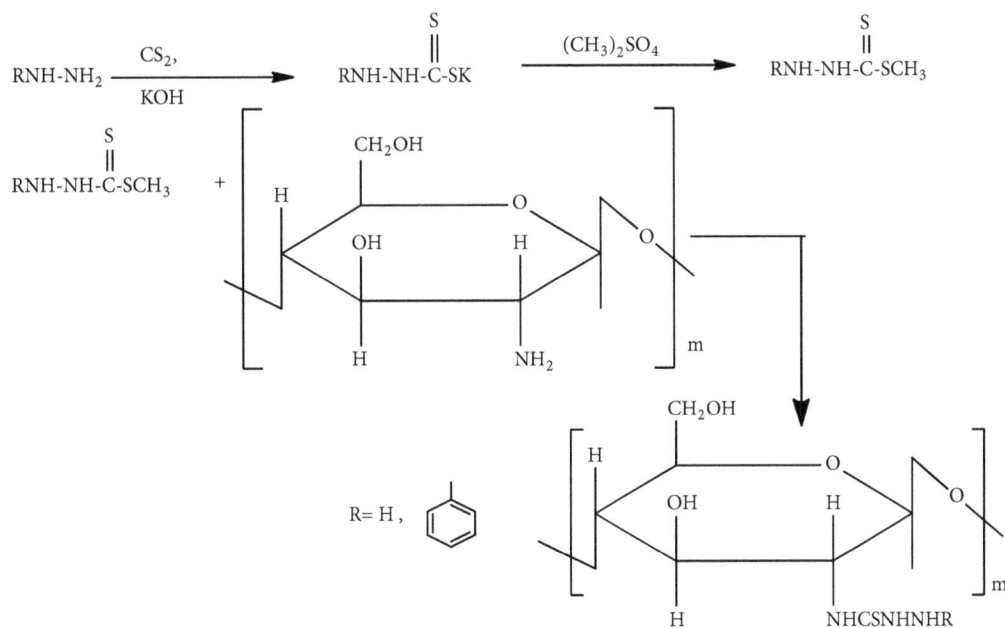

FIGURE 4: Synthetic route to 2-phenylhydrazine (or hydrazine) thiosemicarbazone chitosan.

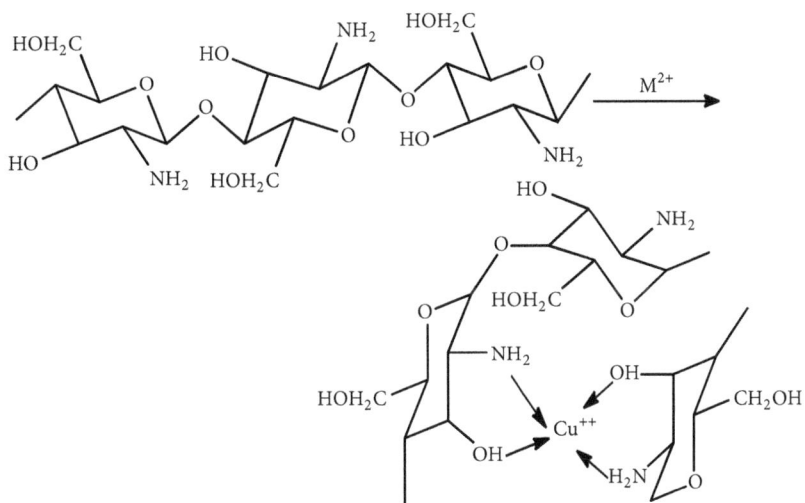

FIGURE 5: Structure of chitosan-metal complex.

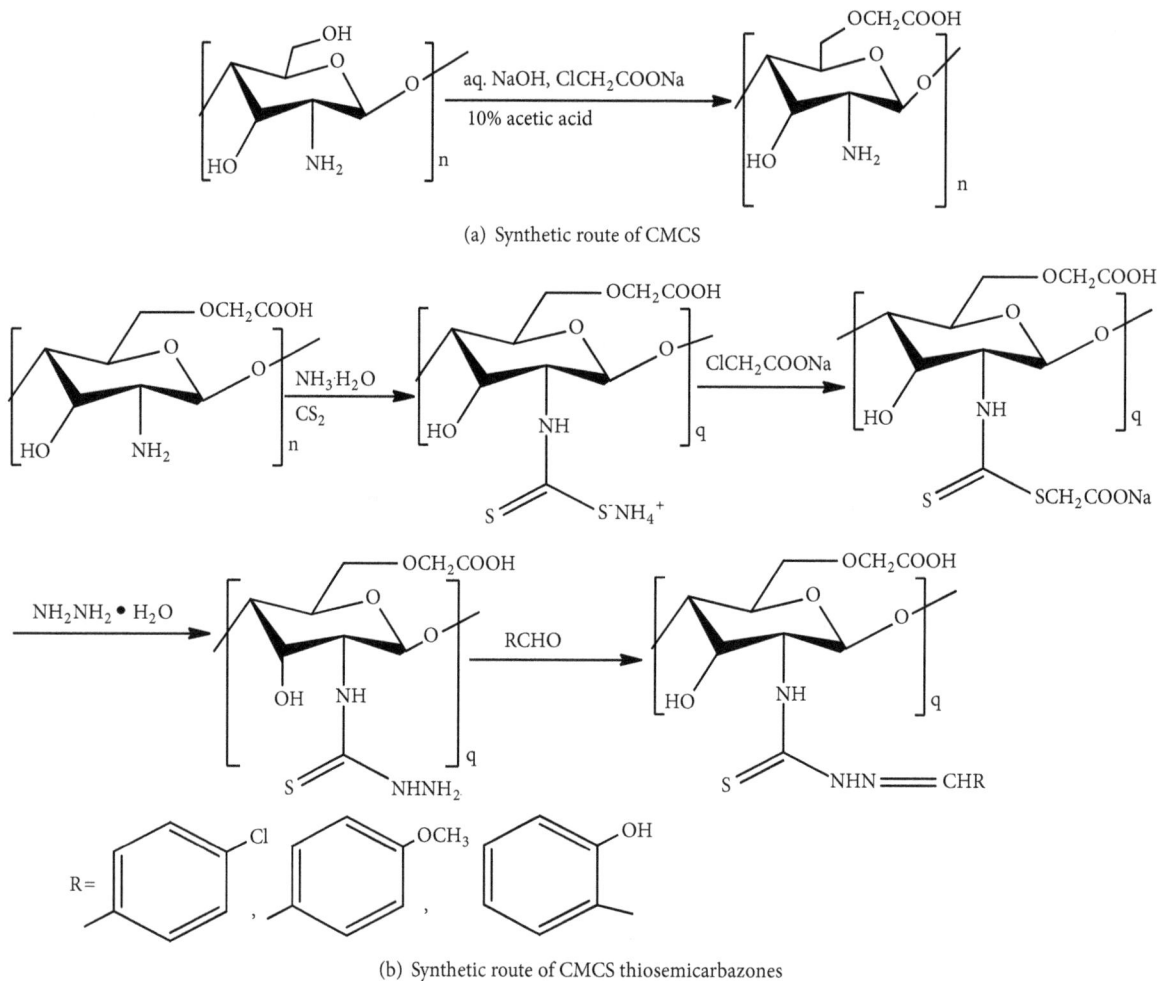

(a) Synthetic route of CMCS

(b) Synthetic route of CMCS thiosemicarbazones

FIGURE 6

10^5 per well were incubated for 24 h at 37°C. Copper-chitosan complexes in 0.1 M HCl were added and further incubated for 48 h. Cell proliferation assays were carried out after adding 2-(2-methoxy-4-nitrophenyl)-3-(4-nitrophenyl)-5-(2,4-disulfo-phenyl)-2H-tetrazolium (WST-8) and 1-methoxyphenazine methosulfate (1-methoxy-PMS). Background control wells also contained the same volume of culture media. Chitosan-copper complexes were found to selectively inhibit HeLa and 293 tumor cell lines but there was no inhibition in the growth of HLF. The IC_{50} values of such a complex with chitosan to copper (II) ratio of 1:0.4 for above cell lines were 48 and 34 μmol/L, respectively [42]. This clearly showed nontoxicity of chitosan-copper complex in noncancerous cells and concentration dependent antitumor activity of chitosan-copper complex in vitro.

3.3. Carboxymethyl Chitosan (CMCS). Carboxymethyl chitosan (CMCS) is an amphoteric, water soluble chitosan derivative [54] that is prepared according to Chen and Park's method (2003) [55] by the reaction of chloroacetic acid with NaOH alkalized chitosan.

CMCS thiosemicarbazones with p-chlorobenzaldehyde, p-methoxybenzaldehyde, and salicylaldehyde are synthesized through one pot synthesis of thiosemicarbazide intermediate and its reaction under reflux at 65°C for 10 h with carboxaldehyde in methanol and acetic acid catalyst [54] (Figures 6(a) and 6(b)).

Treatment of CMCS (0.5 mg/ml, 1 mg/ml, and 1.5 mg/ml) on human umbilical vein endothelial cells (HUVECs) proliferation assessed by 3-(4,5-dimethylthiazol-2-yl)-2,5-diphenyltetrazolium bromide (MTT) assay showed no significant decrease in cell viability (p>0.05) after 24 h and 48 h incubation. So, CMCS was nontoxic to HUVECs at the range of 0.5- 1.5 mg/ml. But, trans well migration assay showed significant inhibition of two-dimensional and three-dimensional HUVECs migration by treatment with CMCS in a concentration dependent manner (p<0.05), confirming the inhibition of angiogenesis in vitro [45]. The in vivo investigation of such effects of CMCS on H22 tumor growth bearing mice model also showed a significant inhibition in tumor growth (p<0.05), in comparison to the control group. The inhibitory rates were found to be 32.63%, 51.43%, and

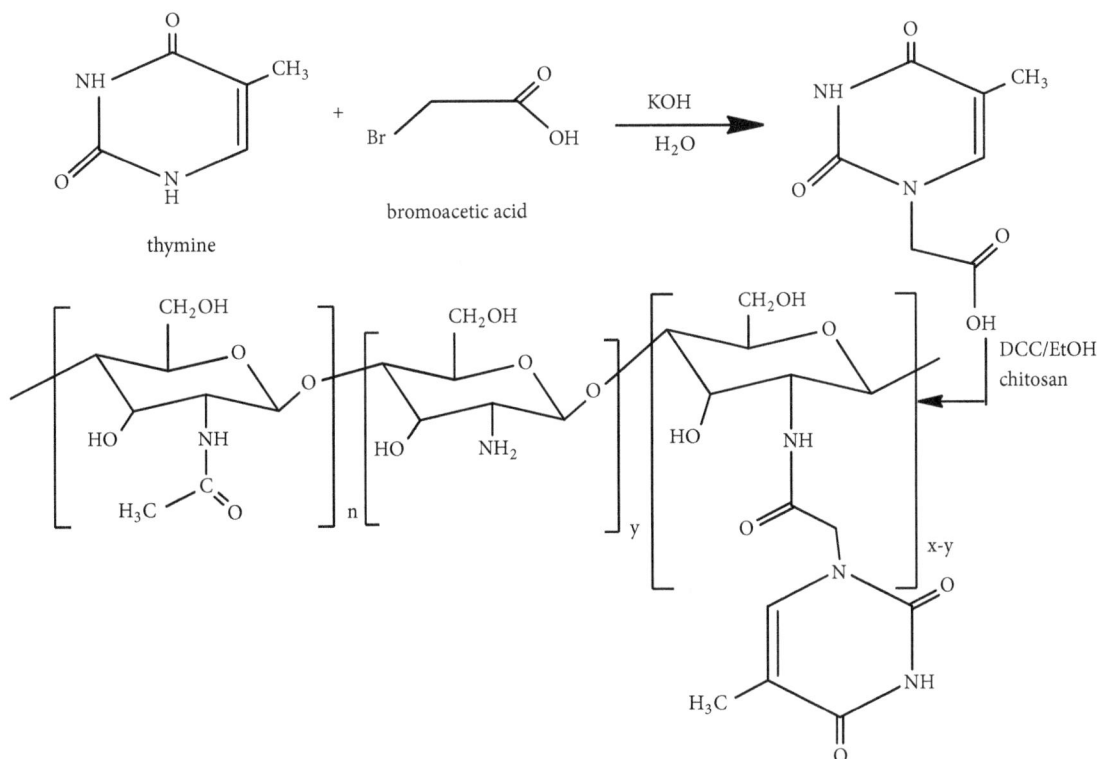

FIGURE 7: Synthetic route of chitosan–thymine conjugate.

29.89% at the doses of 75 mg/kg, 150 mg/kg, and 300 mg /kg, respectively [45]. The effect of CMCS on histopathology of hepatocarcinoma 22 (H22) cells, as examined by HE staining of paraffin sections, showed the necrosis of most of the CMCS treated tumor cells, confirming the repression of H22 cells *in vivo*.

3.4. Chitosan-Thymine Conjugate. Thymine derivatives have been found to show the potent anticancer effect. For instances, nanoparticles (100-250 nm in size) of water-based chitosan thymine conjugate formed by selective binding with Poly(A) inhibit the growth of colon cancer cells *in vitro* [56] and some phosphonotripeptide thymine derivatives show inhibition of human leukemia (HL-60) cell growth *in vitro* [57]. Alpha-methylene-gamma-(4-substituted phenyl)-gamma-butyrolactone bearing thymine, uracil, and 5-bromouracil compounds have also been demonstrated to show inhibition of leukemia cell lines [58]. Ferrocenyl-thymine-3,6-dihydro-2H-thiopyranes have been reported to show *in vitro* antiproliferative activity against human colon carcinoma HT-29, estrogen receptor-responsive human breast adenocarcinoma MCF-7, estrogen-negative human breast adenocarcinoma MDA-MB-231, human promyelocytic leukemia HL-60, and human monocytic MonoMac6 cancer cells [59]. The modification of chitosan with hyaluronic acid and thymine also shows an enhanced anticancer activity [33]. Conjugation of chitosan with thymine appears important in the expansion of biomedical utility. A novel chitosan–thymine conjugate was synthesized

by the reaction of chitosan with thymine-1-yl-acetic acid followed by acylation [46] (Figure 7).

Cellular cytotoxicity, proliferation, and viability assays were carried out with mouse embryonic fibroblast cell line (NIH 3T3) and human liver cancer cell line (HepG2) cultured in DMEM with 10 % (v/v) fetal bovine serum, MEM nonessential amino acids, 50 μM 2-mercaptoethanol, and chitosan thymine conjugate (0, 5,50,100 μM) for seven days at 37°C in a humid atmosphere of 5% carbon dioxide in air. The cells treated with pure thymine or chitosan and untreated cells were taken as control. The assays for cell proliferation and viability of novel chitosan–thymine conjugate were reported to show inhibition (p < 0.05) of HepG2 proliferation in a dose-dependent manner, but no toxicity in noncancerous NIH 3T3 was found [46].

3.5. Sulfated Chitosan (SCS) and Sulfated Benzaldehyde Chitosan (SBCS). Chitosan with an average molecular weight of ~ 1000 k Da that contains one acetamido and two hydroxyl groups in a unit [60] was chosen as a starting compound to make a hybrid sulfated compound with sulfate group as anticancer moiety in glycosyl unit [61]. Hence, sulphonylation of chitosan gives SCS and Schiff's base reaction with benzaldehyde followed by sulphonylation gives SBCS [47] (Figure 8). Human breast cancer (MCF-7) cells culture in DMEM in heat-inactivated fetal bovine, growth inhibition study, western blot, and cell apoptosis evaluation by fluorescence-activated cell sorting (FACS) analysis showed inhibition of MCF-7 cells proliferation and significant induction of

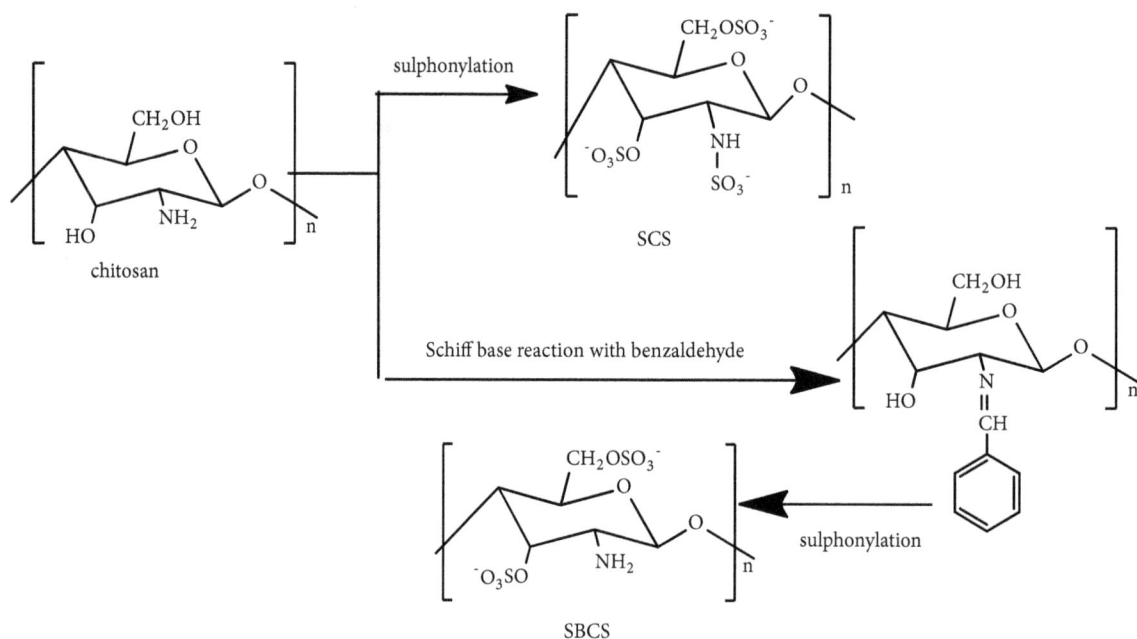

FIGURE 8: Synthetic route of sulfated chitosan (SCS) and sulfated benzaldehyde chitosan (SBCS).

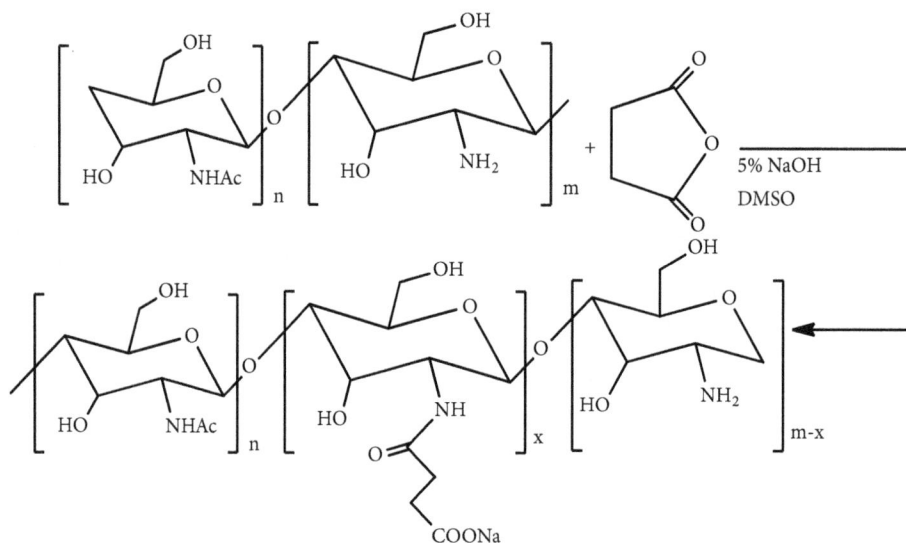

FIGURE 9: Synthetic route of N-succinyl chitosan.

apoptosis by both compounds, SCS and SBCS, obtained in this way [47]. SBCS was investigated to have better inhibitory effects and lower IC$_{50}$ than SCS [47].

3.6. N-Succinyl Chitosan (Suc-Chi) and Glycol Chitosan (GChi). Biocompatibility and cell viability problems with chitosan can be minimized by more deacetylation, depolymerisation, and removal of coexisting ions [62, 63]. Enzymatic degradation of chitosan can be increased by derivatization of its 6-hydroxy group as in glycol-chitosan (GChi) and N-succinyl chitosan (Suc-Chi) [64–68]. These water-soluble derivatives of chitosan have been found both *in vitro* and *in vivo* to efficiently release the drugs to tumor cells [69, 70]. The

synthetic route of N-succinyl chitosan (Figure 9) involved the 24 h reaction of succinic anhydride with DAC-90 in DMSO at 60°C followed by precipitation with 5% aq. NaOH at pH 5. The water dispersion of the precipitate maintained at pH 10-12 with 5% w/v aq. NaOH was dialyzed at room temperature for 2-3 days and the lyophilized samples were recovered [71]. The *in vivo* study, with the single intraperitoneal administration of Suc-Chi-MMC conjugate at 24 hours after the intraperitoneal L1210 tumor inoculation in mice models, showed the increase in antitumor activity with the increase in dose (equivalent MMC /kg). The ILS values of Suc-Chi-MMC conjugate have been reported to be 45.3% at the dose of 5 mg equivalent MMC/kg and 65.3% at the dose of 20 mg equivalent MMC/kg

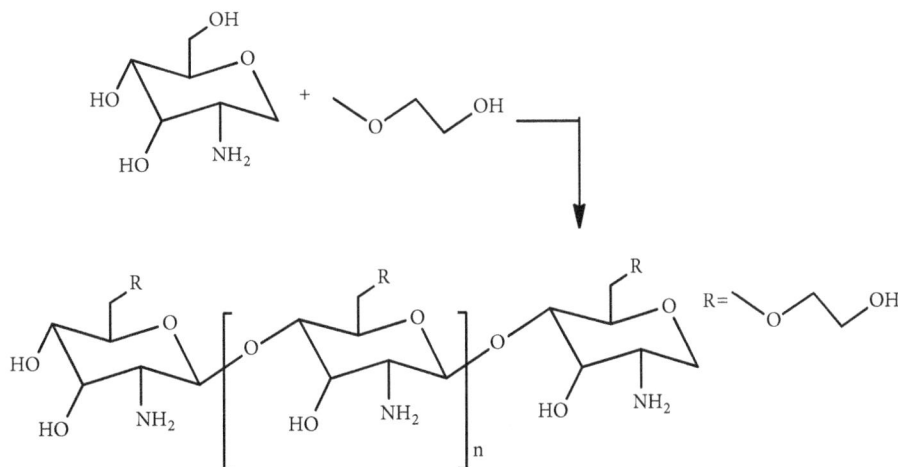

FIGURE 10: Synthetic route of glycol chitosan.

[72]. In addition, Suc-Chi-MMC conjugate has been found effective against solid tumors and metastatic liver cancer [48].

Synthesis of glycol chitosan involves the reaction of ethylene glycol with chitosan [73] (Figure 10). The intravenous *in vivo* study of fluorescein thiocarbamoyl-G-Chi (G-Chi-FTC), a fluorescein labelled derivative of G-Chi with fluorescein isothiocyanate (FITC), in mice showed that G-Chi could have more localization in kidney and longer retention in the blood circulation [48]. The *in vivo* investigation after intraperitoneal administration to mice bearing P388 leukemia showed the decrease in toxic side effects with G-Chi-MMC conjugate, though the therapeutic effect of the conjugate was not found better than MMC [48].

3.7. Furanoallocolchicinoid Chitosan. Use of colchicine as an antitumor agent is limited due to low accumulation in tumor cells. So, conjugation of colchicine with chitosan has been essentially important to decrease the side effects, increase the molecular weight to sequester it from noncancer cells and increase the biodistribution level of colchicine in cancer cells [74].

Furanoallocolchicinoid chitosan conjugate was synthesized by EV Svirshchevskaya et al. [49] by the reaction of furanoallocolchicinoid with succinic anhydride in tetrahydrofuran under an inert atmosphere followed by the extraction with ethyl acetate, addition of 40 k Da chitosan in the presence of acetic acid (pH 6) and methanol, stirring for 24 h with EDC and NHS, and drying and washing with toluene [49, 75, 76] (Figure 11).

Furanoallocolchicinoid chitosan has been found to show tumour growth inhibition as a result of a better accumulation in the tumour tubulin reorganisation and cell cycle arrest [49]. The investigation was made from *in vivo* study of the compound in Wnt-1 breast tumor bearing mice [49].

3.8. Polypyrrole-Chitosan (PPC): Graft Copolymerization. Recently, N. Salahuddin et al. have shown the enhanced *in vitro* inhibitory effect of polypyrrole-chitosan- (PPC-) silver chloride nanocomposite on proliferation of Erlich ascites

carcinoma (EAC) cells after loading of 3-amino-2-phenyl-4(3H)-quinazolinone. The investigation was made from *in vitro* release of PPC nanoparticles in EAC cells at pH 2 [50]. PPC is a polyamine chitosan that can be obtained by graft polymerization of chitosan with pyrrole [50] (Figure 12).

The synthetic routes and activity of chitosan derivatives as anticancer agent have been summarized in Table 1.

4. Mechanism of Anticancer Activity of Chitosan

4.1. Permeation Enhancing Mechanism. Amino group in chitosan leads to protonation in acidic to neutral medium. The positive charge developed in this cationic polysaccharide (pKa ~6.5) makes it water soluble and bioadhesive to bind with and enhance permeation through negatively charged surfaces such as mucosal and basement membranes [4, 77]. Consequently, chitosan facilitates oral bioavailability of polar drugs and their transportation through epithelial surfaces. Due to its biocompatibility and nontoxicity, chitosan finds applications in pharmaceutical and commercial fields like in the preparation of binder in wet granulation, tablets with slow release of drugs, drug carrier in microparticle system, disintegrant, hydrogels, site specific drug delivery, and carrier of vaccine delivery and gene therapy [4]. Its antimetastatic activity both *in vitro* and *in vivo* has been reported due to its permeation enhancing mechanism [4]. It has been found that the treatment of MDA-MB-231 human breast carcinoma cells with increasing concentration of chitosan inhibited the migration of these cells through a matrigel coated membrane [78] because this combination of chitosan and carcinoma cell lines lowered the activity and amount of MMP9 protein and this antimetastatic behavior increased with increase in concentration of chitosan [78].

4.2. Antiangiogenic Mechanism. Chitosan can exhibit antitumor effect by antiangiogenic mechanism. This process interferes with mutual regulation of proangiogenic and antiangiogenic factors under the pathological conditions [45]. Y. Xu and coworkers (2009) showed that chitosan nanoparticles

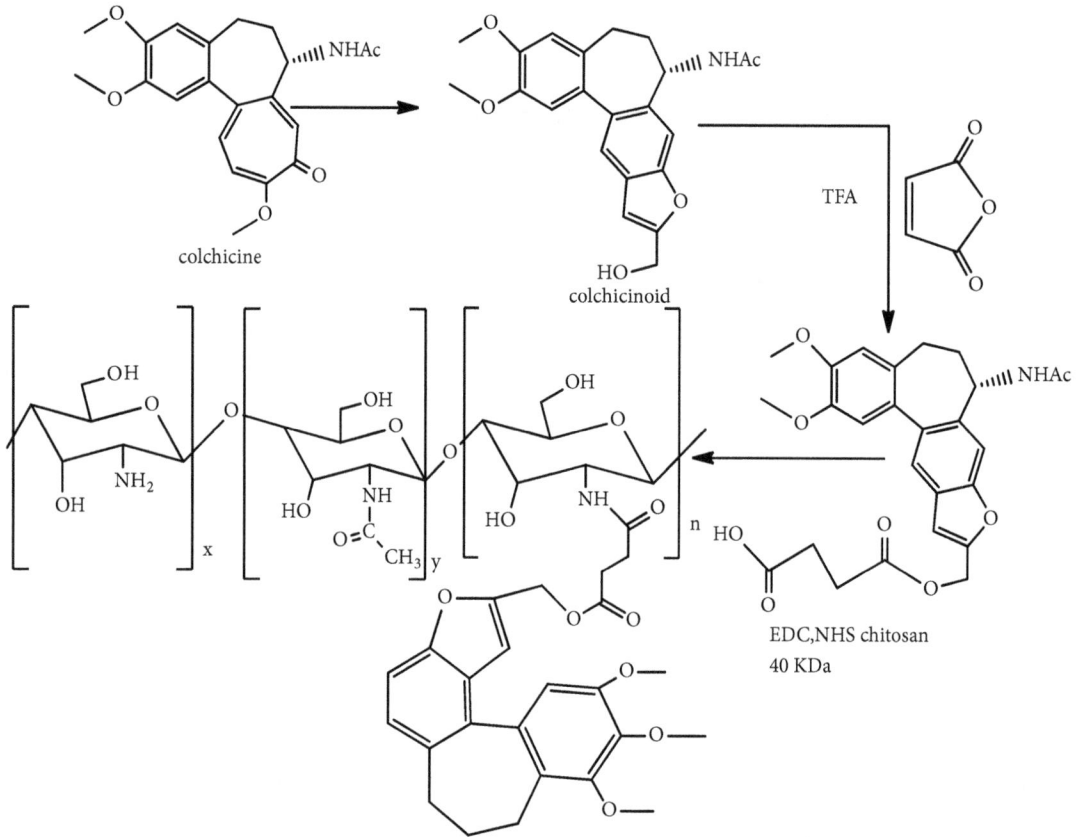

FIGURE 11: Synthetic route to furanoallocolchicinoid chitosan.

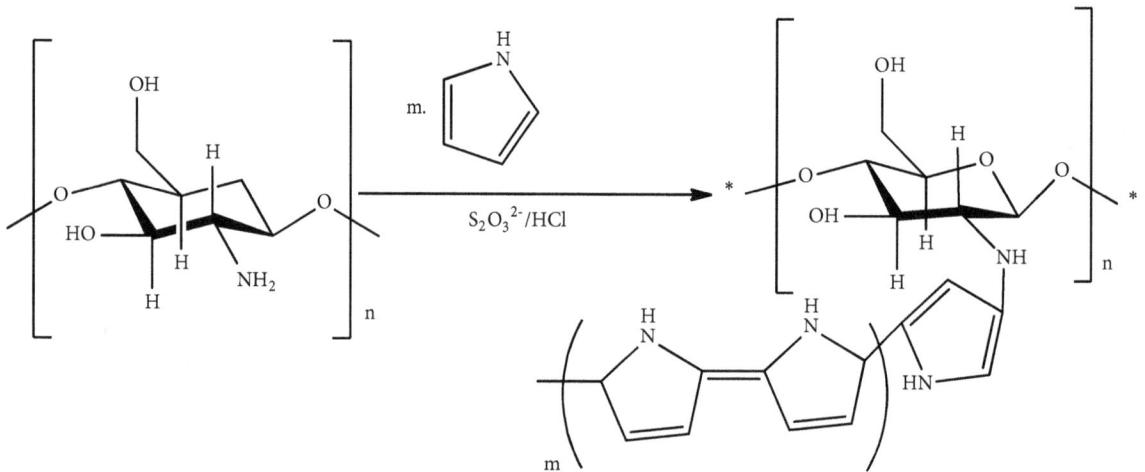

FIGURE 12: Graft copolymerization of polypyrrole-chitosan.

(CNP) could inhibit the growth of human hepatocellular carcinoma through a mechanism of CNP-mediated inhibition of tumor angiogenesis that was associated to impaired levels of vascular endothelial growth factor receptor 2 (VEGFR2) [17].

4.3. Sustained Release Mechanism. A mechanism of anticancer functionality of chitosan is related to its capacity to increase the biodistribution level and accumulation of drug in

tumor cells. Zhang et al. [79] through pharmacokinetic study in vivo have shown that mifepristone (MIF) loaded chitosan nanoparticles (MCNS) ensure controlled drug delivery in a sustained release manner and enhance the oral bioavailability and anticancer activity of the drug [79].

4.4. Immunoenhancement Mechanism. It was also shown that the tumor growth inhibitory mechanism of chitosan involved

TABLE 1: Synthetic routes and activity of chitosan derivatives as anticancer agent.

S. No.	Compound	Method of synthesis	Test	Outcome	Year	Ref.
1	2-Phenyl hydrazine (or hydrazine) thiosemicarbazone chitosan	Reaction of 2-phenylhydrazine (or hydrazine) dithiocarboxylate intermediate with chitosan in DMSO.	Superoxide radical scavengingassay *in vitro*.	Higher superoxide radical scavenging effect than chitosan.	2010	[39]
2	Chitosan copper(II) complex	Reaction of chitosan with 1% acetic acid containing copper sulfate in 1:0.4 molar ratio of chitosan to $CuSO_4.5H_2O$, neutralized by dilute ammonia solution.	Cell proliferation assays after adding WST-8 and 1-methoxy-PMS in chitosan -copper cell well *in vitro*.	Inhibition of the proliferation of HeLa and 293 cells.	2006	[42]
3	CMCS	Reaction of chloroacetic acid with NaOH alkalized chitosan (Chen and Park)	Antitumor angiogenesis effects *in vitro* through MTT, and transwell migration assay in HUVECs and *in vivo* test in H22 bearing mice.	Significant inhibition of the migration of HUVECs *in vitro* and H22 growth inhibition *in vivo*.	2003 2015	[55] [45]
4	Chitosan-thymine conjugate	Reaction of chitosan with thymine-1-yl-acetic acid followed by acylation.	Cellular cytotoxicity, proliferation and viability assays with HepG2 culture in DMEM with fetal bovine serum in suitable seeding conditions.	*In vitro* inhibition of human HepG2 proliferation in a dose-dependent manner.	2012	[46]
5	SCS and SBCS	SCS from Sulphonylation of chitosan and SBCS from Schiff's base reaction with benzaldehyde followed by sulphonylation.	MCF-7 cells culture in DMEM in heat -inactivated fetal bovine, growth inhibition study, western blot and cell apoptosis analysis.	Significant induction of MCF-7 cells apoptosis and inhibition of MCF-7 cells proliferation *in vitro*.	2011	[47]
6	Suc-Chi	Reaction of succinic anhydride with DAC-90 in DMSO followed by precipitation with aq. NaOH at pH 5	Intraperitoneal administration after the intraperitoneal tumor inoculation in mice models.	Increase in antitumor activity with increase in dose in L1210 *in vivo*.	2005 2006 1993	[48] [71] [72]

TABLE 1: Continued.

S. No.	Compound	Method of synthesis	Test	Outcome	Year	Ref.
7.	G-Chi	Reaction of ethylene glycol with chitosan	The intravenous *in vivo* study of fluorescein thiocarbamyl-G-Chi (G-Chi-FTC) in normal mice.	Localization in kidney and longer retention in the blood circulation	2001 2005	[73] [48]
			Intraperitoneal administration of G-Chi-MMC to mice bearing P388 leukemia.	Decrease in toxic side effects	2001 2005	[73] [48]
8.	Furanoallocolchicinoid –chitosan	Reaction of furanoallocolchicinoid with succinic anhydride in tetrahydrofuran under an inert atmosphere followed by the extraction with ethyl acetate, addition of chitosan in the presence of acetic acid (pH 6) and methanol, stirring with EDC and NHS, drying and washing with toluene.	*In vivo* study of the compound in Wnt-1 breast tumor bearing mice.	Decrease in side effects, sequestering of colchicine drug from noncancer cells and increase in its biodistribution in cancer cells, more inhibition of tumor growth than chitosan.	2016 2011 2015 2014	[49] [74] [75] [76]
9.	PPC	Graft copolymerization of chitosan with pyrrole	*In vitro* release of PPC nanoparticles in EAC cells at pH 2.	Enhanced *in vitro* inhibitory effect of PPC silver nanocomposite on EAC cells proliferation after loading of 3-amino -2-phenyl -4(3H)-quinazolinone.	2017 2017	[50] [51]

enhancement of immunological system consisting of tumoricidal immunocytes as cytotoxic lymphocytes natural killer cells as observed in sarcoma 180 bearing mice [19, 80]. Antitumor activity of oligochitosan was suggested to have been related to activation of intestinal immune functions due to enhancement of NK activity in intraepithelial lymphocytes (IELs) or splenic lymphocytes [19]. Microcrystalline chitosan has been found to inhibit cell viability on HT29 colon carcinoma cell line [81] and suppress the tumor growth in HepG2 bearing severe combined immune deficient (SCID) mice [82]. Applications of native chitosan are limited by its higher molecular weight that results in low solubility in nonacidic aqueous media. So, to be absorbed in human body it is converted into low molecular weight COS [83]. Cellulase treated chitosan forms water soluble oligosaccharide product with low molecular weight due to enzymatic hydrolysis followed by degradation of the chain without any modification in chemical structure of the residues [83]. Such water-soluble product has been found to inhibit the growth of tumor cells [84–86]. Tokoro et al. suggested that the mechanism of such tumor growth inhibitory effect of hexa-N-acetylchitohexaose and chitohexaose is associated with higher production of interleukin I and interleukin II to bring about the maturation of splenic T- lymphocytes and killer T-cells [84]. Seo et al. showed that the antitumor activity of low molecular weight chitosan was due to activation of murine peritoneal macrophases to kill the tumor cells in the presence of IFN-γ [85].

Immunoenhancing molecular mechanisms of COS could precede either with direct killing of pathogenic microorganisms or tumor cells because of an immune response or with enhancement of cytotoxic activity to inhibit the production of tumor cells by activation of T-cells and NK-cells with the help of IL-1 and TNF-α cytokines [87, 88]. Synergistic effects shown by TNF-α are critical to bring about the proliferation of Th1 cells together with IL-1 and IL-2 in vitro [88]. So, the innate immune responses shown by COS are associated with upregulation of IL-1, TNF-α, and IFN-γ to increase the immune functions of lymphocytes [85–89]. The antitumor effect of chitosan has also been shown to be due to its antioxidant profile improvement pathway [90].

4.5. Cellular Apoptotic Mechanism. Anticancer activity of chitosan in different cell lines has been found to be due to apoptosis [10, 13, 17] that is initiated by activation of procaspase triggered from outside the cell to accelerate the cleavage of cascade to amplify the death signals [13].

Cytotoxicity of chitosan has been found to depend on its molecular weight and degree of deacetylation (DDA) [10]. Low molecular weight chitosan (LMWC) has been found to exhibit cytotoxic effects on the oral squamous cell carcinoma (SCC) Ca9-22 in vitro through induction of apoptosis by activation of caspase 3 and cell cycle arrest through extrinsic apoptosis by the activation of caspase 8 [13, 91, 92]. Higher cytotoxic effect of LMWC than higher molecular weight chitosan has been found to be the result of difference in mechanism of cytotoxicity. LMWC possesses higher positive charge in amino group and is more attracted to cancer cell membrane that has greater negative charge than in normal

cells [93]. LMWC attacks cancer cells through electrostatic interaction with tumor cell membrane or extracellularly through endocytosis [13, 16].

Antiproliferative effect of chitosan on T24 urinary bladder cancer cell lines as shown by fluorescent activated cell sorbent assay (FACS) and DNA fragmentation assay [80] has been found to be the result of apoptosis. Investigation of cell cycle distribution mechanism of the chitosan induced inhibition of T24 cell growth with the help of flow cytometry showed that there was progressive increase in DNA content up to G2 DNA level with decrease in concentration until the end of S phase. Duration of G1 phase increased with increase in concentration eventually causing the disruption of cell membrane and hence necrosis of chitosan treated cell lines. This effect showed how chitosan could arrest the growth of tumor cells [80, 94].

Chitosan nanoparticles have been shown to inhibit human hepatoma BEL7402 cells proliferation because of cell necrosis by neutralization of its surface charge, permeation through the cell membrane, decrease in MMP, and induction of lipid peroxidation in vitro [95]. They have been proved to inhibit cell viability on HT-29 colon carcinoma cell lines [81]. Chitosan nanoparticles have been reported to target the cancer cells because of their preferential accumulation in tumor cells due to enhanced permeation and retention (EPR) effect and lower the p-glycoprotein induced multidrug resistance [96, 97].

LMWC has been shown to induce S phase arrest in cancer cells [13]. The mechanism of such cell cycle arrest at S phase generally involves cytokine signaling from the environment and subsequent inhibition of DNA synthesis for several hours [98]. Cellular senescence due to permanent arresting of cell cycle is a major cause of aging and a mechanism of anticancer activity [99]. It has been reported that cell senescence due to cell cycle arrest at G1 and S- phase by LMWC is probably associated with higher production of reactive oxygen species (ROS). This process is initiated by higher expression of TGF-β molecules that causes step by step activation of Smads 2/3, Smad 4, p15, and p21 before the ultimate activation of ROS production [99]. Necessity of further research has been pointed out to clarify this mechanism of anticancer activity of chitosan [13].

G1 arrest by LMWC has been reported to be an indicative of the mechanism that involves the changes in protein expression to prevent the cells from entering S phase in a manner independent of p53 [98]. The rate of protein synthesis increases in case there is DNA damage requiring a rapid response without transcription or translation [100]. When there is checkpoint at G1 or S phase, TGF-β molecules induce CKIp15 and p27 to inhibit Cdk-4/Cdk-6-cyclin complex formation and prevent RB phosphorylation in a manner independent of p53 [101, 102]. When the checkpoint is in mid to late G1 phase, the cell cycle arrest takes place as a result of no RB phosphorylation in mid phase and low cyclin E-Cdk-2 activity in late phase [103]. G1 arrest by LMWC has also been reported to probably involve decrease in concentration of Cdc25A and inactivation of cyclin E-Cdk2 due to ubiquitination of Cdc25A. This process of ubiquitination in mammalian cells exposed to UV radiation is the result of Cdc25A phosphorylation through Chk1/Chk2 due to activation of ATM/ATR [103]. Necessity of further investigation has been pointed out to clarify this anticancer pathway of LMWC [13].

Shen et al. discovered that chitosan oligosaccharide (COS) *in vitro* inhibited cell proliferation, lowered the number of cells in S phase, and decreased the rate of DNA synthesis, as a result of increase in the level of p21 and decrease in cyclin A and CDK-2 [82]. MMP-9 that has key role in tumor growth was inhibited by COS in Lewis Lung Carcinoma (LLC) cells [82].

Prolonged survival of nude mice with human pancreatic cancer xenografts upon the treatment of porcine pancreatic enzyme (PPE) extracts [104] was an evidence of proteolytic enzymes as a defense against cancer. Chemo preventive activity of COS in human colorectal adenocarcinoma cell line HT-29 was reported to be the result of increased activity of enzymes QR, GST, and GSH [105]. COS was also found to inhibit proinflammatory cytokinin mediated nitric oxide (NO) production and inducible NO synthase (iNOS) leading to decrease in proliferation of HT-29 [106]. Antiangiogenic activity of COS was hypothesized to be the result of heparanase inhibition [107] and reduction in colorectal adenocarcinoma HT-29 tumor size by COS was attributed to concentration dependent reduction in secretion of zinc dependent proteolytic enzyme MMP-2 [108] as a result of lowering of its induction by cytokines IFN-γ, IL-1α, and TNF-α [106]. COS was found to have inhibitory effects on the types of MMPs gelatinase and matrilysin on HT-29 cells [109]. COS was also found to inhibit ODC activity induced by 12-O–tetradecanoylphorbol-13-acetate (TPA) and TPA induced expression of COX-2 in HT-29 cells [105].

Increase in the expression of iNOS is associated with tumor growth, vascular invasion and metastatic potential [110, 111]. COS has been found to bring about the inhibition of angiogenesis and platelet aggregation effect of NO [112] by inhibition of NO production because of reduction in iNOS expression [106]. COS has been demonstrated to exert inhibitory effect on LPS-induced IL-8 expression in human umbilical vein endothelial cells (HUVECs), LPS-induced HUVECs migration, and U937 monocyte adhesion to HUVECs [113]. COS has been found to induce apoptosis in human colon adenocarcinoma, HT-29 [114], and HL-60 cell lines [115]. Higher concentration of chitosan was found to inhibit the growth of mouse monocyte macrophage in RAW 264.7 cell lines [116] and suppress the colon and gastric cells proliferation [94]. *In vivo* effect of chitosan on Erlich ascites tumor (EAT) cells in EAT bearing mice showed a significant decrease in volume of ascites [18] and there was 25% increase in caspase 3 activity in Caco-2 cells after 24 h incubation with chitosan compared to the control that was not treated with chitosan [117]. Through nucleosomal DNA fragmentation, chitosan induced apoptosis on EAT cells was studied [118].

The effects of molecular weight (M_w) and degree of deacetylation (DDA) of chitosan on its antitumor activity against PC3 (human prostate), A549 (human lung), and HepG2 (human hepatoma) cell lines were demonstrated by the cytotoxic potentials of high molecular weight chitosan (HMWC) and COS fractions with different M_w and DDA.

The results showed that high HMWC was less effective than COS against these cells [10]. The antitumor activity is associated with both molecular size and chemical structure but antitumor mechanism of HMWC has yet been unclear.

Anticancer mechanism of action of chitosan in some potential target cells is summarized in Table 2.

5. Mechanism of Anticancer Activity of Chitosan Derivatives

5.1. 2-Phenylhydrazine (or Hydrazine) Thiosemicarbazone Chitosan. In 1956, Brockman et al. [119] reported pyridine 2-carboxaldehyde thiosemicarbazone as the first heterocyclic thiosemicarbazone (HCT) to show anticancer activity in prolonging the life span of mice bearing L1210 leukemia. Then, many HCT derivatives with anticancer activity were synthesized by modification in the heterocyclic ring system, thiosemicarbazone side chain, and ring substituents [120–122]. In 1979, Klayman et al. [123] showed antineoplastic activity of 2-formylpyridine thiosemicarbazones. Ribonucleotide reductase (RR) is essentially involved in the *de novo* synthesis of deoxyribonucleotides required for DNA replication and repair [124, 125], and the antineoplastic activity of α-(N)-heterocyclic carboxaldehyde thiosemicarbazones was found to be associated with inhibition of RR activity [126].

Chitosan thiosemicarbazones impart more antioxidant ability to scavenge and minimize the formation of free radicals [39] that would cause the immune system decline, brain dysfunction, and cancer [40, 41]. Due to presence of reactive functional groups and cationic nature, chitosan can make tight junctions in cell membrane and it can be biochemically modified into different derivatives of unique properties [4]. Antioxidant behavior of chitosan and its derivatives is due to ability of amino and hydroxyl groups in C-2, C-3 and C-6 positions of pyranose ring to abstract proton from free radicals [127]. When thiosemicarbazone is grafted to chitosan, both intramolecular and intermolecular hydrogen bonds are weakened, N-H and C=S groups interact with free radicals, and there is an increase in its antioxidant capacity [39]. Anticancer effects of chitosan thiosemicarbazones can be inferred from their structural and antioxidant behavior.

5.2. Chitosan–Metal Complexes. Cisplatin is a complex widely used as an antineoplastic drug in solid tumors, but it has limited spectrum of activity and several side effects of dose dependent severity [128, 129]. In an attempt to develop the antitumor compounds with less side effects and wide spectrum of biological activity, platinum and nonplatinum metal complexes with different carrier ligands have been synthesized [130–134].

Owing to the presence of multiple hydroxyl, acetamido, and amino groups in the chain, chitosan shows chelation with many metal ions [42, 135, 136]. Chitosan copper(II) complexes in copper to chitosan mixture ratio of 2:5 have been found to show antitumor activity with 293 cells and HeLa cells [42].

Investigation by sulforhodamine B assay *in vitro* of low-molecular-weight chitosan salicylaldehyde Schiff-base and its zinc(II) complexes have been found to show inhibition of

TABLE 2: Anticancer mechanism of action of chitosan in some potential target cells.

Compound	Target cells	Mechanism of action	Test	Outcome	Year	Ref.
Chitosan	MDA-MB-231	Permeation enhancement, lowering of MMP9 activity	In vitro and in vivo	Antimetastatic effect	2013	[4]
					2009	[78]
	T24 urinary bladder cell lines	Disruption of cell membrane, necrosis	In vitro	Antiproliferative effect	2013	[80]
					2001	[94]
	Human hepato carcinoma	Nano particles mediated antiangiogenic action and impairment of VEGFR2 levels.	In vitro	Antiangiogenic effect	2010	[17]
					2015	[45]
Chitosan nano particles	BEL7402, HT-29	Cell necrosis, decrease in MMP, induction of lipid peroxidation, enhanced permeation and retention (EPR) effect	In vitro	Inhibition of cellular proliferation	2012	[81]
					2007	[95]
					2017	[96]
					2017	[97]
MIF loaded chitosan nano particles	Solid tumor	Sustained release and enhancement of bioavailability of drug	In vivo	Drug accumulation and growth inhibition	2016	[79]
Oligochitosan, (N-Acetyl) chitohexaose	Sarcoma 180, HT- 29, HepG2	Immunoenhancement through increase in activity of NK cells, T cells, killer lymphocytes and cytokines.	In vivo and in vitro	Suppression of tumor growth	2004	[19]
					2013	[80]
					2012	[81]
					2009	[82]

TABLE 2: Continued.

Compound	Target cells	Mechanism of action	Test	Outcome	Year	Ref.
	SCC Ca9-22	Cellular apoptosis, activation of caspase-3 and caspase-8, electrostatic interaction and endocytosis	In vitro	Inhibition of tumor growth and proliferation	2014 2004 2010 2004	[13] [92] [93] [16]
	SCC Ca9-22	Cytokine signaling cell cycle arrest, ROS activation	In vitro	cell senescence, inhibition of cell growth and proliferation	2014 2003 2010	[13] [98] [99]
	LLC cells	Inhibition of MMP-9	In vitro	Cell death and antiproliferation	2009	[82]
		Increased activity of enzymes QR, GST and GSH.	In vitro	Increase in chemo preventive activity	2007	[105]
	HT-29	Inhibition of NO and iNOS	In vitro	Decrease in tumor cells proliferation	2007	[106]
LMWC/COS		Antiangiogenesis by heparanase inhibition	In vitro	Inhibition of tumor growth	2009	[107]
		Cytokines mediated MMP-2 reduction	In vitro	Reduction in tumor size	2007	[106]
		Inhibitory effect on LPS-induced IL-8	In vitro		1999	[108]
	HUVECs	expression, LPS-induced HUVECs migration and U937 monocyte adhesion to HUVECs	In vitro	Tumor growth inhibition	2011	[113]
	EAT cells	Apoptosis through nucleosomal DNA fragmentation	In vivo	Decrease in volume of ascites	2005 2004 2002	[18] [117] [118]

the growth of SMMC-7721 liver cancer cells because of the synergistic effect of chitosan matrix and planar geometry of the complexes [43].

The results of electrophoretic analysis have shown that the zinc complexes are bound to DNA by means of electrostatic interactions and intercalation. Experiments have shown more inhibitory effect of complex than ligand and still more potent antitumor activity of low molecular weight chitosan–zinc complex than high molecular weight analogue [43]. Square planar geometry of chitosan-metal complex favors the reaction of metal ion with free radicals to cause better scavenging of oxidative free radicals [44]. The free donor atoms in the complex molecules can chemically induce the cleavage of DNA to show antitumor activity [42].

The antitumor activity of chitosan copper(II) complex depends on concentration of copper and the possible mechanism of this action is that the positive charge on amino group of chitosan is strengthened due to the chelation with copper(II) ion and the complex develops more interaction with anionic components of cell surface [42, 137]. This complex has been tested to inhibit tumor cell proliferation in 293 cells and HeLa cells in *vitro* [42] and the underlying mechanism of antitumor activity is associated with checkpoint-controlled progression of cell proliferation at S phase [138]. It has been shown that chitosan-loaded copper nanoparticles are biocompatible towards the execution of enhanced retention and permeation (EPR) effect to be preferentially accumulated in cancer cells *in vivo* and their superior anticancer effect has been demonstrated by maximum damage and apoptotic body formation in cancer cells [96]. Oxidative stress, apoptosis, and inflammation to endothelial cells have been noted as the causes of anticancer effect of such nanoparticles [139, 140]. Indeed, the selective accumulation and internalization of these nanoparticles (< 200 nm) on cancer cells have now been a progressive research perspective [96]. Anticancer activity of copper chitosan nanoparticles has been attributed to generation of higher mitochondrial ROS level as a prime hallmark of cellular oxidative damage, DNA fragmentation, and apoptosis [141]. Experimentally, higher apoptotic activity owing to an increase in caspase 3/7 activity has been illustrated by the higher expression of caspase 3 [96].

5.3. Carboxymethyl Chitosan (CMCS). Owing to its solubility in water [142], lower toxicity, better biodegradability, and biocompatibility [143], CMCS has been prepared as a carrier of anticancer drug such as 5- fluorouracil, curcumin, and doxorubicin [144–147]. CMCS has been found to exhibit antitumor activity as a result of antiangiogenic effects *in vitro* and *in vivo* [45]. It showed the concentration and time dependent inhibition of HUVECs migration *in vitro* and a significant decrease in growth rate of mouse hepatocarcinoma (H-22) tissues because of cell necrosis *in vivo* [45]. Most of the CMCS treated H-22 cells were found to undergo necrosis due to distortion in their shape and disintegration of their nuclei [45]. CMCS was also found to inhibit the growth of BEL-7402, SGC-7901, and HeLa cells (p<0.05) [148].

CMCS has been shown to stimulate immune functions and suppress the tumor angiogenesis [45]. Molecular mechanism of tumor angiogenesis involves the formation of new blood vessels from the vascular endothelial cells. So, the method of immune histochemistry to investigate angiogenesis also adopts the way of labeling of such cells as 'marker cells' to reflect the formation of new blood vessels in the tumor. Among many endothelial marker cells, CD34 antigen is selected to study this process in H-22 hepatic tumor cells [149]. J. Zhiwen et al. showed the inhibition in the expression of CD34 (p<0.05) in CMCS (150-300 mg/kg) treated H-22 tumor tissue and this result strongly indicated dose-dependent antiangiogenic activity of CMCS in H-22 hepatic tumor *in vivo* [45].

Tumor angiogenesis is regulated by the proangiogenic and antiangiogenic effects in the cells. VEGF, a specific mitogen for vascular endothelial cells, and its kinase receptors found in many human tumors bring about the proangiogenic effect and TIMPs cause antiangiogenic effect by the inhibition of extracellular matrix degradation and transformation of malignant cells [45]. J. Zhiwen et al. found the decrease in VEGF level and increase in TIMP1 level after 14-day treatment of mouse serum with CMCS *in vivo*. This result clearly showed the inhibition of angiogenesis by CMCS in mouse serum [45]. The mechanism of this antiangiogenic activity may be associated with stimulation effect of key cytokines causing the inhibition of MMP activity that inhibits the extracellular matrix degradation and transformation of malignant cells [45, 82, 95, 106].

Human body can resist infection and cancer through the immune system consisting of the thymus, spleen, lymph nodes, and lymph ducts [150]. TNF-α and IFN-γ are important immune- related cytokines being used in the clinical cancer treatment for many years [151, 152]. TNF- α enhances the immune function [153] and induces apoptosis of tumor cells [154, 155]. IFN- γ, a pleiotropic cytokine with immunomodulatory effects, is produced by activated T cells and NK cells in the immune system to promote apoptosis and kill the tumor cells [156, 157]. J. Zhiwen et al. showed a significant increase in thymus index in mice (p<0.05) upon the treatment of CMCS and in another experiment, through detection by ELISA assay, they showed an enhancement in IFN-γ and TNF-α levels in CMCS treated mouse serum. These results clearly indicated the antitumor effects of CMCS by the regulation of immune-related cytokines induction and improvement in immune system [45].

5.4. Chitosan-Thymine Conjugate. Conjugation of nucleobase with various natural and synthetic biopolymers can form the derivatives with enhanced biological activity. For instance, phenanthridinium–nucleobase conjugates [158], metallocene–nucleobase conjugates [159], symmetrical and unsymmetrical, ω-nucleobase mono- and bis-amide conjugates [160], cyclodextrin–DNA conjugate [161], ferrocene–bis(nucleobase) conjugates [162, 163], neamine–nucleoside conjugates [164], DNA-peptide conjugates [165], peptide–nucleobase conjugates, and nucleobase PNA conjugates [166] have been found to inhibit a specific DNA or mRNA molecular expression as a result of an induced blockade in the transfer of genetic information from DNA to protein. Chitosan–nucleobase conjugate is an analogue of natural nucleobases and its anticancer mechanism is associated with

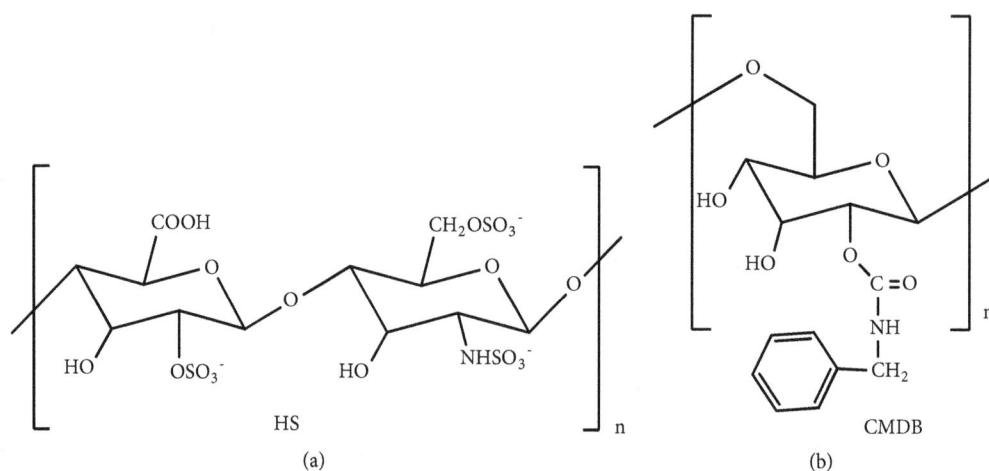

FIGURE 13: (a) Heparan sulfate (HS). (b) Carboxymethyl benzylamide dextran (CMDB).

its incorporation into the nuclear DNA during DNA synthesis and into mRNA during transcription. Incorporation of chitosan–nucleobase into DNA or mRNA induces breakage of the strand as a result of the chain termination leading to cell cycle arrest [167] and this mechanism is attributed to the absence of 3OH group required for the addition of more nucleotides. Cancer cells have shorter cell cycle and hence faster cell division. In comparison to noncancerous or slow dividing cells, these cells are far more affected by the chitosan–nucleobase [168]. Such a selective cytotoxicity against the cancer cells can be increased by conjugation of chitosan with a polynucleotide having a complementary sequence to that of oncogene or its mRNA product so that a specific nucleobase of chitosan–nucleobase conjugate can interact with the DNA or mRNA of tumor cell. This interaction owing to complementary base pairing (Thymine or Uracil with Adenine and Cytosine with Guanine) leads to inhibition of DNA synthesis, mRNA transcription, and translation of the cancer-causing gene [46]. Inhibition of HepG2 proliferation *in vitro* was shown by chitosan-thymine conjugate in a dose-dependent manner [46].

5.5. Sulfated Chitosan (SCS) and Sulfated Benzaldehyde Chitosan (SBCS). Endothelial cell proliferation and angiogenesis in metastatic breast carcinomas are associated with the role of heparin-binding growth factor [169–171]. The interaction of fibroblast growth factor-2 (FGF-2) with a low affinity receptor heparan sulfate (HS) (Figure 13(a)) brings about a suitable conformational change and a subsequent binding of FGF-2 to its high-affinity receptor tyrosine kinase (FGFR). Thus, HS is crucial for storage and regulated release of FGF-2 and other HS-binding growth factors like vascular endothelial growth factor (VEGF) at the cell surface. Evidently the HS alterations during the progression of cancer cause the change in FGF-2 binding and fibroblast growth factor–receptor (FGFR) ternary complex assembly in breast carcinomas [47].

Literature shows that natural sulfated polysaccharides, such as pentosan polysulphate [172–174], tecogalan [175], and fucoidan [176], can bind with FGF-2 and block the

binding of FGF-2 with HS [174] resulting in inhibition of cell proliferation [173] and metastasis [176]. Heparin like binding of carboxymethyl benzylamide dextrans (CMDB) [177–181] and phenylacetate carboxymethyl benzylamide dextran (NaPaC) [182–185] with FGF-2 was also found to alter the cell growth. CMDB inhibited autocrine and paracrine growth of breast tumor cells as a result of formation of a stable 1:1 complex FGFR [186]. NaPaC showed antiproliferative effects and inhibited VEGF binding to VEGFR2 and abolished VEGFR2 activity [187]. Sulfate group in heparin, tecogalan [175], and phenyl group in CMDB and NaPaC [177] were found responsible for the anticancer effects, and, in an attempt to get both functional groups in the same compound, the sulfate group was introduced to the end of the phenyl group of CMDB (Figure 13(b)) and a hybrid compound, carboxymethyl benzylamide sulfonate dextran, was obtained [188].This compound was found to interact strongly with FGF and potentiate the FGF-induced mitogenic activity; but it had no antiproliferative activity [188]. So, chitosan that contains one acetamido and two hydroxyl groups in a unit was chosen as a starting compound to make such a hybrid compound with the sulfate group on other sites of glycosyl unit [61].

Both the sulfated chitosan (SCS) and the sulfated benzaldehyde chitosan (SBCS) were investigated to significantly inhibit cell proliferation through induction of apoptosis and blockade of the FGF-2-induced phosphorylation of extracellular signal-regulated kinases (ERK) in the human breast cancer cell lines MCF-7 cells [47].

5.6. Glycol-Chitosan and N-Succinyl Chitosan. The conjugates of anticancer drug with chitosan have been found to show less adverse effects due to a predominantly higher distribution of such conjugates in cancer cells. Due to such a higher bioavailability in cancer cells, both insoluble and soluble formulations of glycol chitosan (G-Chi) and N-succinyl-chitosan (N-Suc-Chi) MMC conjugates have been found useful polymeric drug carrier in cancer chemotherapy [48].

N-Suc-Chi was found to show a long systemic half-life and a high distribution level in tumor cells [48]. *In vivo* study of activity of G-Chi, using its fluorescein labelled derivative after intravenous administration in normal mice, showed that G-Chi was more distributed in blood and kidneys with a long retention in kidneys [48].

Conjugation of doxifluridine and 1-β-D-arabinofuranosylcytosine (Ara-C) *via* glutaric spacer with chitosan has shown higher antitumor effect against P388-bearing leukemia model mice *in vivo*. The conjugates of mitomycin C (MMC) with both G-Chi and N-Suc-Chi have been found to show a remarkable antitumor activity in solid tumors, leukemia, and metastatic liver cancer [64, 65, 189] by a sustained release mechanism of the free drug from conjugates [48] *in vitro* and *in vivo*. The toxic side effects of MMC–G-Chi conjugate were lower than free MMC, possibly due to high distribution of G-Chi in tumor cells [48].

5.7. Furanoallocolchicinoid Chitosan Conjugates. Colchicine is a small hydrophobic molecule that binds to tubulin in serum albumin and accumulates in leukocytes [190, 191]. It prevents microtubule formation by such binding with tubulin and inhibits cell division [192–195]. However, its use as an antitumor agent is limited due to low accumulation in tumor cells. So, the increase in molecular weight by conjugation of colchicine with chitosan has been found essentially important to sequester colchicine molecules from the noncancer cells. It results in the increase in biodistribution level of colchicine in cancer cells and decrease in the side effects [74].

Furanoallocolchicinoid chitosan is a "smart" Ringsdorf's antitumor drug conjugate [196] that has been found to induce *in vitro* tubulin reorganization, cell cycle arrest, and more effective inhibition of the tumor cell proliferation in Wnt-1 breast tumor bearing mice [49, 75]. Due to better accumulation in tumor cells, furanoallocolchicinoid chitosan conjugate was found more effective (p <0.05) than chitosan towards tumor growth inhibition [49]. Lowering of tumor growth by chitosan was not reported to be associated with tubulin reorganization and cell cycle inhibition [49].

5.8. Polypyrrole Chitosan (PPC). Loading of 3-amino-2-phenyl-4(3H)-quinazolinone on polypyrrole chitosan- (PPC-) silver chloride nanocomposite has shown an increase in bioavailability of chitosan in cancer cells and the mechanism of this activity is associated with sequestering of molecules from noncancer cells and their sustained release to cancer cells [50, 51]. Owing to large surface area to volume ratio and stability [197, 198] polypyrrole chitosan nanoparticles loaded 1,2,4- triazoles have been reported to show higher antitumor activity than 1,2,4- triazole against Ehrlich ascites carcinoma (EAC) cells and breast cancer cell line (MCF-7) [197]. Polypyrrole chitosan loaded nanoparticles exhibit biocompatibility with mammalian cells [199] towards their delivery to targeted cells in a sustained release manner [50, 200]. Nanosized polypyrrole chitosan particles are not easily cleared by phagocytes and can easily make their way through the smallest blood capillaries and penetrate the cells to reach the target organs [50].

The *in vitro* release of PPC nanoparticles in EAC and MCF-7 at pH 2 was found to follow the zero-order kinetics in a gradual release manner [197]. The rapid release of 1,2,4-triazoles from chitosan nanoparticles at pH 2 was attributed to electrostatic repulsion between NH_3^+ and NH_2^+ groups in the chitosan nanoparticles [201]. At basic medium of pH 7.4, hydrogen bonding between S-H of triazole and N-H of NH_2 group in chitosan is strengthened and it causes the decrease in release percentage of triazoles [197].

Anticancer mechanism of action of heterocyclic thiosemicarbazone (HCT), as a precursor of chitosan thiosemicarbazone, and chitosan derivatives in some potential target cells is summarized in Table 3.

6. Nanochitosan and Its Mechanism of Anticancer Activity

Nanoparticles refer to particulate dispersions or solid particles in the range of 10-1000 nm in size [202]. Nanochitosan in this range of particle size can be prepared as biocompatible polymeric nanoparticles. Chitosan is a hydrophilic polymer, and hence nanochitosan lends itself to prolonged circulation in blood with more extravasation and passive targeting [203]. So, nanochitosan is a suitable drug delivery candidate [204, 205].

Khanmohammadi et al. prepared nanochitosan by addition of chitosan gel, obtained by dispersion of chitosan in sodium chloride solution as electrolyte in 3% acetic acid solution on stirring for two hours, in linseed oil with Span 80 as a surfactant on magnetic stirring for 30 min at room temperature, using an optimized spontaneous emulsification method with further addition of acetone and Glutaraldehyde-Saturated Toluene as a chemical cross-linking agent. Nanoparticle size was strongly dependent on synthesis parameters such as sodium chloride, surfactant, and chemical cross-linking agent. These nanoparticles were found to have particle sizes from 33.64 to 74.87 nm in average [206]. Agarwal et al. prepared nanochitosan by ionic gelation method, inducing gelation of chitosan solution with tripolyphosphate (TPP). The sizes of nanochitosan particles were optimized at different concentrations of chitosan and TPP. At chitosan concentration up to 4mg/ml and TPP concentration of less than 1.5 mg/ml, the nanoparticle size was found to range from 168-682nm [207]. Chitosan, being a biodegradable and mucoadhesive cationic polymer, has been widely used in the last few years in target delivery of anticancer chemotherapeutics to tumor cells. The chitosan nanoparticles loaded with therapeutic agent have been found more stable, permeable, and bioactive [208].

Chitosan is easily degraded by the kidney *in vivo*. So, during drug delivery, it appears less cytotoxic to healthy cells [209]. Drug discovery trials with nanochitosan have been more adaptable as it is biocompatible and cheap [210]. Chitosan nanoparticles are easily internalized by the cells [211] and this specificity of nanochitosan has shown its therapeutic significance in different types of cancer [17, 212, 213]. Nanochitosan has been found to show antiangiogenesis by RNA interference [17] and immune enhancement in breast cancer mice model4. Nanochitosan has been found to inhibit

TABLE 3: Anticancer mechanism of action of HCT as a precursor of chitosan thiosemicarbazone and chitosan derivatives in some potential target cells.

Compounds	Target cells	Mechanism of action	Test	Outcome	Year	Ref.
HCT	L1210	Inhibition of RR activity	In vitro	Antineoplastic effect	1956-89	[119–126]
Chitosan copper(II) complex	293 and HeLa cells	Checkpoint-controlled progression of cell proliferation at S phase	In vitro	Inhibition of cellular proliferation	2006 / 2000	[42] / [138]
Copper loaded chitosan nano particles	Osteocarcinoma	nano particles mediated enhanced permeation and retention (EPR) effect, increase in ROS level, DNA fragmentation and apoptosis	In vitro and in vivo	Inhibition of tumor growth	2017 / 2009 / 2013 / 2009	[96] / [139] / [140] / [141]
	HUVECs	Inhibition of extracellular matrix degradation and transformation of malignant cells	In vitro	Suppression of angiogenesis, decrease in VEGF and increase in TIMP1 levels	2015 / 2009 / 2007 / 2007	[45] / [82] / [95] / [106]
	H-22	Necrosis due to cell distortion and disintegration of nuclei	In vivo	Inhibition of tumor growth	2015	[45]
CMCS	Solid tumor	Enhancement in IFN-γ and TNF-α levels, regulation of immune-related cytokines induction and immunoenhancement	In vivo mice model	Increase in thymus index, tumor growth inhibition	2015	[45]
Chitosan thymine conjugate	HepG2	Inhibition of DNA synthesis, mRNA transcription and translation of the cancer-causing gene	In vitro	Inhibition of tumor growth	2012	[46]
SCS and SBCS	MCF-7 cells	Induction of apoptosis and blockade of the FGF-2-induced phosphorylation of ERK	In vitro	Inhibition of cells proliferation	2011	[47]
G-Chi- MMC and N-Suc-Chi -MMC conjugate	Solid tumors, leukemia, metastatic liver cancer	Sustained release of drug from conjugate	In vitro and in vivo	Higher antitumor effect and less side effects	2005	[48]
Furanoallocolchicinoid-chitosan conjugate	Wnt-1 breast tumor bearing mice	Tubulin reorganization, cell cycle arrest, sequestering of colchicine molecules.	In vivo	Inhibition of tumor cell proliferation and less side effects.	2016 / 2015	[49] / [75]
3-Amino-2-phenyl-4(3H)-quinazolinone PPC-silver chloride nano composite	EAC and MCF-7	Sequestering of molecules from noncancer cells and sustained release to cancer cells with zero order kinetics	In vitro	Target delivery of nano particles	2017	[50, 51]

the proliferation of human gastric cancer cells *in vitro* in a sustained release manner [213].

The paclitaxel loaded modified glycol chitosan nanoparticles in the size of 400 nm has been found to show sustained release of paclitaxel to bring about the inhibition of MCF-7 tumor growth due to EPR effect *in vitro* [214]. Encapsulation of paclitaxel and thymoquinone in nanochitosan has been found effective in breast cancer therapy [215]. The target specificity of nanochitosan can be established through the binding of protein with chitosan nanoparticles. For instance, binding of $\alpha v \beta 3$ integrin, with receptors for tumor cells, to nanochitosan has shown inhibition of the ovarian cancer *in vivo* [216]. Nanochitosan has been shown to increase the immune response in murine model by elevation of IgG, IgA, and IgM as well as IL-2, Il-4, and IL-6 receptors [217]. In acidic microenvironment with poor vasculature outside the tumor, amino group of chitosan gets protonated, the nanoparticles swell, and there is faster release of the drug. The EPR effect due to accumulation of nanochitosan macromolecules in the tumor microenvironment [218] and protonation of chitosan are significant in adaptation of nanochitosan-drug system in cancer therapy.

Chitosan-curcumin nanoformulation has been found to show anticancer activity following the apoptotic pathways associated with DNA damage, cell-cycle blockage, and elevation of ROS levels *in vivo* [219]. Nanochitosan has been demonstrated to inhibit the growth of human hepatocellular carcinoma (HCC) cells by cell necrosis and inhibition of tumor angiogenesis. The antiangiogenic activity of nanochitosan is associated to suppression of VEGFR2 gene expression [17]. Nanochitosan can bring about the HCC cell death *in vitro* by disruption of cell membrane, lowering of negative surface charge, decrease in mitochondrial membrane potential, induction of lipid peroxidation, disruption of fatty acid layer of the membrane, and fragmentation of DNA [95]. The mechanism of HCC cell growth inhibition *in vivo* by nanochitosan is associated with increase in apoptosis and decrease in cell proliferation. It has been found that nanochitosan is nontoxic to normal cells, but it has potent and specific cytotoxic effects on tumor cells [17].

Chitosan folate hesperetin nanoparticles (450 nm size) have been found to show apoptosis of HCT15 cells (IC_{50} 28 μM), after passive targeting through the leaky vasculature of tumor environment, more effectively than hesperetin (IC_{50} 28 μM) by proper regulation proapoptotic genes expression. So, chitosan folate hesperetin nanocomposite is suitable carrier of hesperetin to colorectal cancer cells *in vivo* [220].

Han et al. showed Arg-Gly-Asp (RGD) peptide-labeled chitosan nanoparticle (RGD-CH-NP) as a novel tumor targeted delivery system for short interfering RNA (siRNA). The RGD-CH-NP loaded with siRNA was found to significantly increase (i) selective intra tumoral delivery in orthotopic animal models of ovarian cancer, (ii) targeted silencing of multiple growth-promoting genes (POSTN, FAK, and PLXDC1) along with therapeutic efficacy in the SKOV3ip1, HeyA8, and A2780 models, and (iii) *in vivo* delivery of PLXDC1-targeted siRNA into the alphanubeta3 integrin-positive tumor endothelial cells in the A2780 tumor-bearing

mice. Overall, there was a significant inhibition of tumor growth *in vivo* [216].

Mechanism of anticancer activity of nanochitosan in some target cells is summarized in Table 4.

7. Chitosan and Chitosan Derivatives on Anticancer Clinical Study and Trial

Kim et al. in a phase IIb clinical study showed the complete tumor necrosis in 77.5% of the patients with HCC lesions <3 cm and 91.7% of the patients with HCC lesions <2 cm in two months after holmium-166 percutaneous (166Ho)/chitosan complex injection (PHI) therapy. Interfered by the cases of cumulative local recurrences and transient bone marrow depression, the survival rates were observed to be 87.2% for 1 year, 71.8% for 2 years, and 65.3% for 3 years. So, PHI proved a safe and novel local ablative procedure for the treatment of small HCC to be used as a bridge to transplantation and necessity of a phase III randomized active control trial in a larger study population was pointed out [221].

Clinical trials with chitosan and chitosan derivatives are being performed as (i) intervention of drug: chitosan on prostate cancer with the title 'Study of Chitosan for Pharmacologic Manipulation of AGE (Advanced Glycation End Products) Levels in Prostate Cancer Patients' [222], (ii) intervention of morphine, ketamine, placebo, and chitosan on cancer pain with the title 'Comparison of Oral Morphine Versus Nasal Ketamine Spray With Chitosan in Cancer Pain Outpatients' [223], (iii) intervention of the device adhesive barrier on axillary dissection of breast cancer with the title 'Anti-adhesive Effect and Safety of a Mixed Solid of Poloxamer, Gelatin and Chitosan (Mediclore®) After Axillary Dissection for Breast Cancer' [224], (iv) intervention of drug: 1% glycated chitosan and the device: photothermal laser on breast cancer stages IIIA, IIIB, and IV with the title 'Randomized Clinical Trial Evaluating the Use of the Laser-Assisted Immunotherapy (LIT/inCVAX) in Advanced Breast Cancer' [225], and (v) intervention of the implant: bilaminar chitosan scaffold on cerebrospinal fluid (csf) leakage with the title 'Chitosan Scaffold for Sellar Floor Repair in Endoscopic Endonasal Transsphenoidal Surgery' [226].

8. Prospects of Chitosan and Its Derivatives as Anticancer Drugs

Chitosan already finds its uses as a pharmaceutical excipient [5], permeation enhancer [6], and a hemostatic agent [7]. It is being utilized as nonwoven sheet in wound healing, dressing [8], weight loss, and cholesterol management [227]. Study *in vitro and in vivo* has shown that many cancer cells are resistant to the chemotherapeutic drugs like cisplatin, 5-fluorouracil (5-FU), docetaxel, procarbazine, methotrexate, etc. in practice [228]. These chemical compounds of current therapeutic use are associated with acute and chronic, life-threatening toxicity of gastrointestinal lining, bone marrow, reticuloendothelial system, and gonads [229]. Chitosan and its derivatives, specially the chitosan-drug nanocomposites as the leading anticancer formulations, due to their selective antitumor effects, nontoxicity, biocompatibility, and

TABLE 4: Mechanism of anticancer activity of nanochitosan (composite) in some target cells.

Nanochitosan (composite)	Target cell(s)	Mechanism of action	Test	Outcome	Year	Ref.
	Breast cancer mice model 4	Interference to RNA and immunoenhancement	In vivo	Inhibition of angiogenesis and proliferation	2010 2015	[17] [211]
	Human gastric cancer cells	Sustained release manner	In vitro	Inhibition of cells proliferation	2010 2005	[17] [213]
Nano chitosan	Ovarian cancer cells	Binding of $\alpha v\beta 3$ integrin with tumor cell receptors	In vivo	Inhibition of tumor growth	2010	[216]
	HCC cells	Decrease in mitochondrial membrane potential, and fragmentation of DNA, suppression of VEGFR2 gene expression	In vitro	Cell death and inhibition of angiogenesis	2010 2007	[17] [95]
Paclitaxel-glycol chitosan nano composite	MCF-7	sustained release of paclitaxel by EPR effect	In vitro	Tumor growth inhibition	2006	[214]
Chitosan–curcumin nano formulation	Solid tumor	Sustained release manner, DNA damage, cell cycle blockage and elevation of ROS levels	In vitro	Inhibition of tumor growth	2018	[219]
Chitosan folate hesperetin nanoparticles	HCT15 cells	Passive targeting through the leaky vasculature of tumor environment	In vivo	Cellular apoptosis	2018	[220]
Peptide-labeled chitosan nanoparticle	Solid tumors	Tumor targeted delivery for short interfering RNA (siRNA)	In vivo	Inhibition of tumor growth	2010	[216]

biodegradability can be the promising natural alternatives to overcome these problems.

Cancer is a major cause of deaths across the globe and, for several decades, intensive research has been focused on more potent anticancer drug development strategies. Despite this, the clinical intervention options of chitosan are still limited for many types of human cancers [80]. Therapeutic use of chitosan-based compounds with the minimal toxicity on noncancer cells [13] is critically important.

Chitosan, with generally recognized as safe (GRAS) status, has been labelled as a nontoxic and biocompatible polymer by US Food and Drug Administration (FDA) for wound dressing. It has been reported safe for regular oral administration (4.5 g/day) in humans for 12 weeks, after which the side effects such as mild nausea and constipation may be seen [230]. But the *in vivo* toxicity with the change in pharmacokinetic properties may appear in nanoparticles and derivatives upon chemical modification [209]. So, the individual assessment of toxicity profile of the compounds is necessary. In addition, solubility and biological activity of chitosan can be enhanced by increase in deacetylation and chemical modification to give chitosan derivatives [4].

The clinical trials are limited by limited accumulation in the target cells and unfavorable conditions of drug uptake, such as tumor perfusion, arteriovenous shunting, necrotic and hypoxic areas, and a high interstitial fluid pressure work. However, the targeted delivery of nanoparticulated anticancer drug can be made more effective by encapsulation of drug conjugate in chitosan nanoparticles that helps better accumulation of drug in tumor cells by EPR effect. For instance, there is targeted delivery of doxorubicin to cancer cells when its conjugate with dextran is encapsulated in chitosan nanoparticles (100 nm diameter) [231]. A few chitosan formulations on clinical study and trial [222–226] may prove significant in diagnosis, treatment, and pain relief management of cancer.

Chitosan and its derivatives can permeate more effectively through negatively charged tumor-cell membrane to ensure the higher bioavailability in tumor cells [79]. Correlation of their structural behavior with suppression of tumor growth and metastasis in different cellular pathways can lead to further understanding of anticancer mechanism. Chemical modification, complex formation, and graft polymerization of chitosan could open an avenue in tailoring the hybrid materials formulation of anticancer therapeutic application.

Abbreviations

AGE: Advanced Glycation End products
Ara-C: 1-D-Arabinofuranosylcytosine
bFGF: Basic Fibroblast Growth Factor
BM: Basement membrane
CMCS: Carboxymethyl chitosan
CMDB: Carboxymethyl benzylamide dextrans
CNP: Chitosan nanoparticles
COS: Chitosan oligosaccharide
COX: Cyclooxygenase
COX-2:   Cyclooxygenase-2
csf: Cerebrospinal fluid

DDA: Degree of deacetylation
DMSO: Dimethyl sulfoxide
DNA: Deoxyribonucleic acid
EAC: Erlich ascites carcinoma
EDC: N-Ethyl-N′-(3-dimethylaminopropyl) carbodiimide hydrochloride
EAT: Erlich ascites tumor
EPR: Enhanced permeation and retention
ERK: Extracellular Signal Regulated Kinase
FACS: Fluorescent activated cell sorbent assay
FDA: US Food and Drug Administration
FGF-2:   Fibroblast growth factor-2
FGFR: Fibroblast growth factor kinase receptor
5-FU: 5-Fluorouracil
G-Chi: Glycol chitosan
GRAS: Generally recognized as safe
GSH: Glutathione
GST: Glutathione S-transferase
HMWC: High molecular weight chitosan
HS: Heparan sulfate
HUVECs: Human umbilical vein endothelial cells
IELs: Intraepithelial lymphocytes
IFN-α: Interferon cell signaling pathway-α
IFN-γ: Interferon cell signaling pathway-γ
IL: Interleukin
IL-1:   Interleukin 1
IL-2:   Interleukin 2
IL-8:   Interleukin 8
iNOS: Inducible Nitric Oxide Synthase
LLC: Lewis Lung Carcinoma
LMWC: Low molecular weight chitosan
LPS: Lipopolysaccharide
MCNS: Mifepristone loaded chitosan nanoparticles
MIF: Mifepristone
MMC: Mitomycin C
MMP9:   Matrix metalloproteinase 9
MMPs: Matrix metalloproteinases
mRNA: Messenger ribonucleic acid
M_w: Molecular weight
NaPaC: Phenylacetate carboxymethyl benzylamide dextran
NHS: N-Hydroxysuccinimide
NK: Natural killer
NK-cells: Natural killer cells
NO: Nitric oxide
ODC: Ornithine decarboxylase
PDGF: Platelets derived growth factor
PHI: Holmium-166 percutaneous (166Ho)/chitosan complex injection
PNA: Peptide-nucleobase conjugate
PPC: Polypyrrole chitosan
QR: Quinone/quinine reductase
RNA: Ribonucleic acid
ROS: Reactive oxygen species
RR: Ribonucleotide reductase
SBCS: Sulfated benzaldehyde chitosan
SCS: Sulfated chitosan
SCC: Squamous cell carcinoma

SCID: Severe combined immune deficient
siRNA: Short interfering RNA
Suc-Chi: N-succinyl chitosan
TGF-β: Transforming growth factor- β
TIMP 1: Tissue inhibitor of metalloproteinase 1
TIMPs: Tissue inhibitor of metalloproteinases
TNF -α: Tumor necrosis factor-α
TPA: 12-O-tetradecanoylphorbol-13-acetate
TPP: Tripolyphosphate
VEGF: Vascular endothelial growth factor
VEGFR2: Vascular endothelial growth factor receptor 2.

Conflicts of Interest

The authors declare no conflicts of interest.

Acknowledgments

Nepal Academy of Science and Technology (NAST) is gratefully acknowledged for providing the financial assistance (2017) to support our work.

References

[1] R. Ramya, P. N. Sudha, and J. Mahalakshmi, "Preparation and Characterization of Chitosan Binary Blend," *International Journal of Scientific and Research Publications*, vol. 2, pp. 1–9, 2012.

[2] Y. Yuan, B. M. Chesnutt, W. O. Haggard, and J. D. Bumgardner, "Deacetylation of chitosan: Material characterization and in vitro evaluation via albumin adsorption and pre-osteoblastic cell cultures," *Materials* , vol. 4, no. 8, pp. 1399–1416, 2011.

[3] Z. T. Zhang, D. H. Chen, and L. Chen, "Preparation of two different serials of chitosan," *Journal of Dong Hua University (English Edition)*, vol. 19, pp. 36–39, 2002.

[4] Y. N. Gavhane, A. S. Gurav, and A. V. Yadav, "Chitosan and Its Applications: A Review of Literature," *International journal of research in pharmaceutical and biomedical sciences*, vol. 4, no. 1, pp. 312–331, 2013.

[5] S. D. Ray, "Potential aspects of chitosan as pharmaceutical excipient," *Acta Poloniae Pharmaceutica*, vol. 68, no. 5, pp. 619–622, 2011.

[6] A. M. M. Sadeghi, F. A. Dorkoosh, M. R. Avadi et al., "Permeation enhancer effect of chitosan and chitosan derivatives: Comparison of formulations as soluble polymers and nanoparticulate systems on insulin absorption in Caco-2 cells," *European Journal of Pharmaceutics and Biopharmaceutics*, vol. 70, no. 1, pp. 270–278, 2008.

[7] R. Gu, W. Sun, H. Zhou et al., "The performance of a fly-larva shell-derived chitosan sponge as an absorbable surgical hemostatic agent," *Biomaterials*, vol. 31, no. 6, pp. 1270–1277, 2010.

[8] M. Burkatovskaya, G. P. Tegos, E. Swietlik, T. N. Demidova, A. P Castano, and M. R. Hamblin, "Use of chitosan bandage to prevent fatal infections developing from highly contaminated wounds in mice," *Biomaterials*, vol. 27, no. 22, pp. 4157–4164, 2006.

[9] J. H. Park, G. Saravanakumar, K. Kim, and I. C. Kwon, "Targeted delivery of low molecular drugs using chitosan and its derivatives," *Advanced Drug Delivery Reviews*, vol. 62, no. 1, pp. 28–41, 2010.

[10] J. K. Park, M. J. Chung, H. N. Choi, and Y. I. Park, "Effects of the molecular weight and the degree of deacetylation of chitosan oligosaccharides on antitumor activity," *International Journal of Molecular Sciences*, vol. 12, no. 1, pp. 266–277, 2011.

[11] M. Günbeyaz, A. Faraji, A. Özkul, N. Purali, and S. Şenel, "Chitosan based delivery systems for mucosal immunization against bovine herpesvirus 1 (BHV-1)," *European Journal of Pharmaceutical Sciences*, vol. 41, no. 3-4, pp. 531–545, 2010.

[12] Y. S. Wimardhani, D. F. Suniarti, H. J. Freisleben, S. I. Wanadi, and M. A. Ikeda, "Cytotoxic effects of chitosan against oral cancer cell lines is molecular-weight-dependent and cell-type-specific," *International Journal of Oral Research*, vol. 3, p. e1, 2012.

[13] Y. S. Wimardhani, D. F. Suniarti, H. J. Freisleben, S. I. Wanandi, N. C. Siregar, and M.-A. Ikeda, "Chitosan exerts anticancer activity through induction of apoptosis and cell cycle arrest in oral cancer cells," *Journal of oral science*, vol. 56, no. 2, pp. 119–126, 2014.

[14] W. Xia, P. Liu, and J. Liu, "Advance in chitosan hydrolysis by non-specific cellulases," *Bioresource Technology*, vol. 99, no. 15, pp. 6751–6762, 2008.

[15] B. A. Vishu Kumar, M. C. Varadaraj, and R. N. Tharanathan, "Low molecular weight chitosan - Preparation with the aid of pepsin, characterization, and its bactericidal activity," *Biomacromolecules*, vol. 8, no. 2, pp. 566–572, 2007.

[16] M. Huang, E. Khor, and L.-Y. Lim, "Uptake and cytotoxicity of chitosan molecules and nanoparticles: effects of molecular weight and degree of deacetylation," *Pharmaceutical Research*, vol. 21, no. 2, pp. 344–353, 2004.

[17] Y. Xu, Z. Wen, and Z. Xu, "Chitosan nanoparticles inhibit the growth of human hepatocellular carcinoma xenografts through an antiangiogenic mechanism," *Anticancer Research*, vol. 29, no. 12, pp. 5103–5109, 2009.

[18] K. V. Harish Prashanth and R. N. Tharanathan, "Depolymerized products of chitosan as potent inhibitors of tumor-induced angiogenesis," *Biochimica et Biophysica Acta (BBA) - General Subjects*, vol. 1722, no. 1, pp. 22–29, 2005.

[19] Y. Maeda and Y. Kimura, "Antitumor Effects of Various Low-Molecular-Weight Chitosans Are Due to Increased Natural Killer Activity of Intestinal Intraepithelial Lymphocytes in Sarcoma 180-Bearing Mice," *Journal of Nutrition*, vol. 134, no. 4, pp. 945–950, 2004.

[20] C. Qin, Y. Du, L. Xiao, Z. Li, and X. Gao, "Enzymic preparation of water-soluble chitosan and their antitumor activity," *International Journal of Biological Macromolecules*, vol. 31, no. 1–3, pp. 111–117, 2002.

[21] S.-L. Wang, H.-T. Lin, T.-W. Liang, Y.-J. Chen, Y.-H. Yen, and S.-P. Guo, "Reclamation of chitinous materials by bromelain for the preparation of antitumor and antifungal materials," *Bioresource Technology*, vol. 99, no. 10, pp. 4386–4393, 2008.

[22] S. Yamada, T. Ganno, N. Ohara, and Y. Hayashi, "Chitosan monomer accelerates alkaline phosphatase activity on human osteoblastic cells under hypofunctional conditions," *Journal of Biomedical Materials Research Part A*, vol. 83, no. 2, pp. 290–295, 2007.

[23] S. Kumar, J. Dutta, and P. Dutta, "Preparation and characterization of N-heterocyclic chitosan derivative based gels for biomedical applications," *International Journal of Biological Macromolecules*, vol. 45, no. 4, pp. 330–337, 2009.

[24] S. Kumar, N. Nigam, T. Ghosh et al., "Preparation, characterization and optical properties of a novel azo-based chitosan biopolymer," *Materials Chemistry and Physics*, vol. 120, no. 2-3, pp. 361–370, 2010.

[25] S. Kumar, P. K. Dutta, and P. Sen, "Preparation and characterization of optical property of crosslinkable film of chitosan with 2-thiophenecarboxaldehyde," *Carbohydrate Polymers*, vol. 80, no. 2, pp. 564–570, 2010.

[26] S. Kumar, J. Koh, D. K. Tiwari, and P. K. Dutta, "Optical Study of Chitosan-Ofloxacin Complex for Biomedical Applications," *Journal of Macromolecular Science, Part A Pure and Applied Chemistry*, vol. 48, no. 10, pp. 789–795, 2011.

[27] N. M. Alves and J. F. Mano, "Chitosan derivatives obtained by chemical modifications for biomedical and environmental applications," *International Journal of Biological Macromolecules*, vol. 43, no. 5, pp. 401–414, 2008.

[28] R. Jayakumar, K. P. Chennazhi, R. A. A. Muzzarelli, H. Tamura, S. V. Nair, and N. Selvamurugan, "Chitosan conjugated DNA nanoparticles in gene therapy," *Carbohydrate Polymers*, vol. 79, no. 1, pp. 1–8, 2010.

[29] R. Jayakumar, N. Nwe, S. Tokura, and H. Tamura, "Sulfated chitin and chitosan as novel biomaterials," *International Journal of Biological Macromolecules*, vol. 40, no. 3, pp. 175–181, 2007.

[30] R. Jayakumar, M. Prabaharan, S. V. Nair, and H. Tamura, "Novel chitin and chitosan nanofibers in biomedical applications," *Biotechnology Advances*, vol. 28, no. 1, pp. 142–150, 2010.

[31] M. K. S. Batista, L. F. Pinto, C. A. R. Gomes, and P. Gomes, "Novel highly-soluble peptide-chitosan polymers: Chemical synthesis and spectral characterization," *Carbohydrate Polymers*, vol. 64, no. 2, pp. 299–305, 2006.

[32] V. K. Mourya and N. N. Inamdar, "Chitosan-modifications and applications: opportunities galore," *Reactive and Functional Polymers*, vol. 68, no. 6, pp. 1013–1051, 2008.

[33] U. Manna, S. Bharani, and S. Patil, "Layer-by-layer self-assembly of modified hyaluronic acid/chitosan based on hydrogen bonding," *Biomacromolecules*, vol. 10, no. 9, pp. 2632–2639, 2009.

[34] R. Jayakumar, M. Prabaharan, R. L. Reis, and J. F. Mano, "Graft copolymerized chitosan—present status and applications," *Carbohydrate Polymers*, vol. 62, no. 2, pp. 142–158, 2005.

[35] R. Jayakumar, H. Nagahama, T. Furuike, and H. Tamura, "Synthesis of phosphorylated chitosan by novel method and its characterization," *International Journal of Biological Macromolecules*, vol. 42, no. 4, pp. 335–339, 2008.

[36] R. Jayakumar and H. Tamura, "Synthesis, characterization and thermal properties of chitin-g-poly(ε-caprolactone) copolymers by using chitin gel," *International Journal of Biological Macromolecules*, vol. 43, no. 1, pp. 32–36, 2008.

[37] R. Jayakumar, M. Prabaharan, P. T. Sudheesh Kumar, S. V. Nair, and H. Tamura, "Biomaterials based on chitin and chitosan in wound dressing applications," *Biotechnology Advances*, vol. 29, no. 3, pp. 322–337, 2011.

[38] N. S. Rejinold, K. P. Chennazhi, S. V. Nair, H. Tamura, and R. Jayakumar, "Biodegradable and thermo-sensitive chitosan-g-poly(N-vinylcaprolactam) nanoparticles as a 5-fluorouracil carrier," *Carbohydrate Polymers*, vol. 83, no. 2, pp. 776–786, 2011.

[39] Z. Zhong, Z. Zhong, R. Xing, P. Li, and G. Mo, "The preparation and antioxidant activity of 2-[phenylhydrazine (or hydrazine)-thiosemicarbazone]-chitosan," *International Journal of Biological Macromolecules*, vol. 47, no. 2, pp. 93–97, 2010.

[40] J. M. McCord, "The evolution of free radicals and oxidative stress," *American Journal of Medicine*, vol. 108, no. 8, pp. 652–659, 2000.

[41] R. L. Rao, M. Bharani, and V. Pallavi, "Role of antioxidants and free radicals in health and disease," *Advances in Pharmacology And Toxicology*, vol. 7, pp. 29–38, 2006.

[42] Y. Zheng, Y. Yi, Y. Qi, Y. Wang, W. Zhang, and M. Du, "Preparation of chitosan-copper complexes and their antitumor activity," *Bioorganic & Medicinal Chemistry Letters*, vol. 16, no. 15, pp. 4127–4129, 2006.

[43] R.-M. Wang, N.-P. He, P.-F. Song, Y.-F. He, L. Ding, and Z. Lei, "Preparation of low-molecular-weight chitosan derivative zinc complexes and their effect on the growth of liver cancer cells in vitro," *Pure and Applied Chemistry*, vol. 81, no. 12, pp. 2397–2405, 2009.

[44] X. Yin, X. Zhang, Q. Lin, Y. Feng, W. Yu, and Q. Zhang, "Metal-coordinating controlled oxidative degradation of chitosan and antioxidant activity of chitosan-metal complex," *Arkivoc*, vol. 2004, no. 9, pp. 66–78, 2004.

[45] Z. Jiang, B. Han, H. Li, Y. Yang, and W. Liu, "Carboxymethyl chitosan represses tumor angiogenesis in vitro and in vivo," *Carbohydrate Polymers*, vol. 129, pp. 1–8, 2015.

[46] S. Kumar, J. Koh, H. Kim, M. K. Gupta, and P. K. Dutta, "A new chitosan-thymine conjugate: Synthesis, characterization and biological activity," *International Journal of Biological Macromolecules*, vol. 50, no. 3, pp. 493–502, 2012.

[47] M. Jiang, H. Ouyang, P. Ruan et al., "Chitosan derivatives inhibit cell proliferation and induce apoptosis in breast cancer cells," *Anticancer Research*, vol. 31, no. 4, pp. 1321–1328, 2011.

[48] Y. Kato, H. Onishi, and Y. Machida, "Contribution of chitosan and its derivatives to cancer chemotherapy," *In Vivo*, vol. 19, no. 1, pp. 301–310, 2005.

[49] E. V. Svirshchevskaya, I. A. Gracheva, A. G. Kuznetsov, and E. V. Myrsikova, "Antitumor Activity of Furanoallocolchicinoid-Chitosan Conjugate," *Medicinal Chemistry*, vol. 6, no. 9, pp. 571–577, 2016.

[50] N. Salahuddin, A. A. Elbarbary, and H. A. Alkabes, "Quinazolinone derivatives loaded polypyrrole/chitosan core-shell nanoparticles with different morphologies: Antibacterial and anticancer activities," *Nano*, vol. 12, no. 1, Article ID 1750002, 17 pages, 2017.

[51] N. Salahuddin, A. A. Elbarbary, and H. A. Alkabes, "Antibacterial and antitumor activities of 3-amino-phenyl-4(3H)-quinazolinone/polypyrrole chitosan core shell nanoparticles," *Polymer Bulletin*, vol. 74, no. 5, pp. 1775–1790, 2017.

[52] A. Pestov and S. Bratskaya, "Chitosan and its derivatives as highly efficient polymer ligands," *Molecules*, vol. 21, no. 3, article no. 330, 2016.

[53] C. Q. Qin, Y. M. Du, and L. Xiao, "Effect of hydrogen peroxide treatment on the molecular weight and structure of chitosan," *Polymer Degradation and Stability*, vol. 76, no. 2, pp. 211–218, 2002.

[54] N. A. Mohamed, R. R. Mohamed, and R. S. Seoudi, "Synthesis and characterization of some novel antimicrobial thiosemicarbazone O-carboxymethyl chitosan derivatives," *International Journal of Biological Macromolecules*, vol. 63, pp. 163–169, 2014.

[55] X.-G. Chen and H.-J. Park, "Chemical characteristics of O-carboxymethyl chitosans related to the preparation conditions," *Carbohydrate Polymers*, vol. 53, no. 4, pp. 355–359, 2003.

[56] J. Fangkangwanwong, N. Sae-Liang, C. Sriworarat, A. Sereemaspun, and S. Chirachanchai, "Water-Based Chitosan for Thymine Conjugation: A Simple, Efficient, Effective, and Green Pathway to Introduce Cell Compatible Nucleic Acid Recognition," *Bioconjugate Chemistry*, vol. 27, no. 10, pp. 2301–2306, 2016.

[57] X.-J. Liu and R.-Y. Chen, "Synthesis of novel phosphonotripeptides containing uracil or thymine group," *Phosphorus, Sulfur, and Silicon and the Related Elements*, vol. 176, pp. 19–28, 2001.

[58] L. Kuan-Han, B.-R. Huang, and C.-C. Tzeng, "Synthesis and anticancer evaluation of certain α-methylene-γ-(4- substituted phenyl)-γ-butyrolactone bearing thymine, uracil, and 5- bromouracil," *Bioorganic & Medicinal Chemistry Letters*, vol. 9, no. 2, pp. 241–244, 1999.

[59] J. Skiba, R. Karpowicz, I. Szabó, B. Therrien, and K. Kowalski, "Synthesis and anticancer activity studies of ferrocenyl-thymine-3,6-dihydro-2H-thiopyranes - A new class of metallocene-nucleobase derivatives," *Journal of Organometallic Chemistry*, vol. 794, pp. 216–222, 2015.

[60] A. K. Singla and M. Chawla, "Chitosan: some pharmaceutical and biological aspects—an update," *Journal of Pharmacy and Pharmacology*, vol. 53, no. 8, pp. 1047–1067, 2001.

[61] C. K. S. Pillai, W. Paul, and C. P. Sharma, "Chitin and chitosan polymers: chemistry, solubility and fiber formation," *Progress in Polymer Science*, vol. 34, no. 7, pp. 641–678, 2009.

[62] K. Y. Lee, W. S. Ha, and W. H. Park, "Blood compatibility and biodegradability of partially N-acylated chitosan derivatives," *Biomaterials*, vol. 16, no. 16, pp. 1211–1216, 1995.

[63] B. Carreño-Gómez and R. Duncan, "Evaluation of the biological properties of soluble chitosan and chitosan microspheres," *International Journal of Pharmaceutics*, vol. 148, no. 2, pp. 231–240, 1997.

[64] Y. Song, H. Onishi, and T. Nagai, "Synthesis and Drug-Release Characteristics of the Conjugates of Mitomycin C with N-Succinyl-chitosan and Carboxymethyl-chitin," *Chemical & Pharmaceutical Bulletin*, vol. 40, no. 10, pp. 2822–2825, 1992.

[65] Y. Song, H. Onishi, and T. Nagai, "Conjugate of mitomycin C with N-succinyl-chitosan: In vitro drug release properties, toxicity and antitumor activity," *International Journal of Pharmaceutics*, vol. 98, no. 1-3, pp. 121–130, 1993.

[66] Y. Song, H. Onishi, and T. Nagai, "Toxicity and antitumor activity of the conjugate of mitomycin C with carboxymethyl-chitin," *Yakuzaigaku*, vol. 53, pp. 141–147, 1993.

[67] K. Kamiyama, H. Onishi, and Y. Machida, "Biodisposition characteristics of N-succinyl-chitosan and glycol- chitosan in normal and tumor-bearing mice," *Biological & Pharmaceutical Bulletin*, vol. 22, no. 2, pp. 179–186, 1999.

[68] Y. Kato, H. Onishi, and Y. Machida, "Evaluation of N-succinyl-chitosan as a systemic long-circulating polymer," *Biomaterials*, vol. 21, no. 15, pp. 1579–1585, 2000.

[69] J. Hosoda, S. Unezaki, K. Maruyama, S. Tsuchiya, and M. Iwatsuru, "Antitumor activity of doxorubicin encapsulated in poly(ethylene glycol)-coated liposomes," *Biological & Pharmaceutical Bulletin*, vol. 18, no. 9, pp. 1234–1237, 1995.

[70] T. Nakanishi, S. Fukushima, K. Okamoto et al., "Development of the polymer micelle carrier system for doxorubicin," *Journal of Controlled Release*, vol. 74, no. 1-3, pp. 295–302, 2001.

[71] C. Yan, D. Chen, J. Gu, H. Hu, X. Zhao, and M. Qiao, "Preparation of N-succinyl-chitosan and their physical-chemical properties as a novel excipient," *Yakugaku Zasshi*, vol. 126, no. 9, pp. 789–793, 2006.

[72] Y. Song, H. Onishi, and T. Nagai, "Pharmacokinetic characteristics and antitumor activity of the n-succinyl-chitosan-mitomycin C conjugate and the carboxymethyl-chitin-mitomycin C conjugate," *Biological & Pharmaceutical Bulletin*, vol. 16, no. 1, pp. 48–54, 1993.

[73] T. Muslim, M. Morimoto, H. Saimoto, Y. Okamoto, S. Minami, and Y. Shigemasa, "Synthesis and bioactivities of poly(ethylene glycol)-chitosan hybrids," *Carbohydrate Polymers*, vol. 46, no. 4, pp. 323–330, 2001.

[74] B. J. Crielaard, S. van der Wal, T. Lammers et al., "A polymeric colchicinoid prodrug with reduced toxicity and improved efficacy for vascular disruption in cancer therapy." *International Journal of Nanomedicine*, vol. 6, pp. 2697–2703, 2011.

[75] Y. V. Voitovich, E. S. Shegravina, N. S. Sitnikov et al., "Synthesis and biological evaluation of furanoallocolchicinoids," *Journal of Medicinal Chemistry*, vol. 58, no. 2, pp. 692–704, 2015.

[76] R. Mathiyalagan, S. Subramaniyam, Y. J. Kim, Y.-C. Kim, and D. C. Yang, "Ginsenoside compound K-bearing glycol chitosan conjugates: Synthesis, physicochemical characterization, and in vitro biological studies," *Carbohydrate Polymers*, vol. 112, pp. 359–366, 2014.

[77] M. Thanou, J. C. Verhoef, and H. E. Junginger, "Chitosan and its derivatives as intestinal absorption enhancers," *Advanced Drug Delivery Reviews*, vol. 50, no. 1, pp. S91–S101, 2001.

[78] K.-S. Nam and Y.-H. Shon, "Suppression of metastasis of human breast cancer cells by chitosan oligosaccharides," *Journal of Microbiology and Biotechnology*, vol. 19, no. 6, pp. 629–633, 2009.

[79] H. Zhang, F. Wu, Y. Li et al., "Chitosan-based nanoparticles for improved anticancer efficacy and bioavailability of mifepristone," *Beilstein Journal of Nanotechnology*, vol. 7, pp. 1861–1870, 2016.

[80] S. Kuppusamy and J. Karuppaiah, "Screening of Antiproliferative Effect of Chitosan on Tumor Growth and Metastasis in T24 Urinary Bladder Cancer Cell Line," *Austrl-Asian Journal of Cancer*, vol. 12, no. 3, pp. 145–149, 2013.

[81] H. Hosseinzadeh, F. Atyabi, R. Dinarvand, and S. N. Ostad, "Chitosan-Pluronic nanoparticles as oral delivery of anticancer gemcitabine: preparation and in vitro study," *International Journal of Nanomedicine*, vol. 7, pp. 1851–1863, 2012.

[82] K.-T. Shen, M.-H. Chen, H.-Y. Chan, J.-H. Jeng, and Y.-J. Wang, "Inhibitory effects of chitooligosaccharides on tumor growth and metastasis," *Food and Chemical Toxicology*, vol. 47, no. 8, pp. 1864–1871, 2009.

[83] C. Qin, B. Zhou, L. Zeng et al., "The physicochemical properties and antitumor activity of cellulase-treated chitosan," *Food Chemistry*, vol. 84, no. 1, pp. 107–115, 2004.

[84] A. Tokoro, K. Suzuki, T. Matsumoto, T. Mikami, S. Suzuki, and M. Suzuki, "Chemotactic Response of Human Neutrophils to N-Acetyl Chitohexaose in vitro," *Microbiology and Immunology*, vol. 32, no. 4, pp. 387–395, 1988.

[85] W.-G. Seo, H.-O. Pae, N.-Y. Kim et al., "Synergistic cooperation between water-soluble chitosan oligomers and interferon-γ for induction of nitric oxide synthesis and tumoricidal activity in murine peritoneal macrophages," *Cancer Letters*, vol. 159, no. 2, pp. 189–195, 2000.

[86] K. Suzuki, T. Mikami, Y. Okawa, A. Tokoro, S. Suzuki, and M. Suzuki, "Antitumor effect of hexa-N-acetylchitohexaose and chitohexaose," *Carbohydrate Research*, vol. 151, pp. 403–408, 1986.

[87] K. Tsukada, T. Matsumoto, K. Aizawa et al., "Antimetastatic and Growth-inhibitory Effects of N-Acetylchitohexaose in Mice Bearing Lewis Lung Carcinoma," *Japanese Journal of Cancer Research*, vol. 81, no. 3, pp. 259–265, 1990.

[88] A. Tokoro, M. Kobayashi, N. Tatewaki et al., "Protective Effect of N-Acetyl Chitohexaose on Listeria monocytogenes Infection in Mice," *Microbiology and Immunology*, vol. 33, no. 4, pp. 357–367, 1989.

[89] F. Y. Wang and Y. S. He, "Study on antitumor effect of water-soluble chitosan," *Journal of Clinical Biochemistry Drug*, vol. 22, pp. 21-22, 2001.

[90] J. C. Fernandes, J. Sereno, P. Garrido et al., "Inhibition of bladder tumor growth by chitooligosaccharides in an experimental carcinogenesis model," *Marine Drugs*, vol. 10, no. 12, pp. 2661–2675, 2012.

[91] M. Sugano, T. Fujikawa, Y. Hiratsuji, K. Nakashima, N. Fukuda, and Y. Hasegawa, "A novel use of chitosan as a hypocholesterolemic agent in rats," *American Journal of Clinical Nutrition*, vol. 33, no. 4, pp. 787–793, 1980.

[92] H. Takimoto, M. Hasegawa, K. Yagi, T. Nakamura, T. Sakaeda, and M. Hirai, "Proapoptotic effect of a dietary supplement: water soluble chitosan activates caspase-8 and modulating death receptor expression.," *Drug Metabolism and Pharmacokinetics*, vol. 19, no. 1, pp. 76–82, 2004.

[93] J. Zhang, W. Xia, P. Liu et al., "Chitosan modification and pharmaceutical/biomedical applications," *Marine Drugs*, vol. 8, no. 7, pp. 1962–1987, 2010.

[94] M. Hasegawa, K. Yagi, S. Iwakawa, and M. Hirai, "Chitosan induces apoptosis via caspase-3 activation in bladder tumor cells," *Japanese Journal of Cancer Research*, vol. 92, no. 4, pp. 459–466, 2001.

[95] L. Qi, Z. Xu, and M. Chen, "In vitro and in vivo suppression of hepatocellular carcinoma growth by chitosan nanoparticles," *European Journal of Cancer*, vol. 43, no. 1, pp. 184–193, 2007.

[96] J.-W. Ai, W. Liao, and Z.-L. Ren, "Enhanced anticancer effect of copper-loaded chitosan nanoparticles against osteosarcoma," *RSC Advances*, vol. 7, no. 26, pp. 15971–15977, 2017.

[97] T. Ramasamy, H. B. Ruttala, N. Chitrapriya et al., "Engineering of cell microenvironment-responsive polypeptide nanovehicle co-encapsulating a synergistic combination of small molecules for effective chemotherapy in solid tumors," *Acta Biomaterialia*, vol. 48, pp. 131–143, 2017.

[98] K. Vermeulen, D. R. van Bockstaele, and Z. N. Berneman, "The cell cycle: a review of regulation, deregulation and therapeutic targets in cancer," *Cell Proliferation*, vol. 36, no. 3, pp. 131–149, 2003.

[99] S. Senturk, M. Mumcuoglu, O. Gursoy-Yuzugullu, B. Cingoz, K. C. Akcali, and M. Ozturk, "Transforming growth factor-beta induces senescence in hepatocellular carcinoma cells and inhibits tumor growth," *Hepatology*, vol. 52, no. 3, pp. 966–974, 2010.

[100] J. Bartek and J. Lukas, "Mammalian G1- and S-phase checkpoints in response to DNA damage," *Current Opinion in Cell Biology*, vol. 13, no. 6, pp. 738–747, 2001.

[101] G. J. Hannon and D. Beach, "p15INK4B is a potential effector of TGF-β-induced cell cycle arrest," *Nature*, vol. 371, no. 6494, pp. 257–261, 1994.

[102] I. Reynisdottir, K. Polyak, A. Iavarone, and J. Massague, "Kip/Cip and Ink4 Cdk inhibitors cooperate to induce cell cycle arrest in response to TGF-β," *Genes & Development*, vol. 9, no. 15, pp. 1831–1845, 1995.

[103] J. Falck, N. Mailand, R. G. Syljuåsen, J. Bartek, and J. Lukas, "The ATM-Chk2-Cdc25A checkpoint pathway guards against radioresistant DNA synthesis," *Nature*, vol. 410, no. 6830, pp. 842–847, 2001.

[104] M. Saruc, S. Standop, J. Standop et al., "Pancreatic Enzyme Extract Improves Survival in Murine Pancreatic Cancer," *Pancreas*, vol. 28, no. 4, pp. 401–412, 2004.

[105] K.-S. Nam, M.-K. Kim, and Y.-H. Shon, "Chemopreventive effect of chitosan oligosaccharide against colon carcinogenesis," *Journal of Microbiology and Biotechnology*, vol. 17, no. 9, pp. 1546–1549, 2007.

[106] K. S. Nam, M. K. Kim, and Y.-H. Shon, "Inhibition of proinflammatory cytokine-induced invasiveness of HT-29 cells by chitosan oligosaccharide," *Journal of Microbiology and Biotechnology*, vol. 17, no. 12, pp. 2042–2045, 2007.

[107] H. Quan, F. Zhu, X. Han, Z. Xu, Y. Zhao, and Z. Miao, "Mechanism of anti-angiogenic activities of chitooligosaccharides may be through inhibiting heparanase activity," *Medical Hypotheses*, vol. 73, no. 2, pp. 205-206, 2009.

[108] H. Nagaset and J. F. Woessner Jr., "Matrix metalloproteinases," *The Journal of Biological Chemistry*, vol. 274, no. 31, pp. 21491–21494, 1999.

[109] P. D. Brown, "Matrix metalloproteinases in gastrointestinal cancer," *Gut*, vol. 43, no. 2, pp. 161–163, 1998.

[110] J. A. Lagares-Garcia, R. A. Moore, B. Collier, M. Heggere, F. Diaz, and F. Qian, "Nitric oxide synthase as a marker in colorectal carcinoma," *The American Surgeon*, vol. 67, no. 7, pp. 709–713, 2001.

[111] N. Yagihashi, H. Kasajima, S. Sugai et al., "Increased in situ expression of nitric oxide synthase in human colorectal cancer," *Virchows Archiv*, vol. 436, no. 2, pp. 109–114, 2000.

[112] J. Folkman, "Angiogenesis and angiogenesis inhibition: an overview.," *EXS*, vol. 79, pp. 1–8, 1997.

[113] H.-T. Liu, P. Huang, P. Ma, Q.-S. Liu, C. Yu, and Y.-G. Du, "Chitosan oligosaccharides suppress LPS-induced IL-8 expression in human umbilical vein endothelial cells through blockade of p38 and Akt protein kinases," *Acta Pharmacologica Sinica*, vol. 32, no. 4, pp. 478–486, 2011.

[114] Z. Hossain and K. Takahashi, "Induction of permeability and apoptosis in colon cancer cell line with chitosan," *Journal of Food and Drug Analysis*, vol. 16, no. 5, pp. 1–8, 2008.

[115] J. Dou, P. Ma, C. Xiong, C. Tan, and Y. Du, "Induction of apoptosis in human acute leukemia HL-60 cells by oligochitosan through extrinsic and intrinsic pathway," *Carbohydrate Polymers*, vol. 86, no. 1, pp. 19–24, 2011.

[116] S.-M. Hwang, C.-Y. Chen, S.-S. Chen, and J.-C. Chen, "Chitinous materials inhibit nitric oxide production by activated RAW 264.7 macrophages," *Biochemical and Biophysical Research Communications*, vol. 271, no. 1, pp. 229–233, 2000.

[117] M. Silano, O. Vincentini, R. A. A. Muzzarelli, C. Muzzarelli, and M. De Vincenzi, "MP-chitosan protects Caco-2 cells from toxic gliadin peptides," *Carbohydrate Polymers*, vol. 58, no. 2, pp. 215–219, 2004.

[118] K. M. Yamada and K. Clark, "Cell biology: Survival in three dimensions," *Nature*, vol. 419, no. 6909, pp. 790-791, 2002.

[119] R. Wallace, J. Richard Thomson, M. J. Bell, and H. E. Skipper, "Observations on the Antileukemic Activity of Pyridine- 2-carboxaldehyde Thiosemicarbazone and Thiocarbohydrazone," *Cancer Research*, vol. 16, no. 2, pp. 167–170, 1956.

[120] F. A. French and E. J. Blanz Jr., "The Carcinostatic Activity of Thiosemicarbazones of Formyl Heteroaromatic Compounds," *Journal of Medicinal Chemistry*, vol. 9, no. 4, pp. 585–589, 1966.

[121] K. C. Agrawal and A. C. Sartorelli, "Potential Antitumor Agents. II. Effects of Modifications in the Side Chain of 1-Formylisoquinoline Thiosemicarbazone," *Journal of Medicinal Chemistry*, vol. 12, no. 5, pp. 771–774, 1969.

[122] F. A. French and E. J. Blanz Jr., "Chemotherapy studies on experimental mouse tumors X," *Cancer Chemotherapy Reports Part 2*, vol. 2, pp. 199–235, 1971.

[123] D. L. Klayman, J. F. Bartosevich, T. S. Griffin, C. J. Mason, and J. P. Scovill, "2-Acetylpyridine Thiosemicarbazones. 1. A New Class of Potential Antimalarial Agents," *Journal of Medicinal Chemistry*, vol. 22, no. 7, pp. 855–862, 1979.

[124] L. Thelander and P. Reichard, "Reduction of ribonucleotides.," *Annual Review of Biochemistry*, vol. 48, pp. 133–158, 1979.

[125] J. G. Cory and A. Sato, "Regulation of ribonucleotide reductase activity in mammalian cells," *Molecular and Cellular Biochemistry*, vol. 53-54, no. 1-2, pp. 257–266, 1983.

[126] E. C. Moore and A. C. Sartorelli, "The inhibition of ribonucleotide reductase by alpha-(N)-heterocyclic carboxaldehyde thiosemicarbazones," in *Inhibitors of Ribonucleotide Diphosphate Reductase Activity*, J. G. Cory and A. H. Cory, Eds., pp. 203–215, Pergamon Press, New York, NY, USA, 1989.

[127] W. Xie, P. Xu, and Q. Liu, "Antioxidant activity of water–soluble chitosan derivatives," *Bioorganic & Medicinal Chemistry Letters*, vol. 11, no. 13, pp. 1699–1701, 2001.

[128] L. Astolfi, S. Ghiselli, V. Guaran et al., "Correlation of adverse effects of cisplatin administration in patients affected by solid tumours: A retrospective evaluation," *Oncology Reports*, vol. 29, no. 4, pp. 1285–1292, 2013.

[129] M. A. Fuertes, C. Alonso, and J. M. Pérez, "Biochemical modulation of cisplatin mechanisms of action: enhancement of antitumor activity and circumvention of drug resistance," *Chemical Reviews*, vol. 103, no. 3, pp. 645–662, 2003.

[130] H.-T. Arkenau, C. P. Carden, and J. S. de Bono, "Targeted agents in cancer therapy," *Medicine*, vol. 36, no. 1, pp. 33–37, 2008.

[131] F. Arnesano and G. Natile, ""Platinum on the road": Interactions of antitumoral cisplatin with proteins," *Pure and Applied Chemistry*, vol. 80, no. 12, pp. 2715–2725, 2008.

[132] M. J. Hannon, "Metal-based anticancer drugs: From a past anchored in platinum chemistry to a post-genomic future of diverse chemistry and biology," *Pure and Applied Chemistry*, vol. 79, no. 12, pp. 2243–2261, 2007.

[133] D. Steinborn and H. Junicke, "Carbohydrate complexes of platinum-group metals," *Chemical Reviews*, vol. 100, no. 12, pp. 4283–4317, 2000.

[134] U. Kalinowska-Lis, J. Ochocki, and K. Matlawska-Wasowska, "Trans geometry in platinum antitumor complexes," *Coordination Chemistry Reviews*, vol. 252, no. 12-14, pp. 1328–1345, 2008.

[135] A. J. Varma, S. V. Deshpande, and J. F. Kennedy, "Metal complexation by chitosan and its derivatives: a review," *Carbohydrate Polymers*, vol. 55, no. 1, pp. 77–93, 2004.

[136] L. Gritsch, C. Lovell, W. H. Goldmann, and A. R. Boccaccini, "Fabrication and characterization of copper(II)-chitosan complexes as antibiotic-free antibacterial biomaterial," *Carbohydrate Polymers*, vol. 179, pp. 370–378, 2018.

[137] L. Qi, Z. Xu, X. Jiang, C. Hu, and X. Zou, "Preparation and antibacterial activity of chitosan nanoparticles," *Carbohydrate Research*, vol. 339, no. 16, pp. 2693–2700, 2004.

[138] P. Nurse, "A long twentieth century of the cell cycle and beyond," *Cell*, vol. 100, no. 1, pp. 71–78, 2000.

[139] J. H. Niazi and M. B. Gu, "Toxicity of metallic nanoparticles in microorganisms- A review," *Atmospheric and Biological Environmental Monitoring*, pp. 193–206, 2009.

[140] O. Bondarenko, K. Juganson, A. Ivask, K. Kasemets, M. Mortimer, and A. Kahru, "Toxicity of Ag, CuO and ZnO nanoparticles to selected environmentally relevant test organisms and mammalian cells in vitro: a critical review," *Archives of Toxicology*, vol. 87, no. 7, pp. 1181–1200, 2013.

[141] M. P. Murphy, "How mitochondria produce reactive oxygen species," *Biochemical Journal*, vol. 417, no. 1, pp. 1–13, 2009.

[142] M. Kurniasih, P. Purwati, D. Hermawan, and M. Zaki, "Optimum conditions for the synthesis of high solubility carboxymethyl chitosan," *Malaysian Journal of Fundamental and Applied Sciences*, vol. 10, no. 4, pp. 189–194, 2014.

[143] J. Ji, D. Wu, L. Liu, J. Chen, and Y. Xu, "Preparation, evaluation, and in vitro release of folic acid conjugated O-carboxymethyl chitosan nanoparticles loaded with methotrexate," *Journal of Applied Polymer Science*, vol. 125, no. S2, pp. E208–E215, 2012.

[144] A. Anitha, K. P. Chennazhi, S. V. Nair, and R. Jayakumar, "5-Flourouracil loaded N,O-carboxymethyl chitosan nanoparticles as an anticancer nanomedicine for breast cancer," *Journal of Biomedical Nanotechnology*, vol. 8, no. 1, pp. 29–42, 2012.

[145] A. Anitha, S. Maya, N. Deepa, K. P. Chennazhi, S. V. Nair, and R. Jayakumar, "Curcumin-loaded N, O-carboxymethyl chitosan nanoparticles for cancer drug delivery," *Journal of Biomaterials Science, Polymer Edition*, vol. 23, no. 11, pp. 1381–1400, 2012.

[146] Y. H. Jin, H. Hu, M. Qiao et al., "pH-sensitive chitosan-derived nanoparticles as doxorubicin carriers for effective anti-tumor activity: preparation and in vitro evaluation," *Colloids and Surfaces B: Biointerfaces*, vol. 94, pp. 184–191, 2012.

[147] Y. Wang, X. Yang, J. Yang et al., "Self-assembled nanoparticles of methotrexate conjugated O-carboxymethyl chitosan: Preparation, characterization and drug release behavior in vitro," *Carbohydrate Polymers*, vol. 86, no. 4, pp. 1665–1670, 2011.

[148] M. Zheng, B. Han, Y. Yang, and W. Liu, "Synthesis, characterization and biological safety of O-carboxymethyl chitosan used to treat Sarcoma 180 tumor," *Carbohydrate Polymers*, vol. 86, no. 1, pp. 231–238, 2011.

[149] J. Folkman, "New perspectives in clinical oncology from angiogenesis research," *European Journal of Cancer Part A: General Topics*, vol. 32, no. 14, pp. 2534–2539, 1996.

[150] R. B. Effros, "Genetic alterations in the ageing immune system: Impact on infection and cancer," *Mechanisms of Ageing and Development*, vol. 124, no. 1, pp. 71–77, 2003.

[151] W. Aulitzky, G. Gastl, W. E. Aulitzky et al., "Successful treatment of metastatic renal cell carcinoma with a biologically active dose of recombinant interferon-gamma," *Journal of Clinical Oncology*, vol. 7, no. 12, pp. 1875–1884, 1989.

[152] A. M. Eggermont, H. Schraffordt Koops, D. Liénard et al., "Isolated limb perfusion with high-dose tumor necrosis factor-α in combination with interferon-γ and melphalan for non-resectable extremity soft tissue sarcomas: A multicenter trial," *Journal of Clinical Oncology*, vol. 14, no. 10, pp. 2653–2665, 1996.

[153] L. M. Ebert, S. Meuter, and B. Moser, "Homing and function of human skin γδ T cells and NK cells: Relevance for tumor surveillance," *The Journal of Immunology*, vol. 176, no. 7, pp. 4331–4336, 2006.

[154] L. Chang, H. Kamata, G. Solinas et al., "The E3 ubiquitin ligase itch couples JNK activation to TNFα-induced cell death by inducing c-FLIPL turnover," *Cell*, vol. 124, no. 3, pp. 601–613, 2006.

[155] G. M. Hur, J. Lewis, Q. Yang et al., "The death domain kinase RIP has an essential role in DNA damage-induced NF-κB activation," *Genes & Development*, vol. 17, no. 7, pp. 873–882, 2003.

[156] O. Trubiani, D. Bosco, and R. Di Primio, "Interferon-γ (IFN-γ) induces programmed cell death in differentiated human leukemia B cell lines," *Experimental Cell Research*, vol. 215, no. 1, pp. 23–27, 1994.

[157] E. Ahn, G. Pan, S. M. Vickers, and J. M. McDonald, "IFN-gamma upregulates apoptosis-related molecules and enhances Fas-mediated apoptosis in human cholangiocarcinoma," *International Journal of Cancer*, vol. 100, no. 4, pp. 445–451, 2002.

[158] L.-M. Tumir, M. Grabar, S. Tomić, and I. Piantanida, "The interactions of bis-phenanthridinium-nucleobase conjugates with nucleotides: adenine-conjugate recognizes UMP in aqueous medium," *Tetrahedron*, vol. 66, no. 13, pp. 2501–2513, 2010.

[159] A. R. Pike, L. C. Ryder, B. R. Horrocks et al., "Metallocene - DNA: Synthesis, molecular and electronic structure and DNA incorporation of C5-ferrocenylthymidine derivatives," *Chemistry - A European Journal*, vol. 8, no. 13, pp. 2891–2899, 2002.

[160] S. Boncel, M. Mączka, K. K. Koziol, R. Motyka, and K. Z. Walczak, "Symmetrical and unsymmetrical α,ω-nucleobase amide-conjugated systems," *Beilstein Journal of Organic Chemistry*, vol. 6, no. 34, 2010.

[161] T. Ihara, A. Uemura, A. Futamura et al., "Cooperative DNA probing using a β-cyclodextrin DNA conjugate and a nucleobase-specific fluorescent ligand," *Journal of the American Chemical Society*, vol. 131, no. 4, pp. 1386-1387, 2009.

[162] H. Kraatz, "Ferrocene-Conjugates of Amino Acids, Peptides and Nucleic Acids," *Journal of Inorganic and Organometallic Polymers and Materials*, vol. 15, no. 1, pp. 83–106, 2005.

[163] W. A. Wlassoff and G. C. King, "Ferrocene conjugates of dUTP for enzymatic redox labelling of DNA," *Nucleic Acids Research*, vol. 30, no. 12, Article ID e58, 2002.

[164] Y. Xu, H. Jin, Z. Yang, L. Zhang, and L. Zhang, "Synthesis and biological evaluation of novel neamine-nucleoside conjugates potentially targeting to RNAs," *Tetrahedron*, vol. 65, no. 27, pp. 5228–5239, 2009.

[165] T. Kubo, R. Bakalova, H. Ohba, and M. Fujii, "Antisense effects of DNA– peptide conjugates," *Nucleic Acids Research*, no. 3, pp. 179-180, 2003.

[166] G. N. Roviello, E. Benedetti, C. Pedone, and E. M. Bucci, "Nucleobase-containing peptides: An overview of their characteristic features and applications," *Amino Acids*, vol. 39, no. 1, pp. 45–57, 2010.

[167] D. L. Nelson and M. M. Cox, "DNA Replication, in: Lehninger Principles of Biochemistry," in *Lehninger Principles of Biochemistry*, pp. 950–966, 4th edition, 2005.

[168] D. Hanahan and R. A. Weinberg, "Hallmarks of cancer: the next generation," *Cell*, vol. 144, no. 5, pp. 646–674, 2011.

[169] D. G. Fernig and J. T. Gallagher, "Fibroblast growth factors and their receptors: An information network controlling tissue growth, morphogenesis and repair," *Progress in Growth Factor Research*, vol. 5, no. 4, pp. 353–377, 1994.

[170] D. Qiao, K. Meyer, C. Mundhenke, S. A. Drew, and A. Friedl, "Heparan sulfate proteoglycans as regulators of fibroblast growth factor-2 signaling in brain endothelial cells: Specific role for glypican-1 in glioma angiogenesis," *The Journal of Biological Chemistry*, vol. 278, no. 18, pp. 16045–16053, 2003.

[171] C. Mundhenke, K. Meyer, S. Drew, and A. Friedl, "Heparan sulfate proteoglycans as regulators of fibroblast growth factor-2 receptor binding in breast carcinomas," *The American Journal of Pathology*, vol. 160, no. 1, pp. 185–194, 2002.

[172] S. W. McLeskey, L. Zhang, B. J. Trock et al., "Effects of AGM-1470 and pentosan polysulphate on tumorigenicity and metastasis of FGF-transfected MCF-7 cells," *British Journal of Cancer*, vol. 73, no. 9, pp. 1053–1062, 1996.

[173] S. Zaslau, D. R. Riggs, B. J. Jackson et al., "In vitro effects of pentosan polysulfate against malignant breast cells," *The American Journal of Surgery*, vol. 188, no. 5, pp. 589–592, 2004.

[174] J. L. Marshall, A. Wellstein, J. Rae et al., "Phase I trial of orally administered pentosan polysulfate in patients with advanced cancer," *Clinical Cancer Research*, vol. 3, no. 12, pp. 2347–2354, 1997.

[175] M. K. Yunmbam and A. Wellstein, "The bacterial polysaccharide tecogalan blocks growth of breast cancer cells in vivo," *Oncology Reports*, vol. 8, no. 1, pp. 161–164, 2001.

[176] J. M. Liu, J. Bignon, F. Haroun-Bouhedja et al., "Inhibitory effect of fucoidan on the adhesion of adenocarcinoma cells to fibronectin," *Anticancer Reseach*, vol. 25, no. 3 B, pp. 2129–2133, 2005.

[177] M. Di Benedetto, A. Starzec, R. Vassy, G. Y. Perret, M. Crépin, and M. Kraemer, "Inhibition of epidermoid carcinoma A431 cell growth and angiogenesis in nude mice by early and late treatment with a novel dextran derivative," *British Journal of Cancer*, vol. 88, no. 12, pp. 1987–1994, 2003.

[178] R. Bagheri-Yarmand, J.-F. Liu, D. Ledoux, J. F. Morere, and M. Crepin, "Erratum: Inhibition of human breast epithelial HBL100 cell proliferation by a dextran derivative (CMDB7): Interference with the FGF2 autocrine loop (Biochemical and Biophysical Research Communications (1997) 239 (424-428))," *Biochemical and Biophysical Research Communications*, vol. 241, no. 3, p. 804, 1997.

[179] J. F. Morere, D. Letourneur, P. Planchon et al., "Inhibitory effect of substituted dextrans on MCF7 human breast cancer cell growth in vitro," *Anti-Cancer Drugs*, vol. 3, no. 6, pp. 629–634, 1992.

[180] P. Bittoun, T. Avramoglou, J. Vassy, M. Crépin, F. Chaubet, and S. Fermandjian, "Low-molecular-weight dextran derivatives (f-CMDB) enter the nucleus and are better cell-growth inhibitors compared with parent CMDB polymers," *Carbohydrate Research*, vol. 322, no. 3-4, pp. 247–255, 1999.

[181] J. Liu, R. Bagheri-Yarmand, Y. Xia, and M. Crépin, "Modulations of breast fibroblast and carcinoma cell interactions by a dextran derivative (CMDB7)," *Anticancer Reseach*, vol. 17, no. 1 A, pp. 253–258, 1997.

[182] M. D. Benedetto, A. Starzec, B. M. Colombo et al., "Aponecrotic, antiangiogenic and antiproliferative effects of a novel dextran derivative on breast cancer growth in vitro and in vivo," *British Journal of Pharmacology*, vol. 135, no. 8, pp. 1859–1871, 2002.

[183] S. Malherbe, M. Crépin, C. Legrand, and M. X. Wei, "Cytostatic and pro-apoptotic effects of a novel phenylacetate-dextran derivative (NaPaC) on breast cancer cells in interactions with endothelial cells," *Anti-Cancer Drugs*, vol. 15, no. 10, pp. 975–981, 2004.

[184] C. Gervelas, T. Avramoglou, M. Crépin, and J. Jozefonvicz, "Growth inhibition of human melanoma tumor cells by the combination of sodium phenylacetate (NaPA) and substituted dextrans and one NaPA-dextran conjugate," *Anti-Cancer Drugs*, vol. 13, no. 1, pp. 37–45, 2002.

[185] M. Di Benedetto, Y. Kourbali, A. Starzec et al., "Sodium phenylacetate enhances the inhibitory effect of dextran derivative on breast cancer cell growth in vitro and in nude mice," *British Journal of Cancer*, vol. 85, no. 6, pp. 917–923, 2001.

[186] P. Bittoun, R. Bagheri-Yarmand, F. Chaubet, M. Crépin, J. Jozefonvicz, and S. Fermandjian, "Effects of the binding of a dextran derivative on fibroblast growth factor 2: Secondary structure and receptor-binding studies," *Biochemical Pharmacology*, vol. 57, no. 12, pp. 1399–1406, 1999.

[187] M. Di Benedetto, A. Starzec, R. Vassy, G.-Y. Perret, and M. Crépin, "Distinct heparin binding sites on VEGF165 and its receptors revealed by their interaction with a non sulfated glycoaminoglycan (NaPaC)," *Biochimica et Biophysica Acta (BBA) - General Subjects*, vol. 1780, no. 4, pp. 723–732, 2008.

[188] D. Logeart-Avramoglou and J. Jozefonvicz, "Carboxymethyl benzylamide sulfonate dextrans (CMDBS), a family of biospecific polymers endowed with numerous biological properties: A review," *Journal of Biomedical Materials Research Part B: Applied Biomaterials*, vol. 48, no. 4, pp. 578–590, 1999.

[189] Y. Song, H. Onishi, Y. Machida, and T. Nagai, "Drug release and antitumor characteristics of N-succinyl-chitosan-mitomycin C as an implant," *Journal of Controlled Release*, vol. 42, no. 1, pp. 93–100, 1996.

[190] A. Sabouraud, O. Chappey, T. Dupin, and J. M. Scherrmann, "Binding of colchicine and thiocolchicoside to human serum proteins and blood cells," *International Journal of Clinical Pharmacology and Therapeutics*, vol. 32, no. 8, pp. 429–432, 1994.

[191] O. N. Chappey, E. Niel, J. Wautier et al., "Colchicine disposition in human leukocytes after single and multiple oral administration," *Clinical Pharmacology & Therapeutics*, vol. 54, no. 4, pp. 360–367, 1993.

[192] E. Stec-Martyna, M. Ponassi, M. Miele, S. Parodi, L. Felli, and C. Rosano, "Structural comparison of the interaction of tubulin with various ligands affecting microtubule dynamics," *Current Cancer Drug Targets*, vol. 12, no. 6, pp. 658–666, 2012.

[193] J. Seligmann and C. Twelves, "Tubulin: An example of targeted chemotherapy," *Future Medicinal Chemistry*, vol. 5, no. 3, pp. 339–352, 2013.

[194] J. Cortes and M. Vidal, "Beyond taxanes: The next generation of microtubule-targeting agents," *Breast Cancer Research and Treatment*, vol. 133, no. 3, pp. 821–830, 2012.

[195] C. D. Katsetos and P. Dráber, "Tubulins as therapeutic targets in cancer: From bench to bedside," *Current Pharmaceutical Design*, vol. 18, no. 19, pp. 2778–2792, 2012.

[196] H. Ringsdorf, "Structure and properties of pharmacologically active polymers," *Journal of Polymer Science: Polymer Symposia*, vol. 51, no. 1, pp. 135–153, 1975.

[197] N. Salahuddin, A. A. Elbarbary, M. L. Salem, and S. Elksass, "Antimicrobial and antitumor activities of 1,2,4-triazoles/polypyrrole chitosan core shell nanoparticles," *Journal of Physical Organic Chemistry*, vol. 30, no. 12, p. e3702, 2017.

[198] L. Zhang and T. J. Webster, "Nanotechnology and nanomaterials: promises for improved tissue regeneration," *Nano Today*, vol. 4, no. 1, pp. 66–80, 2009.

[199] J. Y. Wong, R. Langer, and D. E. Ingber, "Electrically conducting polymers can noninvasively control the shape and growth of mammalian cells," *Proceedings of the National Acadamy of Sciences of the United States of America*, vol. 91, no. 8, pp. 3201–3204, 1994.

[200] Y. Li, K. G. Neoh, and E. T. Kang, "Controlled release of heparin from polypyrrole-poly(vinyl alcohol) assembly by electrical stimulation," *Journal of Biomedical Materials Research Part A*, vol. 73, no. 2, pp. 171–181, 2005.

[201] M. Shivashankar, B. K. Mandal, and K. Uma, "Chitosan-acryl amide grafted polyethylene glycol interpenetrating polymeric network for controlled release studies of Cefotaxime," *Journal of Chemical and Pharmaceutical Research*, vol. 5, no. 5, pp. 140–146, 2013.

[202] M. N. Koopaei, M. R. Khoshayand, S. H. Mostafavi et al., "Docetaxel loaded PEG-PLGA nanoparticles: Optimized drug loading, in-vitro cytotoxicity and in-vivo antitumor effect," *Iranian Journal of Pharmaceutical Research*, vol. 13, no. 3, pp. 819–833, 2014.

[203] U. Gaur, S. K. Sahoo, T. K. De, P. C. Ghosh, A. Maitra, and P. K. Ghosh, "Biodistribution of fluoresceinated dextran using novel nanoparticles evading reticuloendothelial system," *International Journal of Pharmaceutics*, vol. 202, no. 1-2, pp. 1–10, 2000.

[204] E. Lee, J. Lee, I.-H. Lee et al., "Conjugated chitosan as a novel platform for oral delivery of paclitaxel," *Journal of Medicinal Chemistry*, vol. 51, no. 20, pp. 6442–6449, 2008.

[205] Y. Zhang, M. Huo, J. Zhou, D. Yu, and Y. Wu, "Potential of amphiphilically modified low molecular weight chitosan as a novel carrier for hydrophobic anticancer drug: Synthesis, characterization, micellization and cytotoxicity evaluation," *Carbohydrate Polymers*, vol. 77, no. 2, pp. 231–238, 2009.

[206] M. Khanmohammadi, H. Elmizadeh, and K. Ghasemi, "Investigation of size and morphology of chitosan nanoparticles used in drug delivery system employing chemometric technique," *Iranian Journal of Pharmaceutical Research*, vol. 14, no. 3, pp. 665–675, 2015.

[207] M. Agarwal, M. K. Agarwal, N. Shrivastav, S. Pandey, R. Das, and P. Gaur, "Preparation of Chitosan Nanoparticles and their In-vitro Characterization," *International Journal of Life-Sciences Scientific Research*, vol. 4, no. 2, pp. 1713–1720, 2018.

[208] P. R. Kamath and D. Sunil, "Nano-Chitosan Particles in Anticancer Drug Delivery: An Up-to-Date Review," *Mini-Reviews in Medicinal Chemistry*, vol. 17, no. 15, pp. 1457–1487, 2017.

[209] T. Kean and M. Thanou, "Biodegradation, biodistribution and toxicity of chitosan," *Advanced Drug Delivery Reviews*, vol. 62, no. 1, pp. 3–11, 2010.

[210] A. Grenha, C. I. Grainger, L. A. Dailey et al., "Chitosan nanoparticles are compatible with respiratory epithelial cells in vitro," *European Journal of Pharmaceutical Sciences*, vol. 31, no. 2, pp. 73–84, 2007.

[211] M. Malatesta, S. Grecchi, E. Chiesa, B. Cisterna, M. Costanzo, and C. Zancanaro, "Internalized chitosan nanoparticles persist for long time in cultured cells," *European Journal of Histochemistry*, vol. 59, no. 1, Article ID 2492, pp. 61–65, 2015.

[212] U. Aruna, R. Rajalakshmi, Y. Indira Muzib et al., "Role of Chitosan Nanoparticles in Cancer Therapy," *International Journal of Innovative Pharmaceutical Sciences and Research*, vol. 4, no. 3, pp. 318–324, 2013.

[213] L.-F. Qi, Z.-R. Xu, Y. Li, X. Jiang, and X.-Y. Han, "In vitro effects of chitosan nanoparticles on proliferation of human gastric carcinoma cell line MGC803 cells," *World Journal of Gastroenterology*, vol. 11, no. 33, pp. 5136–5141, 2005.

[214] J.-H. Kim, Y.-S. Kim, S. Kim et al., "Hydrophobically modified glycol chitosan nanoparticles as carriers for paclitaxel," *Journal of Controlled Release*, vol. 111, no. 1-2, pp. 228–234, 2006.

[215] P. Soni, J. Kaur, and K. Tikoo, "Dual drug-loaded paclitaxel–thymoquinone nanoparticles for effective breast cancer therapy," *Journal of Nanoparticle Research*, vol. 17, no. 1, p. 18, 2015.

[216] H. D. Han, L. S. Mangala, J. W. Lee et al., "Targeted gene silencing using RGD-labeled chitosan nanoparticles," *Clinical Cancer Research*, vol. 16, no. 15, pp. 3910–3922, 2010.

[217] X. Li, M. Min, N. Du et al., "Chitin, Chitosan, and Glycated Chitosan Regulate Immune Responses: The Novel Adjuvants for Cancer Vaccine," *Clinical and Developmental Immunology*, vol. 2013, Article ID 387023, 8 pages, 2013.

[218] H. Maeda, "The enhanced permeability and retention (EPR) effect in tumor vasculature: the key role of tumor-selective macromolecular drug targeting," *Advances in Enzyme Regulation*, vol. 41, pp. 189–207, 2001.

[219] P. Yadav, A. Bandyopadhyay, A. Chakraborty, and K. Sarkar, "Enhancement of anticancer activity and drug delivery of chitosan-curcumin nanoparticle via molecular docking and simulation analysis," *Carbohydrate Polymers*, vol. 182, pp. 188–198, 2018.

[220] L. Mary Lazer, B. Sadhasivam, K. Palaniyandi et al., "Chitosan-based nano-formulation enhances the anticancer efficacy of hesperetin," *International Journal of Biological Macromolecules*, vol. 107, pp. 1988–1998, 2018.

[221] J. K. Kim, K.-H. Han, J. T. Lee et al., "Long-term clinical outcome of phase IIb clinical trial of percutaneous injection with holmium-166/chitosan complex (milican) for the treatment of small hepatocellular carcinoma," *Clinical Cancer Research*, vol. 12, no. 2, pp. 543–548, 2006.

[222] Medical University of South Carolina, "Study of Chitosan for Pharmacologic Manipulation of AGE (Advanced Glycation End products) Levels in Prostate Cancer Patients," ClinicalTrials.gov identifier: NCT03712371, 2018, https://clinicaltrials.gov/ct2/show/NCT03668431.

[223] Basel University Hospital, "Comparison of Oral Morphine Versus Nasal Ketamine Spray with Chitosan in Cancer Pain Outpatients," ClinicalTrials.gov Identifier: NCT02591017, 2018, https://ClinicalTrials.gov/ct2/show/NCT02591017.

[224] S. Won Kim and Samsung Medical Center, "Anti-adhesive Effect and Safety of a Mixed Solid of Poloxamer, Gelatin and Chitosan (Medichlore®) After Axillary Dissection for Breast Cancer," ClinicalTrials.gov Identifier: NCT02967146, 2018, https://ClinicalTrials.gov/ct2/show/NCT02967146.

[225] S. A. C. Eske Corporation, "Randomized Clinical Trial Evaluating the Use of the Laser-assisted Immunotherapy (LIT/inCVAX) in Advanced Breast Cancer," ClinicalTrials.gov Identifier: NCT03202446, 2018, https://ClinicalTrials.gov/ct2/show/NCT03202446.

[226] I. Segura Duran, "University of Guadalajara, Chitosan Scaffold for Sellar Floor Repair in Endoscopic Endonasal Transsphenoidal Surgery," ClinicalTrials.gov identifier: NCT03280849, 2018, https://ClinicalTrials.gov/ct2/show/NCT03280849.

[227] K. M. Shields, N. Smock, C. E. McQueen, and P. J. Bryant, "Chitosan for weight loss and cholesterol management," *American Journal of Health-System Pharmacy*, vol. 60, no. 13, pp. 1315-1316, 2003.

[228] C. Andreadis, K. Vahtsevanos, T. Sidiras, I. Thomaidis, K. Antoniadis, and D. Mouratidou, "5-Fluorouracil and cisplatin in the treatment of advanced oral cancer," *Oral Oncology*, vol. 39, no. 4, pp. 380–385, 2003.

[229] A. Remesh, "Toxicities of anticancer drugs and its management," *International Journal of Basic & Clinical Pharmacology*, vol. 1, no. 1, pp. 2–12, 2012.

[230] P. Baldrick, "The safety of chitosan as a pharmaceutical excipient," *Regulatory Toxicology and Pharmacology*, vol. 56, no. 3, pp. 290–299, 2010.

[231] S. Bisht and A. Maitra, "Dextran-doxorubicin/chitosan nanoparticles for solid tumor therapy," *Wiley Interdisciplinary Reviews: Nanomedicine and Nanobiotechnology*, vol. 1, no. 4, pp. 415–425, 2009.

17

Stabilisation of Collagen Sponges by Glutaraldehyde Vapour Crosslinking

Yong Y. Peng, Veronica Glattauer, and John A. M. Ramshaw

CSIRO Manufacturing, Bayview Avenue, Clayton, VIC 3169, Australia

Correspondence should be addressed to Veronica Glattauer; veronica.glattauer@csiro.au

Academic Editor: Kheng-Lim Goh

Glutaraldehyde is a well-recognised reagent for crosslinking and stabilising collagens and other protein-based materials, including gelatine. In some cases, however, the use of solutions can disrupt the structure of the material, for example, by causing rapid dispersion or distortions from surface interactions. An alternative approach that has been explored in a number of individual cases is the use of glutaraldehyde vapour. In this study, the effectiveness of a range of different glutaraldehyde concentrations in the reservoir providing vapour, from 5% to 25% (w/v), has been explored at incubation times from 5 h to 48 h at room temperature. These data show the effectiveness of the glutaraldehyde vapour approach for crosslinking collagen and show that materials with defined, intermediate stability could be obtained, for example, to control resorption rates in vivo.

1. Introduction

Glutaraldehyde (GA) has been used extensively as a crosslinking agent for collagen-based biomedical materials [1]. This includes its use in tissue based devices such as heart valve replacements [2, 3] and for tissue biosynthetic products [4]. Also, it has been used for products based on purified collagen, including collagen pastes [5] and freeze dried collagen sponges [6]. Most recently it has been examined for stabilisation of recombinant collagen products [7].

Despite its extensive use in medical products, concerns remain still about its potential cytotoxicity [8–10] and it being a causative agent of nonspecific tissue calcification [11]. Certainly, nonspecific calcification of biologically derived heart valves is a significant issue and leads to loss of function and the need for revision [12], although catastrophic failure is not a normal issue. Various methods have been examined to reduce this calcification [13, 14]. It has been suggested that the cytotoxicity and calcification arise from the propensity of GA to form reactive polymers, particularly at the neutral pH conditions normally used for tissue and collagen stabilisation. At acidic conditions, for example, around pH 3–pH 4, GA is found predominantly as a monomer but taking the reagent to neutral pH leads to formation of polymeric forms [15]. Crosslinking will occur at pH 4, but it is slow and gives materials of lower thermal stability [16].

Other approaches have looked at ways to minimise the amount of GA polymer present during neutral pH stabilisation. One approach is to stabilise collagen with GA at an elevated temperature, for example, up to 50°C [17], which is less than typical tissue shrinkage temperatures. Examination of GA solutions at elevated temperature, for example, by NMR spectroscopy, shows an increase in free aldehyde content [18].

Another approach is to use GA vapour, which has been used in the preparation of collagen-based biomedical materials [19–21]. In particular, it has been used for samples that are initially hard to handle without damage, including sponges from bacterial collagens which may disperse in solvents [7]. More studies, however, have been done on the denatured form of collagen, gelatine, especially on electrospun gelatine materials [22–24], as well as on a number of protein composite materials that include notionally collagen or gelatine [25, 26]. Electrospun gelatine can also be formed unintentionally, from electrospinning of collagen in harsh solvents [27], although more recently benign solvents have been used for electrospinning of collagen without associated denaturation [28, 29]. Electrospun films of collagen or gelatine are frequently quite fragile, so the use of vapour phase crosslinking has clear advantages over solution methods.

Previous GA vapour stabilisation studies have used a wide variety of conditions with variations in time of exposure,

GA concentration in the reservoir, and reaction temperature, with no preferred procedure emerging. In the present study, we have examined a range of conditions, all at room temperature to understand better the extent and rate of crosslinking that can occur.

2. Materials and Methods

2.1. Collagen Sponge Preparation. Bovine type I collagen was purified from yearling hides obtained from a local abattoir using the well-established method [30] of pepsin solubilisation of minced, unhaired dermis, using 1 mg/ml pepsin (Sigma-Aldrich) in 100 mM acetic acid (Merck). Fractionation of different collagens and purification of type I collagen was by NaCl precipitation in acetic acid and then at pH 7.2, as previously described [30, 31]. Purified bovine type I collagen at 10 mg/ml in 50 mM acetic acid was used to prepare freeze dried collagen sponges, around 3 mm thick. Circular samples 6 mm in diameter for glutaraldehyde (GA) treatment and analysis were cut from the sponge sheets with a biopsy punch. Except where otherwise noted all other chemicals were of the highest grade readily available and obtained from Merck (Victoria).

2.2. Glutaraldehyde Treatments. For GA vapour crosslinking, GA was obtained as a 50% (w/v) stock solution (ProSciTech, Thuringowa, QLD). For crosslinking, >20 ml GA solutions of various concentrations made by diluting the stock GA solution with water, from 5% up to 25% (w/v), were held in the lower part of glass desiccators in 70 mm dishes. Cut sponge disks were held in glass dishes above these solutions, allowing ready access to vapour, and the desiccator lid was placed to seal the chamber. Samples were removed at selected times, up to 48 h, of GA vapour exposure. Control samples, with no GA treatment, were handled in a similar manner but were held over water and had no exposure to GA. Samples for Differential Scanning Calorimetry (DSC) analysis were washed with 40 mM glycine for 3 h and then washed in MilliQ water and air dried. Other samples were held isolated, open to a stream of clean air for at least 12 h prior to any further analysis.

2.3. Differential Scanning Calorimetry. The thermal stability of untreated and GA vapour treated collagen disks were determined by DSC using a Mettler Toledo DSC821 instrument. The collagen sponge disks were between 0.8 and 0.9 mg each and were rehydrated in phosphate buffered saline (PBS) prior to analysis. A heating rate of 5°C/min was used. Data were averaged from separate sample determinations, with at least 2 determinations for each condition. The range of values obtained was typically around 1°C for each condition tested.

2.4. Scanning Electron Microscopy. Samples were examined after Ir coating (30 sec, 60 mA) using a Cressington 208HR sputter coater using a Zeiss Merlin Gemini 2 FESEM instrument.

3. Results and Discussion

GA is a widely used crosslinking agent for collagen, gelatine, and many other proteins that is normally used at a dilute concentration, for example, 0.1 to 2.0% (w/v), in aqueous solution [1] or less frequently in organic solvents [32]. In some cases, the nature of the sample makes it unsuitable for stabilisation in solution. In these instances, using GA vapour has proved a suitable alternative.

Previously a wide range of isolated treatment conditions have been used (Table 1), in which variations in temperature and in the GA concentration in the vapour reservoir have been used. The present study has compared the effects of concentration and time variations on the effectiveness of collagen crosslinking. The effectiveness of the crosslinking has been determined by the increase in the collagen melting temperature (T_m, denaturation temperature). This method allows moderately rapid, reproducible determinations, but the high rate of heating, 5°C/min, can lead to a slight increase in values, especially at lower temperatures, compared with methods that use a lower heating rate, where a T_m around 4°C lower may be observed.

The present study has examined concentrations ranging from 5% (w/v) GA up to 25% GA (w/v) in aqueous solution and incubation time up to 48 h (Figure 1). The temperature used was room temperature, which had most frequently been used by others (Table 1). Previously, higher temperatures have been used in some studies [20, 22], where increased speed of crosslinking is expected, in part from the increase in GA vapour pressure. For example, the GA vapour pressure increases around 7-fold for a 15% (w/v) solution between room temperature (20°C, 32 ppm) and 40°C (226 ppm) [42].

These present data show that essentially full crosslinking, $T_m > 80°C$, can be achieved by using 20% or 25% (w/v) reservoir solutions for 24 h or 48 h. This is consistent with previously report T_m data [20, 21], where a T_m of >80°C was reported for a collagen sponge over 25% (w/v) GA at room temperature for 24 h [21] and similarly a T_m of >80°C was reported for a collagen film over 8% (w/v) GA at 37°C after 8 h [20]. T_m provides a good quantitative measure for crosslinking but is not always reported. Often, the physical appearance and stability of materials are quoted under varying conditions, such as in acid solution. For gelatine samples, a T_m cannot be given as the gelatine is already denatured (from collagen). The efficiency of crosslinking for GA treated gelatine samples can be estimated by examining the stability of the material to proteolysis. However, the conditions used often vary between studies, making comparisons difficult.

At lower reservoir concentrations of GA, full crosslinking did not seem to occur, even after 48 h incubation time (Figure 1). Further, it appeared that for lower concentrations the extent of crosslinking, as shown by T_m values, was appearing to approach a maximum value dependent on the concentration being used (Figure 1). Samples incubated over 5% (w/v) GA showed T_m values in the mid-50°C range after 24 h, and these values did not increase much after 48 h incubation. Samples over 10% and 15% (w/v) GA solutions may also be approaching maximum values which are in the mid-70°C range (Figure 1) and lower than the T_m values of >80°C found with higher GA concentrations. These apparent plateau values dependent on the concentration of GA in the reservoir have the potential to provide sample series of varying crosslinking, for example, for studies on resorption

TABLE 1: Examples of previously reported conditions for GA vapour stabilisation of collagen and gelatine materials.

Substrate	Format	Temperature	% GA	Time	Reference
Collagen (Limed bovine)	Reconstituted Fibrils	RT	25%	24 h, 48 h	Kato et al. [19]
Collagen (Bovine)	Reconstituted Fibrils	RT	25%	96 h	Law et al. [33]
Collagen (Bovine)	Film	37°C	8%	Various 3 h to 72 h	Barbani et al. [20]
Collagen (Bovine)	Electrospun mat	RT	(Not stated)	24 h	Matthews et al. [34]
Collagen (Bovine)	Sponges	RT	25%	4 h, 8 h, 24 h	Lickorish et al. [21]
Collagen (Bovine)	Reconstituted Fibrils	RT	25%	Various, 1 h to 24 h	Rho et al. [35]
Collagen (Bovine)	Electrospun mat	RT	25%	24 h	Yang et al. [36]
Bacterial Collagen	Sponge	RT	20%	18 h	Peng et al. [7]
Collagen (Bovine)	Electrospun mat	RT	25%	8 h	Takeda et al. [37]
Gelatine (Fish skin)	Electrospun mat	37°C	50 vol%	3 h	Songchotikunpan et al. [22]
Gelatine	Electrospun mat		0.5%	19 h	Sisson et al. [38]
Gelatine (Porcine)	Electrospun mat	37°C	"Saturated"	5 min	Dheraprasart et al. [39]
Gelatine (A-type, B-type)	DHT-treated Electrospun mat	4°C	0.06% in acetone/HCl	48 h	Ratanavaraporn et al. [40]
Gelatine (A-type)	Electrospun mat	RT	1.5% in EtOH	48 h	Zha et al. [23]
Gelatine (Fish skin)	Electrospun mat	40°C	5%	Various, 0.5 h to 24 h	Gomes et al. [24]
Gelatine (Fish skin)	Electrospun mat	40°C	5%	5 h	Gomes et al. [41]

RT: Room temperature.

rate, similar to those obtained from the use of different GA concentrations in solution stabilisation [43].

GA vapour crosslinking has the advantage for any porous sample that by avoiding surface tension and repeated freeze drying that are found with solution approaches the use of vapour leads to negligible changes to the collagen organisation and topology. SEM studies (Figure 2), show little if any changes in the collagen sponge structure in control untreated material (Figure 2(a)) and one extensively stabilised (20% (w/v) GA, 24 h) by GA vapour (Figure 2(b)).

In the present study, we have examined purified collagen with no additions. GA vapour stabilisation can also be used for composite materials based on collagen and gelatine. These include, for example, mixtures of collagen or gelatine with other proteins [25, 26], or with other components including carbohydrates [44], polymers [45], and mineral [46].

In addition to mixtures, collagen and gelatine can be used as external coatings during coaxial spinning [47].

Several previous studies have used GA vapour crosslinking collagen-based materials for cell growth, and these studies have consistently shown that the resultant crosslinking is not cytotoxic [7, 21, 33–35, 37], even when higher GA concentrations are used at longer time points [7, 21, 33, 35]. This is consistent with the GA being reactive as the monomer and not allowing a significant build-up of polymers in the stabilised material. Other studies have shown the enhanced mechanical properties arising from GA vapour crosslinking fabricated of materials [19, 35, 36].

4. Conclusion

The present study has demonstrated the effectiveness of GA crosslinking over a range of conditions. It has shown that essentially full crosslinking can be obtained for collagen sponges with treatment with 20% or 25% GA vapour for 24 or

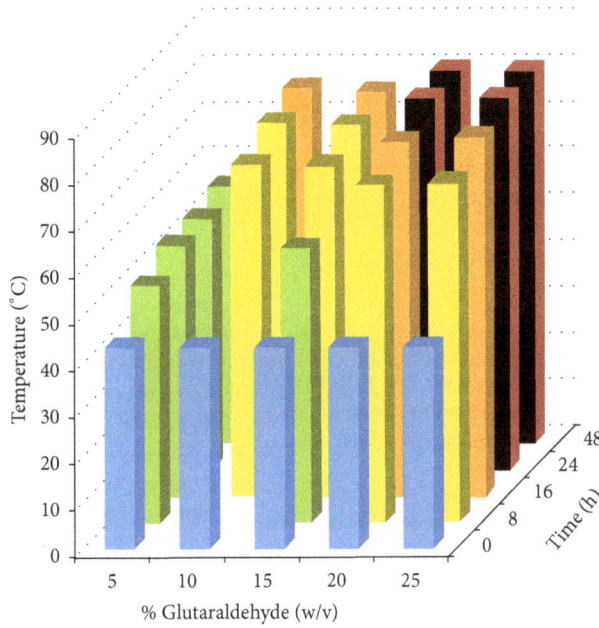

T_m values (°C)

	Time (h)				
	0	8	16	24	48
GA%					
5	43.4	51.1	54.1	54.2	55.4
10	43.4	n.d.	71.5	74.9	76.7
15	43.4	59.1	71.5	74.3	75.8
20	43.4	72.8	76.3	80.1	80.4
25	43.4	73.3	77.3	80.2	80.3

FIGURE 1: Bar diagram, showing the melting temperature, T_m, of collagen sponges treated with varying amounts of aqueous GA vapour for various time periods. Red bars indicate $T_m > 80°C$, orange bars indicate $75°C < T_m < 80°C$, yellow bars indicate $70°C < T_m < 75°C$, green bars indicate $50°C < T_m < 70°C$, and blue bars indicate $T_m < 50°C$, as found for the average value of control untreated sponges. A single average T_m value for control untreated samples was obtained and this average value was used in each of the different % GA treatment series. T_m values are given below the diagram. n.d.: not determined.

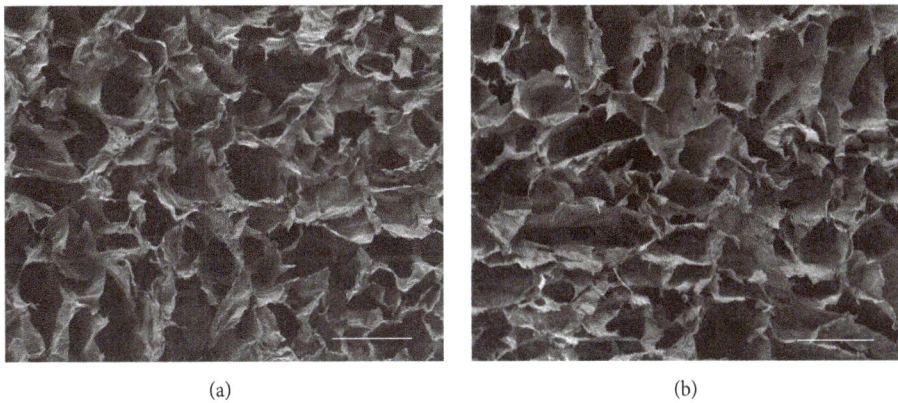

FIGURE 2: FESEM examination of collagen sponges. (a) Control untreated sponges, (b) GA vapour stabilised sponge; 25% (w/v) GA, 24 h at room temperature. Bar = 100 μm.

48 h at room temperature. Intermediate degrees of crosslinking may be obtained by varying the GA concentration. These observations, and the understanding of the variation in crosslinking through changes in GA concentration and time, should be useful in designing the preferred crosslinking characteristics for collagens and gelatines and in composites based on these protein materials. The use of GA vapour crosslinking is particularly useful for porous materials that are not easily handled, providing stability. Subsequently, additional solution based crosslinking could be used to augment the stability if necessary or to introduce chemical modifications while maintaining a stable structure.

Disclosure

John A. M. Ramshaw's present address is Department of Surgery, St. Vincent's Hospital, University of Melbourne, Melbourne, VIC 3065, Australia.

Conflicts of Interest

The authors declare that there are no conflicts of interest regarding the publication of this paper.

Acknowledgments

The authors thank Dr. Wendy Tian for advice on DSC and Mark Greaves for assistance with FESEM.

References

[1] E. Khor, "Methods for the treatment of collagenous tissues for bioprostheses," *Biomaterials*, vol. 18, no. 2, pp. 95–105, 1997.

[2] N. Zuhdi, W. Hawley, V. Voehl, W. Hancock, J. Carey, and A. Greer, "Porcine aortic valves as replacements for human heart valves," *Annals of Thoracic Surgery*, vol. 17, no. 5, pp. 479–491, 1974.

[3] M. I. Ionescu, A. P. Tandon, D. A. S. Mary, and A. Abid, "Heart valve replacement with the Ionescu Shiley pericardial xenograft," *Journal of Thoracic and Cardiovascular Surgery*, vol. 73, no. 1, pp. 31–42, 1977.

[4] J. A. M. Ramshaw, D. E. Peters, J. A. Werkmeister, and V. Ketharanathan, "Collagen organization in mandrel-grown vascular grafts," *Journal of Biomedical Materials Research*, vol. 23, no. 6, pp. 649–660, 1989.

[5] J. M. McPherson, P. W. Ledger, S. Sawamura et al., "The preparation and physicochemical characterization of an injectable form of reconstituted, glutaraldehyde cross-linked, bovine corium collagen," *Journal of Biomedical Materials Research*, vol. 20, no. 1, pp. 79–92, 1986.

[6] R. F. Oliver, R. A. Grant, R. W. Cox, and A. Cooke, "Effect of aldehyde cross-linking on human dermal collagen implants in the rat," *British Journal of Experimental Pathology*, vol. 61, no. 5, pp. 544–549, 1980.

[7] Y. Y. Peng, A. Yoshizumi, S. J. Danon et al., "A Streptococcus pyogenes derived collagen-like protein as a non-cytotoxic and non-immunogenic cross-linkable biomaterial," *Biomaterials*, vol. 31, no. 10, pp. 2755–2761, 2010.

[8] D. P. Speer, M. Chvapil, C. D. Eskelson, and J. Ulreich, "Biological effects of residual glutaraldehyde in glutaraldehyde-tanned collagen biomaterials," *Journal of Biomedical Materials Research*, vol. 14, no. 6, pp. 753–764, 1980.

[9] E. Gendler, S. Gendler, and M. E. Nimni, "Toxic reactions evoked by glutaraldehyde-fixed pericardium and cardiac valve tissue bioprosthesis," *Journal of Biomedical Materials Research*, vol. 18, no. 7, pp. 727–736, 1984.

[10] L. M. Delgado, K. Fuller, and D. I. Zeugolis, "Collagen crosslinking: biophysical, biochemical, and biological response analysis," *Tissue Engineering Part A*, 2017.

[11] R. J. Levy, F. J. Schoen, F. S. Sherman, J. Nichols, M. A. Hawley, and S. A. Lund, "Calcification of subcutaneously implanted type I collagen sponges. Effects of formaldehyde and glutaraldehyde pretreatments," *American Journal of Pathology*, vol. 122, no. 1, pp. 71–82, 1986.

[12] F. J. Schoen and R. J. Levy, "Pathology of substitute heart valves: new concepts and developments," *Journal of Cardiac Surgery*, vol. 9, pp. 222–227, 1994.

[13] M. K. Dewanjee, E. Solis, J. Lanker et al., "Effect of diphosphonate binding to collagen upon inhibition of calcification and promotion of spontaneous endothelial cell coverage on tissue valve prostheses," *ASAIO transactions/American Society for Artificial Internal Organs*, vol. 32, no. 1, pp. 24–29, 1986.

[14] N. Vyavahare, D. Hirsch, E. Lerner et al., "Prevention of bioprosthetic heart valve calcification by ethanol preincubation: efficacy and mechanisms," *Circulation*, vol. 95, no. 2, pp. 479–488, 1997.

[15] K. E. Rasmussen and J. Albrechtsen, "Glutaraldehyde. The influence of pH, temperature, and buffering on the polymerization rate," *Histochemistry*, vol. 38, no. 1, pp. 19–26, 1974.

[16] D. E. Peters, L. J. Stephens, and J. A. M. Ramshaw, "Examination of collagen tanning by glutaraldehyde using isometric tension measurements," *Das Leder*, vol. 41, no. 7, pp. 129–133, 1990.

[17] J. M. Ruijgrok, J. R. de Wijn, and M. E. Boon, "Glutaraldehyde crosslinking of collagen: effects of time, temperature, concentration and presoaking as measured by shrinkage temperature," *Clinical Materials*, vol. 17, no. 1, pp. 23–27, 1994.

[18] C. E. Holloway and F. H. Dean, "^{13}C-NMR study of aqueous glutaraldehyde equilibria," *Journal of Pharmaceutical Sciences*, vol. 64, no. 6, pp. 1078–1079, 1990.

[19] Y. P. Kato, D. L. Christiansen, R. A. Hahn, S.-J. Shieh, J. D. Goldstein, and F. H. Silver, "Mechanical properties of collagen fibres: a comparison of reconstituted and rat tail tendon fibres," *Biomaterials*, vol. 10, no. 1, pp. 38–42, 1989.

[20] N. Barbani, P. Giusti, L. Lazzeri, G. Polacco, and G. Pizzirani, "Bioartificial materials based on collagen: 1. Collagen crosslinking with gaseous glutaraldehyde," *Journal of Biomaterials Science, Polymer Edition*, vol. 7, no. 6, pp. 461–469, 1996.

[21] D. Lickorish, J. A. M. Ramshaw, J. A. Werkmeister, V. Glattauer, and C. R. Howlett, "Collagen-hydroxyapatite composite prepared by biomimetic process," *Journal of Biomedical Materials Research - Part A*, vol. 68, no. 1, pp. 19–27, 2004.

[22] P. Songchotikunpan, J. Tattiyakul, and P. Supaphol, "Extraction and electrospinning of gelatin from fish skin," *International Journal of Biological Macromolecules*, vol. 42, no. 3, pp. 247–255, 2008.

[23] Z. Zha, W. Teng, V. Markle, Z. Dai, and X. Wu, "Fabrication of gelatin nanofibrous scaffolds using ethanol/phosphate buffer saline as a benign solvent," *Biopolymers*, vol. 97, no. 12, pp. 1026–1036, 2012.

[24] S. R. Gomes, G. Rodrigues, G. G. Martins, C. M. R. Henriques, and J. C. Silva, "In vitro evaluation of crosslinked electrospun

fish gelatin scaffolds," *Materials Science and Engineering C, Materials for Biological Applications*, vol. 33, no. 3, pp. 1219–1227, 2013.

[25] I.-S. Yeo, J.-E. Oh, L. Jeong et al., "Collagen-based biomimetic nanofibrous scaffolds: preparation and characterization of collagen/silk fibroin bicomponent nanofibrous structures," *Biomacromolecules*, vol. 9, no. 4, pp. 1106–1116, 2008.

[26] E. Tamimi, D. C. Ardila, D. G. Haskett et al., "Biomechanical comparison of glutaraldehyde-crosslinked gelatin fibrinogen electrospun scaffolds to porcine coronary arteries," *Journal of Biomechanical Engineering*, vol. 138, no. 1, Article ID 011001, 2016.

[27] D. I. Zeugolis, S. T. Khew, E. S. Y. Yew et al., "Electro-spinning of pure collagen nano-fibres—just an expensive way to make gelatin?" *Biomaterials*, vol. 29, no. 15, pp. 2293–2305, 2008.

[28] B. Dong, O. Arnoult, M. E. Smith, and G. E. Wnek, "Electrospinning of collagen nanofiber scaffolds from benign solvents," *Macromolecular Rapid Communications*, vol. 30, no. 7, pp. 539–542, 2009.

[29] A. Elamparithi, A. M. Punnoose, and S. Kuruvilla, "Electrospun type 1 collagen matrices preserving native ultrastructure using benign binary solvent for cardiac tissue engineering," *Artificial Cells, Nanomedicine and Biotechnology*, vol. 44, no. 5, pp. 1318–1325, 2016.

[30] E. J. Miller and R. K. Rhodes, "Preparation and characterization of the different types of collagen," *Methods in Enzymology*, vol. 82, pp. 33–64, 1982.

[31] R. L. Trelstad, V. M. Catanese, and D. F. Rubin, "Collagen fractionation: separation of native types I, II and III by differential preciptitation," *Analytical Biochemistry*, vol. 71, no. 1, pp. 114–118, 1976.

[32] P. F. Gratzer, C. A. Pereira, and J. M. Lee, "Solvent environment modulates effects of glutaraldehyde crosslinking on tissue-derived biomaterials," *Journal of Biomedical Materials Research*, vol. 31, no. 4, pp. 533–543, 1996.

[33] J. K. Law, J. R. Parsons, F. H. Silver, and A. B. Weiss, "An evaluation of purified reconstituted type 1 collagen fibers," *Journal of Biomedical Materials Research*, vol. 23, no. 9, pp. 961–977, 1989.

[34] J. A. Matthews, G. E. Wnek, D. G. Simpson, and G. L. Bowlin, "Electrospinning of collagen nanofibers," *Biomacromolecules*, vol. 3, no. 2, pp. 232–238, 2002.

[35] K. S. Rho, L. Jeong, G. Lee et al., "Electrospinning of collagen nanofibers: effects on the behavior of normal human keratinocytes and early-stage wound healing," *Biomaterials*, vol. 27, no. 8, pp. 1452–1461, 2006.

[36] L. Yang, C. F. C. Fitié, K. O. van der Werf, M. L. Bennink, P. J. Dijkstra, and J. Feijen, "Mechanical properties of single electrospun collagen type I fibers," *Biomaterials*, vol. 29, no. 8, pp. 955–962, 2008.

[37] N. Takeda, K. Tamura, R. Mineguchi et al., "In situ cross-linked electrospun fiber scaffold of collagen for fabricating cell-dense muscle tissue," *Journal of Artificial Organs*, vol. 19, no. 2, pp. 141–148, 2016.

[38] K. Sisson, C. Zhang, M. C. Farach-Carson, D. B. Chase, and J. F. Rabolt, "Evaluation of cross-linking methods for electrospun gelatin on cell growth and viability," *Biomacromolecules*, vol. 10, no. 7, pp. 1675–1680, 2009.

[39] C. Dheraprasart, S. Rengpipat, P. Supaphol, and J. Tattiyakul, "Morphology, release characteristics, and antimicrobial effect of nisin-loaded electrospun gelatin fiber mat," *Journal of Food Protection*, vol. 72, no. 11, pp. 2293–2300, 2009.

[40] J. Ratanavaraporn, R. Rangkupan, H. Jeeratawatchai, S. Kanokpanont, and S. Damrongsakkul, "Influences of physical and chemical crosslinking techniques on electrospun type A and B gelatin fiber mats," *International Journal of Biological Macromolecules*, vol. 47, no. 4, pp. 431–438, 2010.

[41] S. R. Gomes, G. Rodrigues, G. G. Martins et al., "In vitro and in vivo evaluation of electrospun nanofibers of PCL, chitosan and gelatin: a comparative study," *Materials Science and Engineering, C Materials for Biological Applications*, vol. 46, pp. 348–358, 2015.

[42] J. D. Olson, "The vapor pressure of pure and aqueous glutaraldehyde," *Fluid Phase Equilibria*, vol. 150, no. 151, pp. 713–720, 1998.

[43] V. Glattauer, J. F. White, W.-B. Tsai et al., "Preparation of resorbable collagen-based beads for direct use in tissue engineering and cell therapy applications," *Journal of Biomedical Materials Research—Part A*, vol. 92, no. 4, pp. 1301–1309, 2010.

[44] Y.-F. Qian, K.-H. Zhang, F. Chen, Q.-F. Ke, and X.-M. Mo, "Cross-linking of gelatin and chitosan complex nanofibers for tissue-engineering scaffolds," *Journal of Biomaterials Science, Polymer Edition*, vol. 22, no. 8, pp. 1099–1113, 2011.

[45] X. Qiao, S. J. Russell, X. Yang, G. Tronci, and D. J. Wood, "Compositional and in vitro evaluation of nonwoven type I collagen/poly-dl-lactic acid scaffolds for bone regeneration," *Journal of Functional Biomaterials*, vol. 6, no. 3, pp. 667–686, 2015.

[46] V. Thomas, D. R. Dean, M. V. Jose, B. Mathew, S. Chowdhury, and Y. K. Vohra, "Nanostructured biocomposite scaffolds based on collagen coelectrospun with nanohydroxyapatite," *Biomacromolecules*, vol. 8, no. 2, pp. 631–637, 2007.

[47] Y. Lu, H. Jiang, K. Tu, and L. Wang, "Mild immobilization of diverse macromolecular bioactive agents onto multifunctional fibrous membranes prepared by coaxial electrospinning," *Acta Biomaterialia*, vol. 5, no. 5, pp. 1562–1574, 2009.

18

Modeling and Synthesis of Ag and Ag/Ni Allied Bimetallic Nanoparticles by Green Method: Optical and Biological Properties

Anuoluwa Abimbola Akinsiku ⓘ,[1] **Enock Olugbenga Dare ⓘ,**[2] **Kolawole Oluseyi Ajanaku,**[1] **Olayinka Oyewale Ajani,**[1] **Joseph Adebisi O. Olugbuyiro,**[1] **Tolutope Oluwasegun Siyanbola,**[1] **Oluwaseun Ejilude,**[3] **and Moses Eterigho Emetere**[4,5]

[1]*Department of Chemistry, Covenant University, PMB 1023, Ota, Ogun State, Nigeria*
[2]*Department of Chemistry, Federal University of Agriculture, PMB 2240, Alabata Road, Abeokuta, Nigeria*
[3]*Department of Medical and Parasitology, Sacred Heart Hospitals, Lantoro, Abeokuta, Nigeria*
[4]*Department of Physics, Covenant University, PMB 1023, Ota, Ogun State, Nigeria*
[5]*Department of Mechanical Engineering Science, University of Johannesburg, Auckland Park Kingsway Campus, Johannesburg 2006, South Africa*

Correspondence should be addressed to Anuoluwa Abimbola Akinsiku; anu.akinsiku@covenantuniversity.edu.ng

Academic Editor: Vijaya Kumar Rangari

In the quest for environmental remediation which involves eco-friendly synthetic routes, we herein report synthesis and modeling of silver nanoparticles (Ag NPs) and silver/nickel allied bimetallic nanoparticles (Ag/Ni NPs) using plant-extract reduction method. Secondary metabolites in the leaf extract of *Canna indica* acted as reducing agent. Electronic transitions resulted in emergence of surface plasmon resonance in the regions of 416 nm (Ag NPs) and 421 nm (Ag/Ni NPs) during optical measurements. Further characterizations were done using TEM and EDX. Antimicrobial activity of the nanoparticles against clinical isolates was highly significant as $P < 0.05$. These findings suggest application of Ag NPs as antibacterial agent against *E. coli*, *S. pyogenes,* and antifungal agent against *C. albicans*. Possible antibacterial drugs against *S. pyogenes* and *E. coli* can also be designed using Ag/Ni nanohybrid based on their strong inhibition activities. Similarly, the enhanced SPR in the nanoparticles is suggested for applications in optical materials, as good absorbers and scatters of visible light. Theoretical model clarified that the experiment observation on the relationship between metallic nanoparticles penetration through peptidoglycan layers and the activeness of microbial species depends on the nature of the nanoparticles and pore size of the layer.

1. Introduction

Recently, paradigm shift in technology has led to the synthetic protocols involving application of green chemistry, part of environmental remediation, encompassing use of biomaterials because of their eco-friendliness. The use of biological materials as sensors, information storage devices, and bimolecular array is on the increase. No doubt, novel characteristics are possessed by materials on nanometre scale, as this creates special interest which is applicable virtually to every aspect of life including medicine, agriculture, and polymer industry among others [1]. There is a growing interest in magnetic NPs of nickel origin due to their superior magnetic characteristics which has a useful application in medicine (as magnetic drug delivery) and therapeutics [2–4]. Imran Din and Rani [5] reviewed progress in the "green" protocols for the syntheses and stabilization of nickel and nickel oxide nanoparticles when *Azadirachta indica* and *Psidium guajava* leaves were utilized in the synthesis of NiO and Ni nanoparticles with an average size of 17–77 nm. Face-centered cubic Ni NPs were reported by Chen et al. in which Ni $(NO_3)_2$ was reduced with *Medicago sativa* (alfalfa) extract. Ni NPs were also synthesized by Chen et al. with the leaf extract of *Ocimum sanctum* as reducing and stabilizing agents: hydrated electrons of *O. sanctum* aqueous leaf extract were considered to have reduced Ni(II) ions into Ni(0) [6, 7].

Biosynthesized conjugated bimetallic nanoparticles are now used in biomedical field, imaging, luminescence tagging, labeling, and drug delivery due to their compatibility in *in vivo* screening [8]. Glucose-capped nickel nanoparticles (G-Ni NPs) were synthesized by Vaseem et al. via aqueous solution method in which glucose bifunctioned as capping agent and a reducing agent [9]. *Canna indica* (Linn.), commonly known as Indian shot belonging to the family Cannaceae is a medicinal plant of diverse uses. The herb consists of rhizomatous root stocks; reddish or yellowish showy flowers, which encompass a variable number of rounds; and shiny black seeds. In folkloric medicine, root decoction is used for the treatment of fever, dropsy, and dyspepsia [10]. Seed juice is used to relieve ear aches. Many varieties of *C. indica* are grown in the gardens and around houses for beautification. Ethnomedical use of *C. indica* leaves include antimicrobial, analgesic, and anthelmintic activities [11]. Research has shown that extraction of *C. indica* rhizomes using water was effective for HIV-1 reverse transcriptase inhibitory activity [12].

From survey, biological syntheses of nanomaterials make use of bacteria, yeasts, fungi, and algae (microorganisms). The use of plants or plant extracts for metal and metal hybrid nanoparticles synthesis is currently a new research focus that has gained wide acceptance [13]. As the demands for commercial nanoparticles are on the increase due to their wider applications in many fields, plant mediated green synthesis presents a cost-effective alternative method that is eco-friendly and sustainable [14]. The technique provides stable nanoparticles dispersions that resist aggregation in biological media and have high resistance to oxidation which is of significant importance [15]. Consequently, there is a need to solve public health problem of drug resistance by the disease causing microbes; new drug procedure needs to be adopted. It is noteworthy that Nigeria is endowed with natural biodiversity whose potential for developing novel health care and active drug candidates through green method has been underutilized. Plant extract with adequate phytochemicals has been confirmed to be faster in initiating bioreduction compared to microbes and the conventional chemical methods [16].

Quite a number of literatures have reported syntheses of allied silver-nickel nanoparticles using various chemical methods. Adekoya et al. [17] synthesized optically active fractal seed mediated silver-nickel bimetallic nanoparticles by chemical method; nevertheless, no investigations were carried out using *Canna indica* green-mediated route technique for the synthesis of Ag/Ni bimetallic nanoparticles. In view of the potential of Ag and Ni NPs having good antibacterial activity against some pathogenic organisms [18] and their optical properties as a result of shift of surface plasmon resonance wavelength (λ_{SPR}) to a longer wavelength based on particle size [19], we report facile green synthetic methodology and theoretical modeling of silver and silver/nickel hybrid nanoparticles that entails in situ reduction of aqueous

Ag(I) and Ni(II) ions. The phytochemicals present in the aqueous leaf extract of *Canna indica* were considered to act as reducing/capping agents. UV-Visible spectrophotometer was used to monitor the optical properties; morphological characterization, size determination, elemental analysis, and antimicrobial screening of the biosynthesized nanoparticles were also carried out.

2. Experimental Details

2.1. Materials. *Canna indica* leaf extract (Indian shot), Whatman number 1 filter paper, distilled deionized water, silver nitrate ($AgNO_3$), and nickel nitrate hexahydrate $Ni(NO_3)_2 \cdot 6H_2O$, commercially obtained from Sigma-Aldrich Company, UK, were used as obtained.

2.1.1. Test Microorganisms. Freshly cultured clinical isolates of *Escherichia coli*, *Pseudomonas aeruginosa* (gram-negative bacteria), *Staphylococcus aureus*, *Streptococcus pyogenes* (gram-positive bacteria), *Candida albicans,* and *Trichophyton rubrum* (Fungi) were collected from the Department of Medical Microbiology and Parasitology, Sacred Heart Hospital, Lantoro, Abeokuta, in Nigeria.

2.1.2. Preparation of Leaf Extract. Indian shot plant was collected from a garden at Atan-Iju, Ogun State, Nigeria. Plant identification and authentication were carried out at Forest Research Institute of Nigeria (FRIN); voucher specimen FHI 109928 was deposited at the herbarium headquarters, Ibadan, Nigeria. Fresh leafy part of the plant was washed with distilled water, finely cut, and ground using mortar and pestle. It was then extracted at a ratio of 1 : 5 wt/v using distilled deionized water, filtered with Whatman number 1 filter paper, and then kept at 4°C. The filtrate was used for phytochemical screening and nanoparticle synthesis. The procedure is modified from previous work [20].

2.1.3. Phytochemical Screening of the Plant Extracts. Plant extract was screened to identify the phytochemicals present according to literature [21].

2.1.4. Syntheses of Ag and Ag/Ni Bimetallic Nanoparticles. Metallic nanoparticles were prepared by plant-extract reduction method with modification to previous work [22]. For the synthesis of Ag NPs, 10 mL of the 0.2 g/mL aqueous filtrate of *C. indica* extract was added to 100 mL of varied concentrations of aqueous silver nitrate solution (0.5–2.0 mM). The reaction mixture was continuously stirred and gradually heated to 70°C on a hotplate. In the case of Ag/Ni bimetallic nanoparticle synthesis, 20 mL of the plant extract was added to equal molar concentration mixture of 100 mL $AgNO_3$ and 100 mL $Ni(NO_3)_2 \cdot 6H_2O$ in a beaker. Precursor concentrations were varied between 0.5 and 3.0 mM. Initial colour of the mixture was noted. The resulting mixture was continuously stirred and gradually heated to 70°C until there was a change in colour. Bioreductions of Ag(I) ions to Ags(0) and Ni(II) to Ni(0) were monitored by taking samples at varied time intervals, using UV-Vis spectrophotometer (double beam Thermo Scientific GENESYS 10S model), starting from the 5th minute until a noticeable colour change and appearance of surface plasmon resonance band (SPRB). Sample was placed in quartz cuvette, operated at a resolution of 1 nm so as to measure the absorbance.

2.1.5. Isolation of Metallic Ag and Ag/Ni Nanoparticles.

The biosynthesized nanoparticles were collected by centrifugation using centrifuge model 0508-1, operated at 5000 rpm for 30 minutes. For purification, the nanoparticles suspension was redispersed in distilled deionized water so as to remove the unbounded organics and finally centrifuged at 5,000 rpm for 10 minutes. The suspension was oven dried and kept in Eppendorf tubes for further characterizations.

2.2. Characterization

2.2.1. Optical Characterization.
Optical properties of the prepared metallic nanoparticles were determined using a double beam Thermo Scientific GENESYS 10S UV-Vis spectrophotometer between 200 and 800 nm wavelength ranges. Absorbance measurement was carried out by placing each aliquot sample taken at time intervals in quartz cuvette (1 cm path length), operated at a resolution of 1 nm, using distilled deionized water as blank.

2.2.2. Structural Characterization.
Structural, morphological characteristics and size determination of the particles were verified with Technai G2 transmission electron microscope (TEM) coupled with an energy-dispersive X-ray spectrometer (EDX), operated at an accelerating voltage of 200 KeV and 20 μA current. Samples for TEM analysis were prepared by drop-coating Ag and Ag/Ni suspensions onto carbon-coated copper TEM grids. The films on the TEM grids were allowed to dry prior to measurement.

2.3. Antimicrobial Activity

2.3.1. Turbidity Standard for Inocula Preparation.
McFarland standard on laboratory guidance was used for the standardization of organisms for susceptibility testing, using a modified method by British Society for Antimicrobial Chemotherapy. BaSO$_4$ turbidity standard equivalent to 0.5 McFarland standards or its optical equivalent was used. The 0.5 McFarland standard was prepared by adding 0.5 mL of 0.048 M BaCl$_2$ of (1.175% w/v) BaCl$_2$ in 2H$_2$O to 99.5 mL of 0.18 M H$_2$SO$_4$ with constant stirring to maintain a suspension. The correct density of the turbidity standard was verified by a pg instrument UV-Vis spectrophotometer model T90+, with 1 cm light path, and matched cuvette to determine the optical density at a wavelength of 625 nm. The acceptable range for the standard is 0.08–0.13 for 0.5 McFarland standard which is equivalent to 1.5 × 10^8 bacterial cells per mL. The standard was distributed into screw cap tubes of the same size and volume, similar to method of growing or diluting the bacterial inocula. The tubes were tightly sealed to prevent loss by evaporation. They were then stored in the dark at room temperature. The turbidity standard was vigorously agitated on a vortex mixer before use. The standard remains potent

for six months; appearance of large particles in the standard is an indication of expiration [23].

2.3.2. Preparation of Inocula.
The microbial strains were propagated in Mueller Hinton broth, prepared by dispersing 5 mL of the prepared broth medium into each screw capped test tube, sterilized by autoclaving at 12°C for 15 minutes. The test tubes were cooled and kept in an incubator for 24 hours at 37°C in order to determine the sterility. The isolates were inoculated into the sterilized test tubes containing the medium and placed in an incubator overnight at 37°C. Appearance of turbidity in broth culture was adjusted equivalent to 0.5 McFarland standards. This was done to obtain standardized suspension. Sterile normal saline was added in order to obtain turbidity optically comparable to that of the 0.5 McFarland standards or against a white background with contrasting black line. The McFarland 0.5 standard provided turbidity comparable to bacterial suspension containing 1.5 × 10^8 cfu/mL [24]. The suspension was used within 5 minutes so as to avoid population increase.

2.3.3. Sensitivity of Test Organisms.
Antimicrobial properties of the biosynthesized nanoparticles were investigated in the form of sensitivity testing, using modified version of the method described by Aida [25]. The test organisms were collected on sterile agar slant and incubated at 37°C for 24 hours. The following biochemical analyses were carried out on the bacteria test organisms: sugar fermentation, citrate utilization, oxidase reaction, Voges-Proskauer, methyl red, capsule staining, spore staining, motility, indole test, urease test, hydrogen sulphide test, gelatin liquefaction, and gram staining. Conversely, the fungus Candida albicans was identified by gram staining, germ tube test, sugar fermentation, and assimilation tests. Trichophyton rubrum (fungus) was identified macroscopically and microscopically using lactophenol cotton blue stains. These were then kept as stock culture on slant in the refrigerator at 4°C. The procedure was in agreement with recommended standards of National Committee for Clinical Laboratory Standards (NCCLS) [24].

2.3.4. Agar Well Diffusion Method.
Antibacterial activity of synthesized nanoparticles was evaluated by the well plate agar diffusion method as described in the Aida modified method [25]. The microbial cultures were adjusted to 0.5 McFarland turbidity standards and inoculated on Muller-Hinton agar plate of diameter 9 cm. The plate was flooded with each of the standardized test organisms (1 mL) and then swirled. Excess inoculum was carefully decanted. A sterile cork borer was used to make wells (6 mm in diameter) on the agar plates. Aliquots of the nanoparticle dilutions (0.1 mL) were reconstituted in 50% DMSO at concentrations of 100 mg/mL and applied on each of the wells in the culture plates previously inoculated with the test organisms. However, each extract was tested in duplicate with 0.1 mL of 5 μg/mL ciprofloxacin as positive control for bacteria and fluconazole as positive control for fungi. These were then left on the bench for 1 hour for proper diffusion of the nanoparticles [24]. Thereafter, the plates were incubated at 37°C for 24 hours for bacteria and

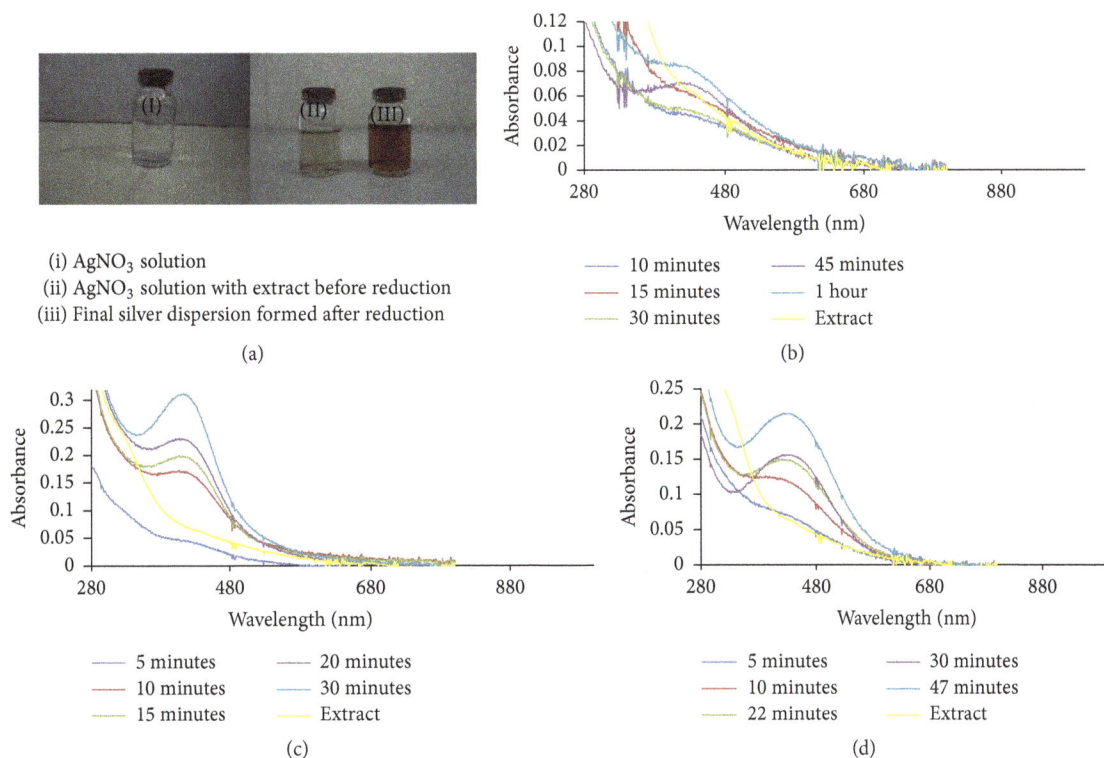

(i) AgNO$_3$ solution
(ii) AgNO$_3$ solution with extract before reduction
(iii) Final silver dispersion formed after reduction

(a)

(b)

(c)

(d)

FIGURE 1: (a) Colour dispersion before and after nanoparticles formation, UV-Vis spectra of Ag NPs prepared by reducing (b) 0.5 mM, (c) 1.0 mM, and (d) 2.0 mM precursor solutions using the extract of *C. indica* at 70°C.

yeast and at 28°C for 72 hours for *T. rubrum*. Antimicrobial activity was determined by measuring the zone of inhibition around each well (excluding the diameter of the well) for nanoparticles obtained from the plant extract. Duplicate tests were conducted against each organism and significant growth inhibitions were found using analysis of variance (ANOVA), SPSS statistical tool.

2.3.5. Minimum Inhibitory Concentration (MIC).

Serial dilution method was employed according to CLSI guidelines. Sterile test tubes (12) were arranged in a rack. 1 mL of sterile nutrient broth was added to tube labeled 2 to 10. 1 mL of known nutrients broth concentration was added to tubes 1 and 2. Afterwards, serial doubling dilution from tube 2 to tube 10 was made, while the remaining 1 mL was discarded. 1 mL of ciprofloxacin was added to tube 11 (positive control) and water to tube 12 (negative control). 1 mL of 0.5 McFarland was added overnight and broth culture to all the tubes and then covered. The experiment was incubated overnight at 37°C and observed for the highest dilution showing no turbidity. The zone of inhibition was then verified and interpreted according to CLSI guidelines [26]; the MIC was determined.

2.3.6. Minimum Bactericidal Concentration (MBC) and Minimum Fungicidal Concentration (MFC).

MBC, the lowest concentration of antibiotic agent that kills at least 99.9% of the organisms, was determined by using Doughari et al. method. 0.5 mL of the sample was removed from those tubes from MIC which did not show any visible sign of

growth and inoculated on sterile Mueller Hinton agar by streaking. The plates were then incubated at 37°C for 24 hours. The concentration at which no visible growth was seen was recorded as the minimum bactericidal concentration (MBC). For MFC, 0.5 mL of the sample which showed no visible sign of growth during MIC screening was taken from the test tubes and then inoculated on sterile potato dextrose agar by streaking. The plates were then incubated at 37°C for 24 hours. The concentration at which no visible growth was seen was recorded as the minimum fungicidal concentration [27].

3. Results and Discussion

3.1. Optical Properties of the Metallic Nanoparticles.

UV/Visible spectra of the biosynthesized silver nanoparticles (Ag NPs) and silver-nickel (Ag/Ni) bimetallic nanoparticles (Ag/Ni NPs) as a result of photon absorption by their solutions are displayed in Figures 1–3. There was a noticeable colour change from light brown to deep brown which signalled formation of nanoparticles (Figure 1(a)). This is as a result of electronic transitions within the structures of Ag and Ag/Ni nanoclusters as they interacted with light. The electronic transitions within the structures of metallic nanoparticles resulted in emergence of surface plasmon resonance (SPR) which increased in peak intensity and confirmed Ag and Ag/Ni NPs formation [28]; bioreduction of Ag$^+$ to Ag^0 is an indication for potential application as excellent absorbers of visible light scatter and absorbers.

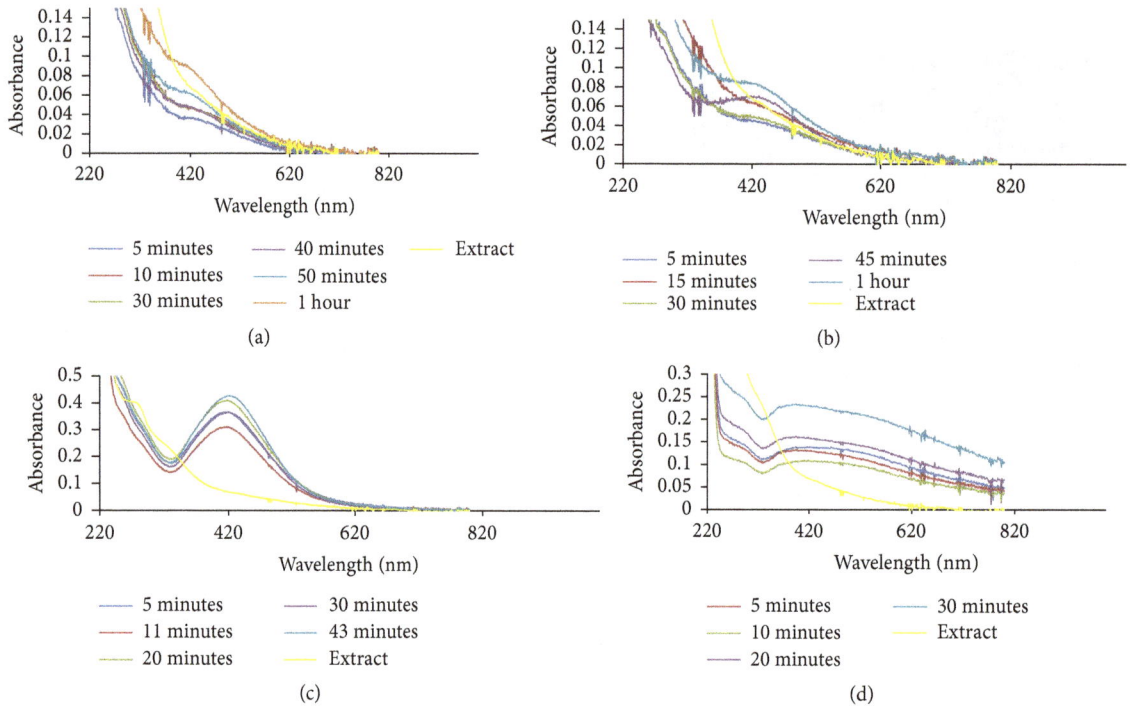

FIGURE 2: UV-Vis spectra of Ag/Ni bimetallic nanoparticles prepared by reducing (a) 0.5 mM, (b) 1.0 mM, (c) 2.0 mM, and (d) 3.0 mM solutions using the extract of *C. indica* leaves at 70°C. The blue curve represents surface plasmon resonance after 30 minutes.

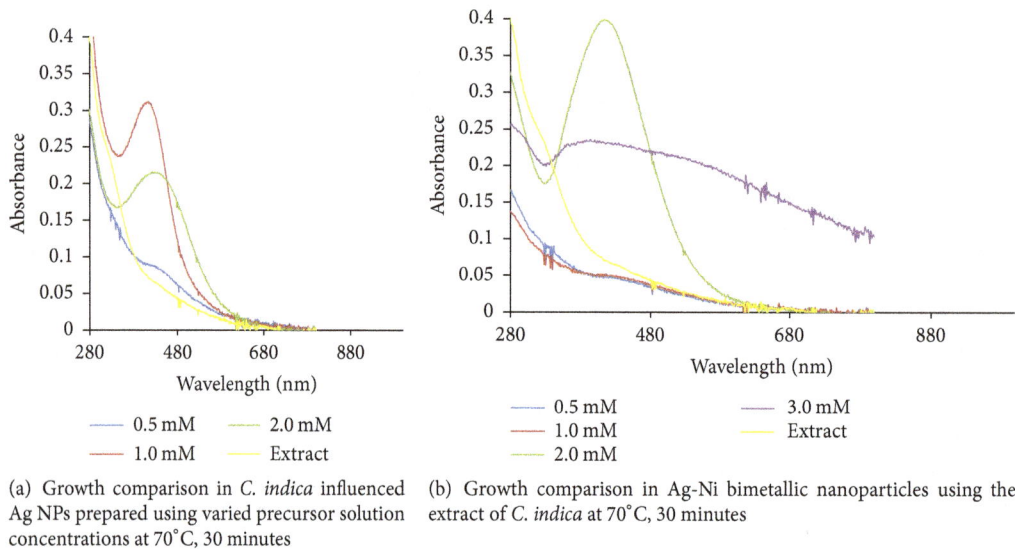

(a) Growth comparison in *C. indica* influenced Ag NPs prepared using varied precursor solution concentrations at 70°C, 30 minutes

(b) Growth comparison in Ag-Ni bimetallic nanoparticles using the extract of *C. indica* at 70°C, 30 minutes

FIGURE 3

Ag NPs formation had maximum absorption in the visible region with absorption wavelength of 416 nm and maximum intensity of 0.312 a.u. There was electron confinement effect when 1.0 mM AgNO$_3$ was reduced, and this culminated in sharp peaks and strong intensity observed. Intensity of absorption in the hybrid nanoparticles prepared from 2.0 mM precursor solution was in contrast to the maximum intensity of absorption noticed in the corresponding monometallic Ag NPs. In the bimetallic Ag/Ni NPs, there were narrow absorption spectra which increased in peak intensity without any shift in wavelength (421 nm). Hence, this signified presence of spherically shaped nanoparticles. Moreover, surface of the hybrid nanoparticles in Figure 2(c) is proposed to be enriched with silver, optically enhanced by nickel which is in line with previous work [29]. The observed narrow peak also depicted confinement of excitons in the nanoparticles as shown in Figures 1(c) and 2(c). Broad band in absorption wavelength between 400 and 450 nm in other nanobimetallic solutions of 0.5, 1.0, and 3.0 mM precursor mixture suggested aggregation and polydispersed structures [30] and this could be as a result of interaction between solute and solvent, hereby reducing the structural resolution and maximum energy of the reaction. The observed spectra overlap at 20th and 30th minute with no further changes in intensity indicating reaction completion. In Ag/Ni hybrid nanoparticles formation, an unprecedented bioreduction feature was observed, resulting in nucleation and growth of the nanoparticles within 5 minutes of reaction in all the concentrations. This observation was different in the case of monometallic Ag NPs in which nucleation and formation were delayed till 20 minutes in 0.5 mM metal precursor but was relatively faster at higher precursor solution concentrations (1.0 and 2.0 mM). The belated onset growth and reduction of 0.5 mM precursor suggested diverse mechanistic character in the formation of nanoparticles [31].

However, presence of nickel (Ni) in the hybrid synthesis of course led to a red shift in the absorbance wavelength from 416.0 to 421 nm, as observed in the reduced 2.0 mM precursor solution of Ag/Ni NPs. There was an obvious increase in intensity of absorption when compared with the corresponding Ag NPs. Growth comparison and optimum concentration for Ag NPs and Ag/Ni bimetallic synthesized at 70°C are displayed in Figures 3(a) and 3(b), respectively. Size increase due to red shift of absorption wavelength with an enhanced surface plasmon resonance was noted for application in biodiagnostic, optical materials, optoelectronics, and good absorbers of visible light and scatters [32].

Biomolecules which acted as the reducing and capping/stabilizing agents for the newly formed nanoparticles were considered to be adequate as a result of unprecedented fast and successful bioreduction [33]. It is noteworthy that *Canna indica* leaf extracts contained the following secondary metabolite: alkaloids, glycosides, and terpenoids were identified in the water extract (Table 1). Tannins, flavonoids, and saponins were also detected in addition when leaf part of the plant was extracted with methanol (Table 1). These were confirmed thorough phytochemical screening. Despite choice of water as the extraction solvent ("green" part of the study) with limited amount of phytochemicals observed compared with methanol extract, the bioreduction and nanoparticles formation were successful. Proposed mechanisms of nanoparticles formation are presented in Schemes 1–6.

3.2. Proposed Mechanisms of Reactions. See Schemes 1–6.

3.3. Morphology of the Metallic Nanoparticles.
Particle size distribution histogram and TEM image of the biosynthesized Ag/Ni bimetallic nanoparticles are depicted in Figures 4(a) and 4(b), respectively. Other representative TEM images are shown in Figures 5(a) and 5(b). TEM image revealed quasi-spherical shapes of average diameter of 9.10 ± 1.12 nm for Ag NPs, while micrograph of the bimetallic Ag/Ni NPs revealed different shapes: cube with truncated/irregular edges plausibly due to the effect of Ostwald ripening and multiply twinned hybrid after 30 minutes of the reaction with a mean particle size of 9.86 ± 2.37 nm [34]. Structural elucidation from TEM image also revealed formation of core-shell Ag/Ni

nanoparticles. The denser silver particles were distinctly visible in the TEM image. The Ag nanoparticles appeared as a dark core with Ni particles appearing less dark on the surface (Figure 2(d)). EDX analysis showing elemental compositions confirmed presence of nickel in the nanohybrid which was silver/nickel enriched with organic capping agent of carbon content which originated from plant extract (Figure 6(b)).

This finding is unique in biosynthesis of nanoparticles. Related finding was reported by Mntungwa et al. [35]; nevertheless, chemical method was applied to obtain the core-shell structure. Mechanism of the process could be explained as the ability of the plant extract to reduce the metal ions, followed by nucleation of metal atoms. Ostwald's ripening took place as a result of redissolution of high solubility and surface energy of smaller particles in the solution as the growth of larger particles continued more, as described by Lifshic and Slezov [36]. The growth pattern is considered to be anisotropic due to size increase of the NPs. Furthermore, in the mechanism, Ag (I) got reduced first because of its higher positive electrochemical potential than nickel which led to the formation of silver core [37].

3.4. Antimicrobial Activity

3.4.1. Antimicrobial Assay. Activity of the biosynthesized Ag NPs and Ag/Ni bimetallic nanoparticles based on size of zones of inhibition in millimetre (mm) is shown in Figure 7. Agar diffusion test revealed that the prepared nanoparticles possessed both antibacterial and antifungal properties. The screened nanoparticles exhibited higher activities on all the test organisms at higher concentration of 3.0 mM except *P. aeruginosa* in which low activity was recorded. Sensitivity testing of organisms (Agar diffusion test) in triplicate showed zones of inhibition. Mean zone of inhibition diameter (mm) ± standard deviation is shown in Table 2. One-way analysis of variance (ANOVA) using SPSS statistical tool indicated that growth inhibition by the nanoparticles was significant at $P < 0.05$, F-value 34.06 (Table 3).

Zones of inhibition recorded in agar well diffusion test led to the conduction of minimum inhibitory concentration (MIC), minimum bactericidal concentration (MBC) and minimum fungicidal concentration (MFC) tests. Results of MIC, MBC, and MFC tests are presented in Table 3. Interestingly, all prepared nanoparticles showed concentration-dependent inhibitory effects on the *in vitro* antimicrobial assay [38]. Highest activity of Ag NPs was on *E. coli* and *C. albicans* with MIC value of 12.5 mg/mL and 2.5 mg/mL (same value for MBC and MFC), followed by *S. aureus* and *S. pyogenes* (12.5 mg/mL value of MIC and MBC). The activity was least on *T. rubrum* (50 mg/mL MIC, 100 mg/mL MFC). However, no detectable activity was found against *P. aeruginosa*. Higher activity was observed in Ag/Ni nanoparticles against *S. pyogenes* with MIC value of 6.25 mg/mL and 12.5 mg/mL MBC.

In hybrid Ag/Ni nanoparticles, analysis of variance (ANOVA) using SPSS statistical tool indicated no significant difference in the concentrations as $P > 0.05$, yet possessing better activity than its corresponding Ag NPs which was observed at different levels in the test organisms. Moreover,

TABLE 1: Phytochemical analysis of *Canna indica* leaf extract.

	Proteins	Carbohydrates	Phenols	Tannins	Flavonoids	Saponins	Glycoside	Steroids	Terpenoids	Alkaloids
					Phytochemical					
Aqueous extract	–	–	–	–	–	–	++	–	+	+++
Methanolic extract	–	–	–	+++	++	+	++	–	++	+++

Weak presence +; strong presence ++; stronger presence +++; absent –.

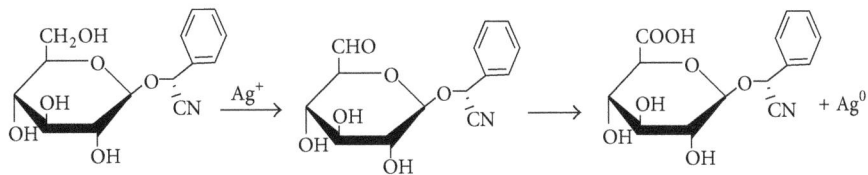

SCHEME 1: Bioreduction of silver ion to silver nanoparticles by glycosides.

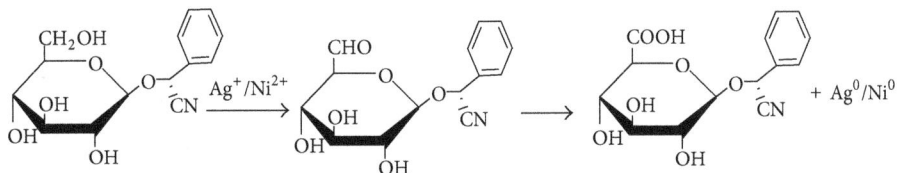

SCHEME 2: Bioreduction of silver/nickel ions to silver/nickel nanoparticles by glycosides.

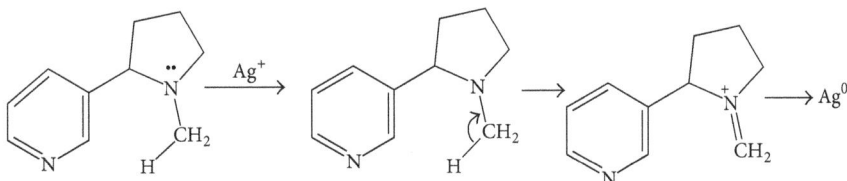

SCHEME 3: Bioreduction of silver ion to silver nanoparticles by alkaloids.

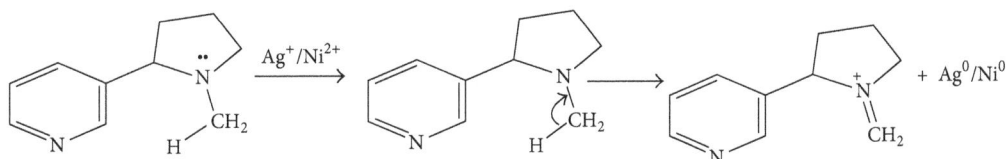

SCHEME 4: Bioreduction of silver/nickel ions to silver/nickel nanoparticles by alkaloids.

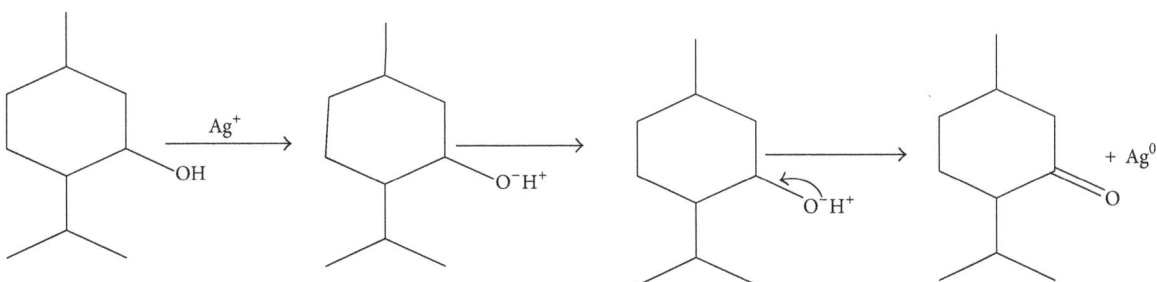

SCHEME 5: Bioreduction of silver ion to silver nanoparticles by terpenoids.

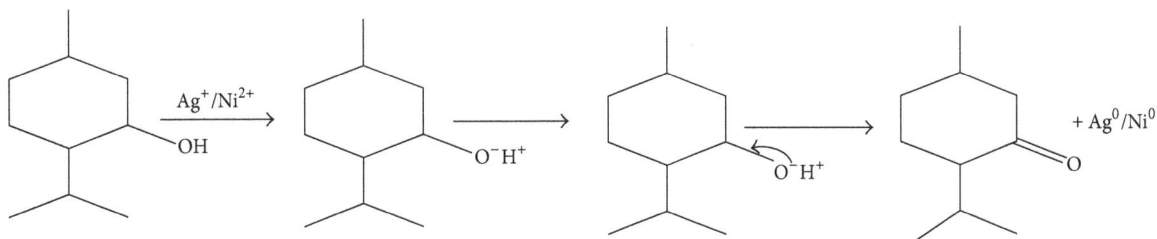

SCHEME 6: Bioreduction of silver/nickel ions to silver/nickel nanoparticles by terpenoids.

(a)

(b)

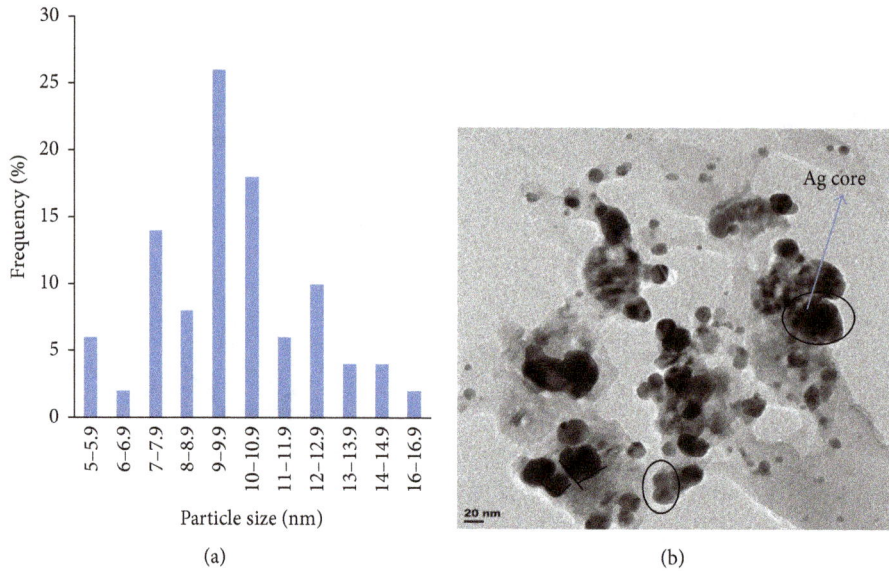

FIGURE 4: (a) Particle size distribution histogram of Ag/Ni determined from TEM images. (b) Representative TEM image of the bimetallic Ag/Ni NPs under *C. indica* influenced synthesis.

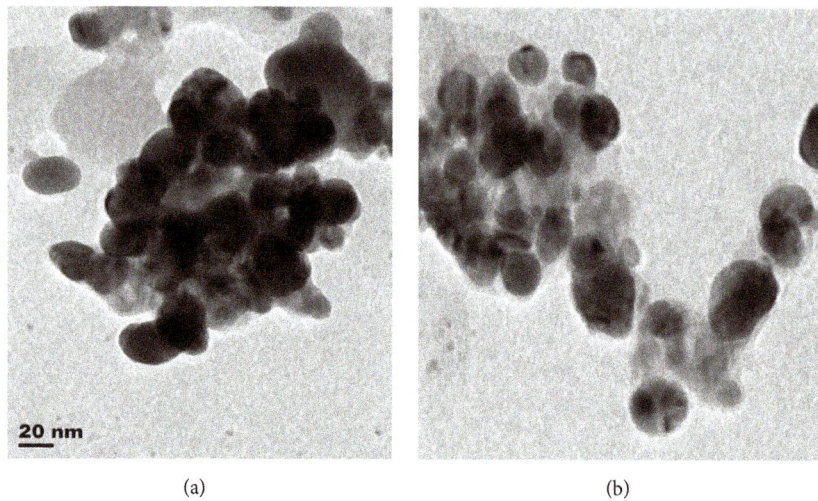

(a)

(b)

FIGURE 5: Representative TEM images of Ag/Ni bimetallic NPs derived from *C. indicia* leaf extract.

these nanoparticles could not hinder the growth of *P. aeruginosa*. None of the as-synthesized nanoparticles was able to compete with ciprofloxacin and fluconazole (standards) in terms of activity. *S. aureus, S. pyogenes,* and *E. coli* showed similar behaviour relative to MIC and MBC, while *P. aeruginosa* was highly resistant. The metallic nanoparticles were more active on *C. albicans* than *T. rubrum* possibly because they were able to penetrate the thin peptidoglycan layer of the fungus with the outer membrane composed of phospholipids and lipopolysaccharides (LPS) of the *E. coli* (complex gram-negative bacterium). Not only were the bionanoparticles considered to have passed through thicker peptidoglycan cell wall layer which is accountable for rigidity and low activity in gram-positive bacteria, as detected in the

MIC test carried out on *S. aureus* and *S. pyogenes* [39], but, according to Marini et al., the observed growth inhibition in bacteria can also be related to the reaction of thiol groups present in bacteria protein with the release of Ag^+ which slowed down or changed the replication of DNA [40].

From the above, there is a need to affirm if the penetrations of the nanoparticle through the microbial samples influence its experimental result. Tan et al. [41] did the characterization of nanoparticle dispersion through red blood cell using the nanoparticle model. Through experimentation, the nanoparticle dispersion model was given as

$$D_r = \frac{D - D_0}{d_{\text{layer}}^2 \eta}. \tag{1}$$

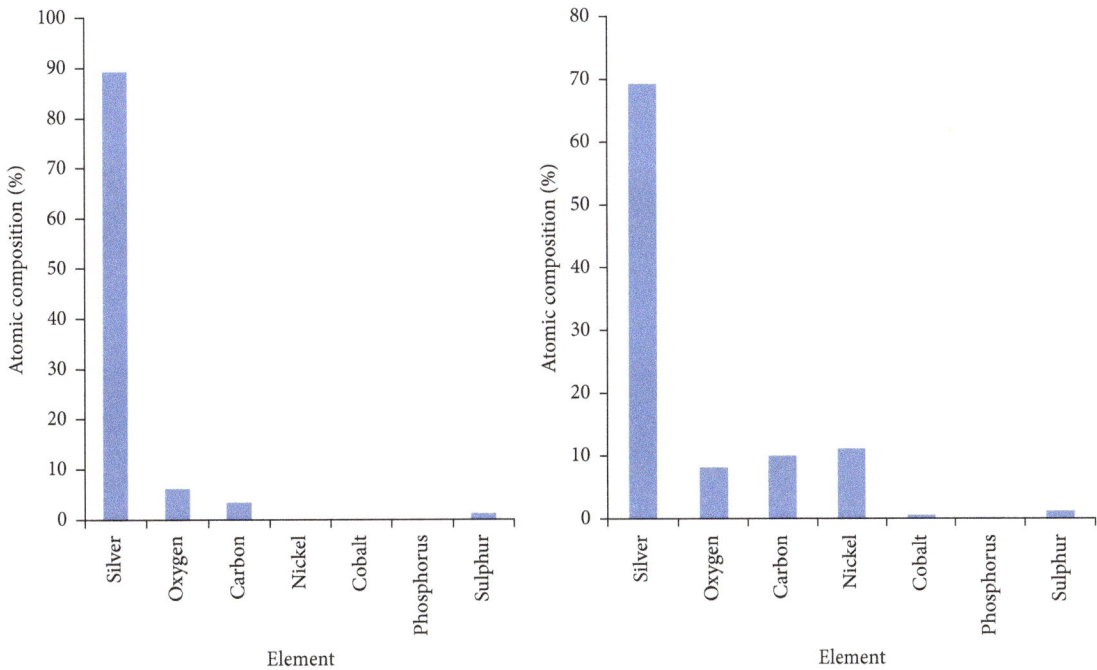

(a) EDX showing atomic composition of elements present in Ag NPs

(b) EDX showing atomic composition of elements present in Ag/Ni bimetallic NPs

Figure 6

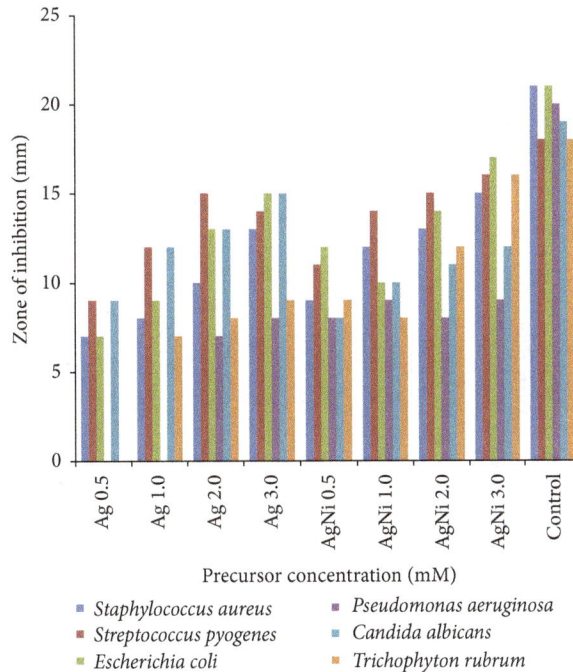

Figure 7: Comparison of inhibition zones between Ag NPs and Ag-Ni bimetallic nanoparticles synthesized using *C. indica* leaf extract.

TABLE 2: Sensitivity testing of organisms with standard deviation in zones of inhibition (agar diffusion test).

NPs	Organisms/mean zone diameter (mm) ± SD					
	Staphylococcus aureus	*Streptococcus pyogenes*	*Escherichia coli*	*Pseudomonas aeruginosa*	*Candida albicans*	*Trichophyton rubrum*
Ag 0.5	7 ± 0.2	9 ± 0.4	7 ± 0.1	Nil	9 ± 0.2	Nil
Ag 1.0	8 ± 0.5	12 ± 0.8	9 ± 0.2	Nil	12 ± 0.3	7 ± 0.2
Ag 2.0	10 ± 0.4	15 ± 0.5	13 ± 0.4	7 ± 0.2	13 ± 0.4	8 ± 0.1
Ag 3.0	13 ± 1	14 ± 0.2	15 ± 0.3	8 ± 0.1	15 ± 0.2	9 ± 0.1
Stat	*P > 0.05*	*P > 0.05*	*P > 0.05*	*P > 0.05*	*P > 0.05*	*P > 0.05*
Ag-Ni 0.5	9 ± 0.2	11 ± 0.4	12 ± 0.5	8 ± 0.3	8 ± 0.1	9 ± 0.2
Ag-Ni 1.0	12 ± 0.3	14 ± 0.6	10 ± 0.2	9 ± 0.2	10 ± 0.2	8 ± 0.1
Ag-Ni 2.0	13 ± 0.1	15 ± 0.2	14 ± 0.4	8 ± 0.1	11 ± 0.2	12 ± 0.3
Ag-Ni 3.0	15 ± 0.4	16 ± 0.6	17 ± 0.6	9 ± 0.1	12 ± 0.1	16 ± 0.2
Control	21 ± 0.8	18 ± 0.3	21 ± 0.2	20 ± 0.4	19 ± 0.6	18 ± 0.3
Stat	$P > 0.05$	$P > 0.05$	$P > 0.05$	$P > 0.05$	$P > 0.05$	$P > 0.05$

Control: ciprofloxacin (bacteria) and fluconazole (fungi); mean zone inhibition (mm) ± standard deviation of triplicate measurements. Ag = silver nanoparticles of specified precursor concentration using *C. indica* leaf extract; Ag/Ni = silver-nickel bimetallic nanoparticles of specified precursor concentration using *C. indica* leaf extract.

TABLE 3: Minimum inhibitory concentration (MIC), minimum bactericidal concentration (MBC), and minimum fungicidal concentration (MFC).

Nanoparticles	Organisms/MIC, MBC & MFC (mg/mL)					
	Staphylococcus aureus MIC, MBC	Streptococcus pyogenes MIC, MBC	Escherichia coli MIC, MBC	Pseudomonas aeruginosa MIC, MBC	Candida albicans MIC, MFC	Trichophyton rubrum MIC, MFC
Ag 0.5	100, 100	50, 100	100, 100	100, 100	50, 50	100, 100
Ag 1.0	100, 100	25, 50	50, 100	100, 100	25, 25	100, 100
Ag 2.0	50, 100	12.5, 25	12.5, 25	100, 100	12.5, 25	100, 100
Ag 3.0	12.5, 25	12.5, 25	12.5, 12.5	100, 100	12.5, 12.5	50, 100
Statistics	*P < 0.05*	*P < 0.05*	*P < 0.05*	*P < 0.05*	*P < 0.05*	*P < 0.05*
Ag/Ni 0.5	50, 100	25, 50	12.5, 25	100, 100	50, 100	50, 100
Ag/Ni 1.0	12.5, 25	12.5, 12.5	25, 50	50, 100	25, 50	100, 100
Ag/Ni 2.0	12.5, 25	12.5, 12.5	12.5, 12.5	100, 100	12.5, 25	12.5, 12.5
Ag/Ni 3.0	12.5, 12.5	6.25, 12.5	6.25, 12.5	100, 100	12.5, 25	12.5, 25
Statistics	*P < 0.05*	*P < 0.05*	*P < 0.05*	*P < 0.05*	*P < 0.05*	*P < 0.05*
Control	3.13	6.25	6.25	6.25	6.25	6.25

D_r is the dimensionless dispersion rate, D is the dispersion rate, D_0 is the dispersion tendency, d is the cell/layer, and η is the shear rate. Tan et al. [41] worked with the shear rate below $40\,\text{s}^{-1}$ and above $200\,\text{s}^{-1}$.

Kleinstreuer and Xu [42] hence gave D_0 for metallic nanoparticles penetration through the thin peptidoglycan layer as

$$D_0 = \frac{k_B T}{3\pi\mu_{\text{bf}}d_p}, \qquad (2)$$

where k_B is the Stefan Boltzmann constant, T is the local temperature, μ_{bf} is the base fluid viscosity, and d_p is the particle diameter. From past experimentation, Benakashani et al. [43] found average particle size of Ag NPs using the Debye-Scherrer equation:

$$d_{\text{nano}} = \frac{K\lambda}{\beta\cos\theta}, \qquad (3)$$

where d is the size of the Ag NPs which is about 20 nm, K is the Scherrer constant that ranges between 0.9 and 1, λ is the wavelength of the X-ray source, in this case it ranges between 280 and 780 nm, β is the full width at the half maximum of the diffraction peak, and θ is Bragg's angle.

The basic condition for the metallic nanoparticle to go through the polymer matrix of the thin peptidoglycan layer is $d_{\text{nano}} < d_{\text{layer}}$. Hence,

$$d_{\text{layer}} = d_{\text{nano}} + d_o. \qquad (4)$$

The dimensionless dispersion of the metallic nanoparticle through the peptidoglycan layer is given as

$$D_r = \frac{D}{\left(K^2\lambda^2/\beta^2\cos^2(\theta) + 2K\lambda d_0/\beta\cos(\theta) + d_0{}^2\right)\eta}$$
$$- \left(\frac{1}{\left(K^2\lambda^2/\beta^2\cos^2(\theta) + 2K\lambda d_0/\beta\cos(\theta) + d_0{}^2\right)\eta}\right) \qquad (5)$$
$$\cdot \left(\frac{k_B T}{3\pi\mu_{\text{bf}}d_p}\right),$$

where

$$D = \begin{cases} \dfrac{1}{4}\cos\left(\dfrac{\pi d_p}{4}\right) & d_p < 10^{-9} \\ 0 & \text{otherwise.} \end{cases} \qquad (6)$$

When $D = 0$, then the dimensionless dispersion becomes

D_r

$$= \left(\frac{1}{\left(K^2\lambda^2/\beta^2\cos^2(\theta) + 2K\lambda d_0/\beta\cos(\theta) + d_0{}^2\right)\eta}\right) \qquad (7)$$
$$\cdot \left(\frac{k_B T}{3\pi\mu_{\text{bf}}d_p}\right).$$

Hence, the illustration of the two cases is shown in Figures 8(a)–8(d) and 9(a)–9(d).

The first case (i.e., (5)) was when the diameter of the metallic particle is equal to or less than 10^{-9} m. At a constant Bragg's angle of 45°, the feature of the dispersion/wavelength trend (Figure 8(a)) is the same as results shown in Figures 2 and 3. This is the first evidence that the penetration of metallic nanoparticles through the thin peptidoglycan layer does not necessarily influence the efficiency of its constituents. Then in a case when the sizes of the nanoparticle are heterogeneous, we assumed Bragg's angle of the nanoparticle ranges between −30° and 30° (Figure 8(b)). Five peaks appeared with the maximum peak at wavelength 340 nm. The significance of this result can be seen in its effect in Figures 2(a)–2(d). A further analysis was conducted to see the features of the dimensionless dispersion of the nanoparticles when Bragg's angle of the nanoparticle ranges between −45° and 45° (Figure 8(c)). Fifteen peaks were observed showing the response or sensitivity of Bragg's angle to nanoparticle transport within multilayers. Then Bragg's angle when the nanoparticle range is between −60° and 60° was considered (Figure 8(d)). Only two peaks were observed. Hence, the penetration of nanoparticle does not depict the activeness of the microbial samples in all cases which largely depend on the nature and pore size of the peptidoglycan layer.

The second case (i.e., (7)) was when the diameter of the metallic particle is greater than 10^{-9} m (Figures 9(a)–9(d)). It was observed that the features of Figures 8 and 9 were the same. However, the dimensionless dispersion is very low.

4. Conclusion

Rapid, facile, and environmental-friendly syntheses of monometallic Ag NPs and Ag/Ni bimetallic nanoparticles using *Canna indica* leaf extract were successful via reduction of $AgNO_3$ and $Ni(NO_3)_2\cdot 6H_2O$ metal precursors. No doubt, the leaf extract acted as the reducing/capping agent. The metallic nanoparticles were characterized for their optical and morphological properties. Nucleation and onset growth which were considered to be by diffusion control and Ostwald's ripening commenced as early as 5 minutes. Ag NPs had maximum absorption in the visible region with absorption wavelength of 416 nm. However, presence of nickel (Ni) in the hybrid synthesis led to a red shift in the absorbance wavelength from 416 to 421 nm after 30 minutes of the reaction. TEM image revealed quasi-spherical shapes of average diameter 9.10 ± 1.12 nm in Ag NPs while micrograph of the bimetallic Ag/Ni cluster revealed different

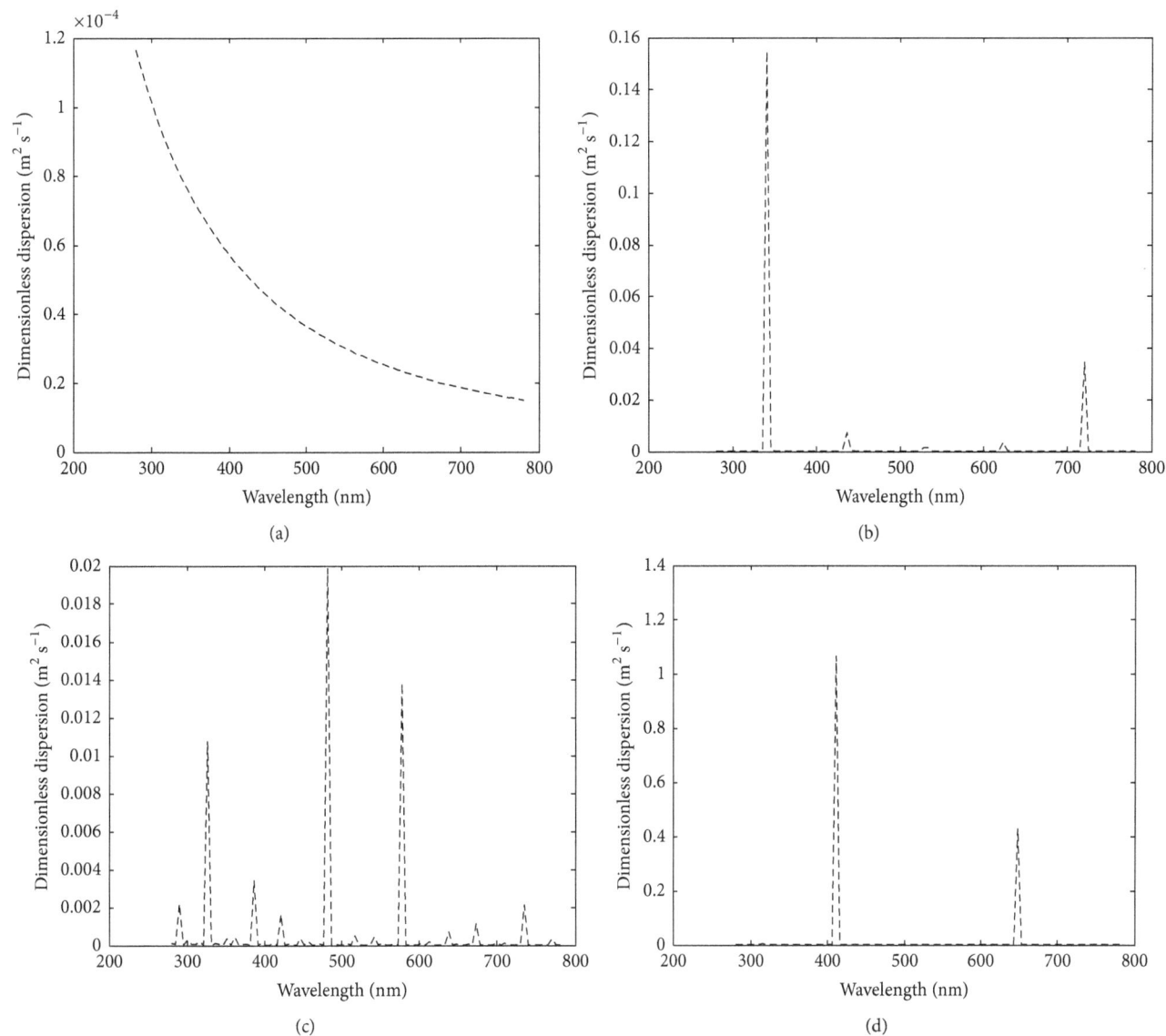

FIGURE 8: The dimensionless dispersion of the metallic nanoparticle under varying conditions (a) when Bragg's angle of the nanoparticle is strictly 45°; (b) when Bragg's angle of the nanoparticle ranges between −30° and 30°; (c) when Bragg's angle of the nanoparticle ranges between −45° and 45°; (d) when Bragg's angle of the nanoparticle ranges between −60° and 60°.

shapes: cube with truncated/irregular edges plausibly due to the effect of Ostwald ripening and multiply twinned hybrid after 30 minutes of the reaction with a mean particle size of 9.86 ± 2.37 nm. Structural elucidation from the TEM image also revealed formation of core-shell Ag/Ni nanoparticles. The denser silver particles were distinctly visible in the TEM image. The Ag nanoparticles appeared as a dark core with Ni particles appearing less dark on the surface (Figure 2(d)). EDX analysis showing elemental compositions of the nanoparticles indicated that the nanohybrid was silver enriched with organic capping due to the composition of carbon content which originated from plant extract. Furthermore, Ag/Ni bimetallic nanocluster exhibited better antimicrobial activity against the test pathogens than its corresponding monometallic Ag NPs. Hence, from the findings, Ag/Ni NPs are potential antibacterial agents against *E. coli* and possible antifungal agents against *C. albicans*. Possible antibacterial drugs against *S. pyogenes* and *E. coli* can be designed using Ag-Ni nanohybrid, based on their strong inhibition activities observed. The observed enhanced SPR in the nanoclusters is noted for applications in optical materials, also as good absorbers of visible light absorber and scatters.

Conflicts of Interest

The authors declare no conflicts of interest in this research work.

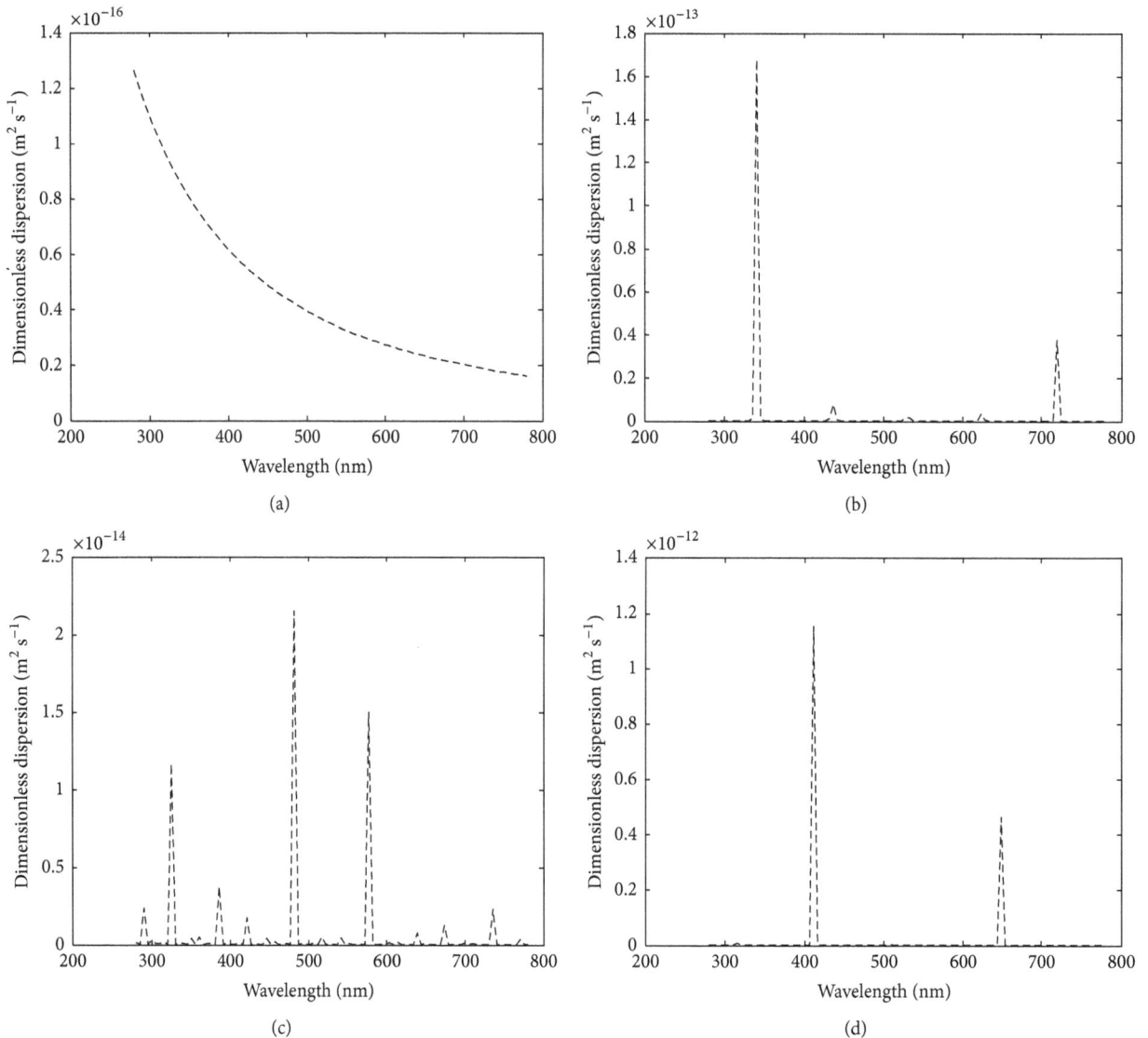

FIGURE 9: The dimensionless dispersion of the metallic nanoparticle under varying conditions at diameter greater than 10^{-9} nm (a) when Bragg's angle of the nanoparticle is strictly 45°; (b) when Bragg's angle of the nanoparticle ranges between −30° and 30°; (c) when Bragg's angle of the nanoparticle ranges between −45° and 45°; (d) when Bragg's angle of the nanoparticle ranges between −60° and 60°.

Acknowledgments

The authors are grateful to Mr. Olusola Rotimi of the University of Western Cape, Bellville Campus, Cape-Town, Mr. Shitole Joseph of iThemba Labs, and Mr. Olufemi Olaofe in South Africa for the TEM and EDX characterizations. The authors also acknowledge Covenant University for funding this publication.

References

[1] S. Ahmed, M. Ahmad, B. L. Swami, and S. Ikram, "A review on plants extract mediated synthesis of silver nanoparticles for antimicrobial applications: a green expertise," *Journal of Advanced Research*, vol. 7, no. 1, pp. 17–28, 2016.

[2] I. Brigger, C. Dubernet, and P. Couvreur, "Nanoparticles in cancer therapy and diagnosis," *Advanced Drug Delivery Reviews*, vol. 54, no. 5, pp. 631–651, 2002.

[3] A. K. Gupta and M. Gupta, "Cytotoxicity suppression and cellular uptake enhancement of surface modified magnetic nanoparticles," *Biomaterials*, vol. 26, no. 13, pp. 1565–1573, 2005.

[4] R. Hao, R. J. Xing, Z. C. Xu, Y. L. Hou, S. Goo, and S. H. Sun, "Synthesis, functionalization, and biomedical applications of multifunctional magnetic nanoparticles," *Advanced Materials*, vol. 22, no. 25, pp. 2729–2742, 2010.

[5] M. Imran Din and A. Rani, "Recent advances in the synthesis and stabilization of nickel and nickel oxide nanoparticles: A green adeptness," *International Journal of Analytical Chemistry*, vol. 2016, Article ID 3512145, 14 pages, 2016.

[6] H. Chen, J. Wang, D. Huang et al., "Plant-mediated synthesis of size-controllable Ni nanoparticles with alfalfa extract," *Materials Letters*, vol. 122, pp. 166–169, 2014.

[7] A. A. Mariam, M. Kashif, J. S. Arokiyara et al., "Bio-synthesis of Nio and Ni nanoparticles and their characterization," *Digest Journal of Nanomaterials and Biostructures*, vol. 9, no. 3, pp. 1007–1019, 2014.

[8] P. K. Jain, X. Huang, I. H. El-Sayed, and M. A. El-Sayed, "Noble metals on the nanoscale: Optical and photothermal properties and some applications in imaging, sensing, biology, and medicine," *Accounts of Chemical Research*, vol. 41, no. 12, pp. 1578–1586, 2008.

[9] M. Vaseem, N. Tripathy, G. Khang, and Y.-B. Hahn, "Green chemistry of glucose-capped ferromagnetic hcp-nickel nanoparticles and their reduced toxicity," *RSC Advances*, vol. 3, no. 25, pp. 9698–9704, 2013.

[10] I. P. Darsini, S. Shamshad, and M. John Paul, "Canna Indica (L.): A plant with potential healing powers: A review," *International Journal of Pharma and Bio Sciences*, vol. 6, no. 2, pp. 1–8, 2015.

[11] E. Abdullah, R. A. Raus, and P. Jamal, "Extraction and evaluation of antibacterial activity from selected flowering plants," *American Medical Journal*, vol. 3, no. 1, pp. 27–32, 2012.

[12] W. Woradulayapinij, N. Soonthornchareonnon, and C. Wiwat, "*In vitro* HIV type 1 reverse transcriptase inhibitory activities of Thai medicinal plants and *Canna indica*L. rhizomes," *Journal of Ethnopharmacology*, vol. 101, no. 1–3, pp. 84–89, 2005.

[13] P. Rauwel, S. Küünal, S. Ferdov, and E. Rauwel, "A review on the green synthesis of silver nanoparticles and their morphologies studied via TEM," *Advances in Materials Science and Engineering*, vol. 2015, Article ID 682749, 9 pages, 2015.

[14] A. P. Kulkarni, A. A. Srivastava, P. M. Harpale, and R. S. Zunjarrao, "Plant Mediated Synthesis of Silver Nanoparticles - Tapping The Unexploited Sources," *The Journal of Natural Product and Plant Resources*, vol. 1, no. 4, pp. 100–107, 2011.

[15] S. C. Boca and S. Astilean, "Detoxification of gold nanorods by conjugation with thiolated poly(ethylene glycol) and their assessment as SERS-active carriers of Raman tags," *Nanotechnology*, vol. 21, no. 23, Article ID 235601, 2010.

[16] M. Rai, A. Yadav, and A. Gade, "CRC 675—current trends in phytosynthesis of metal nanoparticles," *Critical Reviews in Biotechnology*, vol. 28, no. 4, pp. 277–284, 2008.

[17] J. A. Adekoya, E. O. Dare, M. A. Mesubi, and N. Revaprasadu, "Synthesis and characterization of optically active fractal seed mediated silver nickel bimetallic nanoparticles," *Journal of Materials*, vol. 2014, pp. 1–9, 2014.

[18] S. Prabhu and E. K. Poulose, "Silver nanoparticles: mechanism of antimicrobial action, synthesis, medical applications, and toxicity effects," *International Nano Letters*, vol. 2, no. 32, pp. 1–10, 2012.

[19] N. Krishnamurthy, P. Vallinayagam, and D. Madhavan, *Engineering Chemistry*, PHI Learning Pvt. Ltd., 2014.

[20] A. A. Akinsiku, E. O. Dare, K. O. Ajanaku, J. A. Adekoya, S. O. Alayande, and A. O. Adeyemi, "Synthesis of silver nanoparticles by plant-mediated green method: optical and biological properties," *Journal of Bionanoscience*, vol. 10, no. 3, pp. 171–180, 2016.

[21] N. Ahmad and S. Sharma, "Green synthesis of silver nanoparticles using extracts of Ananas comosus," *Green and Sustainable Chemistry*, vol. 2, no. 4, pp. 141–147, 2012.

[22] A. A. Akinsiku, K. O. Ajanaku, J. A. Adekoya, and E. O. Dare, "Green synthesis, characterization of silver nanoparticles using Canna indica and Senna occidentalis leaf extracts," in *Proceedings of 2nd Covenant University International Conference on African Development Issues*, pp. 154–157, 2015.

[23] British Society for Antimicrobial Chemotherapy (BSAC, 1990).

[24] National Committee for Clinical Laboratory Standards (NCCLS), (1993).

[25] P. Aida, V. Rosa, F. Blamea, A. Thomas, and C. Salvador, "Paraguyan plants used in raditional medicine," *Journal of Ethnopharmacology*, vol. 16, pp. 93–98, 2001.

[26] Clinical and Laboratory Standards Institute (CLSI), 2006.

[27] J. H. Doughari, M. S. Pukuma, and N. De, "Antibacterial effects of Balanites aegyptiaca L. Drel. and Moringa oleifera Lam. on Salmonella typhi," *African Journal of Biotechnology*, vol. 6, no. 19, pp. 2212–2215, 2007.

[28] U. Guler, V. M. Shalaev, and A. Boltasseva, "Nanoparticle plasmonics: going practical with transition metal nitrides," *Materials Today*, vol. 18, no. 4, pp. 227–237, 2015.

[29] K. Sridharan, T. Endo, S.-G. Cho, J. Kim, T. J. Park, and R. Philip, "Single step synthesis and optical limiting properties of Ni-Ag and Fe-Ag bimetallic nanoparticles," *Optical Materials*, vol. 35, no. 5, pp. 860–867, 2013.

[30] S. Ponarulselvam, C. Panneerselvam, K. Murugan, N. Aarthi, K. Kalimuthu, and S. Thangamani, "Synthesis of silver nanoparticles using leaves of *Catharanthus roseus* Linn. G. Don and their antiplasmodial activities," *Asian Pacific Journal of Tropical Biomedicine*, vol. 2, no. 7, pp. 574–580, 2012.

[31] A. M. Smith, H. Duan, M. N. Rhyner, G. Ruan, and S. Nie, "A systematic examination of surface coatings on the optical and chemical properties of semiconductor quantum dots," *Physical Chemistry Chemical Physics*, vol. 8, no. 33, pp. 3895–3903, 2006.

[32] O. Kvítek, J. Siegel, V. Hnatowicz, and V. Švorčík, "Noble metal nanostructures influence of structure and environment on their optical properties," *Journal of Nanomaterials*, vol. 2013, Article ID 743684, p. 8, 2013.

[33] K. Raja, A. Saravanakumar, and R. Vijayakumar, "Efficient synthesis of silver nanoparticles from Prosopis juliflora leaf extract and its antimicrobial activity using sewage," *Spectrochimica Acta Part A: Molecular and Biomolecular Spectroscopy*, vol. 97, pp. 490–494, 2012.

[34] O. V. Kharissova, H. V. R. Dias, B. I. Kharisov, B. O. Pérez, and V. M. J. Pérez, "The greener synthesis of nanoparticles," *Trends in Biotechnology*, vol. 31, no. 4, pp. 240–248, 2013.

[35] N. Mntungwa, V. S. R. Pullabhotla, and N. Revaprasadu, "Facile synthesis of organically capped CdTe nanoparticles," *Journal of Nanoscience and Nanotechnology*, vol. 12, no. 3, pp. 2640–2644, 2012.

[36] I. M. Lifshic and V. V. Slezov, "The kinetics of precipitation from supersaturated solid solutions," *Journal of Physics and Chemistry of Solids*, vol. 19, no. 1-2, pp. 35–50, 1961.

[37] R. G. Haverkamp and A. T. Marshall, "The mechanism of metal nanoparticle formation in plants: limits on accumulation," *Journal of Nanoparticle Research*, vol. 11, no. 6, pp. 1453–1463, 2009.

[38] C. R. Andrighetti-Fröhner, K. N. de Oliveira, D. Gaspar-Silva et al., "Synthesis, biological evaluation and SAR of sulfonamide 4-methoxychalcone derivatives with potential antileishmanial activity," *European Journal of Medicinal Chemistry*, vol. 44, no. 2, pp. 755–763, 2009.

[39] F. Nazzaro, F. Fratianni, L. De Martino, R. Coppola, and V. De Feo, "Effect of essential oils on pathogenic bacteria," *Pharmaceuticals*, vol. 6, no. 12, pp. 1451–1474, 2013.

[40] M. Marini, S. De Niederhausern, R. Iseppi et al., "Antibacterial activity of plastics coated with silver-doped organic-inorganic hybrid coatings prepared by sol-gel processes," *Biomacro-molecules*, vol. 8, no. 4, pp. 1246–1254, 2007.

[41] J. Tan, W. Keller, S. Sohrabi, J. Yang, and Y. Liu, "Characterization of nanoparticle dispersion in red blood cell suspension by the lattice boltzmann-immersed boundary method," *Nanoma-terials*, vol. 6, no. 2, article no. 30, 2016.

[42] C. Kleinstreuer and Z. Xu, "Mathematical modeling and computer simulations of nanofluid flow with applications to cooling and lubrication," *Fluids*, vol. 1, no. 2, p. 16, 2016.

[43] F. Benakashani, A. R. Allafchian, and S. A. H. Jalali, "Biosynthesis of silver nanoparticles using Capparis spinosa L. leaf extract and their antibacterial activity," *Karbala International Journal of Modern Science*, vol. 2, no. 4, pp. 251–258, 2016.

19

Emulsion Cross-Linking Technique for Human Fibroblast Encapsulation

Watcharaphong Chaemsawang,[1] **Weerapong Prasongchean,**[2]
Konstantinos I. Papadopoulos ⓘ **,**[3] **Suchada Sukrong** ⓘ **,**[4]
W. John Kao,[5] **and Phanphen Wattanaarsakit** ⓘ[1]

[1] Department of Pharmaceutics and Industrial Pharmacy, Faculty of Pharmaceutical Sciences, Chulalongkorn University, Bangkok, Thailand
[2] Department of Biochemistry and Microbiology, Faculty of Pharmaceutical Sciences, Chulalongkorn University, Bangkok, Thailand
[3] THAI StemLife Co., Ltd., Bangkok, Thailand
[4] Department of Pharmacognosy and Pharmaceutical Botany, Faculty of Pharmaceutical Sciences, Chulalongkorn University, Bangkok, Thailand
[5] Chemistry and Biology Centre, Li Ka Shing Faculty of Medicine and Faculty of Engineering, The University of Hong Kong, Hong Kong SAR, Hong Kong

Correspondence should be addressed to Phanphen Wattanaarsakit; aphanphe@chula.ac.th

Academic Editor: Nicholas Dunne

Microencapsulation with biodegradable polymers has potential application in drug and cell delivery systems and is currently used in probiotic delivery. In the present study, microcapsules of human fibroblast cells (CRL2522) were prepared by emulsion cross-linking technique. Tween 80 surfactant at a 2% concentration through phase inversion resulted in the most efficient and stable size, morphology, and the cells survival at least 50% on day 14. Emulsion cross-linking microcapsule preparation resulted in smaller and possibly more diverse particles that can be developed clinically to deliver encapsulated mammalian cells for future disease treatments.

1. Introduction

Microencapsulation using biodegradable polymers has potential application in drug and cell delivery systems [1–4]. Coacervation, solvent evaporation, and spray drying are examples of technique used in producing microcapsules that are robust enough to withstand external forces and allowing them to be implanted using needles or catheters for drug delivery to target organs. Another commonly used encapsulation method is ionic gelation that does not require heat or organic solvents but the size of particles is larger and difficult to use in the clinic [5–7]. High shear speed cutter and ultrasonic and spray guns can be used to reduce particle size of ionic gelation microcapsules. Microencapsulation by emulsion cross-linking is an easier technique to achieve microcapsule size reduction. Emulsion cross-linking is a favored method in probiotic and other biological product encapsulation [5]. The objective of this study is to develop emulsion cross-linking encapsulation for cell delivery. How surfactant type and concentration and oil aqueous phase ratio impacted particle size, stability, and number as well as cell viability in the resulting encapsulation was examined.

2. Materials and Methods

2.1. Chemicals and Reagents. Sodium alginate was purchased from Sigma–Aldrich (CAS number 9005-38-3). Calcium chloride was purchased from Merck. Lecithin (phosphatidyl choline S75) was obtained as a gift sample from Lipoid. Tween 80 and Span 80 were purchased from Srichand United Dispensary.

2.2. Cell Culture.
Human fibroblast cells (CRL-2522ATCC) were cultured in high glucose Dulbecco's modified Eagle's medium (DMEM: containing 10 % fetal bovine serum and 1% penicillin-streptomycin- amphotericin B). Cells were cultured at 7°C with 5% CO_2 and the medium was replenished every three days. Cells were dissociated with trypsin–EDTA and were enumerated with trypan blue under microscope. The cells passage of 20-30 was used for microencapsulation. All of medium materials were purchased from Invitrogen.

2.3. Development of Alginate Microcapsules by Emulsion Cross-Linking Technique.
Water-in-oil (W/O) emulsion was prepared by mixing 1% sodium alginate solution with rice bran oil. A 1% sodium alginate solution was generated by dissolving 1 g of alginate (Sigma–Aldrich) in 100 ml cell culture medium. Rice bran oil and surfactant were mixed with a magnetic stirrer. The alginate solution was mixed in the rice bran oil solution and stirred for 10 minutes. In the second stage the primary emulsion was rinsed in 2% calcium chloride solution and continuously stirred for 20 minutes. Then, the microcapsules were centrifuged at 2,000 rpm for 20 minutes and washed three times with PBS pH 7.4. The effect of rice bran oil, aqueous phase ratio, surfactant type, and concentration of surfactant on the resulting microcapsules, was examined.

2.4. Cell Encapsulation in Alginate Microcapsule by Emulsion Cross-Linking Technique.
Human fibroblast cells at a concentration of $5x10^5$ cells/ml were placed in 1 % (w/v) sodium alginate solution and transferred to the oil solution and mixed for 10 minutes. The cell suspension was placed into calcium chloride bath and stirred for 20 minutes at room temperature. The resulting fibroblast-containing microcapsules were kept in DMEM medium and incubated at 37°C with 5% CO_2. The cell culture medium was washed and changed every 3rd day.

2.5. Characterization of Microcapsules.
The size and morphology of the microcapsules were determined under inverted microscope and Malvern mastersizer. Percent living cell entrapment was calculated from

$$Living\ cell\ entrapment$$
$$= \frac{Number\ of\ living\ cell\ in\ microcapsule}{Number\ of\ living\ cell\ loading} x100 \quad (1)$$

2.6. Assessment of Encapsulated Cells Viability.
Number of living cell in microcapsule was determined by fluorimetric quantitative PrestoBlue[5] assay. PrestoBlue reagent, a solution of resazurin base, is rapidly taken up by living cells. The reducing environment within viable cells converts PrestoBlue reagent to an intensely red-fluorescent dye which was analyzed using microplate reader (Perkin Elmer) at excitation 560, emission 590 nm.

3. Results and Discussion

3.1. Development of Microencapsulation.
The microcapsules were prepared by emulsion crosslink using variable types and concentrations of surfactants. We used Tween 80, Span 80, and Lecithin to investigate the effects of surfactants and while no surfactant preparation was used as the control. Microencapsulation using 2% Tween 80 (Figure 1(A)) led to a turbid calcium chloride solution from suspended microcapsules in the solution. In the absence of surfactant (Figure 1(B)) no microcapsules could be produced as the emulsion was not stable and thus the calcium chloride solution remained clear. The aqueous phase separated directly from emulsion when the stir is stopped due to the lack of surfactant and the non-homogenous hydrophilic and hydrophobic portions.

When Span 80 and Lecithin were used the results were the same compared to absence of surfactant and no formation of microcapsules (data not shown). The emulsion was more stable than without surfactant but despite increasing the concentrations of both types of surfactants we were not able to prepare the particles. Span 80 and Lecithin are low hydrophilic lipophilic balance (HLB) type surfactants that can produce stable water-in-oil (w/o) emulsion but after mixing with calcium chloride solution microcapsules cannot form as phase inversion cannot be induced. On the other hand, Tween 80 has a high HLB value and readily forms an oil-in-water (o/w) emulsion and inverted from a w/o emulsion when calcium chloride solution was mixed [8–10]. Phase inversion is an important mechanism to change the internal phase in the calcium chloride solution. In our study, when phase inversion occurred, sodium alginate was pushed out from the w/o emulsion to the calcium chloride solution. Calcium ions diffuse through the sodium alginate polymer and the cross-linking between the carboxylic groups of the polymer chains with calcium ions hardens the polymer shell and solidifies it to a microcapsule (Figure 2) [11–14].

The effects of different surfactant concentrations on microcapsule formation were studied. At 1% Tween 80 (Figure 1(A1)) microcapsules had a small size but the particles occurred rarely. This might be due to the fact that there are too few surfactants to surround the internal phase and thus particle preparation was similar to the instance when no surfactant was used. Using 2% and 3 % Tween 80 surfactant concentrations ((Figures 1(A2)-1(A3)) drop-shaped particles could be prepared without any size difference between the two concentrations. As the concentration of the surfactant increased surface tension was reduced and a stable emulsion could be produced. When the concentration of surfactant increased further to 5% (Figure 1(A4)) the particles became smaller with a more spherical shape compared to 2% and 3% Tween 80 at similar particle numbers. At higher surfactant concentrations, the surfactant will evenly surround the internal phase and reduce surface tension to maintain thermodynamic balance. As the aim in the present study was to encapsulate living cells and a high surfactant concentration is known to be toxic to encapsulated cells, the lowest effective concentration for particle preparation was chosen at 2% Tween 80 [6, 15].

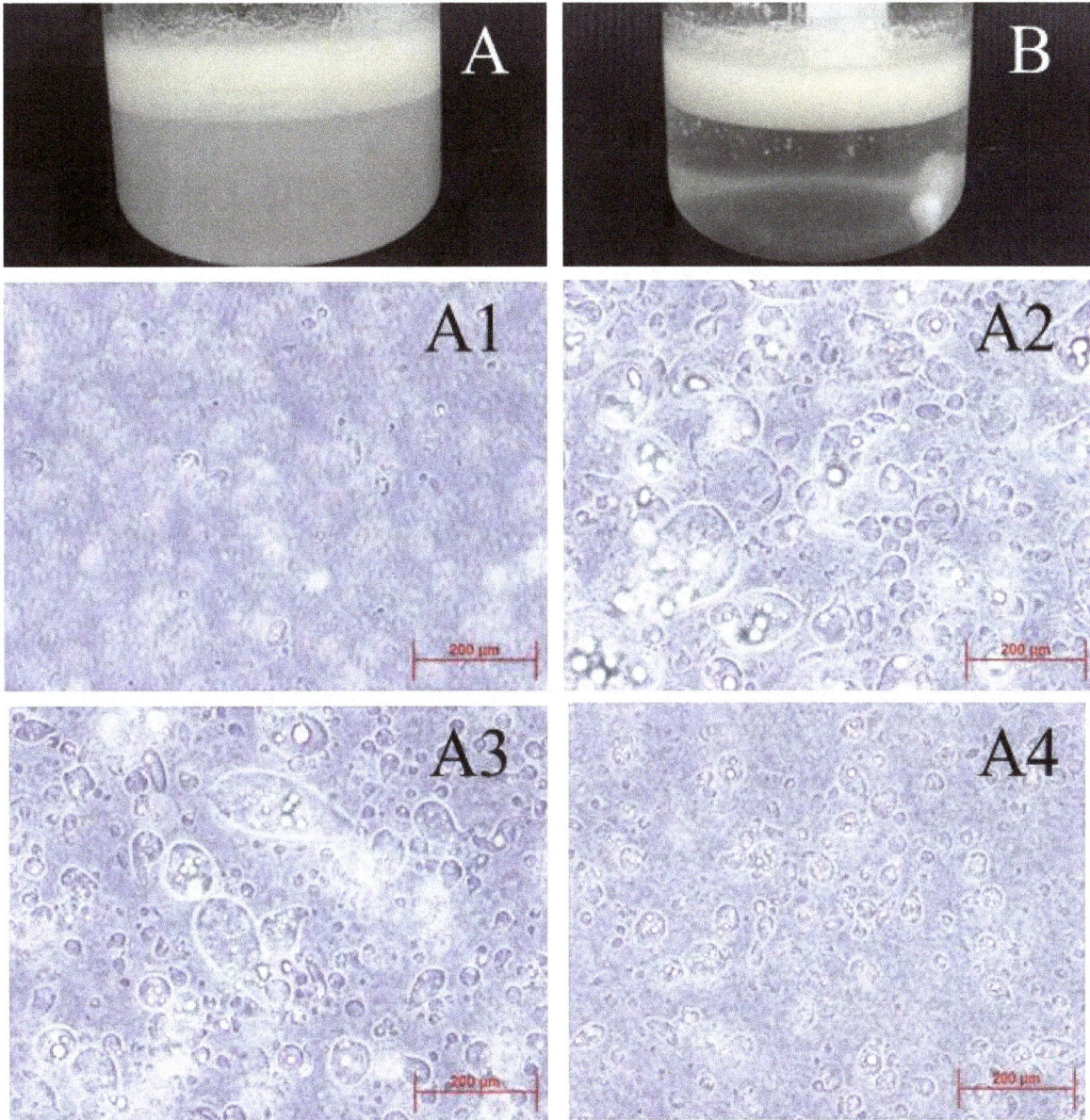

FIGURE 1: Microencapsulation. Prepared microcapsules with (A) Tween 80 and (B) without surfactant. The emulsion or oil layer is seen in the upper white opaque area while the calcium chloride solution is the clearer area under the oil layer. The picture shows the effect on microcapsule morphology under microscope of (A1) 1% Tween 80, (A2) 2% Tween 80 (A3) 3% Tween 80, and (A4) 5% Tween 80.

Oil and aqueous phase ratios, at 1:1 and 1:2, showed similar results and were both unable to form microcapsules gel layer formation around the emulsion was seen instead around microcapsules (Table 1). When we reduced the aqueous phase to 2:1 and 4:1, these ratios could prepare microcapsules. However, the ratio 2:1 resulted in a higher number of particles compared to the 4:1 ratio as there was more oil phase has a higher viscosity that cause hard to produce the particles and the particles are less precipitated. Microcapsule size and morphology were not different with particles drop shaped at both ratios and an average size of about 300 micrometers.

The percentage of entrapped living cells determined by the PrestoBlue assay was calculated at 52.4%.

Then we randomized the microcapsules and observed them at various dates of incubation confirming microcapsule stability for at least 30 days with unaltered morphology and statistically non-significant size difference (Figure 3).

Furthermore, we evaluated the stability of the microcapsules in a push pass test through a no. 18 needle and found no breakage of the microcapsules at microscopy observation (Figure 4). We noticed though that as the particles had a high size distribution (Poly Disperse Index: PDI), some were

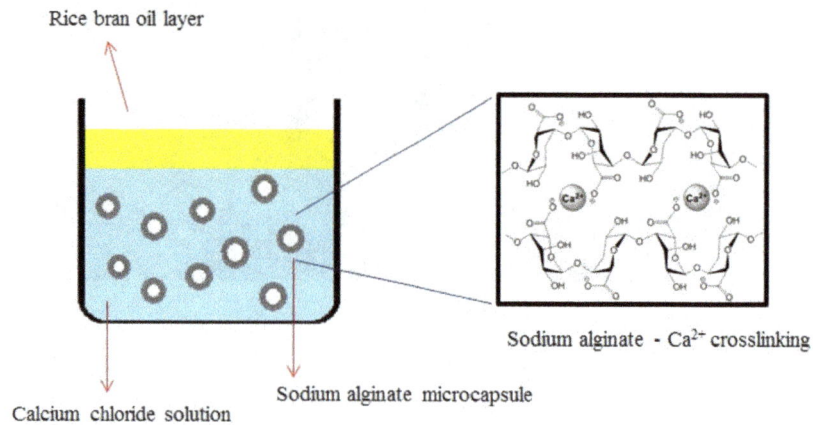

FIGURE 2: Sodium alginate crosslinking.

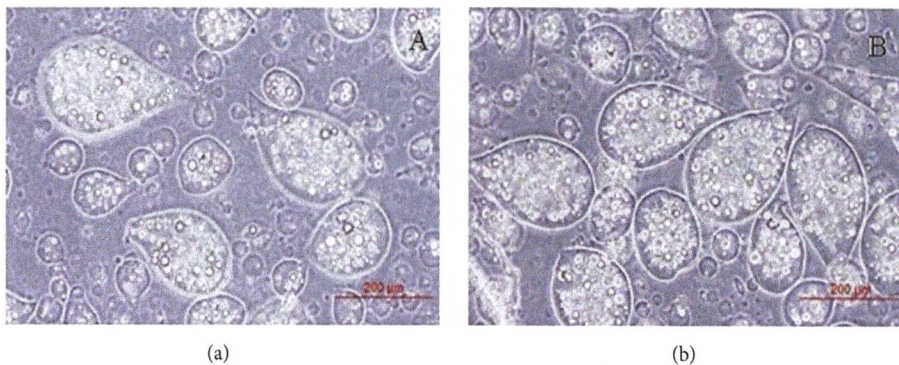

(a)   (b)

FIGURE 3: Microcapsule morphology under light microscope incubated in cell culture medium at 37°C on day 0 (a) and day 30 (b).

TABLE 1: Effect of oil and aqueous phase ratio on microcapsule formation.

Ratio (oil: aqueous)	Size (μm) $\pm$ SD	Poly disperse index (PDI)
1:2	No microcapsule formed	No microcapsule formed
1:1	No microcapsule formed	No microcapsule formed
2:1	328 $\pm$14.7	4.12
4:1	294 $\pm$15.3	3.94

too large to pass through. This observation warrants further studies to reduce the particle size distribution for a more convenient clinical use.

3.2. Cell Viability. Emulsion cross-linking encapsulated cells showed continuously decreasing viability that became statistically significant on day 7 (Figure 5(a)). On day 7 some broken microcapsules were encountered and a very small number of cells growing on the well plate could be seen (Figure 5(b)) that disappeared on the following day possibly due to cell damage caused by the microcapsule.

Cell viability continued to decline until day 30 when less than 20% of the encapsulated cells were still viable. A possible reason leading to cell death is a diffusion barrier created by the microcapsule, this being an important factor as cells in the microcapsule need to exchange nutrition, growth factor, and waste and impediment of these processes might be a reason for increased cell death [6, 7, 16–18]. Another potential reason for cell death could be the microcapsule being too densely populated thus waste accumulation along with a diffusion barrier could lead to an increased apoptosis rate [6, 19]. When comparing emulsion cross-linking microcapsule preparation to ionic gelation, the smaller cell proliferation area and longer preparation time of the former could lead to cell weakness and injury. Nevertheless, cell survival was still at 50% on day 14 but rapidly declined thereafter. Reducing particle size and improving particle quality, nutrient exchange and waste product control may increase cell viability and facilitate the use of this method in the clinic.

4. Conclusion

In conclusion, emulsion cross-linking was used to prepare microcapsules for human fibroblast encapsulation. Tween 80 surfactant was employed at a 2% concentration resulting

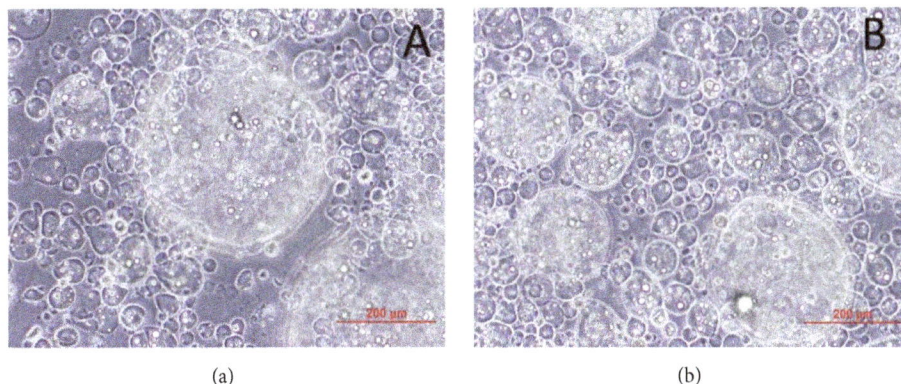

(a)   (b)

FIGURE 4: Encapsulated cells under microscope (a) before and (b) after no. 18 needle push pass test. No microcapsule breakage was observed after the test.

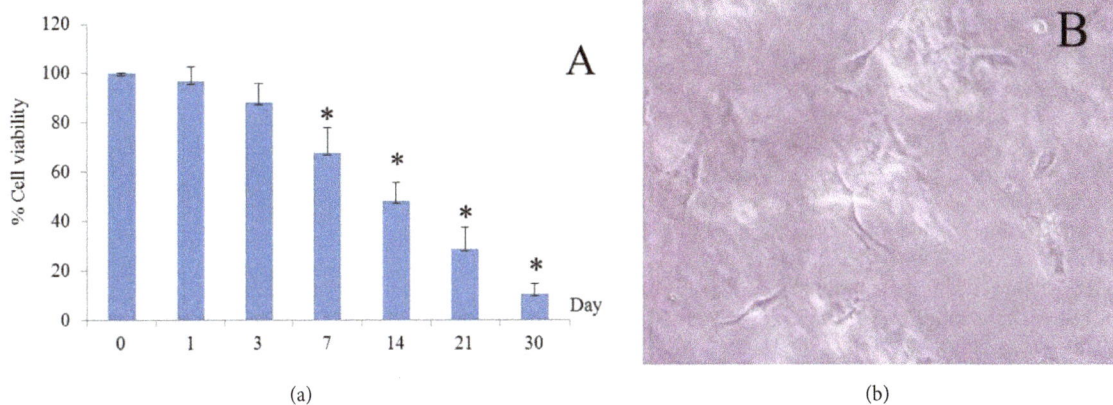

(a)   (b)

FIGURE 5: Cell viability.

from the phase inversion phenomenon in the most efficient and stable size, morphology, and particle numbers without being toxic to the encapsulated cells that showed at least 50% survival on day 14. Emulsion cross-linking microcapsule preparation results in smaller and possibly more diverse particles that can be developed clinically to deliver encapsulated mammalian cells for future disease treatments.

Conflicts of Interest

The authors declare the absence of any conflicts of interest.

Acknowledgments

The authors express their gratitude to the Faculty of Pharmaceutical Sciences, Chulalongkorn University, for providing the research fund (Grant no. Phar2560-RG03) to Dr. Phanphen Wattanaarsakit, Ph.D.

References

[1] B. P. Chan, T. Y. Hui, C. W. Yeung, J. Li, I. Mo, and G. C. F. Chan, "Self-assembled collagen-human mesenchymal stem cell microspheres for regenerative medicine," Biomaterials, vol. 28, no. 31, pp. 4652–4666, 2007.

[2] A. Moshaverinia, X. Xu, C. Chen, K. Akiyama, M. L. Snead, and S. Shi, "Dental mesenchymal stem cells encapsulated in an alginate hydrogel co-delivery microencapsulation system for cartilage regeneration," Acta Biomaterialia, vol. 9, no. 12, pp. 9343–9350, 2013.

[3] N. Wang, G. Adams, L. Buttery, F. H. Falcone, and S. Stolnik, "Alginate encapsulation technology supports embryonic stem cells differentiation into insulin-producing cells," Journal of Biotechnology, vol. 144, no. 4, pp. 304–312, 2009.

[4] X. Yuan, H. Zhang, Y.-J. Wei, and S.-S. Hu, "Embryonic stem cell transplantation for the treatment of myocardial infarction: Immune privilege or rejection," Transplant Immunology, vol. 18, no. 2, pp. 88–93, 2007.

[5] S. Benita, "Microencapsulation: methods and industrial applications," in Drugs and the pharmaceutical sciences, vol. 18, p. 756, Taylor & Francis, New York, NY, USA, 2nd edition, 2006.

[6] R. M. Hernández, G. Orive, A. Murua, and J. L. Pedraz, "Microcapsules and microcarriers for in situ cell delivery," Advanced Drug Delivery Reviews, vol. 62, no. 7-8, pp. 711–730, 2010.

[7] A. Murua, A. Portero, G. Orive, R. M. Hernández, M. de Castro, and J. L. Pedraz, "Cell microencapsulation technology: towards clinical application," Journal of Controlled Release, vol. 132, no. 2, pp. 76–83, 2008.

[8] D. Myers, *Surfactant Science and Technology*, John Wiley & Sons, Inc., Hoboken, NJ, USA, 3rd edition, 2005.

[9] M. Fanun, "Colloids in drug delivery," in *Surfactant science series*, vol. 26, p. 626, CRC Press/Taylor & Francis, Boca Raton, FL, USA, 2010.

[10] A. N. Martin, P. J. Sinko, and Y. Singh, *Martin's physical pharmacy and pharmaceutical sciences: physical chemical and biopharmaceutical principles in the pharmaceutical sciences*, vol. 8, Lippincott Williams & Wilkins, Baltimore, MD, USA, 6th edition, 2011.

[11] S. Cai, M. Zhao, Y. Fang, K. Nishinari, G. O. Phillips, and F. Jiang, "Microencapsulation of Lactobacillus acidophilus CGMCC1.2686 via emulsification/internal gelation of alginate using Ca-EDTA and CaCO3 as calcium sources," *Food Hydrocolloids*, vol. 39, pp. 295–300, 2014.

[12] Z.-G. Cui, K.-Z. Shi, Y.-Z. Cui, and B. P. Binks, "Double phase inversion of emulsions stabilized by a mixture of CaCO3 nanoparticles and sodium dodecyl sulphate," *Colloids and Surfaces A: Physicochemical and Engineering Aspects*, vol. 329, no. 1-2, pp. 67–74, 2008.

[13] P. M. Kruglyakov and A. V. Nushtayeva, "Phase inversion in emulsions stabilised by solid particles," *Advances in Colloid and Interface Science*, vol. 108-109, pp. 151–158, 2004.

[14] S. Mayer, J. Weiss, and D. J. McClements, "Vitamin E-enriched nanoemulsions formed by emulsion phase inversion: Factors influencing droplet size and stability," *Journal of Colloid and Interface Science*, vol. 402, pp. 122–130, 2013.

[15] H. Uludag, P. De Vos, and P. A. Tresco, "Technology of mammalian cell encapsulation," *Advanced Drug Delivery Reviews*, vol. 42, no. 1-2, pp. 29–64, 2000.

[16] W. J. Kao, "Evaluation of protein-modulated macrophage behavior on biomaterials: Designing biomimetic materials for cellular engineering," *Biomaterials*, vol. 20, no. 23-24, pp. 2213–2221, 1999.

[17] M. Endres, N. Wenda, H. Woehlecke et al., "Microencapsulation and chondrogenic differentiation of human mesenchymal progenitor cells from subchondral bone marrow in Ca-alginate for cell injection," *Acta Biomaterialia*, vol. 6, no. 2, pp. 436–444, 2010.

[18] T. Turajane, U. Chaveewanakorn, W. Fongsarun, J. Aojanepong, and K. I. Papadopoulos, "Avoidance of Total Knee Arthroplasty in Early Osteoarthritis of the Knee with Intra-Articular Implantation of Autologous Activated Peripheral Blood Stem Cells versus Hyaluronic Acid: A Randomized Controlled Trial with Differential Effects of Growth Factor Addition," *Stem Cells International*, vol. 2017, pp. 1–10, 2017.

[19] G. Orive, R. M. Hernández, A. Rodríguez Gascón et al., "History, challenges and perspectives of cell microencapsulation," *Trends in Biotechnology*, vol. 22, no. 2, pp. 87–92, 2004.

20

Influence of Vanadium 4+ and 5+ Ions on the Differentiation and Activation of Human Osteoclasts

Matthias A. König,[1] **Oliver P. Gautschi,**[2] **Hans-Peter Simmen,**[1]
Luis Filgueira,[3] **and Dieter Cadosch**[4]

[1]*Department of Traumatology, University Hospital Zurich, Zurich, Switzerland*
[2]*Département de Neurosciences Cliniques, Geneva University Hospital, Geneva, Switzerland*
[3]*School of Anatomy and Human Biology, University of Western Australia, Perth, WA, Australia*
[4]*Department of General and Trauma Surgery, Triemlispital, Zurich, Switzerland*

Correspondence should be addressed to Matthias A. König; matthias.a.koenig@gmail.com

Academic Editor: Rosalind Labow

Background. In the pathophysiology of implant failure, metal ions and inflammation-driven osteoclasts (OC) play a crucial role. The aim of this study was to investigate whether vanadium (V) ions induce differentiation of monocytic OC precursors into osteoresorptive multinucleated cells. In addition, the influence of V ions on the activation and function of *in vitro* generated OC was observed. *Methods.* Human monocytes and osteoclasts were isolated from peripheral blood monocytic cells (PBMCs). Exposition with increasing concentrations (0–$3\,\mu$M) of $V4^{+}/V5^{+}$ ions for 7 days followed. Assessment of OC differentiation, cell viability, and resorptional ability was performed by standard colorimetric cell viability assay 3-(4,5-dimethylthiazol-2-yl)-5-(3-carboxymethoxyphenyl)-2-(4-sulfophenil)-2H-tetrazolium (MTS), tartrate-resistant acid phosphatase (TRAP) expression, and functional resorption assays on bone slides during a period of 21 days. *Results.* No significant differences were noted between $V4^{+}/V5^{+}$ ions ($p > 0.05$). MTS showed significant reduction in cellular viability by V concentrations above $3\,\mu$M ($p < 0.05$). V concentrations above $0.5\,\mu$M showed negative effects on OC activation/differentiation. Higher V concentrations showed negative effects on resorptive function (all $p < 0.05$) without affecting cell viability. $V4^{+}/V5^{+}$ concentrations below $3\,\mu$M have negative effects on OC differentiation/function without affecting cell survival. *Conclusion.* Vanadium-containing implants may reduce implant failure rate by influencing osteoclast activity at the bone-implant interface. V-ligand complexes might offer new treatment options by accumulating in the bone.

1. Introduction

Metal-based implants have become essential and very successful treatment tools in dental and orthopedic trauma surgery. Implant failure rates are usually well below 5% for the first two years. The rates may increase up to 10–20% under local inflammatory conditions, including periodontitis and allergic reactions against the metal implant [1]. To date, surgical revision is the gold standard in treating implant failure. With an increasingly ageing population, implant failure with the consequent need of revision surgery will have a clinical and economic significance in the future.

Aseptic loosening (AL) is the leading cause of failure of total joint arthroplasty [2]. Aside from the well investigated and recognized role of wear debris (in the nanometer range) in the initiation and development of AL, over the last years, there has been increasing evidence that involvement of metal ions released by biocorrosion might influence AL by enhancing osteolysis and decreasing osseointegration [3, 4].

Various metal alloys used in metal implants contain vanadium (V), which gives the alloy favorable physicochemical and mechanical properties. Several *in vitro* and *in vivo* studies have investigated the bioactivity of V and different V compounds. It has been demonstrated that V compounds have an insulin and growth factor mimicking action [5, 6]. In addition, V compounds largely influence phosphatases, a group of enzymes mainly associated with the cell membrane [7]. Phosphatases, specifically the tartrate-resistant acid phosphatase (TRAP), play a paramount role in osteoclastic activities. Inhibiting TRAP might reduce the risk of AL. *In vivo* studies carried out to investigate the biodistribution of V compounds showed that V accumulates predominantly in the bone, kidney, spleen, and liver after 24 h of administration [8]. A recently published study has also shown that V compounds are able to regulate osteoblastic growth [9]. In particular, V ions released by biocorrosion from metal implants and accumulated in the bone could exert specific effects on bone turnover and more specifically on OC. Little is known about the potential biological effects of V ions on OC. In this study, we investigate the hypothesis that $V4^+$ and $V5^+$ ions may interfere with OC differentiation and activation using a well-established *in vitro* human OC model, as well as an *in vitro* bone resorption model.

2. Methods

2.1. Isolation of Peripheral Blood Monocytic Cells and Generation of Osteoclasts. The study protocol established by Lionetto et al. was used in the experiments [10]. Ficoll-gradient centrifugation (Amersham Biosciences, Uppsala, Sweden) was used to isolate pooled peripheral blood monocytic cells (PBMCs) from buffy coats of healthy blood donors. RPMI-1640 plus GlutaMAX™ medium (RPMI) (Gibco/Invitrogen, Auckland, New Zealand) supplemented with 5% human serum and 1% antibiotics (10,000 units/mL penicillin G sodium, 10,000 μg/mL streptomycin sulfate, and 25 μg/mL amphotericin B (Gibco standard medium)) was the culture medium for PBMCs at 37°C (humified, 5% CO_2, in 25 cm^2 tissue culture flasks (Sarstedt)). The nonadherent PBMCs were discarded and the adherent PBMCs were washed twice using 0.1 M phosphate-buffered saline (PBS, pH = 7.2), scraped off, and resuspended in standard medium after 1 h of culturing. These cells were used as monocytic cells (MCs) as described below.

To generate the OC, cell cultures of adherent PBMCs were supplemented with OC differentiation cytokines (10 ng/mL recombinant human macrophage-colony stimulating factor (hM-CSF) and 10 ng/mL recombinant human receptor activator of NF-κB ligand (hRANKL) (ReproTech, Rocky Hill, NJ, USA)).

2.2. Cell Culture Conditions. The MC, MC supplemented with OC differentiation cytokines, and mature OC were subsequently exposed to increasing concentrations (0 to 3 μM) of bis(maltolato)oxovanadium(IV) (BMOV) $V4^+$ or oxidized bis(maltolato)oxovanadium(V) (oxBMOV) $V5^+$ ions for 7 days. Cell viability was assessed using 3-(4,5-dimethylthiazol-2-yl)-5-(3-carboxymethoxyphenyl)-2-(4-sulfophenyl)-2H-tetrazolium (MTS) colorimetric assay (Promega, Madison, WI, USA). The assays were repeated four times for each experimental condition, and the mean optical light density was recorded using a microplate reader (Labsystems, Helsinki, Finland) at an absorbance of 492 nm.

2.3. Detection of TRAP Using ELF97 and Flow Cytometry. The endogenous tartrate-resistant acid phosphatase (TRAP) activity was detected by using the phosphatase substrate ELF97 (Molecular Probes, Eugene, OR, USA) as described previously [11], to assess OC differentiation in the presence of the different concentrations of $V4^+$ or $V5^+$ ions. Cell fixation was performed after 7 days, using a culture medium containing 1% paraformaldehyde. After being washed twice with distilled water, the fixed cells were subsequently incubated with 200 μM ELF97 in a 110 mM acetate buffer staining solution for a period of 20 min. Cells were characterized for expression of surface markers by using fluorescence-labeled mouse monoclonal antibodies binding specifically to human HLA-DR (phycoerythrin (PE) fluorescent labeled; Becton Dickinson Biosciences, San Jose, CA, USA) and CD45 (PerCP fluorescent labeled; Becton Dickinson Biosciences). To account for the background signal and autofluorescence, isotype antibodies and unstained controls were used.

A FACS Vantage Cell Sorter with a UV laser was used for sample analysis. Approximately 10^5 gated cells were analyzed. The ELF97 was excited at 350 nm, and the TRAP-related signal was collected using a 530/30 band pass filter (515–545 nm). The PE and PerCP fluorochromes were excited using a 488 nm laser. The signal was collected using a 575/26 and a 675/20 band pass filter. The data was further analyzed using the Flowjo v8.5.3 software package (Treestar, Ashland, OR, USA).

2.4. Osteoclast Resorptive Function on the Dentine Slices. Evidence of OC function and bone resorptive activity was assessed using lacunar resorption assays on the dentine slices, as previously described [3]. Whale dentine slices (diameter 15 mm, thickness 0.5 mm) were cut from a sperm whale tooth, purchased from Kaempf (Osborne Park, Western Australia, collected prior to 1972, before the ban on whale hunting in Australia). Adherent cells (4×10^5 cells/well) were cultured on dentine slices. The slices were washed with water and left overnight in a 0.25% ammonium hydroxide solution after a culturing period of 21 days to remove all cellular material. Staining was performed with 1% toluidine blue dye in 1% sodium borate for 10 min. A Nikon Inverted Microscope Eclipse TE 300 (Nikon Instruments, Melville, NY, USA) and a Nikon CCD digital camera were used for documentation. The whole surface of the slice was examined. The DXM 1200F-ACT-1 image processing software (Nikon) was used to analyze the images. A dark-blue excavation on the dentine surface with a clear rim of unchanged and unstained original surface located between neighboring pits was defined as a resorption pit. The extent of resorption was determined from the number of pits formed and by calculating the total planar surface area of the resorption pits. The diameter of 30 pits/slice was measured and the mean value used to calculate the surface of the number of pits counted. The number of pits was counted using a single blinded observer.

2.5. Calculations and Statistical Analysis. The data was analyzed using the SPSS for Windows software package (v15.0; SPSS Inc., Chicago, IL). Independent sample *t*-tests were conducted to determine whether the mean ion concentrations

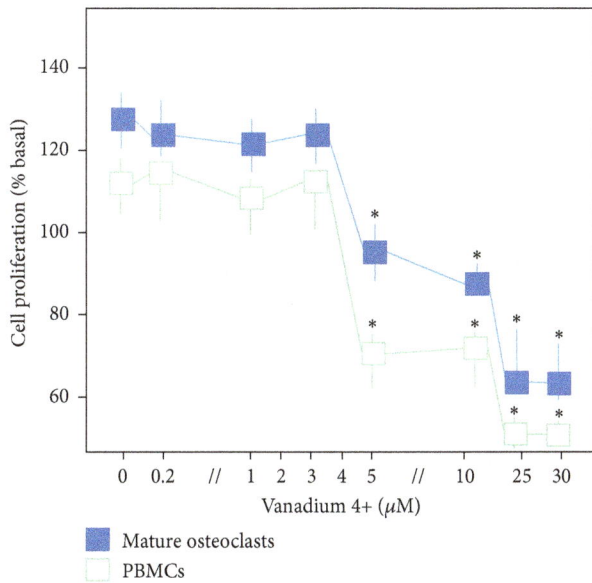

FIGURE 1: Assessment of cell viability. Cell proliferation (percentage of basal proliferation) of monocytic cells and osteoclasts under increasing concentration of vanadium 4^+ ions. No differences were noted in the same concentrations of vanadium 5^+ ions ($p > 0.05$). $^*p < 0.001$.

were significantly greater than the minimum detection limits. A one-way ANOVA test was used to test the differences in the mean in the expression of the TRAP, as well as differences in the mean number of resorption pits and the surface area across the various cell cultures. A p value of $p < 0.05$ was considered to be statistically significant.

3. Results

3.1. Cell Viability. Nontoxic concentrations of $V4^+$ and $V5^+$ ions that did not greatly decrease the MC and OC viability were defined by MTS colorimetric assays. A significant reduction in MC and *in vitro* generated OC was seen at concentrations of $V4^+$ or $V5^+$ ions greater than $3 \mu M$. No significant change in MC and OC viability was found when using concentrations up to $3 \mu M$, indicating the toxic concentrations for MC and OC (Figure 1). Toxicity of V4+ and V5+ concentrations above $3 \mu M$ was additionally confirmed by Annexin-5 staining. No significant difference was found between $V4^+$ and $V5^+$ ions (data not shown) ($p > 0.05$). Subsequently, increasing concentrations (0 to $3 \mu M$ of $V4^+$ or $V5^+$) were defined as standard conditions for the experiments.

3.2. Detection of TRAP Using ELF97 and Flow Cytometry (FACS). All OC cultures were TRAP-positive while all the MC cultures supplemented with OC differentiation cytokines showed a significantly reduced expression of TRAP, indicating OC differentiation inhibition in the presence of $V4^+$ or $V5^+$ for all the concentrations (Figure 2). All cultures of MC exposed only to $V4^+$ or $V5^+$ (no supplementation with OC differentiation cytokines) showed no expression of TRAP.

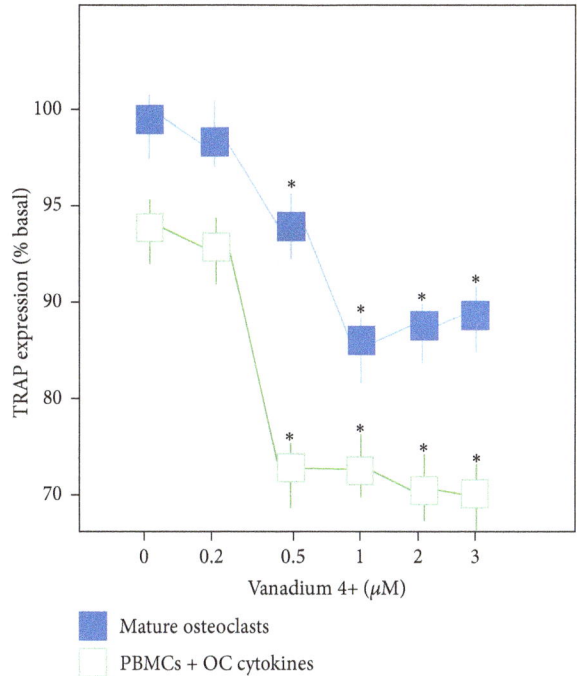

FIGURE 2: Quantitative analysis of TRAP expression by osteoclasts and monocytes supplemented with osteoclast differentiation cytokines in the presence of increasing concentrations (0 to $3 \mu M$) of $V4^+$ ions. Mean fluorescence intensity (MFI) indicates the average TRAP expression per cell. The MFI value indicates relative TRAP expression when compared with the control (MFI = 121). No differences were noted between the same concentrations of $V4^+$ and $V5^+$ ions ($p > 0.05$) ($^*p < 0.05$).

Quantitative FACS analysis showed that TRAP expression was significantly decreased in the OC cultures exposed to increasing concentrations of $V4^+$ or $V5^+$ ions compared with OC cultured without any V ions ($p < 0.05$) (Figure 2).

3.3. Qualitative and Quantitative Assessment of Osteoclastic Resorptive Function on the Dentine Slices. The resorptive function was assessed on the dentin slides after the discovery that $V4^+$ and $V5^+$ ions had an essential effect on OC differentiation for the selected concentrations.

OC were cultured with and without $V4^+$ or $V5^+$ ions on dentin slides for a period of 21 days. The extent of lacunar resorption was analyzed (Figure 3). Between the different conditions, qualitative and quantitative differences in the patterns of absorption were seen. Compared to the control samples, exposition to increased V4+ or V5+ ions led to smaller sized resorption pits, while the resorption pits of untreated OC were larger. Osteoclasts incubated with the different concentrations of $V4^+$ or $V5^+$ ions showed a significantly decreased number of resorption pits and resorption area of the total dentine surface compared to untreated OC (Figure 4). Monocytes, as a negative control, showed only very few resorption pits (22 ± 3.9 pits/mm^2) with small dentine area resorption (0.28%, Figure 4). Untreated OC showed an average of 801 ± 13.2 resorption pits/mm^2 with a dentine resorption area of 10.19% ($p < 0.001$). Regarding the resorption

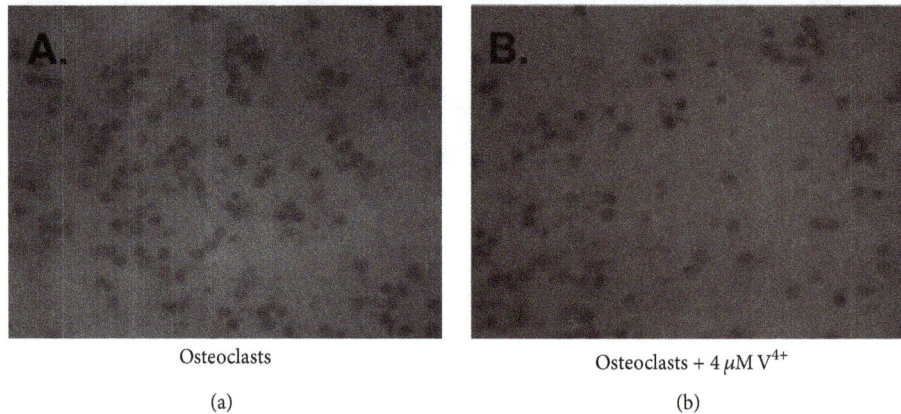

Figure 3: Assessment of bone resorption. The images show a representative example of resorption pits (dark spots) on dentine slides after 21 days of incubation under different culture conditions. (a) Untreated osteoclasts. (b) Osteoclasts + 4 μM vanadium 4^+ ions.

area, no difference was detected between $V4^+$ and $V5^+$ ions ($p > 0.05$). Resorption features were not seen on the dentine slices incubated without cells.

4. Discussion

The hypothesis that $V4^+$ and $V5^+$ ions are able to interfere with OC differentiation and bone resorptive activity was clearly supported in this study. The hypothesis was based on previous data showing the strong effects of different metal ions on bone metabolism as well as the tendency of V ions/compounds to accumulate in bone cells [3, 4, 8].

The integrity of the skeleton as well as the osseous integration of intercortical metal implants (e.g., total joint arthroplasty) requires the regulated activity of bone-forming cells (osteoblasts) and bone-resorbing cells (OC). Numerous growth factors and cytokines with pro- and antiapoptotic effects are produced in the bone, regulating the balance between osteoformation and resorption [9, 11, 12]. Thus, imbalanced bone cell activity results in severe bone alterations and tissue loss leading in the worst cases to implant loosening without osseointegration. The term "osseointegration" is defined as incorporation of nonvital components in a predictive and reliable way into vital bone. This anchorage mechanism is specifically important if cementless implants are used [13, 14]. In addition, altered TRAP levels in bone pathologies like osteoporosis might compromise osseointegration [15].

Metal ions have been shown to affect bone turnover through a variety of both direct and indirect mechanisms. It has been already demonstrated that $Ti4^+$ ions act directly on bone cells by enhancing OC differentiation [3]. In line with this, $Co2^+$ and $Cr3^+$ ions affected human osteoblasts by inhibiting the release of alkaline phosphatase, enhancing apoptosis, RANKL, and osteoprotegerin (OPG) expression, with enhanced osteoclastogenesis and osteolysis [16, 17]. Furthermore, metal ions induce elevated proinflammatory cytokine secretion such as interleukin- (IL-) $1\alpha/\beta$, IL-6, and tumor necrosis factor- (TNF-) α [4, 18]. TNF-α has a direct influence on OC precursors, whereas IL-6 and IL-$1\alpha/\beta$ act indirectly by

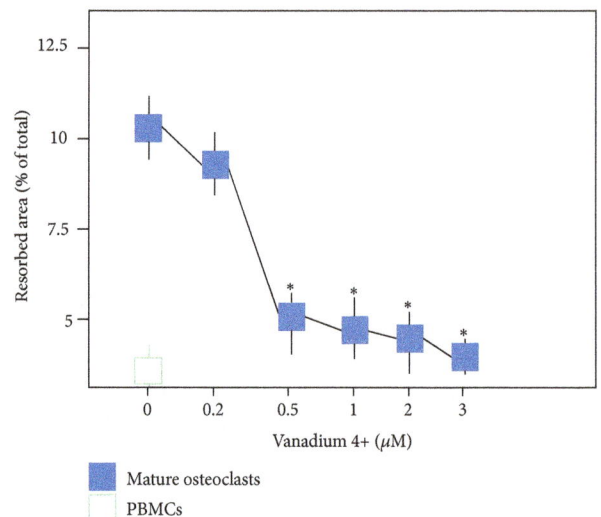

Figure 4: Histogram representing the mean area (percentage of total area $\pm$ SE) of resorption on dentine slides after 21 days in culture under different conditions. Note the significant decrease of dentin resorption by osteoclasts exposed to $V4^+$ ions. No differences were noted between the same concentrations of $V4^+$ and $V5^+$ ions ($p > 0.05$) ($^*p > 0.05$).

increasing osteoblast RANKL and M-CSF expression, which directly drives osteoclastogenesis [4]. Several studies have reported the bioactivity of vanadate, vanadyl, and different $V4^+$ complexes on bone metabolism [19–22]. Low concentrations ($<25\,\mu$M) of $V4^+$ compounds influence the proliferation and differentiation of osteoblast-like cells in culture, while higher concentrations have inhibitory and cytotoxic effects on osteoblasts. V compounds stimulate the protein tyrosine phosphorylation through the inhibition of protein tyrosine phosphatases (PTPases) [23]. The presented results confirm the cytotoxicity of high concentrations ($>3\,\mu$M) of $V4^+$ and $V5^+$ ions on MC and OC in culture. This study demonstrates the inhibitory action of $V4^+$ and $V5^+$ ions at concentration $< 3\,\mu$M on OC differentiation and activity in vitro. This is based on several observations: (a) significantly decreased expression of TRAP in cultures of MC supplemented with OC differentiation cytokines and exposed to V4$^+$ or V5$^+$ ions, (b) significantly decreased TRAP expression in cell cultures of mature and functional OC exposed to V4$^+$ or V5$^+$ ions, and (c) significantly reduced bone resorption by OC exposed to V4$^+$ or V5$^+$.

5. Conclusion

V-containing implants may reduce OC activity at the bone-implant interface and thereby decrease bone resorption and ultimately the rate of implant failure. Primary osseointegration is supported by suppressing local TRAP activity. Additionally, V complex development with different ligands might be an alternative strategy of use in skeletal tissue engineering especially when considering their tendency to accumulate in bone.

Abbreviations

AL: Aseptic loosening
ARCBS: Australian Red Cross Blood Service
Co: Cobalt
Cr: Chrome
MC: Monocytic cells
MTS: 3-(4,5-Dimethylthiazol-2-yl)-5-(3-carboxymethoxyphenil)-2-(4-sulfophenil)-2H-tetrazolium
μM: Micromol
nm: Nanometer
OC: Osteoclasts
PBMC: Peripheral blood monocytic cell
PE: Phycoerythrin
TRAP: Tartrate-resistant acid phosphatase
V: Vanadium
BMOV: Bis(maltolato)oxovanadium(IV)
oxBMOV: Oxidized bis(maltolato)oxovanadium(V).

Disclosure

The results of this study were presented at the 100th Annual Congress of the Swiss Society of Surgery.

Conflicts of Interest

All the authors confirm that there are no conflicts of interest.

Acknowledgments

This study was supported by the International Team for Implantology (ITI) Center, Peter Merian-Weg 10, Basel, Switzerland.

References

[1] W. H. Harris, "Osteolysis and particle disease in hip replacement: A review," *Acta Orthopaedica*, vol. 65, no. 1, pp. 113–123, 1994.

[2] D. Cadosch, E. Chan, O. P. Gautschi, and L. Filgueira, "Metal is not inert: role of metal ions released by biocorrosion in aseptic loosening—current concepts," *Journal of Biomedical Materials Research A*, vol. 91, no. 4, pp. 1252–1262, 2009.

[3] D. Cadosch, E. Chan, O. P. Gautschi, J. Meagher, R. Zellweger, and L. Filgueira, "Titanium IV ions induced human osteoclast differentiation and enhanced bone resorption in vitro," *Journal of Biomedical Materials Research - Part A*, vol. 91, no. 1, pp. 29–36, 2009.

[4] D. Cadosch, M. Sutanto, E. Chan et al., "Titanium uptake, induction of RANK-L expression, and enhanced proliferation of human T-lymphocytes," *Journal of Orthopaedic Research*, vol. 28, no. 3, pp. 341–347, 2010.

[5] A. K. Srivastava and M. Z. Mehdi, "Insulino-mimetic and anti-diabetic effects of vanadium compounds," *Diabetic Medicine*, vol. 22, no. 1, pp. 2–13, 2005.

[6] K. Thompson and C. Orvig, "Vanadium compounds in the treatment of diabetes," *Metal Ions in Biological Systems*, vol. 42, pp. 221–252, 2004.

[7] D. C. Crans, "Antidiabetic, Chemical, and Physical Properties of Organic Vanadates as Presumed Transition-State Inhibitors for Phosphatases," *Journal of Organic Chemistry*, vol. 80, no. 24, pp. 11899–11915, 2015.

[8] A. M. Cortizo, M. S. Molinuevo, D. A. Barrio, and L. Bruzzone, "Osteogenic activity of vanadyl(IV)-ascorbate complex: evaluation of its mechanism of action," *International Journal of Biochemistry and Cell Biology*, vol. 38, no. 7, pp. 1171–1180, 2006.

[9] R. S. Tuan, "Biology of developmental and regenerative skeletogenesis," *Clinical Orthopaedics and Related Research*, vol. 427, pp. S105–S117, 2004.

[10] S. Lionetto, A. Little, G. Moriceau et al., "Pharmacological blocking of the osteoclastic biocorrosion of surgical stainless steel in vitro," *Journal of Biomedical Materials Research - Part A*, vol. 101, no. 4, pp. 991–997, 2013.

[11] S. C. Manolagas, "Birth and death of bone cells: basic regulatory mechanisms and implications for the pathogenesis and treatment of osteoporosis," *Endocrine Reviews*, vol. 21, no. 2, pp. 115–137, 2000.

[12] S. Kousteni, J.-R. Chen, T. Bellido et al., "Reversal of bone loss in mice by nongenotropic signaling of sex steroids," *Science*, vol. 298, no. 5594, pp. 843–846, 2002.

[13] R. Brånemark, P-I. Brånemark, B. Rydevik, and R. R. Myers, "Osseointegration in skeletal reconstruction and rehabilitation," *Journal of Rehabilitation Research & Development*, vol. 38, pp. 1–4, 2001.

[14] G. Piarulli, A. Rossi, and G. Zatti, "Osseointegration in the elderly," *Aging Clinical and Experimental Research*, vol. 25, no. S1, pp. 59-60, 2013.

[15] G. Resmini, S. Migliaccio, L. D. Carbonare et al., "Differential characteristics of bone quality and bone turnover biochemical markers in patients with hip fragility fractures and hip osteoarthritis: Results of a clinical pilot study," *Aging Clinical and Experimental Research*, vol. 23, no. 2, pp. 99–105, 2011.

[16] M. Dai, T. Zhou, H. Xiong, W. Zou, P. Zhan, and W. Fu, "Effect of metal ions Co2+ and Cr3+ on osteoblast apoptosis, cell cycle distribution, and secretion of alkaline phosphatase," *Journal of Reparative and Reconstructive Surgery*, vol. 25, no. 1, pp. 56–60, 2011 (Chinese).

[17] M. Dai, X. Yuan, H. Fan, M. Cheng, and J. Ai, "Expression of receptor activator of nuclear factor kappaB ligand and

osteoprotegerin of mice osteoblast induced by metal ions," *Journal of Reparative and Reconstructive Surgery*, vol. 24, no. 3, pp. 292–295, 2010 (Chinese).

[18] J. Y. Wang, B. H. Wicklund, R. B. Gustilo, and D. T. Tsukayama, "Titanium, chromium and cobalt ions modulate the release of bone-associated cytokines by human monocytes/macrophages in vitro," *Biomaterials*, vol. 17, no. 23, pp. 2233–2240, 1996.

[19] D. Barrio, M. Braziunas, S. Etcheverry, and A. Cortizo, "Maltol complexes of vanadium (IV) and (V) Regulate in vitro alkaline phosphatase activity and osteoblast-like cell growth," *Journal of Trace Elements in Medicine and Biology*, vol. 11, no. 2, pp. 110–115, 1997.

[20] S. B. Etcheverry, D. A. Barrio, A. M. Cortizo, and P. A. M. Williams, "Three new vanadyl(IV) complexes with non-steroidal anti-inflammatory drugs (Ibuprofen, Naproxen and Tolmetin). Bioactivity on osteoblast-like cells in culture," *Journal of Inorganic Biochemistry*, vol. 88, no. 1, pp. 94–100, 2002.

[21] S. B. Etcheverry, P. A. M. Williams, V. C. Sálice, D. A. Barrio, E. G. Ferrer, and A. M. Cortizo, "Biochemical properties and mechanism of action of a vanadyl(IV) - Aspirin complex on bone cell lines in culture," *BioMetals*, vol. 15, no. 1, pp. 37–49, 2002.

[22] D. A. Barrio, P. A. M. Williams, A. M. Cortizo, and S. B. Etcheverry, "Synthesis of a new vanadyl(IV) complex with trehalose (TreVO): insulin-mimetic activities in osteoblast-like cells in culture," *Journal of Biological Inorganic Chemistry*, pp. 459-68, 2003.

[23] A. S. Tracey and M. J. Gresser, "Interaction of vanadate with phenol and tyrosine: implications for the effects of vanadate on systems regulated by tyrosine phosphorylation.," *Proceedings of the National Academy of Sciences*, vol. 83, no. 3, pp. 609–613, 1986.

21

Enhancement of the Mechanical Properties of Hydroxyapatite/Sulphonated Poly Ether Ether Ketone Treated Layer for Orthopaedic and Dental Implant Application

Roohollah Sharifi ⓘ,[1] Davood Almasi ⓘ,[2] Izman Bin Sudin ⓘ,[3]
Mohammed Rafiq Abdul Kadir ⓘ,[4] Ladan Jamshidy,[5] Seyed Mojtaba Amiri,[6]
Hamid Reza Mozaffari ⓘ,[7,8] Maliheh Sadeghi ⓘ,[4]
Fatemeh Roozbahani ⓘ,[4] and Nida Iqbal[9]

[1] Department of Endodontics, School of Dentistry, Kermanshah University of Medical Sciences, Kermanshah, Iran
[2] School of Dentistry, Kermanshah University of Medical Sciences, Kermanshah, Iran
[3] Department of Manufacturing and Industrial Engineering, Faculty of Mechanical Engineering,
  Universiti Teknologi Malaysia, 81310 Skudai, Johor, Malaysia
[4] Medical Devices & Technology Group (MEDITEG), Faculty of Biosciences and Medical Engineering,
  Universiti Teknologi Malaysia, 81310 Johor Bahru, Johor, Malaysia
[5] Department of Prosthodontics, School of Dentistry, Kermanshah University of Medical Sciences, Kermanshah, Iran
[6] Department of Biostatistics and Epidemiology, School of Health, Kermanshah University of Medical Sciences, Kermanshah, Iran
[7] Department of Oral and Maxillofacial Medicine, School of Dentistry, Kermanshah University of Medical Sciences, Kermanshah, Iran
[8] Medical Biology Research Center, Kermanshah University of Medical Sciences, Kermanshah, Iran
[9] Bio-Medical Engineering Center, University of Engineering and Technology (UET), Lahore Kala Shah Kaku (KSK) Campus, Pakistan

Correspondence should be addressed to Hamid Reza Mozaffari; mozaffari@kums.ac.ir

Academic Editor: Vijaya Kumar Rangari

The mechanical properties of coated layers are one of the important factors for the long-term success of orthopeadic and dental implants. In this study, the mechanical properties of the porous coated layer were examined via scratch and nanoindentation tests. The effect of compression load on the porous coated layer of sulphonated poly ether ether ketone/Hydroxyapatite was studied to determine whether it changes its mechanical properties. The water contact angle and surface roughness of the compressed coated layer were also measured. The results showed a significant increase in elastic modulus, with mean values ranging from 0.464 GPa to 1.199 GPa (p<0.05). The average scratch hardness also increased significantly from 69.9 MPa to 95.7 MPa after compression, but the surface roughness and wettability decreased significantly (p<0.05). Simple compression enhanced the mechanical properties of the sulphonated poly ether ether ketone/hydroxyapatite coated layer, and the desired mechanical properties for orthopaedic and dental implant application can be achieved.

1. Introduction

Success in orthopaedic and dental implant depends on several parameters that may be improved by considering both biologic and mechanical criteria [1]. The use of synthetic polymers and composites for biomaterial applications has continued to expand. Fiber-reinforced polymers offer advantages because they can be designed to match tissue properties, can be anisotropic with respect to mechanical properties, can be coated for attachment tissues, and can be fabricated at relatively low cost. Expanded future applications for orthopaedic and dental implant systems are anticipated as interest in combination synthetic and biological composites increases. The more inert polymeric biomaterials include polytetrafluoroethylene (PTFE), polyethylene terephthalate (PET), polymethylmethacrylat (PMMA), ultra-high molecular weight polyethylene (UHMW-PE), polypropylene (PP), polysulfone (PSF), and poly ether ether ketone (PEEK). These are thermal and electrical insulators, and when constituted as a high molecular weight system without plasticizers, they are relatively resistant to biodegradation [2]. The low elasticity modulus, excellent chemical stability, transparency to radio waves, and compatibility with reinforcing agents (such as carbon fiber) make poly ether ether ketone (PEEK) an ideal choice for medical applications, such as dental implants [3–5]. Despite these excellent properties, PEEK is still categorized as bioinert due to its very low reaction with the surrounding bone tissue [6]. There are several methods to improve the bioactivity of PEEK, such as selective wet-chemistry [7], grafting [8, 9], and hydroxyapatite (HA) coating [10]. In our previous studies, the bioactivity of PEEK was increased via sulphonation and deposition of HA crystalline particles on the sulphonated layer [11, 12]. The main advantage of our new method was the fact that the deposition process took place at room temperature, which caused no damage to the heat-sensitive PEEK. Our new method consists of two steps including sulphonation for activating the surface of PEEK and deposition of HA particles on the activated sulphonated PEEK (SPEEK) layer. The diffusion of sulphuric acid in the PEEK caused a porous coated layer [11]. The porous coated layer is a potential for bone interlocking; however, the modulus is expected to be relatively low for load-bearing applications such as orthopaedic and dental applications.

An adequate elastic modulus of the coating layer is important as it affects load distribution and stress shielding at the interface layer between a coated implant and the surrounding bone [13]. Also, the success of a particular implant *in vivo* depends on adequate adhesion of the coating layer [14]. Some standards must be considered for the HA coatings on medical implants, such as the thickness, porosity, roughness, pore size, and adhesion of the coating layer, to ensure the quality of their performance [15, 16]. However, the coated layer in our new type of coated layer consists of SPEEK and HA, and the existing standards are not suitable to evaluate its performance qualification.

In our previous study, we studied the feasibility of changing the mechanical property of the porous coated layer of SPEEK/HA via compression [17]. In this study, mechanical and bioactivity properties of the porous coated layer of SPEEK/HA were examined at different sulphonation times and enhanced via compression using a hydraulic press to ensure that it qualified for orthopaedic and dental implant application. Microscratch and nanoindentation tests were conducted to evaluate the mechanical properties of the coated layer. The water contact angle and surface roughness were measured. The wettability and surface roughness of the coated layer were also examined to determine the level of bioactivity of a particular material.

2. Materials and Methods

The PEEK substrates (Optima® Invibio) discs were ground by 400-grit silicon carbide paper. The PEEK samples were immersed in concentrated sulphuric acid (95-97%) for three different durations, 3, 5, and 10 min, in ambient temperature. The samples were then immersed with distilled water at room temperature until no traces of acid were obtained, and the SPEEK samples were left to dry at room temperature overnight. The SPEEK disc samples were then immersed in a 10% wt/v suspension of hydroxyapatite 21223 (Sigma Aldrich, USA) in water for 5 h and continuously stirred via a magnetic stirrer. After 5 h, the samples were removed, washed, and ultrasonically cleaned with deionized water for 10 min to remove any excess HA particles that were not chemically connected to the samples. The samples were then dried at room temperature overnight [11]. A compressive load of 15 MPa, equal to the ultimate compressive strength of cancellous bone [18], was applied to the surface of the coated layer via a hydraulic press for 10 minutes.

2.1. Nanoindentation Test. The Hysitron TI 750D Ubi nanomechanical test system with a Berkovich indenter tip was used for the nanoindentation test. Three nanoindentation tests were conducted per sample. Based on Buckle's one-tenth rule for the evaluation of the mechanical properties of a coating layer, the maximum indentation depth must be less than one-tenth of the thickness of the coating layer to prevent the effect of the substrate in the resultant force curve [19]. The maximum applied load of 200 μN was chosen based on a preliminary experiment considering the one-tenth indentation rule [19]. The loading rate was chosen as 0.5 μN/s, and the holding time was 300 s.

The Oliver-Pharr model was used to calculate Young's modulus of the coated layer from the indentation force curve. In nanoindentation studies of polymeric materials, the elastic properties of the indenter can be ignored due to the large differences in the elastic modulus between the tip and the sample [20]. The following Oliver-Pharr equations were based on this assumption [21]. For the Berkovich indenter, which was used in the nanomechanical test system, the projected contact area (A_c) was obtained from the contact depth (δ_c) via [22]

$$A_c = 24.5\delta_c^2 \tag{1}$$

The contact depth (δ_c) was obtained from (2) at the peak load (δ_{max}), where the stiffness (S) was calculated by measuring the slope of the unloading part of the force curve at δ_{max}.

$$\delta_c = \delta_{max} - \varepsilon\frac{F_{max}}{S}, \tag{2}$$

where F_{max} is the load at the maximum indentation depth. The geometric constant of ε is 0.75 for the Berkovich indenter tip [23]. Finally, to calculate Young's modulus of the coated layer, (3) was utilized:

$$E = \frac{S\sqrt{\pi}}{2\sqrt{A_c}} \tag{3}$$

2.2. Scratch Test. The progressive scratch test was carried out using microscratch test equipment from Micro Materials, Ltd.

One scratch test was made per sample. The stylus speed was set at 2 μm/s, with a chosen scratch length of 600 μm. Normal load was not applied to the stylus for the first 60 μm. The load was then increased linearly from 0 to 500 mN, with a loading rate of 2 mN/s between 60 and 560 μm scratch length and remained constant in the last 40 μm of the scratch length. A conical spherical Rockwell stylus with a radius of 25 μm and a conical angle of 90° was used in this test. The scratch tracks were analyzed using optical microscope images during which the scratch width was measured and the coating failure point determined. The point at which the stylus reached the substrate was determined through manual observation from the optical microscope images by the changes in color from white (coating layer) to beige (substrate).

Based on the literature, there are more than 250 methods available to determine the adhesion between a coating and the coated layer [24], of which the scratch test is the most effective technique [25, 26]. The test consists of applying a continuously increasing load on the coating surface by a stylus scratching point, while the sample is displaced at a constant speed. The scratching point causes increasing elastic and plastic deformation until damage occurs in the surface region.

Based on the American Society for Testing Materials (ASTM) [27], scratch hardness (H_s) is defined as "the normal load of the stylus over the load-bearing area". To calculate the scratch hardness, (4) was used [27]:

$$H_s = \frac{4Fq}{\pi w^2},\qquad(4)$$

where F was the normal load (in Newtons), w was the width of the scratch (in millimeters), and q was a dimensionless parameter which varied between one (for full elastic recovery of the sample) and two (for samples with no recovery). In this study, the q parameter was assumed to be 2 because of plastic deformation of the sample during the scratch test [27].

For the graphical determination of the scratch hardness, the plot of F-$(\pi w^2/4)$ for the sample was drawn, and the slope of the linear fit to the graph gives the scratch hardness of the material [27]. To calculate the scratch hardness of the coated layer, the data of the scratch test for the scratch distance before the critical point must be used.

Nanoscale morphology of the coated layer before and after compression and the morphology of the scratches were probed using a scanning electron microscope (SEM) (Hitachi Tabletop, TM-3000).

An atomic force microscope (AFM) (SPA-300 HV, Seiko) was used to analyze the surface roughness. The AFM was run in the force-curve mode, and a scan size of 5 μm × 5 μm was used to calculate the arithmetic mean of the surface roughness (Ra). Thirty lines, each with a length of 3 μm, were used to calculate Ra [11].

The Sessile method was used to measure the water contact angle of the surface of the modified PEEK. Contact angle goniometer equipment (OCA 15 plus, Data Physics) was used for the measurement. The ASTM D7334-08 standard, in which deionized water was used as the liquid and the chosen drop size was 0.5 ± 0.1 μl, was used in this test. For each sample, 10 points were randomly chosen from the sample's surface to measure the contact angle [11]. The data of this study were analyzed by SPSS Statistics 22 (IBM, USA) using one-way analysis of variance (ANOVA) and Tukey's test followed by post hoc least significant difference (LSD), with a significance level set at $p < 0.05$.

3. Results and Discussion

3.1. Surface Morphology. Figure 1 shows the surface morphology of the coated layer with a 3 min sulphonation time before and after a 15 MPa compressive load. The load caused an increase in the density of the coated layer. Agglomeration of HA particles in the samples before compression could still be found after the load was removed, but with a reduction in size.

3.2. Surface Roughness. Figure 2 shows the three-dimensional height image of the surface of the samples after applying the compression load via AFM. The surface morphology of all three different samples was almost the same, indicating the independence of sulphonation time on surface roughness after compression. Before compression, the surface roughness increased with increasing sulphonation time [11]. The line of the AFM tip is visible on the sample image due to the soft properties of the coated layer and the force mode which was used for this analysis. The calculated arithmetic mean of the surface roughness obtained via AFM for the compressed samples was 20.4, 22.2, and 20.9 nm for the 3, 5, and 10 min sulphonation times, respectively. These three mean values did not show significant difference ($p > 0.05$); however, they showed significant decrease in comparison with uncompressed samples (from 34.1, 30.9, and 45.2 nm for 3, 5, and 10 min sulphonation times, respectively) [11]. This decrease in surface roughness was due to the compaction of the soft and porous coated layer under the compression load. This condition is less desirable for cell attachment and could not provide a scaffold for mechanical interlocking between mineralized bone and the implant [28].

3.3. Scratch Study Results and Discussions. Figure 3 shows a graph of the penetration depth-normal load against the scratch distance of the samples. As explained above, the scratch normal load begins after 60 μm from the beginning of the scratch length from 0 linearly to 500 mN at the 560 μm length of the scratch and remains fixed for the last 40 μm of the scratch.

Different criteria may be used to probe the failure point in the scratch test. The point of failure may be defined as the onset of microcracking, crazing, fish-scale formation, ploughing, or the point at which the coating is penetrated, revealing the underlying substrate [27]. For brittle coating layers, such as ceramics, failure occurs at two critical loads; the first is the cohesive failure at the coating layer (LC1), and the second is the adhesive failure where the load causes the coating to peel off, exposing the substrate (LC2) [29]. Cohesive failure may not occur in some soft coating layers, in which the stylus reaches straight to the substrate without any cracking, fish scaling, ploughing, etc. [27].

The penetration depth must increase with increasing normal load. However, the accumulation of the detached

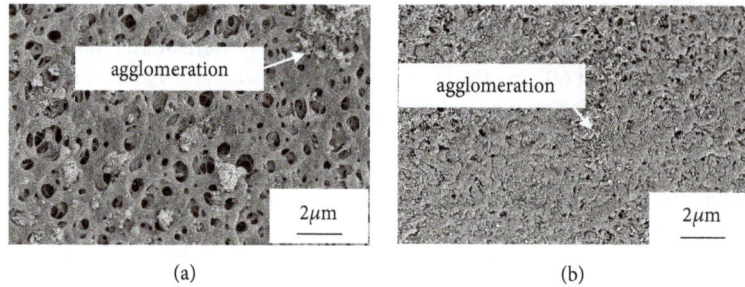

FIGURE 1: SEM image of the surface of the treated layer (a) before and (b) after applying the compression load.

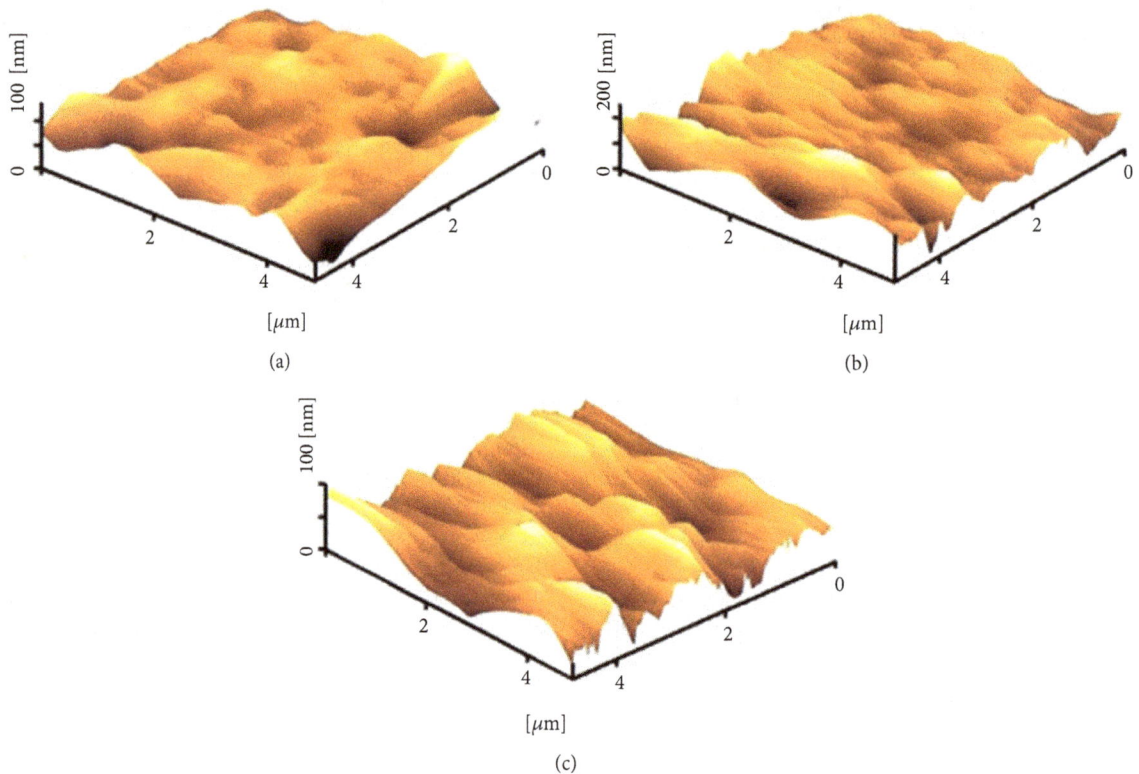

FIGURE 2: AFM 3D height images of the treated PEEK with (a) 3-, (b) 5-, and (c) 10-minute sulphonation times after applying the compression load.

coating and substrate material ahead and under the stylus (macro-chip) pushes the stylus up, affecting the penetration depth. For most of the samples, the upward force caused by the macro-chip formed under the stylus (Figure 4) overcomes the effect of increasing the normal load due to the reduction of the penetration depth of the stylus at the last part of the scratch distance. Once the ultimate shear strength of the coating layer material was exceeded, plastic deformation occurred causing the materials to slip on each other, forming macro-chips. Due to the compliant property of the substrate, the macro-chips surrounding the stylus continued to remove the substrate material after full delamination of the coating. Wrinkles are also visible at the edges of the scratch track. Similar results were observed for samples after compression. The sample with 5 min sulphonation time shows a continuous increase in penetration depth as the normal load increases

due to the formation of macro-chips ahead of the stylus (Figure 4). These results would be different if tested on a brittle coating layer as there would be no macro-chip material ahead or under the stylus and the only factor affecting the penetration depth would be the normal load [30, 31]. Macro-chips can only be found for the scratch testing of compliant coating layers [32].

Figure 5 shows the scratch hardness measurement results on the coated layer at different sulphonation times before and after being compressed. The scratch hardness increased from 75.3, 78.1, and 56.2 MPa to 92.1, 99.6, and 95.3 MPa, respectively, after being compressed for the samples at 3, 5, and 10 min sulphonation time. The mean value of the scratch hardness of compressed samples significantly increased (36.9%) in comparison with uncompressed samples ($p < 0.05$). The scratch hardness results indicate that compression load

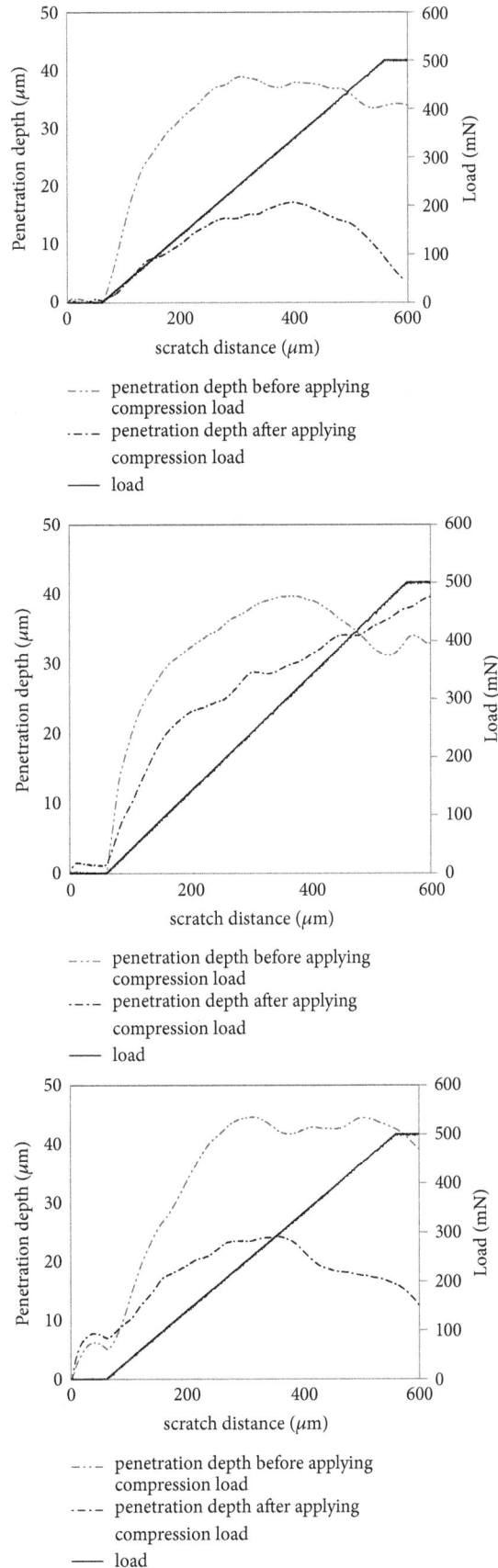

FIGURE 3: Penetration depth versus scratch distance for (a) 3-, (b) 5-, and (c) 10-minute sulphonation time without and with applying the compression load.

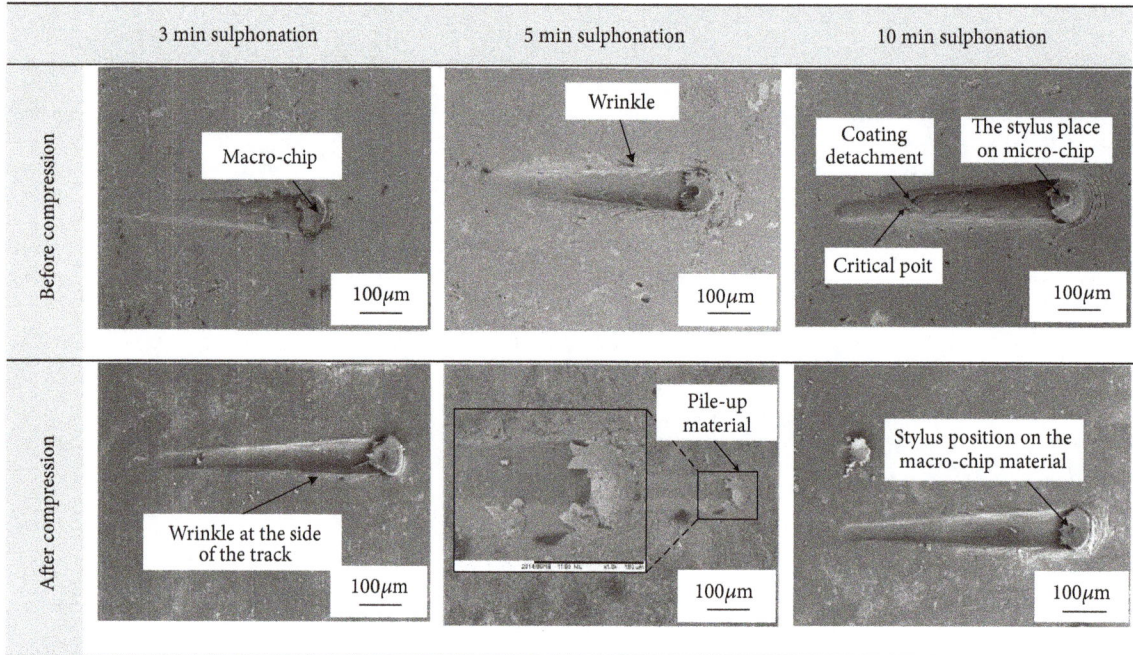

FIGURE 4: Scratch track images on the surface of the treated layer.

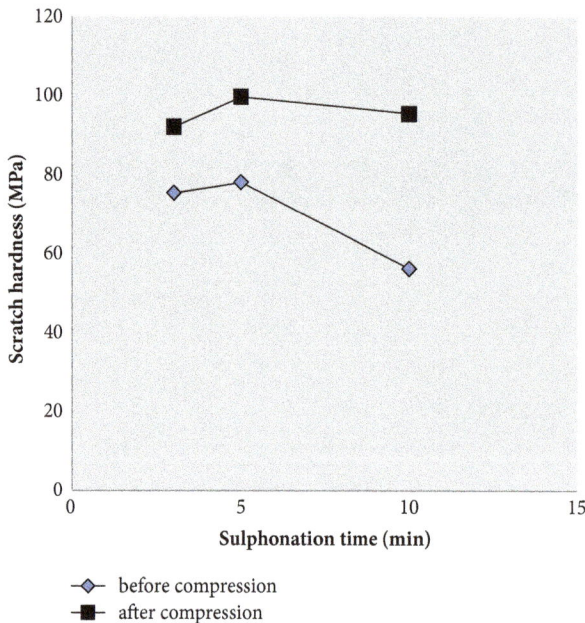

FIGURE 5: Scratch hardness of the treated layer for different sulphonation time, without and with applying compression load.

can be used for improving the mechanical properties of the coated layer.

Figure 6 shows the horizontal load versus scratch distance for different sulphonation times. The horizontal load increased as the applied normal load increased during the scratch test and showed an abrupt drop in load when delamination occurred on the substrate surface. In some samples, the critical point was reached with only the penetration of the stylus into the coated layer without any delamination. In these cases, no abrupt changes were observed in the horizontal load graph. The amount of the horizontal load of the first part of the horizontal load/scratch distance graph, before the critical point (before penetration of the stylus into the sample which is $=200\,\mu m$), is very important because it can be used to calculate the coefficient of the friction. The results showed that the horizontal load of the samples increased after compression of the coated layer. Two important factors affect the horizontal load of soft and rubbery materials, such as our coated layer: first surface roughness [33] and second elastic modulus of the coated layer [34]. As shown above, the surface roughness of the samples decreased with the applied compression which caused a reduction in the horizontal load. However, the compression changed the mechanical properties of the coated layer comprising the elastic modulus, which affected the horizontal load and resulted in the increase in the horizontal load after compression.

3.4. Nanoindentation Study Results and Discussions. Figure 7 shows the modulus of elasticity of the coated layer with different sulphonation times with and without compression. The elastic modulus of the coated layer significantly increased (threefold) for 5 min compared to a 3-min sulphonation time ($p<0.05$). However, increasing the sulphonation time to 10 min did not increase the elastic modulus further ($p>0.05$). After the compression process, all samples, irrespective of the sulphonation time, produced similar magnitudes of elastic modulus. Compression caused the coated layer to become more compact. The applied compression resulted in a significant increase of the mean elasticity modulus of the coated layer from 0.464 GPa to 1.199 GPa ($p<0.05$).

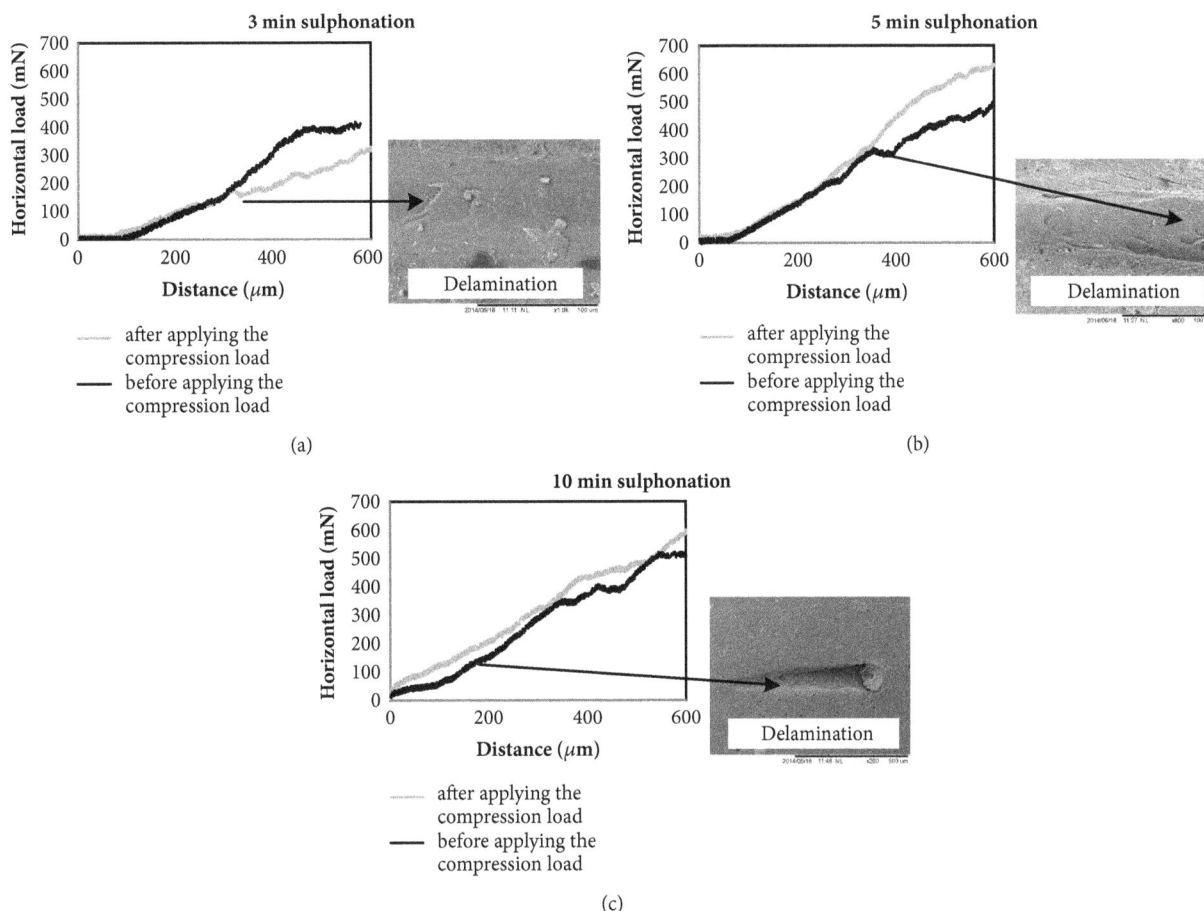

FIGURE 6: The effect of compression on the horizontal load/scratch distance with (a) 3-, (b) 5-, and (c) 10-minute sulphonation time.

The improvement in mechanical properties can be varied to produce coated layers with different elastic moduli based on the requirements for the orthopaedic and dental implant application.

3.5. Water Contact Angle Analysis. Figure 8 shows the effect of sulphonation time and compression load on the water contact angle of the coated PEEK samples. The water contact angle results for the samples before applying the compression load showed a low water contact angle and the expected improvement in the wettability due to the surface treatment [11]. After applying compression load on the coated layer, the mean of the water contact angle significantly increased (59%) ($p<0.05$). However, this is still lower than the contact angle of bare PEEK, which is 72° [11]. It was also noticed that the water contact angle was constant with sulphonation time variations from 3 to 10 min before and after compression ($p>0.05$). The two main parameters of surface roughness and surface chemistry affect the water contact angle. The absorption of the water droplet traces in the compact layer (after compression), in comparison to porous surface layers before applying the compression load that permeated the water droplet, can reduce the water contact angle. However,

the compression can also change the surface chemistry by increasing the water contact angle due to the soft properties of the SPEEK in comparison to HA particles. The increasing the water contact angle leads to reduced osseointegration [35], which is undesirable.

4. Conclusion

The nanoindentation and scratch hardness study revealed the sulphonation time did not have a uniform trend in mechanical properties of coated layer. The elastic modulus of coated layer increased via increasing the sulphonation time from 3 to 5 minutes; however, increasing the sulphonation time to 10 minutes did not increase the elastic modulus further. The scratch hardness of the coated layer increased via increasing the sulphonation time from 3 to 5 minutes; however, increasing the sulphonation time to 10 minutes decreased the scratch hardness. The applied compression resulted in a significant increase of mean elasticity modulus of coated layer from 0.464 GPa to 1.199 GPa, and enhanced the mean scratch hardness of the samples from 69.9 MPa to 95.7 MPa, as the porosity was reduced through compaction. After compression, the surface roughness decreased due

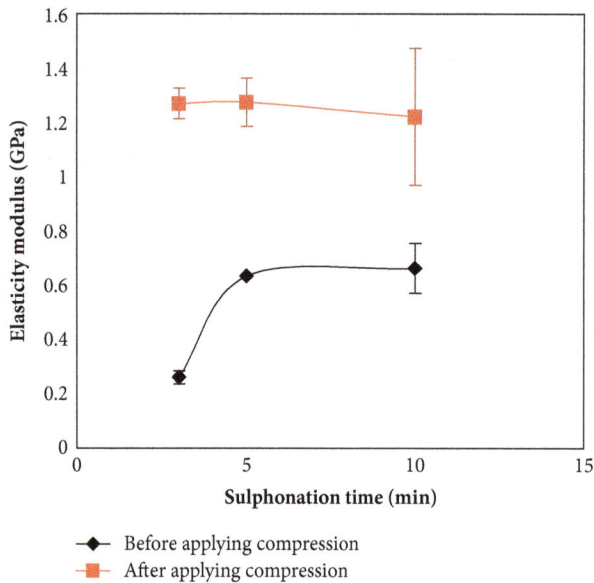

FIGURE 7: The effect of compression on the elastic modulus of a treated layer with different sulphonation times.

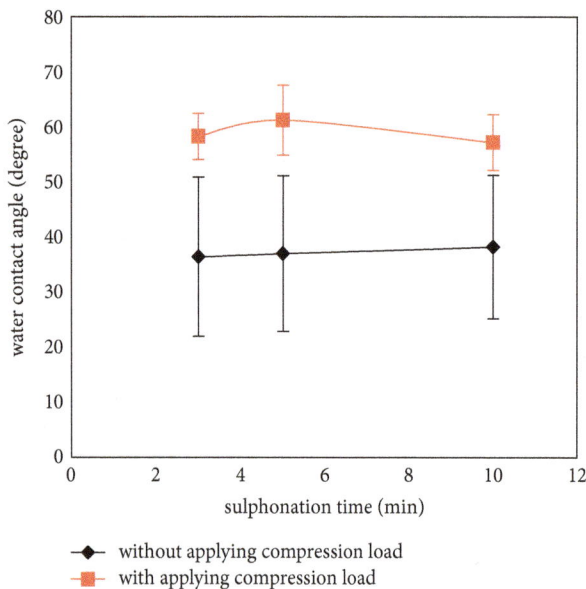

FIGURE 8: The effect of the compression load on the water contact angle of the samples with different sulphonation times [11].

to the compaction of the porous coated layer. The mean elastic modulus of the coated layer for different sulphonation times increased from 0.464 GPa to 1.199 GPa. The water contact angle for 3, 5, and 10 min sulphonation times after compression increased from 37.2° to 58.9°, which is lower than the value for the bare PEEK of 72°. The mechanical properties of PEEK with chemical deposition of HA on its surface can be enhanced through simple compression and reach the requirements for orthopaedic and dental implant application. However, the improvement comes at the expense of achieving lower wettability.

Conflicts of Interest

The author declares that there are no conflicts of interest regarding the publication of this paper.

Acknowledgments

This article was a result of a study conducted at Kermanshah University of Medical Sciences in Kermanshah, Iran.

References

[1] C. E. Misch and H. A. Abbas, *Contemporary implant dentistry*, Elsevier Health Sciences, mosby, 3rd edition, 2008.

[2] M. Goutam, G. S. Chandu, S. K. Mishra, M. Singh, and B. S. Tomar, "Factors affecting osseointegration: a literature review," *Journal of Orofacial Research*, vol. 3, pp. 197–201, 2013.

[3] D. Almasi, N. Iqbal, M. Sadeghi, I. Sudin, M. R. Abdul Kadir, and T. Kamarul, "Preparation methods for improving PEEK's bioactivity for orthopedic and dental application: a review," *International Journal of Biomaterials*, vol. 2016, Article ID 8202653, 12 pages, 2016.

[4] K. Marya, J. Dua, S. Chawla et al., "Polyetheretherketone (PEEK) dental implants: a case for immediate loading," *International Journal of Oral Implantology & Clinical Research*, vol. 2, pp. 97–103, 2011.

[5] A. D. Schwitalla and W. D. Muller, "PEEK dental implants: a review of the literature," *The Journal of Oral Implantology*, 2011.

[6] S. M. Kurtz and J. N. Devine, "PEEK biomaterials in trauma, orthopedic, and spinal implants," *Biomaterials*, vol. 28, no. 32, pp. 4845–4869, 2007.

[7] O. Noiset, Y.-J. Schneider, and J. Marchand-Brynaert, "Adhesion and growth of CaCo2 cells on surface-modified PEEK substrata," *Journal of Biomaterials Science, Polymer Edition*, vol. 11, no. 7, pp. 767–786, 2000.

[8] Y. Kawasaki and Y. Iwasaki, "Surface modification of poly(ether ether ketone) with methacryloyl- functionalized phospholipid polymers via self-initiation graft polymerization," *Journal of Biomaterials Science, Polymer Edition*, vol. 25, no. 9, pp. 895–906, 2014.

[9] O. Noiset, Y. J. Schneider, and J. Marchand-Brynaert, "Fibronectin adsorption or/and covalent grafting on chemically modified PEEK film surfaces," *Journal of Biomaterials Science, Polymer Edition*, vol. 10, no. 6, pp. 657–677, 1999.

[10] A. Rabiei and S. Sandukas, "Processing and evaluation of bioactive coatings on polymeric implants," *Journal of Biomedical Materials Research Part A*, vol. 101, no. 9, pp. 2621–2629, 2013.

[11] D. Almasi, S. Izman, M. Assadian, M. Ghanbari, and M. R. Abdul Kadir, "Crystalline ha coating on peek via chemical deposition," *Applied Surface Science*, vol. 314, pp. 1034–1040, 2014.

[12] D. Almasi, S. Izman, M. Sadeghi et al., "*In vitro* evaluation of bioactivity of chemically deposited hydroxyapatite on polyether ether ketone," *International Journal of Biomaterials*, vol. 2015, Article ID 475435, 5 pages, 2015.

[13] A. Í. S. Antonialli and C. Bolfarini, "Numerical evaluation of reduction of stress shielding in laser coated hip prostheses," *Materials Research*, vol. 14, no. 3, pp. 331–334, 2011.

[14] W. Jiang, J. Cheng, and K. Dinesh, "Improved mechanical properties of nanocrystalline hydroxyapatite coating for dental and orthopedic implants," in *Materials Research Society*, pp. 1140–HH1103-1103, 2009.

[15] ASTM, "Standard test method for stereological evaluation of porous coatings on medical implants," *ASTM F1854-15*, 2015.

[16] ISO, "Implants for surgery-hydroxyapatite-part 4: determination of coating adhesion strength," *ISO 13779-4*, 2002.

[17] D. Almasi, S. Izman, F. Roozbahani, and M. Sadeghi, "Enhancement of Mechanical Properties of Hydroxyapatite/Sulphonated Poly Ether Ether Ketone Treated Layer," in *Proceedings of the International Conference on Experimental Solid Mechanics*, 2016.

[18] J. K. Sherwood, L. G. Griffith, and S. Brown, "omposites for tissue regeneration and methods of manufacture thereof," *Google Patents*, 2002.

[19] M. Ilze and M. Janis, "Effect of substrate hardness and film structure on indentation depth criteria for film hardness testing," *Journal of Physics D: Applied Physics*, vol. 41, article 074010, 2008.

[20] C. A. Clifford and M. P. Seah, "Quantification issues in the identification of nanoscale regions of homopolymers using modulus measurement via AFM nanoindentation," *Applied Surface Science*, vol. 252, no. 5, pp. 1915–1933, 2005.

[21] G. M. Pharr, W. C. Oliver, and F. R. Brotzen, "On the generality of the relationship among contact stiffness, contact area, and elastic modulus during indentation," *Journal of Materials Research*, vol. 7, no. 3, pp. 613–617, 1992.

[22] A. C. Fischer-Cripps, "Nanoindentationv Testing," in *Introduction to Contact Mechanics*, pp. 20–35, Springer, New York, NY, USA, 2000.

[23] S.-R. Jian, G.-J. Chen, and T.-C. Lin, "Berkovich nanoindentation on AlN thin films," *Nanoscale Research Letters*, vol. 5, no. 6, pp. 935–940, 2010.

[24] K. L. Mittal, *Adhesion Measurement of Films and Coatings*, Taylor & Francis, 1995.

[25] D. S. Rickerby, "A review of the methods for the measurement of coating-substrate adhesion," *Surface and Coatings Technology*, vol. 36, no. 1-2, pp. 541–557, 1988.

[26] J. Valli, "A review of adhesion test methods for thin hard coatings," *Journal of Vacuum Science & Technology A*, vol. 4, no. 6, pp. 3007–3014, 1986.

[27] ASTM, D7027-13, Standard Test Method for Evaluation of Scratch Resistance of Polymeric Coatings and Plastics Using an Instrumented Scratch Machine, in, ASTM standard, 2013.

[28] L. Le Guéhennec, A. Soueidan, P. Layrolle, and Y. Amouriq, "Surface treatments of titanium dental implants for rapid osseointegration," *Dental Materials*, vol. 23, no. 7, pp. 844–854, 2007.

[29] ASTM, C1624-05, Standard Test Method for Adhesion Strength and Mechanical Failure Modes of Ceramic Coatings by Quantitative Single Point Scratch Testing, 2010.

[30] F. X. Liu, F. Q. Yang, Y. F. Gao et al., "Micro-scratch study of a magnetron-sputtered Zr-based metallic-glass film," *Surface and Coatings Technology*, vol. 203, no. 22, pp. 3480–3484, 2009.

[31] S. Xu, X. Ma, H. Wen, G. Tang, and C. Li, "Effect of annealing on the mechanical and scratch properties of BCN films obtained by magnetron sputtering deposition," *Applied Surface Science*, vol. 313, pp. 823–827, 2014.

[32] M. Barletta, A. Gisario, and G. Rubino, "Scratch response of high-performance thermoset and thermoplastic powders deposited by the electrostatic spray and "hot dipping" fluidised bed coating methods: the role of the contact condition," *Surface and Coatings Technology*, vol. 205, no. 21-22, pp. 5186–5198, 2011.

[33] P. L. Menezes, Kishore, and S. V. Kailas, "Effect of directionality of unidirectional grinding marks on friction and transfer layer formation of Mg on steel using inclined scratch test," *Materials Science and Engineering: A Structural Materials: Properties, Microstructure and Processing*, vol. 429, no. 1-2, pp. 149–160, 2006.

[34] B. N. J. Persson, "Theory of rubber friction and contact mechanics," *The Journal of Chemical Physics*, vol. 115, no. 8, pp. 3840–3861, 2001.

[35] S. C. Sartoretto, A. T. N. N. Alves, R. F. B. Resende, J. Calasans-Maia, J. M. Granjeiro, and M. D. Calasans-Maia, "Early osseointegration driven by the surface chemistry and wettability of dental implants," *Journal of Applied Oral Science*, vol. 23, pp. 279–287, 2015.

22

Experimental Investigations into the Mechanical, Tribological, and Corrosion Properties of Hybrid Polymer Matrix Composites Comprising Ceramic Reinforcement for Biomedical Applications

Mohammed Yunus ⓘ **and Mohammad S. Alsoufi**

Department of Mechanical Engineering, College of Engineering and Islamic Architecture, Umm Al-Qura University, Al-Abdiah, Makkah 24231, Saudi Arabia

Correspondence should be addressed to Mohammed Yunus; yunus.mohammed@rediffmail.com

Academic Editor: Anna Maria Piras

Hybrid polymer matrix composites (HPMC) are prominent material for the formation of biomaterial and offer various advantages such as low cost, high strength, and the fact that they are easy to manufacture. However, they are associated with low mechanical (low hardness) and tribological properties (high wear rate). The average hip joint load fluctuates between three to five times of the body weight during jumping and jogging and depends on various actions relating to body positions. Alternate bone and prosthesis material plays a critical role in attaining strength as it determines the method of load transferred to the system. The material property called modulus of elasticity is an important design variable during the selection of the geometry and design methodology. The present work is demonstrated on how to improve the properties of high-density polyethylene (HDPE) substantially by the addition of bioceramic fillers such as titanium oxide (TiO_2) and alumina (Al_2O_3). The volume fractions of Al_2O_3 and TiO_2 are limited to 20% and 10%, respectively. Samples were fabricated as per ASTM standards using an injection moulding machine and various properties such as mechanical (tensile, flexural, and impact), tribological (hardness, wear), and corrosion including SEM, density, and fractography analysis studied. Experimental results revealed that an injection moulding process is suitable for producing defect-free mould HPMC. HPMC comprising 70% HDPE/20% Al_2O_3/10% TiO_2 has proved biocompatible and a substitute for biomaterial. A substantial increase in the mechanical and tribological properties and full resistance to corrosion makes HPMC suitable for use in orthopaedic applications such as human bone replacement, bone fixation plates, hip joint replacement, bone cement, and bone graft in bone surgery.

1. Introduction

Polymer and ceramic oxide metal matrix composites are seeing applications as a biomaterial because their combination acts to replace human tissues and anatomical elements to treat or improve and also medical devices for implants. The mechanical properties are important factors that determine the progress of potential biomaterial. Polyethylene is one of the readily available low-cost polymer materials and can be processed at temperatures of 150-250°C.

Polyethylene is hybrid-linked and is available in the form of linear low, low, and high-density polyethylene (LLDPE, LDPE, and HDPE), respectively. Alternate bone materials are

required to fill the gap or portion of the bone missing as bone is a natural composite mainly consisting of organic as well as mineral matrices called collagen fibers (which introduce the mechanical properties of toughness and viscoelasticity) and hydroxyapatite (HAP) forms a bonding gel. However, the challenge for alternative bone materials is to maintain a balance between biological and biomechanical properties to act as a biomaterial. The hybrid linking nature of polyethylene develops a thick chain of high molecular weight leading to the branched structure to improve the mechanical properties such as impact strength, crack, creep, and abrasion resistance without much change in tensile strength and density [1]. Alumina (Al_2O_3) and titanium oxides (TiO_2) are used to improve the mechanical as well as tribological properties (wear properties) of high-density polyethylene (HDPE) further [2]. Alumina is a hard-ceramic oxide (up to 1800 HV) thermodynamically stable (up to 2000°C) which is available in both ionic and covalent bonds [3]. Titanium, meanwhile, due to its outstanding characteristics of biocompatibility, has no reaction with the tissue surrounding the implant, decreases the electronic exchange process so that there is no corrosion, has no problem with cardiac and cardiovascular applications, and has an isoelectric point to maintain a pH value between 5 and 7 of the human body [4]. Both materials are biologically inert as they remain intact after implantation into human bodies and can be retained as foreign materials by accommodation in fibrous tissues in order to isolate them from a human body. There are thus no adverse reactions yet they are endured well by tissues [5].

In the literature study, mostly the work is on the biomaterials such as stainless steel (316L), titanium alloy (Ti-6AL-4V), cobalt-chromium alloy, HAP, UHMWPE, Al_2O_3, TiO_2, and silicon carbide (Sic). The above materials are used for the replacement of knee, hip, and ankle joints as well as for dental implants [6]. In the present work, we study the essential properties of bone alternate or replacement material made of high-density polyethylene. This paper focuses on the study of fundamental attributes needed for other biomaterials based on hybrid polymer matrix composites to accommodate the types of bones and joints fractured [7].

Mostly alumina and zirconia-based ceramics have used bioceramics mainly due to their biocompatibility for an implant, offering a high mechanical strength with no reaction and being nontoxic to tissues along with having a blood compatibility characteristic [8]. The mechanical properties of the 12, 24, and 36% of hybrid fiber (Sisal, Jute, and Hemp) polymer composite material when compared with the femur bone strengths are seen to be increased by increasing the percentage of the fiber [9]. The tensile and compression strength of 30% of Sisal natural fiber reinforcement epoxy composite materials was found to be maximum out of the composition of 10, 20 and 30% [10]. Variation of % of fiber weight of banana fiber in glass reinforced hybrid polypropylene composites improved various mechanical properties such as tensile, flexural, impact at 10% fiber fraction [11]. Hybrid reinforced composites formed by bamboo fiber with fly ash filler prepared by hand lay-up technique produced excellent mechanical properties [12]. Coconut shell fiber reinforced hybrid composites produced by coconut shell fiber compacting epoxy resin matrix with 10 to 30% volume fraction showed increased tensile strength with increase in coconut shell fiber content [13]. Carbon fiber based hybrid polymer composite matrix with +/- 0 to 90° orientations was used as implant material and compared with mechanical properties of a femur bone [14]. Synthesis of hybrid biopolymer matrix composites uses low-density polyethylene as matrix material with reinforcing material; namely, alumina and titanium oxide showed improvement in mechanical properties [15].

An attempt has been made to develop hybrid biopolymer matrix composites using high-density poly ethylene as the matrix material with titanium oxide/titania (TiO_2) and

alumina/aluminium oxide (Al_2O_3) particles as the reinforcement material with varying percentages using an extrudal injection moulding machine. The different testing, namely, tensile, hardness, flexural strength, density, fractography, corrosion, and wear test, was conducted on the standard samples prepared [16]. Substantial improvements are found in the mechanical and tribological properties of the hybrid polymer matrix composite, which can be used for a variety of applications in human body bone replacement [17]. In this case, their application in orthopaedics as an implantable material in bone surgery has been considered and studied. These composite materials have found extensive use in orthopaedic applications, particularly in bone fixation plates, hip joint replacement, bone cementing, and bone graft [18].

In the literature review, most of the work has been conducted on biomaterials such as stainless steel 316L, titanium alloy (Ti-6AL-4V), cobalt-chromium alloy, hydroxyl apatite (HAP), ultra-high molecular weight polyethylene (UHMWPE), alumina (Al_2O_3), titanium oxide (TiO_2), and silicon carbide (sic) as the material for replacement of knee joints, hip joints, and ankle joints as well as dental implants. In this work, we study the essential properties of bone alternate or replacement materials for bone [19, 20].

This paper highlights the study of the fundamental properties required to replace bone materials for various types of bones and joints fractured by the synthesis of biocompatible, hybrid polymer matrix composites [21]. Polymer matrix composite is a material consisting of polymer (resin) matrix combined with a fibrous reinforcing dispersed phase. Polymers make ideal matrix material; they can be processed, i.e., being fabricated more easily, processing light weight, and offering desirable mechanical properties. The reasons for the selection of these composites are their low cost, high strength, and simple manufacturing principles [22].

2. Materials and Methods

Three grades of commercially available polyethylene are low-density, high-density, and ultra-high molecular weight polyethylene (UHMWPE). UHMWPE produces low ductility and fractures toughened material more than those of other classes of polyethylene. However, high-density polyethylene (HDPE) materials have better packing of linear chains and high branching levels which results in their increased crystallinity and enhanced mechanical, tribological properties [23]. Thus, in the present work, HDPE in granule form (transition/softening temperature of 125°C and melt flow index of 0.22 g/min) is used as polymer matrix material and further their properties can be improved by adding metallic (TiO_2) as a coupling agent and ceramic reinforcement (Al_2O_3) materials in the polymer matrix of 325 mesh size were used for the synthesis of polymer composites supplied by alumina Ceramic Manufacturers India, Gujarat, INDIA. Hence, Al_2O_3 and TiO_2 of purists grade have a melting temperature of >350°C. To obtain the various levels of properties [24], different compositions are used as shown in Table 1 by varying percentage by weight of each matrix material; Al_2O_3 and TiO_2 powder materials along with a

TABLE 1: Composition of composites prepared.

Sample No.	HDPE in weight %	TiO$_2$ in weight %	Al$_2$O$_3$ in weight %
1	85	10	05
2	80	10	10
3	75	10	15
4	70	10	20

surfactant (non-inphinoethoxylate) material were used for the synthesis of composite material.

2.1. Production of HPMC Composites. Samples were prepared using 75 tonnage vertical injection moulding machine in which raw material is injected into a mould via a hot barrel to take the inverse shape. Multiple cavity mould is preferred over the single cavity mould to save raw material and the time. Finally the compound is then taken for the specimen preparation. Standard test specimens were prepared as per the ASTM standard for tensile (ASTM D638), flexural (ASTM D790), and impact tests (ASTM D256).

2.2. Mechanical and Tribological Tests. Tension tests were performed according to ASTM D 638 standard using Instron UTM (Universal Testing Machine) 4302 having a load capacity ranging from 0 to 10000N at the crosshead speed of 0.0166 mm/Sec. Tensile test specimen has dimensions 63 mm x 9.53 mm x 3.5 mm with cross section width 9.53 mm and radius of 12.7 mm. Flexural test was conducted in accordance with ASTM D790 having dimensions of 127 mm x 13 mm x 3.5 mm using Instron UTM 3365 possessing a load cell capacity of 50000N with centre loading three-point load system. The crosshead speed of 0.05 mm/sec and span length of 70 mm were set. The impact test from 0 to 10 J at an interval of 0.0001 J on the specimen a 64 mm x 13 mm x 3.5 mm and hardness of the synthesised polymer composites were measured using shore hardness D-scale. Durometer hardness tester (Shore Instrument and MFG Co., Freeport, NY) as per the ASTM D2240 standard was used for testing hardness of samples. Pin-on-disc sliding wear testing machine [25] was used to understand the dry sliding wear characteristics of the hybrid composite specimens. After each test, the coefficient of friction (COF) and height loss was recorded. As per ASTM G99-95 standards, the pin of Ø8 × 32 mm length was used for the tribology test. The pin was cleaned with acetone and its initial mass was measured using a digital electronic balance and then held pressed against the rotating EN-32 steel disc (counter face) with a hardness of 65 HRC during the test. The tribological test was carried out with the normal load varying from 10 N to 30 N, a track diameter of 75 mm, and sliding speed of 500 rpm, and the entire test was carried out for 15 minutes duration. Further, a corrosion test was carried out on the composites as per a salt spray test according to the ASTM B117 standard.

3. Results and Discussion

In this section, the mechanical and tribological characteristics of HDPE /TiO$_2$ /Al$_2$O$_3$ hybrid composites are discussed

FIGURE 1: Photographs of samples prepared.

by carrying out various tests according to ASTM standards on the specimens prepared using an injection moulding machine/process. The results of tensile strength at ultimate point, flexural strength at 7.8 mm deflection, impact strength at breaking point, shore hardness D-scale, and wear and coefficient of test depict/represent the mechanical and tribological characterization of developed HPMC including the corrosion test.

3.1. Mechanical Properties

3.1.1. Ultimate Tensile Strength. The tensile specimens were tested, calculated, and plotted for the ultimate tensile strength of the composite material (refer to Figure 1). Its tensile strength increases substantially with the increase of Al$_2$O$_3$ (from 5 to 20% at the interval of 5%), at fixed 10% of TiO$_2$ which in turn increases the load carrying capacity due to the presence of hard and stiff alumina particles in the composite material. Hence, the load carrying capacity of the composite material increases. It reaches a maximum value of 17 MPa which is 30% higher than the polyethylene with an elastic modulus of 500 MPa for the combination of 70% HDPE/20% Al$_2$O$_3$ /10% TiO$_2$ (which remains constant) of the composite specimens as seen in the stress-strain curve of Figures 2 and 3. The elongation (ductility) of the HDPE is also increased. The presence of alumina along with TiO$_2$ has increased ductility as well as little brittleness which improved both tensile strength and elongation capability. Sample 04 exhibits good yield strength strain but poor elongation due to brittleness provided by alumina reinforcements as seen in Table 2.

3.1.2. Flexural Strength. Figure 4 shows that the variation in flexural strength of the composite specimen also increases with an increasing percentage of Al$_2$O$_3$ (from 5 to 20%) as they resist the deformation of the composite material. The flexural strengths of the joint were found to be about 25% higher than that of polyethylene. The plastic region of sample depends on ductility of the HDPE matrix and

TABLE 2: Comparison of yield strength, strain at failure, and elongation of HPMC samples.

Sample	Yield strength (MPa)	Strain at Failure (%)	Elongation (mm)
01	28.5	3.5	4.2
02	30.2	3.8	4
03	31.25	4	3.9
04	33.355	4.7	3.75

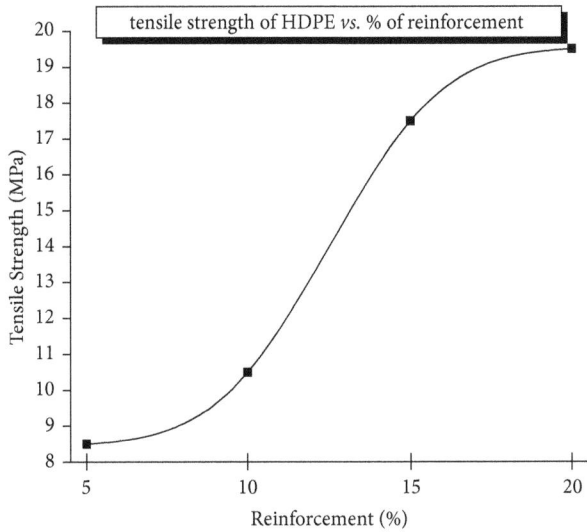

FIGURE 2: Effect of Al_2O_3 loading on the tensile strength of HPMC.

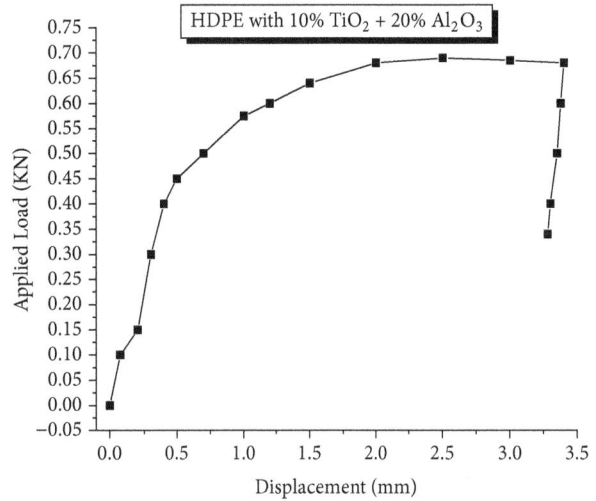

FIGURE 3: Applied load versus displacement curve of HPMC (sample 4).

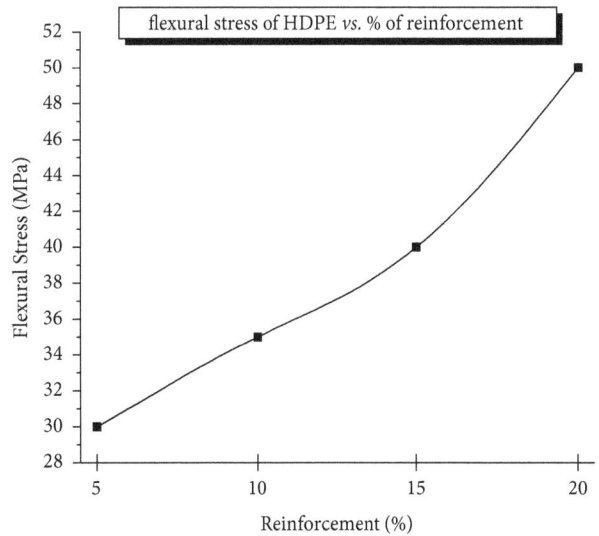

FIGURE 4: Variation of bending stress with % reinforcement.

interparticle distance of reinforcements. At higher % Al_2O_3, the interparticle distance increases make matrix hard to undergo yielding, increasing the flexure strength. When the Al_2O_3 content was less than 5% weight, the interparticle distance reaches a minimum which can change into plastic yield to decrease the flexure strength. At very low % Al_2O_3, the interparticle distance changes HPMC to behave brittle. The highest flexural strength reached a value of 50 MPa for 70% HDPE /10% TiO_2 /20% Al_2O_3 and showed good bend properties. There were no visible cracks in the midsection of the composite specimen.

3.1.3. Hardness. Figures 5 and 6 show the hardness profile measured and it is clearly evident that there is a substantial increase in the HDPE with an increment of Al_2O_3 percentage reinforcement in alumina-titania-HDPE composite compared to the unfilled system. The highest average shore D hardness number was found to be 60 measured at various sections due to uniform distribution of hard alumina particles as well as titanium oxide being bonded together and increasing the resistance to plastic deformation. The mean values of hardness against different TiO_2 and Al_2O_3 of varying concentrations by weight are shown in Figure 6. The shape S indicates hardness increases rapidly but it is limited by interparticle bonding, distance, and nonuniform distribution of reinforcements. Variation of alumina and titania contributes to the hardness of HPMC.

3.1.4. Density Test. The density of composite material depends on parent metal and constituents. Density increased with percentage of alumina and titania constituents but increasing percentage alumina contributed towards density increase due to good bonding with polyethylene as well as titania without changing the cross-linked or branched structure of the composite material. Interparticle distance,

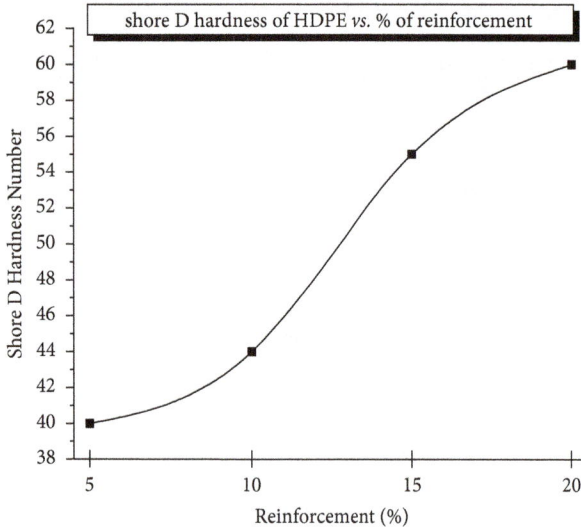

FIGURE 5: Variation of hardness with % reinforcement.

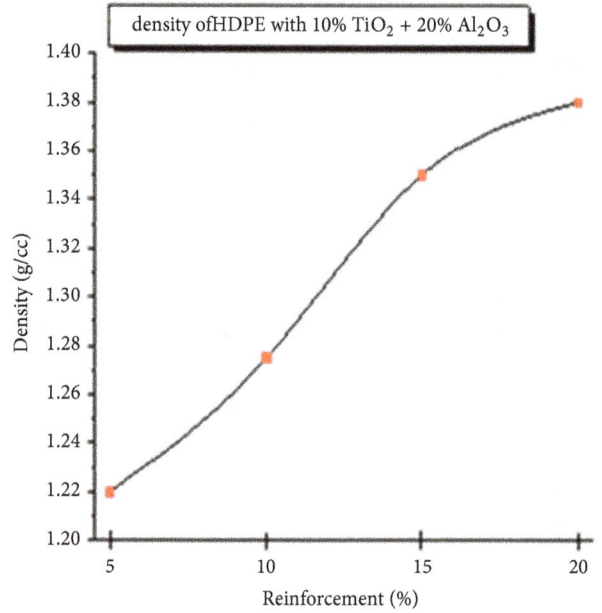

FIGURE 7: Density varying with % reinforcement.

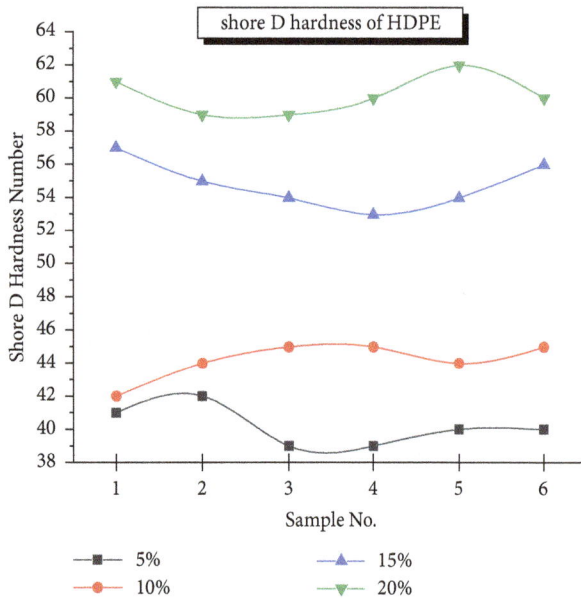

FIGURE 6: Variation of hardness with % reinforcement.

have been reported elsewhere [26–28]. Figure 8 shows the fractured surface observed at 200X magnification. It is clear from the SEM image that there is a homogeneous distribution of reinforcing particles in the matrix of the polymer and there are no casting defects observed. Also, there is a proper bonding between the matrix and reinforcing particles. This enhances the mechanical properties of the composite materials. Further, the image shows that the composite fails by brittle fracture. The image of the fractured surface shows a uniform composite without moulding imperfections. Beautiful populated dimples were observed at the higher magnification which resembles the brittle mode of the failure.

3.1.6. Corrosion Test. As per ASTM B117, a standard corrosion test is carried out by using salt spray test involving a solution of 5% NaCl (AR Grade) with 1.41 ml volume of solution collected per hour on area of 80 Cm^2 at a temperature of 33 to 37°C in distilled water (7.08 pH value). After cleaning with running water, no signs of corrosion or reaction were found on any of the specimens of HPMC during the observations made after a period of 24 hours using the procedure and results are tabulated in Table 3.

3.1.7. Impact Strength. The amount of energy absorbed evaluated at the breaking point of the composites or toughness of material decreases drastically with an increasing percentage of alumina at a fixed value of TiO_2, because of the brittle and hard nature of alumina particles which provide ductility along with titania under the static or slow rate of loading, whereas, for suddenly applied loads, bonding between alumina and titania particles breaks soon without allowing further transfer of load due to higher percentage of alumina and breaking of the bonding that exists between reinforcement constituents. In case of low percentage of alumina, load transfer happens between polyethylene and

distribution of reinforcement particles, weight density difference of reinforcements and matrix polymer, and so on decide the final density. S distribution indicates density value accelerates rapidly with shallow growth during lower and higher % Al_2O_3 as the interparticle distance becomes too large and small, this leads to deboning. Figure 7 shows the variation of density with the percentage of reinforcement to bring the structure to a specific weight which is necessary to use as bone or hip joint replacement material.

3.1.5. Fractography Study. The hybrid composite HDPE/20% Al_2O_3/10% TiO_2 was scanned using SEM (scanning electron microscopy) to get an image of the distribution of the reinforcing particles before the test and the fracture type after the tensile test. More details of the SEM machine procedure

TABLE 3: Corrosion test results.

S. No.	% of Al_2O_3	Time in Hours	Observation
01	5	24	No corrosion was observed
02	10	24	No corrosion was observed
03	15	24	No corrosion was observed
04	20	24	No corrosion was observed

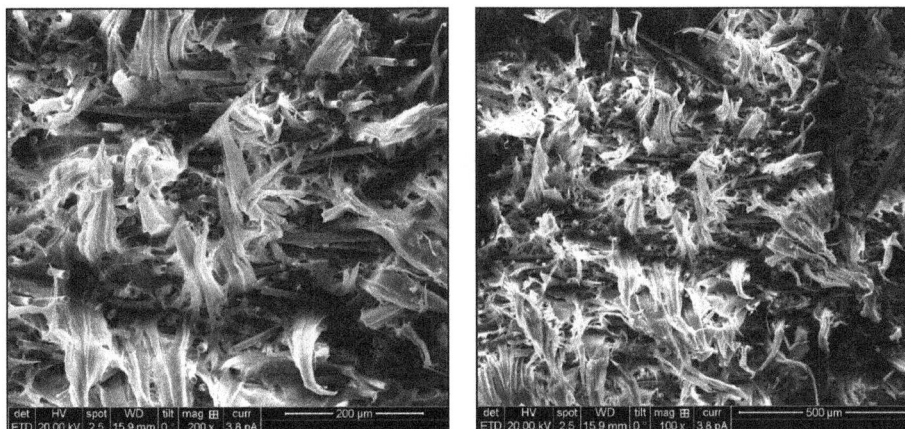

FIGURE 8: Fractured surface after tensile strength test for HDPE/20% Al_2O_3.

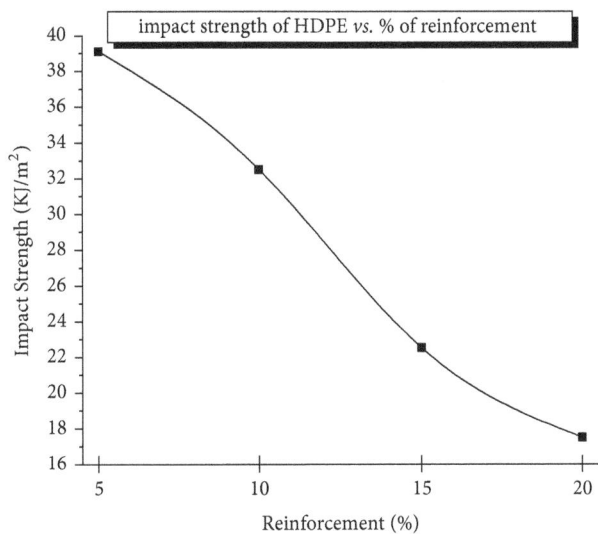

FIGURE 9: Impact strength varying with % reinforcement.

titania rather than alumina particles. Hence, the impact strength can also be increased by refining alumina particle size or intensifying bonding structure between TiO_2 and Al_2O_3. The maximum impact strength obtained was $39kJ/m^2$ for HDPE/5% Al_2O_3/10% TiO_2 which are depicted in Figure 9. The toughness of the alumina particles depends on ductility of the HDPE matrix and interparticle distance of reinforcements. These introduced stress concentrations lead to deboning of the filler particles and in turn void formation. The interparticle distance depends on particle content and the HDPE matrix stress state around the voids. At higher

% Al_2O_3, the interparticle distance increases make matrix hard to undergo yielding, decreasing the impact strength. When the Al_2O_3 content was less than 5% weight, the interparticle distance reaches a minimum which can change into plastic yield to improve the impact strength. Lesser % Al_2O_3 decreases the interparticle distance which makes HPMC behave like brittle.

3.2. Tribological Properties. The wear loss regarding height measurement using a pin-on-disc tester is taken down under different loads (10 N, 20 N, and 30 N) at a constant speed of 500 rpm for different pin specimens prepared with an ambient temperature of $20\pm1°C$ and a relative humidity of greater than $40\pm5\%$ RH. The various observations were made in the wear analysis test. Firstly, the increase of the load on the specimen and sliding time increases the wear loss but decreases with the increase of % of alumina at fixed 10% of TiO_2 as shown in Figure 10. Secondly, the alumina particles which are solid offer resistance to wear as long as bonding is maintained and uniform dispersion is achieved. Finally, frictional force decreases with the increase of alumina particles as they bear higher load without loss of weight or grain size and in turn the coefficient of friction also decreases as shown in Figures 11 and 12.

4. Conclusions

In the present study, the various observations are drawn based on the investigations conducted on hybrid polymer matrix composites (polymeric biocomposite) for orthopaedic applications. There are a variety of applications with the human body as implantable materials for bone surgery especially

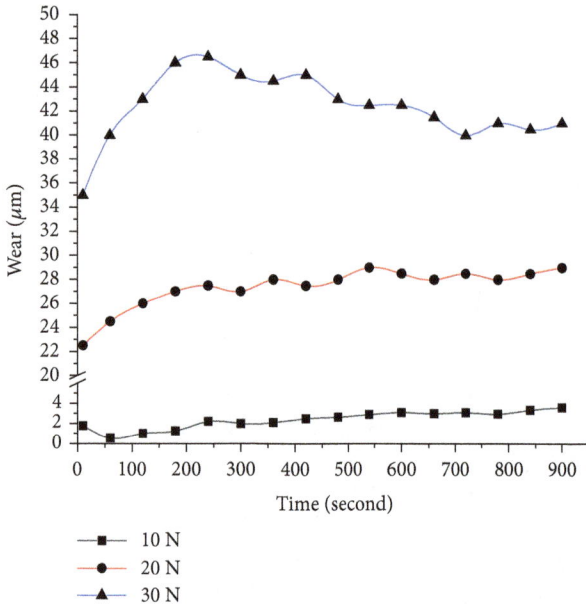

FIGURE 10: Wear varying with contact time at various loads and % reinforcements.

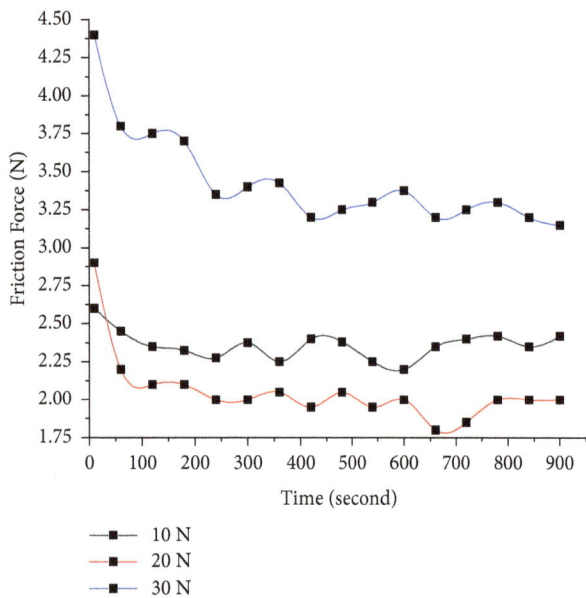

FIGURE 11: Frictional force varying with time of contact for 20% Al_2O_3.

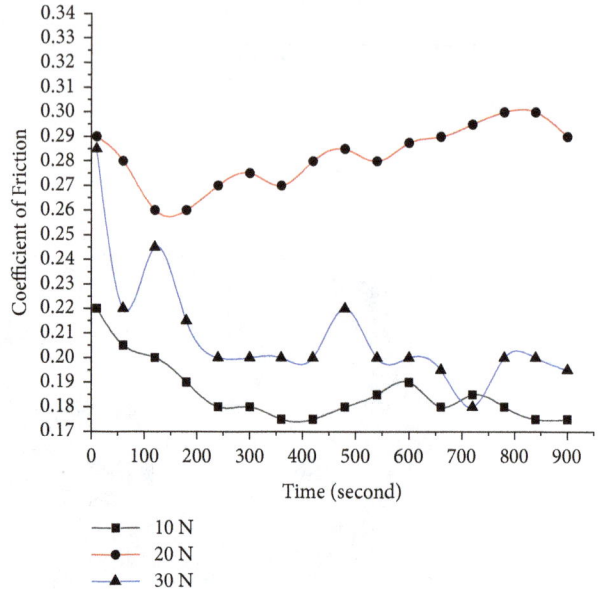

FIGURE 12: Variation of coefficient of friction at various loads and for 20% Al_2O_3.

TiO2 /20% Al_2O_3 combination produces suitable biomaterial in orthopaedic applications during the processing of various tests. No corrosion was observed in the samples after a period of 48 hours at a pH value of 7. Injection moulding process is a successful fabrication technology for preparing biomaterial without any casting defects.

Conflicts of Interest

The authors have no conflicts of interest.

Authors' Contributions

The authors participated in the conceptualization of the research. Mohammed Yunus and Mohammad S. Alsoufi prepared the data for experimentation and analysis. The authors participated in the analysis and interpretation of the data. Mohammed Yunus drafted the manuscript and figures are sketched by Mohammad S. Alsoufi. The authors revised the manuscript for intellectual content and approved the final manuscript.

References

[1] C. X. Dong, S. J. Zhu, M. Mizuno, and M. Hashimoto, "Fatigue Behavior of HDPE Composite Reinforced with Silane Modified TiO2," *Journal of Materials Science and Technology*, vol. 27, no. 7, pp. 659–667, 2011.

[2] C. Y. Chee, N. L. Song, L. C. Abdullah, T. S. Y. Choong, A. Ibrahim, and T. R. Chantara, "Characterization of mechanical

alternate for hard and soft tissues, bone cement, grafting, fixation plates, and hip joint replacement. With the increasing percentage of Al_2O_3 in the HPMC, the tensile, flexural strength, and hardness increased. Density has improved with reinforcing particles. Impact strength, coefficient of friction, frictional force, and wear decrease with an increase in the alumina percentage in the composite material. SEM image analysis has shown a homogeneous distribution of reinforcing particles with proper bonding between matrix and reinforcement without moulding imperfections. 70% HDPE/10%

properties: Low-density polyethylene nanocomposite using nanoalumina particle as filler," *Journal of Nanomaterials*, vol. 2012, Article ID 215978, 6 pages, 2012.

[3] M. S. Alsoufi, M. W. Alhazmi, H. A. Ghulman, S. M. Munshi, and S. Azam, "Surface roughness and knoop indentation micro-hardness behavior of aluminium oxide (Al2O3) and polystyrene (C8H8)n materials," *International Journal of Mechanical and Mechatronics Engineering*, vol. 16, no. 6, pp. 43–49, 2016.

[4] S. W. Lee, C. Morillo, J. Lira-Olivares et al., "Tribological and microstructural analysis of Al2O3/TiO2 nanocomposites to use in the femoral head of hip replacement," *Wear*, vol. 255, no. 7-12, pp. 1040–1044, 2003.

[5] S. M. Tamboli, S. T. Mhaske, and D. D. Kale, "Crosslinked polyethylene," *Indian Journal of Chemical Technology*, vol. 11, pp. 853–864, 2004.

[6] X. Liu, P. K. Chu, and C. Ding, "Surface modification of titanium, titanium alloys, and related materials for biomedical applications," *Materials Science and Engineering: R: Reports*, vol. 47, no. 3, pp. 49–121, 2004.

[7] S. A. R. Alavi, M. T. Angaji, and Z. Gholami, "Twin-Screw Extruder and Effective Parameters on the HDPE Extrusion Process," *World Academy of Science, Engineering and Technology, International Journal of Chemical, Molecular, Nuclear, Materials and Metallurgical Engineering*, vol. 3, no. 1, pp. 60–63, 2009.

[8] T. V. Thamaraiselvi and S. Rajeswari, "Biological evaluation of bioceramic materials - a review," *Trends in Biomaterials & Artificial Organs*, vol. 18, no. 1, pp. 9–17, 2004.

[9] D. A. Gouda, J. S P, D. K. Dinesh, V. Gouda H, and D. N. Prashanth, "Characterization and Investigation of Mechanical Properties of Hybrid Natural Fiber Polymer Composite Materials Used As Orthopaedic Implants for Femur Bone Prosthesis," *IOSR Journal of Mechanical and Civil Engineering (IOSR-JMCE)*, vol. 11, no. 4, pp. 40–52, 2014.

[10] K. R. Dinesh, S. P. Jagadish, A. Thimmanagouda, and N. Hatapaki, "Characterization and Investigation of Tensile and Compression Test on Sisal Fibre Reinforcement Epoxy Composite Materials Used as Orthopaedic Implant," *International Journal of Application or Innovation in Engineering & Management (IJAIEM)*, vol. 2, no. 12, pp. 376–389, 2013.

[11] N. R. Kumar, G. R. Prasad, and B. R. Rao, "Investigation on mechanical properties of banana fiber glass reinforced hybrid thermoplastic composites," *International Journal of Engineering Research & Technology (IJERT)*, vol. 2, no. 11, pp. 3701–3706, 2013.

[12] T. V. Rao, K. Venkatarao, and L. K. Ch, "Mechanical Properties of Bamboo Fibre filled with Fly Ash filler R einforced Hybrid C omposites," *International Journal of Engineering Research & Technology (IJERT)*, vol. 3, no. 9, pp. 725–732, 2014.

[13] J. O. Akindapo, A. Harrison, and O. M. Sanusi, "Evaluation of mechanical properties of coconut shell fibres as reinforcement material in epoxy matrix," *International Journal of Engineering Research & Technology (IJERT)*, vol. 3, no. 2, pp. 2337–2348, 2014.

[14] S. P. Jagadish, A. Thimmana Gouda, K. R. Dinesh et al., "Analysis of mechanical properties of carbon fiber polymer Composite materials used as orthopaedic implants," *International Journal Of Modern Engineering Research (IJMER)*, vol. 5, no. 8, pp. 9–16, 2015.

[15] S. Visbal, J. Lira-Olivares, T. Sekino, K. Niihara, B. K. Moon, and S. W. Lee, "Mechanical properties of Al2O3-TiO2-SiC

[16] nanocomposites for the femoral head of hip joint replacement," *Materials Science Forum*, vol. 486-487, pp. 197–200, 2005.

[16] R. Dhabale and V. S. Jatti, "A bio-material: mechanical behaviour of LDPE-Al2O3 -TiO2," *IOP Conference Series: Materials Science and Engineering*, vol. 149, no. 1, article 012043, 2016.

[17] S. Mazurkiewicz, "The Methods of Evaluating Mechanical Properties of Polymer Matrix Composites," *Archives of Foundry Engineering Journal*, vol. 40, no. 3, pp. 209–212, 2010.

[18] M. Haneef, J. Fazlur Rahman, M. Yunus, S. Zameer, S. patil, and T. Yezdanil, "Hybrid Polymer Matrix Composites for Biomedical Applications," *International Journal of Modern Engineering Research (IJMER)*, vol. 3, no. 2, pp. 970–979, 2013.

[19] K. Van de Velde and P. Kiekens, "Biopolymers: overview of several properties and consequences on their applications," *Polymer Testing*, vol. 21, no. 4, pp. 433–442, 2002.

[20] M. Altan and H. Yildirim, "Mechanical and morphological properties of polypropylene and high density polyethylene matrix composites reinforced with surface modified nano sized TiO2 particles," *World Academy of Science, Engineering and Technology*, vol. 4, no. 10, pp. 654–659, 2010.

[21] K. Okubo, T. Fujii, and Y. Yamamoto, "Development of bamboo-based polymer composites and their mechanical properties," *Composites Part A: Applied Science and Manufacturing*, vol. 35, no. 3, pp. 377–383, 2004.

[22] X. Wang, T. Wang, F. Jiang, and Y. Duan, "The hip stress level analysis for human routine activities," *Biomedical Engineering: Applications, Basis and Communications*, vol. 17, no. 03, pp. 153–158, 2005.

[23] Santavirta S., M. Böhler, W. H. Harris, Y. T. Konttinen, R. Lappalainen, O. Muratoglu et al., "Alternative Materials to Improve Total Hip Replacement Tribology," *Acta Scandinavia*, vol. 74, no. 4, pp. 380–388, 2003.

[24] J. Black and G. Hastings, *Handbook of Biomaterial Properties*, vol. 590, Springer, New York, NY, USA, 1st edition, 1998.

[25] M. S. Alsoufi, "A high dynamic response micro-tribometer measuring-head," in *School of Engineering*, University of Warwick, Coventry, England, 2011.

[26] M. S. Alsoufi, D. K. Suker, A. S. Alsabban, and S. Azam, "Experimental Study of Surface Roughness and Micro-Hardness Obtained by Cutting Carbon Steel with Abrasive WaterJet and Laser Beam Technologies," *American Journal of Mechanical Engineering*, vol. 4, no. 5, pp. 173–181, 2016.

[27] M. S. Alsoufi, "Tactile perception of passenger vehicle interior polymer surfaces: an investigation using fingertip blind observations and friction properties," *International Journal of Science and Research (IJSR)*, vol. 5, no. 5, pp. 1447–1454, 2016.

[28] T. M. Bawazeer, M. S. Alsoufi, D. Katowah, and W. S. Alharbi, "Effect of Aqueous Extracts of Salvadora Persica "Miswak" on the Acid Eroded Enamel Surface at Nano-Mechanical Scale," *Materials Sciences and Applications*, vol. 7, no. 11, pp. 754–771, 2016.

23

Bioceramic-based Root Canal Sealers

Afaf AL-Haddad and Zeti A. Che Ab Aziz

Department of Restorative Dentistry, Faculty of Dentistry, University of Malaya, 50603 Kuala Lumpur, Malaysia

Correspondence should be addressed to Afaf AL-Haddad; afaf_haddad3@yahoo.com

Academic Editor: Traian V. Chirila

Bioceramic-based root canal sealers are considered to be an advantageous technology in endodontics. The aim of this review was to consider laboratory experiments and clinical studies of these sealers. An extensive search of the endodontic literature was made to identify publications related to bioceramic-based root canal sealers. The outcome of laboratory and clinical studies on the biological and physical properties of bioceramic-based sealers along with comparative studies with other sealers was assessed. Several studies were evaluated covering different properties of bioceramic-based sealers including physical properties, biocompatibility, sealing ability, adhesion, solubility, and antibacterial efficacy. Bioceramic-based sealers were found to be biocompatible and comparable to other commercial sealers. The clinical outcomes associated with the use of bioceramic-based root canal sealers are not established in the literature.

1. Introduction

The main functions of root canal sealers are (i) sealing off of voids, patent accessory canals, and multiple foramina, (ii) forming a bond between the core of the filling material and the root canal wall, and (iii) acting as a lubricant while facilitating the placement of the filling core and entombing any remaining bacteria [1]. Due to the relative biological and technical importance of sealers, their chemical and physical properties have been the subject of considerable attention since their initial development in the early twentieth century [2]. Sealers are categorised according to their main chemical constituents: zinc oxide eugenol, calcium hydroxide, glass ionomer, silicone, resin, and bioceramic-based sealers.

Root canal sealers have been reviewed across a number of studies, either collectively [2] or based on their composition, including zinc oxide eugenol [3], calcium hydroxide [4], glass ionomer [5], and resin-based sealers [6]. However, no extensive review of bioceramic-based sealers has been conducted.

Bioceramic-based sealers have only been available for use in endodontics for the past thirty years, their rise to prominence corresponding to the increased use of bioceramic technology in the fields of medicine and dentistry. Bioceramics are ceramic materials designed specifically for medical and dental use. They include alumina, zirconia, bioactive glass, glass ceramics, hydroxyapatite, and calcium phosphates [7]. The classification of bioceramic materials into bioactive or bioinert materials is a function of their interaction with the surrounding living tissue [8]. Bioactive materials, such as glass and calcium phosphate, interact with the surrounding tissue to encourage the growth of more durable tissues [9]. Bioinert materials, such as zirconia and alumina, produce a negligible response from the surrounding tissue, effectively having no biological or physiological effect [8]. Bioactive materials are further classified according to their stability as degradable or nondegradable. Bioceramics are commonly used for orthopaedic treatments, such as joint or tissue replacements, and for coating metal implants to improve biocompatibility. Additionally, porous ceramics, such as calcium phosphate-based materials, have been used as bone graft substitutes [10].

Calcium phosphate was first used as bioceramic restorative dental cement by LeGeros et al. [11]. However, the first documented use of bioceramic materials as a root canal sealer was not until two years later when Krell and Wefel [12] compared the efficacy of experimental calcium phosphate cement with Grossman's sealer in extracted teeth, finding no significant difference between both sealers in terms of apical occlusion, adaptation, dentinal tubule occlusion, adhesion, cohesion, or morphological appearance. Nonetheless, the

experimental calcium phosphate sealer failed to provide apical sealing as effectively as Grossman's sealer [13]. Chohayeb
et al. [14] later evaluated the use of calcium phosphate as a
root canal sealer in adult dog teeth. They reported that the calcium phosphate-based sealer made for a more uniform and
tighter adaptation to the dentinal walls as compared to gutta-
percha [14]. Calcium phosphate cement has subsequently
been used successfully in endodontic treatments, including
pulp capping [15], apical barrier formation, periapical defect
repairs [16], and bifurcation perforation repairs [17].

There are two major advantages associated with the
use of bioceramic materials as root canal sealers. Firstly,
their biocompatibility prevents rejection by the surrounding
tissues [9]. Secondly, bioceramic materials contain calcium
phosphate which enhances the setting properties of bioceramics and results in a chemical composition and crystalline
structure similar to tooth and bone apatite materials [18],
thereby improving sealer-to-root dentin bonding. However,
one major disadvantage of these materials is in the difficulty
in removing them from the root canal once they are set for
later retreatment or post-space preparation [19].

The exact mechanism of bioceramic-based sealer bonding to root dentin is unknown; however, the following mechanisms have been suggested for calcium silicate-based sealers:

(1) Diffusion of the sealer particles into the dentinal
tubules (tubular diffusion) to produce mechanical
interlocking bonds [20].

(2) Infiltration of the sealer's mineral content into the
intertubular dentin resulting in the establishment of
a mineral infiltration zone produced after denaturing
the collagen fibres with a strong alkaline sealer [21, 22].

(3) Partial reaction of phosphate with calcium silicate
hydrogel and calcium hydroxide, produced through
the reaction of calcium silicates in the presence of the
dentin's moisture, resulting in the formation of hydroxyapatite along the mineral infiltration zone [23].

While various branded bioceramic-based root canal sealers are available on the market, others are still experimental, requiring further laboratory and clinical testing to
ascertain their efficacy. A number of commercially available
bioceramic-based root canal sealers, classified according
to their major constituents, are identified in Table 1. The
biological and physical properties of bioceramic-based root
canal sealers were reviewed based on the ideal root canal
sealer properties as described by Grossman [24], as in the
following list:

(1) It should be tacky when mixed to provide good
adhesion between it and the canal wall when set.

(2) It should make a hermetic seal.

(3) It should be radiopaque so that it can be visualized on
the radiograph.

(4) The particles of powder should be very fine so that
they can mix easily with liquid.

(5) It should not shrink upon setting.

(6) It should not discolour tooth structure.

(7) It should be bacteriostatic or at least not encourage
bacterial growth.

(8) It should set slowly.

(9) It should be insoluble in tissue fluids.

(10) It should be well tolerated by the periapical tissue.

(11) It should be soluble in common solvents if it is
necessary to remove the root canal filling.

2. Ideal Root Canal Sealer Properties

2.1. Biocompatibility. Biocompatibility is an essential requirement of any root canal sealer as the root filling material
constitutes a true implant coming into direct contact with
the vital tissue at the apical and lateral foramina of the root
or indirectly via surface restoration [2]. Biocompatibility is
defined as the ability of a material to achieve a proper and
advantageous host response in specific applications [25]. In
other words, a material is said to be biocompatible when
the material coming into contact with the tissue fails to
trigger an adverse reaction, such as toxicity, irritation, inflammation, allergy, or carcinogenicity [26]. Most studies assess
biocompatibility through investigations of cytotoxicity, in
reference to the effect of the material on cell survival [27]. The
cytotoxicity of bioceramic-based sealers has been evaluated
in vitro using mouse and human osteoblast cells [28, 29] and
human periodontal ligaments cells [30]. Most bioceramic-
based root canal sealers have subsequently been found to
be biocompatible. This biocompatibility is attributed to the
presence of calcium phosphate in the sealer itself. Calcium
phosphate also happens to be the main inorganic component
of the hard tissues (teeth and bone). Consequently, the literature notes that many bioceramic sealers have the potential to
promote bone regeneration when unintentionally extruded
through the apical foramen during root canal filling or repairs
of root perforations [30, 31].

Sankin apatite has been shown by Telli et al. [32] to be
biocompatible in *in vitro* studies. However, Kim et al. [33]
showed that Sankin apatite exerts a tissue response when
implanted subcutaneously in rats and that this response
began to subside within two weeks. The biocompatibility
of Sankin apatite root canal sealer was also evaluated in
comparison to an experimental calcium phosphate-based
sealer composed of tetracalcium phosphate, dicalcium phosphate dihydrate, and modified McIlvaine's buffer solution.
Yoshikawa et al. [34] found that Sankin apatite caused severe
inflammatory reactions in both the dorsal subcutaneous and
the periapical tissue of rats. However, the experimental sealer
produced no inflammatory response in the subcutaneous
tissue and only a mild reaction in the periapical tissue [34].
The cytotoxicity of the Sankin apatite root canal sealer is
the result of the presence of iodoform and polyacrylic acids
in the sealer [33]. However, Sankin apatite type II and type
III were found to be more biocompatible than either type
I or Grossman's sealer [35]. EndoSequence BC, iRoot SP,
and MTA-Fillapex showed moderate toxicity when freshly
mixed; however, cytotoxicity reduced over time until being
completely set [29, 36, 37]. Although *in vitro* evaluations
of biocompatibility can be an indicator of the cytotoxicity

TABLE 1: Examples of bioceramic-based root canal sealers.

Type	Brand name	Manufacturer	Components
Calcium silicate-based sealer	iRoot SP	Innovative BioCeramix Inc., Vancouver, Canada	Zirconium oxide, calcium silicates, calcium phosphate, calcium hydroxide, filler, and thickening agents
	EndoSequence BC Sealer	Brasseler USA, Savannah, GA, USA	
	MTA-Fillapex	Angelus, Londrina, PR, Brazil	Salicylate resin, diluting resin, natural resin, bismuth trioxide, nanoparticulate silica, MTA, and pigments
	Endo CPM sealer	Egeo, Buenos Aires, Argentina	Silicon dioxide, calcium carbonate, bismuth trioxide, barium sulfate, propylene glycol alginate, sodium citrate, calcium chloride, and active ingredients
MTA-based sealer	MTA-Angelus	Angelus, Londrina, PR, Brazil	Tricalcium silicate, dicalcium silicate, tricalcium aluminate, tetracalcium aluminoferrite, bismuth oxide, iron oxide, calcium carbonate, magnesium oxide, crystalline silica, and residues (calcium oxide, free magnesium oxide, and potassium and sodium sulphate compounds)
	ProRoot Endo Sealer	DENTSPLY Tulsa Dental Specialties	Powder: tricalcium silicate, dicalcium silicate, calcium sulphate, bismuth oxide, and a small amount of tricalcium aluminate. Liquid: viscous aqueous solution of a water-soluble polymer
Calcium phosphate-based sealer	Sankin apatite root canal sealer (I, II, and III)	Sankin Kogyo, Tokyo, Japan	Powder: alpha-tricalcium phosphate and hydroxy-Sankin apatite in type I, iodoform added to powder in type II (30%) and type III (5%). Liquid: polyacrylic acid and water
	Capseal (I and II)	Experimental [45]	Powder: tetracalcium phosphate (TTCP) and dicalcium phosphate anhydrous (DCPA), Portland cement (gray cement in type I and white cement in type II), zirconium oxide, and others as powder. liquid: hydroxypropyl methyl cellulose in sodium phosphate solution

of a material, *in vitro* immunological deficiencies should be taken into consideration. Some sealers have been shown to have severe cytotoxicity *in vitro*, such as zinc oxide eugenol-based sealers; however, such toxicity is not necessarily clinically significant [38].

Capseal I and Capseal II sealers have been shown to produce less tissue irritation and less inflammation compared to other sealers [30, 31, 33]. Shon et al. [39] studied the effects of Capseal I and Capseal II in comparison to Sankin apatite root sealer (type I and type III) and a zinc oxide eugenol-based sealer (Pulp Canal Sealer). Investigators exposed human periodontal fibroblast cells to the various sealers before measuring the inflammatory response by way of inflammatory mediators and the viability and osteogenic potential of osteoblast MG63 cells. They found Capseal I and Capseal II to possess low cytotoxicity and to facilitate periapical dentoalveolar healing by regulating cellular mediators from periodontal ligaments cells and osteoblast differentiation. MTA-Fillapex was found to have a severe cytotoxic effect on fibroblast cells when freshly mixed. Furthermore, this effect did not decrease with time. The level of cytotoxicity remained moderate even five weeks after mixing [40].

2.2. Setting Time.

The ideal root canal sealer setting time should permit adequate working time. However, a slow setting time can result in tissue irritation, with most root canal sealers producing some degree of toxicity until being completely set. According to the manufacturers of EndoSequence BC Sealer or iRoot SP, the setting reaction is catalysed by the presence of moisture in the dentinal tubules. While the normal setting time is four hours, in patients with particularly dry canals, the setting time might be considerably longer [41]. The amount of moisture present in the dentinal tubules of the canal walls can be affected by absorption with paper points [42], the presence of smear plugs, or tubular sclerosis [43].

Loushine et al. [28] reported that EndoSequence BC Sealer requires at least 168 hours before being completely set under different humidity conditions, as evaluated using the Gilmore needle method. Zhou et al. [44], on the other hand, reported a setting time of 2.7 hours. The setting reaction of EndoSequence BC Sealer is a two-phase reaction. In phase I, monobasic calcium phosphate reacts with calcium hydroxide in the presence of water to produce water and hydroxyapatite. In phase II, the water derived from the dentin humidity, as well as that produced by the phase I reaction, contributes to the hydration of calcium silicate particles to trigger a calcium silicate hydrate phase [28].

The manufacturer of MTA-Fillapex claims that their product will set in a minimum of two hours and this setting time has been confirmed in at least two studies [44, 51]. However, even shorter setting times for MTA-Fillapex (66 min) have been reported [52]. The setting reaction of MTA material is complicated and has been discussed by Darvell and Wu [53]; however, the setting reaction of MTA-based sealers has not been described in the literature.

2.3. Flow.

Flow is an essential property that allows the sealer to fill difficult-to-access areas, such as the narrow irregularities of the dentin, isthmus, accessory canals, and voids between the master and accessory cones [54]. According to ISO 6786/2001 [55], a root canal sealer should have a flow rate of not less than 20 mm. Factors that influence the flow rate of the sealer include particle size, temperature, shear rate, and time from mixing [4]. The internal diameter of the tubes and rate of insertion are considered when assessing flow rate via the Rheometer method [2]. The flow rate for EndoSequence BC Sealer has been variously reported as 23.1 mm and 26.96 mm [44, 54]. Similarly, the flow rate of MTA-Fillapex has been variously reported as 22 mm, 24.9 mm, and 29.04 mm [44, 51, 52]. While most of the bioceramic-based root sealer manufacturers included in Table 1 claim that the flow rate of their sealers meets ISO requirements, the literature does not support such claims.

2.4. Retreatability.

Root filling materials provide a mechanical barrier for the isolation of necrotic tissue or bacteria responsible for the persistence of periapical inflammation or postoperative pain [56, 57]. Wilcox et al. [58] observe that most of the remaining material during retreatment is sealer. Therefore, the complete removal of the sealer is essential during endodontic retreatment to establish healthy periapical tissues. EndoSequence BC Sealer is difficult to remove from the root canal using conventional retreatment techniques, including heat, chloroform, rotary instruments, and hand files. A number of cases have been reported in which obstruction of the apical foramen has resulted in a loss of patency [59]. By contrast, Ersev et al. [60] reported that the removability of EndoSequence BC Sealer from the root canal is comparable to AH Plus. Sankin apatite root canal sealer is easily removed during retreatment with and without the use of solvents [61]. Retreatability with MTA-Fillapex is comparable to that of AH Plus in terms of material remaining in the canal, dentin removal, and time taken to reach working length [62].

2.5. Solubility.

Solubility is the mass loss of a material during a period of immersion in water. According to ANSI/ADA Specification 57 [63], the solubility of a root canal sealer should not exceed 3% by mass. A highly soluble root canal sealer would invariably permit the formation of gaps within and between the material and the root dentin, thereby providing avenues for leakage from the oral cavity and periapical tissues [2].

Both iRoot SP and MTA-Fillapex are highly soluble, 20.64% and 14.89%, respectively, which does not meet ANSI/ADA requirements [52, 64]. This high solubility is the result of hydrophilic nanosized particles being present in both sealers which increases their surface area and allows more liquid molecules to come into contact with the sealer. However, the literature contains conflicting accounts, with Viapiana et al. [52] finding MTA-Fillapex to be highly soluble and Vitti et al. [51] reporting the solubility of MTA-Fillapex to be <3%, consistent with ISO 6876/2001. Similarly, the solubility of EndoSequence BC is reported to be consistent with ISO 6876/2001 [44]. This discrepancy between the findings of these studies might be attributed to variations in the methods used to dry the samples after having subjected them to solubility testing. The low solubility of MTA-Angelus, consistent with ANSI/ADA requirements [64], is the result of an insoluble matrix of crystalline silica present within the sealer that maintains its integrity even in the presence of water [65].

2.6. Discolouration of Tooth Structure. For reasons of aesthetic appearance, a root canal sealer should not stain the tooth. The chromogenic effects of root sealers are increased when excess sealer is not removed from the coronal dentin of the pulp chamber [66]. Partovi et al. [67] observe that Sankin apatite III results in the least discolouration nine months after application as compared with AH26, Endofill, Tubli-Seal, and zinc oxide eugenol sealers. The greatest degree of discolouration was observed following treatment of the cervical third of the crown [67]. MTA-Fillapex was found to cause the least crown discolouration to the extent of not being clinically perceptible [66].

2.7. Radiopacity. Root canal sealers should be sufficiently radiopaque so as to be distinguishable from adjacent anatomical structures [68]. This allows the quality of the root filling to be evaluated through radiographic examination. According to ISO 6876/2001, the minimum radiopacity for a root canal sealer is based on a reference standard of 3.00 mm of aluminium. Candeiro et al. [54] reported the radiopacity of EndoSequence BC Sealer to be 3.83 mm. Endo CPM sealer was found to have a radiopacity of 6 mm due to the presence of bismuth trioxide and barium sulphate [69]. Similarly, the presence of bismuth trioxide in MTA-Fillapex gives it a radiopacity of 7 mm [52, 70].

2.8. Antimicrobial Properties. The antimicrobial activity of a root canal sealer increases the success rate of endodontic treatments by eliminating residual intraradicular infections that might have survived root canal treatment or have invaded the canal later through microleakage [71, 72]. According to the literature, the key antimicrobial properties of root canal sealers lie in their alkalinity and release of calcium ions [4] which stimulates repair via the deposition of mineralised tissue [73].

Two methods are commonly used to evaluate the antibacterial activity of bioceramic-based root canal sealers: the agar diffusion test [74, 75] and direct contact testing [23, 75]. EndoSequence BC Sealer has been shown to have high pH (>11) as well as high tendency to release calcium ions [54]. Zhang et al. [23] tested the antibacterial activity of iRoot SP sealer *in vitro* against *Enterococcus faecalis* through a modified direct contact test, finding that iRoot SP sealer had a high pH value (11.5) even after setting but that its antibacterial effect was greatly diminished after seven days. The investigators suggested two additional mechanisms associated with the antibacterial efficacy of iRoot SP: hydrophilicity and active calcium hydroxide diffusion [23]. Hydrophilicity reduces the contact angle of the sealer and facilitates penetration of the sealer into the fine areas of the root canal system to enhance the antibacterial effectiveness of iRoot SP *in vivo* [23].

Morgental et al. [75] evaluated the antibacterial activity of MTA-Fillapex and Endo CPM against *Enterococcus faecalis*

using an agar diffusion test after mixing and a direct contact test after setting. The pH of the Endo CPM suspension was greater than that of MTA-Fillapex (>11); however, the bacterial inhibition zone produced by MTA-Fillapex was greater than that produced by Endo CPM [75]. The investigators attributed the antibacterial activity of MTA-Fillapex to the presence of resin as a core ingredient. Nevertheless, neither sealer was able to sustain its antibacterial activity after setting despite their initial high pH levels [75].

Enterococcus faecalis is the most common intraradicular microbe isolated from periapical periodontitis [76, 77] and is therefore commonly used to test the antibacterial activity of root canal sealers. Other microorganisms, such as *Micrococcus luteus*, *Staphylococcus aureus*, *Escherichia coli*, *Pseudomonas aeruginosa*, *Candida albicans*, and *Streptococcus mutans*, have also been used to test the antibacterial effects of bioceramic-based sealers [74, 78]. Freshly mixed Endo CPM exhibits antibacterial activity against *Staphylococcus aureus* and *Streptococcus mutans* with no significant reduction of the inhibition zone after setting. Nevertheless, the antibacterial effect is less than that of AH-26 [78]. MTA-Angelus has an antibacterial effect against *Micrococcus luteus*, *Staphylococcus aureus*, *Escherichia coli*, *Pseudomonas aeruginosa*, and *Candida albicans* [74].

2.9. Adhesion. Root canal sealer adhesion is defined as its capacity to adhere to the root canal dentin and promote GP cone adhesion to each other and the dentin [79]. Tagger et al. [80] argued that the term *adhesion* should be replaced with *bonding* in the case of root canal sealers because the attachment between the substances involves mechanical interlocking forces rather than molecular attraction. There is no standard method used to measure the adhesion of a sealer to the root dentin; therefore, the adhesion potential of the root filling material is commonly tested using microleakage and bond strength tests [81].

The sealing ability of a sealer is related to its solubility and to its bonding to the dentin and root canal filling cones [4]. Several studies have evaluated the sealing abilities of different bioceramic-based sealers *in vitro*. These studies are summarised in Table 2. Regardless of the different methodologies used, the sealing ability of bioceramic-based sealers has been found to be satisfactory and comparable to other commercially available sealers. However, until recently, there had been a paucity of literature concerning the long-term sealing ability or clinical outcomes associated with bioceramic-based sealers.

Bond strength is the force per unit area required to debond the adhesive material from the dentin [81]. Although no correlation has been identified between leakage and bond strength [82], the bond strength test has received significant attention due to the development of the "monoblock" concept in which a sealer bonds to both the core material and the dentinal wall to create a singular unit that enhances sealing and strengthens the root-filled tooth against fracture [83]. A strong bond between the root canal sealer and the root dentin is essential for maintaining the integrity of the sealer-dentin interface during the preparation of post-spaces and during tooth flexure [84]. Bioceramic-based sealers have the ability

TABLE 2: Sealing ability of bioceramic-based root canal sealers.

Tested sealers	Compared sealers	Obturation technique	Microleakage site	Microleakage test	Duration	Finding	References
Capseal I and Capseal II	Sankin apatite, AH Plus, Sealapex, and zinc oxide-based sealer (Pulp Canal Sealer-Kerr)	Lateral condensation technique	Apical	Anaerobic bacterial leakage	Every day for a period of 90 days	Capseal I and Capseal II especially Capseal II showed good sealing ability, comparable to that of AH Plus	Yang et al. [45]
Endo CPM and Experimental MTA-based sealer	AH Plus, Sealer 26, Epiphany SE, Sealapex, Activ GP, and Endofill	Cold lateral condensation	Coronal	Bacterial leakage	Every 24 h for a period of 120 days	Activ GP, Endo CPM sealer, and MTAS were less resistant to leakage	Oliveira et al. [46]
Experimental calcium phosphate injectable sealer	Sealapex	Lateral condensation for Sealapex group and single silver cone for experimental sealer	Apical	Poly-R dye penetration test	5 days	Sealing ability was satisfactory in both groups with no significant difference	Cherng et al. [19]
Experimental MTA and fluorodoped MTA (FMTA) sealers	AH Plus	Warm vertical compaction	Apical	Fluid filtration method	24 hours, 48 hours, 1 week, 2 weeks, 1 month, 3 months, and 6 months	FMTA and AH Plus had a significantly better sealing ability than MTA	Gandolfi and Prati [47]
iRoot SP	AH Plus	Continuous wave condensation technique with both sealers and single cone technique with iRoot SP	Apical	Fluid filtration method	24 hours, 1 week, 4 weeks, and 8 weeks	iRoot SP was equivalent to AH Plus sealer in apical sealing ability	Zhang et al. [20]
MTAS	Pulp Canal Sealer	Warm vertical compaction technique	Coronal	Fluid filtration method	1 day and 28 days	MTAS had sealing ability comparable to Pulp Canal Sealer	Camilleri et al. [48]
ProRoot Endo Sealer (DENTSPLY Tulsa Dental Specialties)	Pulp Canal Sealer and AH Plus	Warm vertical compaction technique	Coronal	Fluid filtration method	7 days and 35 days	Sealing ability of ProRoot Endo Sealer and AH Plus is better than that of Pulp Canal Sealer	Weller et al. [49]
Sankin apatite root sealer types I, II, and III	Roth's sealer, Sealapex, and Kerr root canal sealer	Lateral-vertical condensation technique	Apical	Dye penetration using silver nitrate	Not stated	Sealing ability of Sankin apatite II was second better after Sealapex	Barkhordar et al. [50]
Sankin apatite types I, II, and III	Grossman's sealer	Lateral condensation technique	Apical	Methylene blue dye	48 hours	All sealers showed minimal leakage with no significant difference	Bilginer et al. [35]

to create bonds between the dentin and core filling materials [9]. The bonding of iRoot SP to root dentin is comparable to that of AH Plus and stronger than either Sealapex or EndoREZ sealers [85]. Shokouhinejad et al. [86] evaluated the bond strength of EndoSequence BC Sealer compared to AH Plus in the presence and absence of a smear layer, finding that the dislocation resistance of EndoSequence BC Sealer was equal to that of AH Plus and with no significant effect on the smear layer. Nagas et al. [87] studied the bond strengths of several sealers under various moisture conditions present in the root canal, concluding that a sealer's bond strength is greatest in moist and wet canals, the presence of residual moisture positively affecting the adhesion of the root canal sealers to radicular dentin. As compared with AH Plus, Epiphany, and MTA-Fillapex, iRoot SP had the highest dislodgment resistance from the root dentin [87]. Moreover, the prior placement of intracanal calcium hydroxide improved the bonding of iRoot SP to the root dentin; however, the bonding was less than that of AH Plus and comparable to MTA-Fillapex in the absence of calcium hydroxide [88]. This improvement in bonding is explained by way of the chemical interaction between calcium hydroxide and the iRoot SP sealer increasing the frictional resistance and/or micromechanical retention of the sealer [88]. Endo CPM has a significantly higher bond strength compared to MTA-Fillapex or AH Plus [89].

Testing the bond strength at the coronal third of the root canal shows no significant difference between MTA-Fillapex, iRoot SP, and AH Plus. However, in middle and apical thirds, iRoot SP and AH Plus have equivalent bond strengths superior to MTA-Fillapex [90]. Huffman et al. [84] tested the dislocation resistance of ProRoot Endo Sealer, AH Plus Jet, and Pulp Canal Sealer from root dentin with and without immersion in a simulated body fluid (SBF). The investigators concluded that ProRoot Endo Sealer possesses greater bond strength than the other two sealers, especially after SBF immersion. According to Huffman et al. [84], the greater bonding of the ProRoot Endo Sealer is due to the presence of spherical amorphous calcium phosphate and apatite-like phases enhancing frictional resistance. There was no negative effect of the iRoot SP root canal sealer on the push-out bond strength of fibre posts cemented with self-adhesive resin cement [91]. Compared to Activ GP sealer (glass ionomer-based sealer, Brasseler USA, Savanah, GA), iRoot SP was found to increase the fracture resistance of endodontically treated roots *in vitro*, a potential indicator of the high bond strength of the sealer [92].

3. Conclusion

Bioceramic-based root canal sealers show promising results as root canal sealers. However, discrepancies in the results of these studies reveal that these sealers do not fulfil all of the requirements demanded of the ideal root sealer. The biocompatibility and biomineralization effect of these sealers might avail them for alternative uses in direct pulp capping and root end filling. Further studies are required to clarify the clinical outcomes associated with the use of these sealers.

Competing Interests

The authors declare that they have no competing interests.

References

[1] A. Kaur, N. Shah, A. Logani, and N. Mishra, "Biotoxicity of commonly used root canal sealers: a meta-analysis," *Journal of Conservative Dentistry*, vol. 18, no. 2, pp. 83–88, 2015.

[2] D. Orstavik, "Materials used for root canal obturation: technical, biological and clinical testing," *Endodontic Topics*, vol. 12, no. 1, pp. 25–38, 2005.

[3] K. Markowitz, M. Moynihan, M. Liu, and S. Kim, "Biologic properties of eugenol and zinc oxide-eugenol. A clinically oriented review," *Oral Surgery, Oral Medicine, Oral Pathology*, vol. 73, no. 6, pp. 729–737, 1992.

[4] S. Desai and N. Chandler, "Calcium hydroxide-based root canal sealers: a review," *Journal of Endodontics*, vol. 35, no. 4, pp. 475–480, 2009.

[5] R. A. Buck, "Glass ionomer endodontic sealers—a literature review," *General Dentistry*, vol. 50, no. 4, pp. 365–368, 2002.

[6] Y. K. Kim, S. Grandini, J. M. Ames et al., "Critical review on methacrylate resin–based root canal sealers," *Journal of Endodontics*, vol. 36, no. 3, pp. 383–399, 2010.

[7] L. L. Hench, "Bioceramics: from concept to clinic," *Journal of the American Ceramic Society*, vol. 74, no. 7, pp. 1487–1510, 1991.

[8] S. M. Best, A. E. Porter, E. S. Thian, and J. Huang, "Bioceramics: past, present and for the future," *Journal of the European Ceramic Society*, vol. 28, no. 7, pp. 1319–1327, 2008.

[9] K. Koch and D. Brave, "A new day has dawned: the increased use of bioceramics in endodontics," *Dentaltown*, vol. 10, pp. 39–43, 2009.

[10] K. C. Saikia, T. D. Bhattacharya, S. K. Bhuyan, D. J. Talukdar, S. P. Saikia, and P. Jitesh, "Calcium phosphate ceramics as bone graft substitutes in filling bone tumor defects," *Indian Journal of Orthopaedics*, vol. 42, no. 2, pp. 169–172, 2008.

[11] R. LeGeros, A. Chohayeb, and A. Shulman, "Apatitic calcium phosphates: possible dental restorative materials," *Journal of Dental Research*, vol. 61, article 343, 1982.

[12] K. F. Krell and J. S. Wefel, "A calcium phosphate cement root canal sealer—scanning electron microscopic analysis," *Journal of Endodontics*, vol. 10, no. 12, pp. 571–576, 1984.

[13] K. V. Krell and S. Madison, "Comparison of apical leakage in teeth obturated with a calcium phosphate cement or Grossman's cement using lateral condensation," *Journal of Endodontics*, vol. 11, no. 8, pp. 336–339, 1985.

[14] A. A. Chohayeb, L. C. Chow, and P. J. Tsaknis, "Evaluation of calcium phosphate as a root canal sealer-filler material," *Journal of Endodontics*, vol. 13, no. 8, pp. 384–387, 1987.

[15] A. Jean, B. Kerebel, L.-M. Kerebel, R. Z. Legeros, and H. Hamel, "Effects of various calcium phosphate biomaterials on reparative dentin bridge formation," *Journal of Endodontics*, vol. 14, no. 2, pp. 83–87, 1988.

[16] E. Pissiotis and L. S. W. Spngberg, "Biological evaluation of collagen gels containing calcium hydroxide and hydroxyapatite," *Journal of Endodontics*, vol. 16, no. 10, pp. 468–473, 1990.

[17] J. Y. M. Chau, J. W. Hutter, T. O. Mork, and B. K. Nicoll, "An in vitro study of furcation perforation repair using calcium phosphate cement," *Journal of Endodontics*, vol. 23, no. 9, pp. 588–592, 1997.

[18] M. P. Ginebra, E. Fernández, E. A. P. De Maeyer et al., "Setting reaction and hardening of an apatitic calcium phosphate cement," *Journal of Dental Research*, vol. 76, no. 4, pp. 905–912, 1997.

[19] A. M. Cherng, L. C. Chow, and S. Takagi, "In vitro evaluation of a calcium phosphate cement root canal filler/sealer," *Journal of Endodontics*, vol. 27, no. 10, pp. 613–615, 2001.

[20] W. Zhang, Z. Li, and B. Peng, "Assessment of a new root canal sealer's apical sealing ability," *Oral Surgery, Oral Medicine, Oral Pathology, Oral Radiology and Endodontics*, vol. 107, no. 6, pp. e79–e82, 2009.

[21] L. Han and T. Okiji, "Uptake of calcium and silicon released from calcium silicate-based endodontic materials into root canal dentine," *International Endodontic Journal*, vol. 44, no. 12, pp. 1081–1087, 2011.

[22] A. R. Atmeh, E. Z. Chong, G. Richard, F. Festy, and T. F. Watson, "Dentin-cement interfacial interaction: calcium silicates and polyalkenoates," *Journal of Dental Research*, vol. 91, no. 5, pp. 454–459, 2012.

[23] H. Zhang, Y. Shen, N. D. Ruse, and M. Haapasalo, "Antibacterial activity of endodontic sealers by modified direct contact test against *Enterococcus faecalis*," *Journal of Endodontics*, vol. 35, no. 7, pp. 1051–1055, 2009.

[24] L. Grossman, "Obturation of root canal," in *Endodontic Practice*, L. Grossman, Ed., p. 297, Lea and Febiger, Philadelphia, Pa, USA, 10th edition, 1982.

[25] D. F. Williams, *Definitions in Biomaterials: Proceedings of a Consensus Conference of the European Society for Biomaterials, Chester, England, March 3–5, 1986*, vol. 4 of *Progress in Biomedical Engineering*, Elsevier, Amsterdam, The Netherlands, 1987.

[26] Z. L. Sun, J. C. Wataha, and C. T. Hanks, "Effects of metal ions on osteoblast-like cell metabolism and differentiation," *Journal of Biomedical Materials Research*, vol. 34, no. 1, pp. 29–37, 1997.

[27] G. Schmalz, "Use of cell cultures for toxicity testing of dental materials—advantages and limitations," *Journal of Dentistry*, vol. 22, no. 2, pp. S6–S11, 1994.

[28] B. A. Loushine, T. E. Bryan, S. W. Looney et al., "Setting properties and cytotoxicity evaluation of a premixed bioceramic root canal sealer," *Journal of Endodontics*, vol. 37, no. 5, pp. 673–677, 2011.

[29] L. P. Salles, A. L. Gomes-Cornélio, F. C. Guimarães et al., "Mineral trioxide aggregate-based endodontic sealer stimulates hydroxyapatite nucleation in human osteoblast-like cell culture," *Journal of Endodontics*, vol. 38, no. 7, pp. 971–976, 2012.

[30] W.-J. Bae, S.-W. Chang, S.-I. Lee, K.-Y. Kum, K.-S. Bae, and E.-C. Kim, "Human periodontal ligament cell response to a newly developed calcium phosphate-based root canal sealer," *Journal of Endodontics*, vol. 36, no. 10, pp. 1658–1663, 2010.

[31] T. E. Bryan, K. Khechen, M. G. Brackett et al., "In vitro osteogenic potential of an experimental calcium silicate-based root canal sealer," *Journal of Endodontics*, vol. 36, no. 7, pp. 1163–1169, 2010.

[32] C. Telli, A. Serper, A. L. Dogan, and D. Guc, "Evaluation of the cytotoxicity of calcium phosphate root canal sealers by MTT assay," *Journal of Endodontics*, vol. 25, no. 12, pp. 811–813, 1999.

[33] J.-S. Kim, S.-H. Baek, and K.-S. Bae, "In vivo study on the biocompatibility of newly developed calcium phosphate-based root canal sealers," *Journal of Endodontics*, vol. 30, no. 10, pp. 708–711, 2004.

[34] M. Yoshikawa, S. Hayami, I. Tsuji, and T. Toda, "Histopathological study of a newly developed root canal sealer containing tetracalcium-dicalcium phosphates and 1.0% chondroitin sulfate," *Journal of endodontics*, vol. 23, no. 3, pp. 162–166, 1997.

[35] S. Bilginer, T. Esener, F. Söylemezoglu, and A. M. Tiftik, "The investigation of biocompatibility and apical microleakage of tricalcium phosphate based root canal sealers," *Journal of Endodontics*, vol. 23, no. 2, pp. 105–109, 1997.

[36] K. Zoufan, J. Jiang, T. Komabayashi, Y.-H. Wang, K. E. Safavi, and Q. Zhu, "Cytotoxicity evaluation of Gutta Flow and Endo Sequence BC sealers," *Oral Surgery, Oral Medicine, Oral Pathology, Oral Radiology and Endodontology*, vol. 112, no. 5, pp. 657–661, 2011.

[37] D. Mukhtar-Fayyad, "Cytocompatibility of new bioceramic-based materials on human fibroblast cells (MRC-5)," *Oral Surgery, Oral Medicine, Oral Pathology, Oral Radiology and Endodontology*, vol. 112, no. 6, pp. e137–e142, 2011.

[38] T. P. Cotton, W. G. Schindler, S. A. Schwartz, W. R. Watson, and K. M. Hargreaves, "A retrospective study comparing clinical outcomes after obturation with Resilon/Epiphany or Gutta-Percha/Kerr Sealer," *Journal of Endodontics*, vol. 34, no. 7, pp. 789–797, 2008.

[39] W.-J. Shon, K.-S. Bae, S.-H. Baek, K.-Y. Kum, A.-R. Han, and W.-C. Lee, "Effects of calcium phosphate endodontic sealers on the behavior of human periodontal ligament fibroblasts and MG63 osteoblast-like cells," *Journal of Biomedical Materials Research Part B: Applied Biomaterials*, vol. 100, no. 8, pp. 2141–2147, 2012.

[40] E. J. Silva, C. C. Santos, and A. A. Zaia, "Long-term cytotoxic effects of contemporary root canal sealers," *Journal of Applied Oral Science*, vol. 21, no. 1, pp. 43–47, 2013.

[41] Q. Yang and D. Lu, "Premixed biological hydraulic cement paste composition and using the same," Google Patents, 2013.

[42] N. Hosoya, M. Nomura, A. Yoshikubo, T. Arai, J. Nakamura, and C. F. Cox, "Effect of canal drying methods on the apical seal," *Journal of Endodontics*, vol. 26, no. 5, pp. 292–294, 2000.

[43] F. Paqué, H. U. Luder, B. Sener, and M. Zehnder, "Tubular sclerosis rather than the smear layer impedes dye penetration into the dentine of endodontically instrumented root canals," *International Endodontic Journal*, vol. 39, no. 1, pp. 18–25, 2006.

[44] H.-M. Zhou, Y. Shen, W. Zheng, L. Li, Y.-F. Zheng, and M. Haapasalo, "Physical properties of 5 root canal sealers," *Journal of Endodontics*, vol. 39, no. 10, pp. 1281–1286, 2013.

[45] S.-E. Yang, S.-H. Baek, W. Lee, K.-Y. Kum, and K.-S. Bae, "In vitro evaluation of the sealing ability of newly developed calcium phosphate-based root canal sealer," *Journal of Endodontics*, vol. 33, pp. 978–981, 2007.

[46] A. C. M. Oliveira, J. M. G. Tanomaru, N. Faria-Junior, and M. Tanomaru-Filho, "Bacterial leakage in root canals filled with conventional and MTA-based sealers," *International Endodontic Journal*, vol. 44, no. 4, pp. 370–375, 2011.

[47] M. G. Gandolfi and C. Prati, "MTA and F-doped MTA cements used as sealers with warm gutta-percha. Long-term study of sealing ability," *International Endodontic Journal*, vol. 43, no. 10, pp. 889–901, 2010.

[48] J. Camilleri, M. G. Gandolfi, F. Siboni, and C. Prati, "Dynamic sealing ability of MTA root canal sealer," *International Endodontic Journal*, vol. 44, no. 1, pp. 9–20, 2011.

[49] R. N. Weller, K. C. Y. Tay, L. V. Garrett et al., "Microscopic appearance and apical seal of root canals filled with gutta-percha and ProRoot Endo Sealer after immersion in a phosphate-containing fluid," *International Endodontic Journal*, vol. 41, no. 11, pp. 977–986, 2008.

[50] R. A. Barkhordar, M. M. Stark, and K. Soelberg, "Evaluation of the apical sealing ability of apatite root canal sealer," *Quintessence International*, vol. 23, no. 7, pp. 515–518, 1992.

[51] R. P. Vitti, C. Prati, E. J. N. L. Silva et al., "Physical properties of MTA fillapex sealer," *Journal of Endodontics*, vol. 39, no. 7, pp. 915–918, 2013.

[52] R. Viapiana, D. L. Flumignan, J. M. Guerreiro-Tanomaru, J. Camilleri, and M. Tanomaru-Filho, "Physicochemical and mechanical properties of zirconium oxide and niobium oxide modified Portland cement-based experimental endodontic sealers," *International Endodontic Journal*, vol. 47, no. 5, pp. 437–448, 2014.

[53] B. W. Darvell and R. C. Wu, "'MTA'—an hydraulic silicate cement: review update and setting reaction," *Dental Materials*, vol. 27, no. 5, pp. 407–422, 2011.

[54] G. T. D. M. Candeiro, F. C. Correia, M. A. H. Duarte, D. C. Ribeiro-Siqueira, and G. Gavini, "Evaluation of radiopacity, pH, release of calcium ions, and flow of a bioceramic root canal sealer," *Journal of Endodontics*, vol. 38, no. 6, pp. 842–845, 2012.

[55] International Organization for Standardization, "Dental root canal sealing materials," ISO 6876, International Organization for Standardization, Geneva, Switzerland, 2001.

[56] L. R. Wilcox, "Endodontic retreatment: ultrasonics and chloroform as the final step in reinstrumentation," *Journal of Endodontics*, vol. 15, no. 3, pp. 125–128, 1989.

[57] J. F. Schirrmeister, K. T. Wrbas, K. M. Meyer, M. J. Altenburger, and E. Hellwig, "Efficacy of different rotary instruments for gutta-percha removal in root canal retreatment," *Journal of Endodontics*, vol. 32, pp. 469–472, 2006.

[58] L. R. Wilcox, K. V. Krell, S. Madison, and B. Rittman, "Endodontic retreatment: evaluation of gutta-percha and sealer removal and canal reinstrumentation," *Journal of Endodontics*, vol. 13, no. 9, pp. 453–457, 1987.

[59] D. Hess, E. Solomon, R. Spears, and J. He, "Retreatability of a bioceramic root canal sealing material," *Journal of Endodontics*, vol. 37, no. 11, pp. 1547–1549, 2011.

[60] H. Ersev, B. Yilmaz, M. E. Dinçol, and R. Dağlaroğlu, "The efficacy of ProTaper Universal rotary retreatment instrumentation to remove single gutta-percha cones cemented with several endodontic sealers," *International Endodontic Journal*, vol. 45, no. 8, pp. 756–762, 2012.

[61] A. Erdemir, N. Adanir, and S. Belli, "In vitro evaluation of the dissolving effect of solvents on root canal sealers," *Journal of Oral Science*, vol. 45, no. 3, pp. 123–126, 2003.

[62] P. Neelakantan, D. Grotra, and S. Sharma, "Retreatability of 2 mineral trioxide aggregate-based root canal sealers: a cone-beam computed tomography analysis," *Journal of Endodontics*, vol. 39, no. 7, pp. 893–896, 2013.

[63] ANSI/ADA, *Specification No 57 Endodontic Sealing Material*, ADA Publishing, Chicago, Ill, USA, 2000.

[64] R. P. Borges, M. D. Sousa-Neto, M. A. Versiani et al., "Changes in the surface of four calcium silicate-containing endodontic materials and an epoxy resin-based sealer after a solubility test," *International Endodontic Journal*, vol. 45, no. 5, pp. 419–428, 2012.

[65] M. Fridland and R. Rosado, "Mineral trioxide aggregate (MTA) solubility and porosity with different water-to-powder ratios," *Journal of Endodontics*, vol. 29, no. 12, pp. 814–817, 2003.

[66] K. Ioannidis, I. Mistakidis, P. Beltes, and V. Karagiannis, "Spectrophotometric analysis of crown discoloration induced by MTA- and ZnOE-based sealers," *Journal of Applied Oral Science*, vol. 21, no. 2, pp. 138–144, 2013.

[67] M. Partovi, A. H. Al-Havvaz, and B. Soleimani, "In vitro computer analysis of crown discolouration from commonly used endodontic sealers," *Australian Endodontic Journal*, vol. 32, no. 3, pp. 116–119, 2006.

[68] Y. Imai and T. Komabayashi, "Properties of a new injectable type of root canal filling resin with adhesiveness to dentin," *Journal of Endodontics*, vol. 29, no. 1, pp. 20–23, 2003.

[69] J. M. Guerreiro-Tanomaru, M. A. H. Duarte, M. Gonçalves, and M. Tanomaru-Filho, "Radiopacity evaluation of root canal sealers containing calcium hydroxide and MTA," *Brazilian Oral Research*, vol. 23, no. 2, pp. 119–123, 2009.

[70] E. J. N. L. Silva, T. P. Rosa, D. R. Herrera, R. C. Jacinto, B. P. F. A. Gomes, and A. A. Zaia, "Evaluation of cytotoxicity and physicochemical properties of calcium silicate-based endodontic sealer MTA fillapex," *Journal of Endodontics*, vol. 39, no. 2, pp. 274–277, 2013.

[71] G. S. P. Cheung, "Endodontic failures—changing the approach," *International Dental Journal*, vol. 46, no. 3, pp. 131–138, 1996.

[72] U. Sjögren, D. Figdor, S. Persson, and G. Sundqvist, "Influence of infection at the time of root filling on the outcome of endodontic treatment of teeth with apical periodontitis," *International Endodontic Journal*, vol. 30, no. 5, pp. 297–306, 1997.

[73] T. Okabe, M. Sakamoto, H. Takeuchi, and K. Matsushima, "Effects of pH on mineralization ability of human dental pulp cells," *Journal of Endodontics*, vol. 32, no. 3, pp. 198–201, 2006.

[74] M. Tanomaru-Filho, J. M. G. Tanomaru, D. B. Barros, E. Watanabe, and I. Y. Ito, "In vitro antimicrobial activity of endodontic sealers, MTA-based cements and Portland cement," *Journal of Oral Science*, vol. 49, no. 1, pp. 41–45, 2007.

[75] R. D. Morgental, F. V. Vier-Pelisser, S. D. Oliveira, F. C. Antunes, D. M. Cogo, and P. M. P. Kopper, "Antibacterial activity of two MTA-based root canal sealers," *International Endodontic Journal*, vol. 44, no. 12, pp. 1128–1133, 2011.

[76] E. T. Pinheiro, B. P. F. A. Gomes, C. C. R. Ferraz, E. L. R. Sousa, F. B. Teixeira, and F. J. Souza-Filho, "Microorganisms from canals of root-filled teeth with periapical lesions," *International Endodontic Journal*, vol. 36, no. 1, pp. 1–11, 2003.

[77] C. H. Stuart, S. A. Schwartz, T. J. Beeson, and C. B. Owatz, "*Enterococcus faecalis*: its role in root canal treatment failure and current concepts in retreatment," *Journal of Endodontics*, vol. 32, no. 2, pp. 93–98, 2006.

[78] Z. Mohammadi, L. Giardino, F. Palazzi, and S. Shalavi, "Antibacterial activity of a new mineral trioxide aggregate-based root canal sealer," *International Dental Journal*, vol. 62, no. 2, pp. 70–73, 2012.

[79] M. D. Sousa-Neto, F. I. Silva Coelho, M. A. Marchesan, E. Alfredo, and Y. T. C. Silva-Sousa, "Ex vivo study of the adhesion of an epoxy-based sealer to human dentine submitted to irradiation with Er: YAG and Nd: YAG lasers," *International Endodontic Journal*, vol. 38, no. 12, pp. 866–870, 2005.

[80] M. Tagger, E. Tagger, A. H. L. Tjan, and L. K. Bakland, "Measurement of adhesion of endodontic sealers to dentin," *Journal of Endodontics*, vol. 28, no. 5, pp. 351–354, 2002.

[81] R. S. Schwartz, "Adhesive dentistry and endodontics. Part 2: bonding in the root canal system—the promise and the problems: a review," *Journal of Endodontics*, vol. 32, no. 12, pp. 1125–1134, 2006.

[82] A. Wennberg and D. Orstavik, "Adhesion of root canal sealers to bovine dentine and gutta-percha," *International Endodontic Journal*, vol. 23, no. 1, pp. 13–19, 1990.

[83] F. B. Teixeira, E. C. N. Teixeira, J. Y. Thompson, and M. Trope, "Fracture resistance of roots endodontically treated with a new resin filling material," *Journal of the American Dental Association*, vol. 135, no. 5, pp. 646–652, 2004.

[84] B. Huffman, S. Mai, L. Pinna et al., "Dislocation resistance of ProRoot Endo Sealer, a calcium silicate–based root canal sealer, from radicular dentine," *International Endodontic Journal*, vol. 42, no. 1, pp. 34–46, 2009.

[85] S. Ersahan and C. Aydin, "Dislocation resistance of iRoot SP, a calcium silicate-based sealer, from radicular dentine," *Journal of Endodontics*, vol. 36, no. 12, pp. 2000–2002, 2010.

[86] N. Shokouhinejad, H. Gorjestani, A. A. Nasseh, A. Hoseini, M. Mohammadi, and A. R. Shamshiri, "Push-out bond strength of gutta-percha with a new bioceramic sealer in the presence or absence of smear layer," *Australian Endodontic Journal*, vol. 39, no. 3, pp. 102–106, 2013.

[87] E. Nagas, M. O. Uyanik, A. Eymirli et al., "Dentin moisture conditions affect the adhesion of root canal sealers," *Journal of Endodontics*, vol. 38, no. 2, pp. 240–244, 2012.

[88] S. A. Wanees Amin, R. S. Seyam, and M. A. El-Samman, "The effect of prior calcium hydroxide intracanal placement on the bond strength of two calcium silicate-based and an epoxy resin-based endodontic sealer," *Journal of Endodontics*, vol. 38, pp. 696–699, 2012.

[89] E. Assmann, R. K. Scarparo, D. E. Böttcher, and F. S. Grecca, "Dentin bond strength of two mineral trioxide aggregate-based and one epoxy resin-based sealers," *Journal of Endodontics*, vol. 38, no. 2, pp. 219–221, 2012.

[90] B. Sagsen, Y. Ustün, S. Demirbuga, and K. Pala, "Push-out bond strength of two new calcium silicate-based endodontic sealers to root canal dentine," *International Endodontic Journal*, vol. 44, no. 12, pp. 1088–1091, 2011.

[91] E. Özcan, İ. Çapar, A. R. Çetin, A. R. Tunçdemir, and H. A. Aydinbelge, "The effect of calcium silicate-based sealer on the push-out bond strength of fibre posts," *Australian Dental Journal*, vol. 57, no. 2, pp. 166–170, 2012.

[92] A. G. Ghoneim, R. A. Lutfy, N. E. Sabet, and D. M. Fayyad, "Resistance to fracture of roots obturated with novel canal-filling systems," *Journal of Endodontics*, vol. 37, no. 11, pp. 1590–1592, 2011.

Permissions

All chapters in this book were first published in IJB, by Hindawi Publishing Corporation; hereby published with permission under the Creative Commons Attribution License or equivalent. Every chapter published in this book has been scrutinized by our experts. Their significance has been extensively debated. The topics covered herein carry significant findings which will fuel the growth of the discipline. They may even be implemented as practical applications or may be referred to as a beginning point for another development.

The contributors of this book come from diverse backgrounds, making this book a truly international effort. This book will bring forth new frontiers with its revolutionizing research information and detailed analysis of the nascent developments around the world.

We would like to thank all the contributing authors for lending their expertise to make the book truly unique. They have played a crucial role in the development of this book. Without their invaluable contributions this book wouldn't have been possible. They have made vital efforts to compile up to date information on the varied aspects of this subject to make this book a valuable addition to the collection of many professionals and students.

This book was conceptualized with the vision of imparting up-to-date information and advanced data in this field. To ensure the same, a matchless editorial board was set up. Every individual on the board went through rigorous rounds of assessment to prove their worth. After which they invested a large part of their time researching and compiling the most relevant data for our readers.

The editorial board has been involved in producing this book since its inception. They have spent rigorous hours researching and exploring the diverse topics which have resulted in the successful publishing of this book. They have passed on their knowledge of decades through this book. To expedite this challenging task, the publisher supported the team at every step. A small team of assistant editors was also appointed to further simplify the editing procedure and attain best results for the readers.

Apart from the editorial board, the designing team has also invested a significant amount of their time in understanding the subject and creating the most relevant covers. They scrutinized every image to scout for the most suitable representation of the subject and create an appropriate cover for the book.

The publishing team has been an ardent support to the editorial, designing and production team. Their endless efforts to recruit the best for this project, has resulted in the accomplishment of this book. They are a veteran in the field of academics and their pool of knowledge is as vast as their experience in printing. Their expertise and guidance has proved useful at every step. Their uncompromising quality standards have made this book an exceptional effort. Their encouragement from time to time has been an inspiration for everyone.

The publisher and the editorial board hope that this book will prove to be a valuable piece of knowledge for researchers, students, practitioners and scholars across the globe.

List of Contributors

Liliane Pimenta de Melo and Carlos Rodrigo de Mello Roesler
LEBm Biomechanics Engineering Laboratory, University Hospital (HU), Federal University of Santa Catarina, 88040-900 Florianópolis, SC, Brazil

Gean Vitor Salmoria
LEBm Biomechanics Engineering Laboratory, University Hospital (HU), Federal University of Santa Catarina, 88040-900 Florianópolis, SC, Brazil
NIMMA Laboratory of Innovation on Additive Manufacturing and Molding, Federal University of Santa Catarina, 88040-900 Florianópolis, SC, Brazil

Eduardo Alberto Fancello
LEBm Biomechanics Engineering Laboratory, University Hospital (HU), Federal University of Santa Catarina, 88040-900 Florianópolis, SC, Brazil
GRANTE, Department of Mechanical Engineering, Federal University of Santa Catarina, 88040-900 Florianópolis, SC, Brazil

Bakhtawar Ghafoor, Murtaza Najabat Ali, Umar Ansari, Muhammad Faraz Bhatti, Mariam Mir, Hafsah Akhtar and Fatima Darakhshan
Biomedical Engineering and Sciences Department, School of Mechanical and Manufacturing Engineering (SMME), National University of Sciences and Technology (NUST), Islamabad, Pakistan

Vivekjot Brar and Gurpreet Kaur
Department of Pharmaceutical Sciences and Drug Research, Punjabi University, Patiala, Punjab 147002, India

Melisa A. Quinteros
IMBIV, CONICET, Departamento de Farmacia, Facultad de Ciencias Químicas, Universidad Nacional de Córdoba, Ciudad Universitaria, 5000 Córdoba, Argentina

Ivana M. Aiassa Martínez and Paulina L. Páez
UNITEFA, CONICET, Departamento de Farmacia, Facultad de Ciencias Químicas, Universidad Nacional de Córdoba, Ciudad Universitaria, 5000 Córdoba, Argentina

Pablo R. Dalmasso
CITSE, CONICET, Universidad Nacional de Santiago del Estero, RN 9, Km 1125, 4206 Santiago del Estero, Argentina

Carlos Rodrigo de Mello Roesler
Biomechanical Engineering Laboratory (LEBm), University Hospital (HU), Federal University of Santa Catarina, 88040-900 Florianópolis, SC, Brazil

Liliane Pimenta de Melo and Gean Vitor Salmoria
Biomechanical Engineering Laboratory (LEBm), University Hospital (HU), Federal University of Santa Catarina, 88040-900 Florianópolis, SC, Brazil
Laboratory of Innovation on Additive Manufacturing and Molding (NIMMA), Federal University of Santa Catarina, 88040-900 Florianópolis, SC, Brazil

Eduardo Alberto Fancello
Biomechanical Engineering Laboratory (LEBm), University Hospital (HU), Federal University of Santa Catarina, 88040-900 Florianópolis, SC, Brazil
GRANTE, Department of Mechanical Engineering, Federal University of Santa Catarina, 88040-900 Florianópolis, SC, Brazil

Nosheen Fatima
Department of Biomedical Engineering and Sciences, National University of Sciences & Technology, Islamabad, Pakistan

Sundus Riaz
Department of Biomedical Engineering and Sciences, National University of Sciences & Technology, Islamabad, Pakistan
Pakistan Agricultural Research Council, FQSRI, SARC, Karachi, Pakistan

Faiza Anwar
Pakistan Agricultural Research Council, FQSRI, SARC, Karachi, Pakistan

Ahmed Rasheed
PhD. Scholar, Sun Yat-Sen University (East Campus), Higher Education Mega Centre North, Guangzhou, China

Mehvish Riaz
MPH, London South Bank University, UK

Yamna Khatoon
Postgraduate Scholar, Department of Agriculture and Agribusiness Management, University of Karachi, Karachi, Pakistan

Alexandra Vinagre, João Ramos, Sofia Alves and Ana Messias
Faculty of Medicine, University of Coimbra, Avenida Bissaya Barreto, Blocos de Celas, 3000-075 Coimbra, Portugal

Nélia Alberto and Rogério Nogueira
Instituto de Telecomunicações (IT), Campus Universitário de Santiago, 3810-193 Aveiro, Portugal

Loreto M. Valenzuela
Department of Chemical and Bioprocess Engineering, Research Center for Nanotechnology and Advanced Materials "CIEN-UC", Pontificia Universidad Católica de Chile, Vicuña Mackenna 2860, Macul, 7820436 Santiago, Chile

Doyle D. Knight
Department of Mechanical and Aerospace Engineering, Rutgers, The State University of New Jersey, New Brunswick, NJ 08854-8087, USA

Joachim Kohn
New Jersey Center for Biomaterials, Rutgers, The State University of New Jersey, 145 Bevier Road, Piscataway, NJ 08854, USA

K. Z. M. Abdul Motaleb and Md Shariful Islam
Department of Textile Engineering, BGMEA University of Fashion and Technology, Dhaka, Bangladesh

Mohammad B. Hoque
Department of Textile Engineering, World University of Bangladesh, Dhaka, Bangladesh

J. R. Anusha and Albin T. Fleming
Department of Advanced Zoology and Biotechnology, Loyola College, Chennai, Tamil Nadu 600 034, India

Nopparuj Soomherun, Narumol Kreua-ongarjnukool and Saowapa Thumsing
Department of Industrial Chemistry, Faculty of Applied Science, King Mongkut's University of Technology North Bangkok, Bangkok, Thailand

Sorayouth Chumnanvej
Neurosurgery Unit, Surgery Department, Faculty of Medicine Ramathibodi Hospital, Mahidol University, Bangkok, Thailand

Carlo Galli, Giuseppe Pedrazzi and Stefano Guizzardi
Dep. of Medicine and Surgery, University of Parma, Italy

Monica Mattioli-Belmonte
DISCLIMO, Department of Clinical and Molecular Sciences, Polytechnic University of Marche, Ancona, Italy

Ho Hieu Minh, Nguyen Thi Hiep and Vo Van Toi
Tissue Engineering and Regenerative Medicine Laboratory, Department of Biomedical Engineering, International University of Vietnam National Universities, Ho Chi Minh City 700000, Vietnam

Nguyen Dai Hai
Institute of Applied Materials Science, Vietnam Academy of Science and Technology, 01 Mac Dinh Chi, District 1, Ho Chi Minh City, Vietnam
Graduate University of Science and Technology, Vietnam Academy of Science and Technology, Hanoi, Vietnam

Helo-sa A. B. Guimarães, Paula C. Cardoso, Rafael A. Decurcio, Lúcio J. E. Monteiro, Let-cia N. de Almeida, Wellington F. Martins and Ana Paula R. Magalhães
Restorative Dentistry, Brazilian Dental Association, Goiânia 74325-110, Brazil

Deepak M. Kalaskar
UCL Centre for Nanotechnology & Regenerative Medicine, University College London, Royal Free London NHS Foundation Trust, Pond Street, London NW3 2QG, UK

Michelle Griffin and Peter E. M. Butler
UCL Centre for Nanotechnology & Regenerative Medicine, University College London, Royal Free London NHS Foundation Trust, Pond Street, London NW3 2QG, UK
The Charles Wolfson Center for Reconstructive Surgery, Royal Free London NHS Foundation Trust Hospital, London, UK
Department of Plastic Surgery, Royal Free London NHS Foundation Trust, Pond Street, London NW3 2QG, UK

Naghmeh Naderi
UCL Centre for Nanotechnology & Regenerative Medicine, University College London, Royal Free London NHS Foundation Trust, Pond Street, London NW3 2QG, UK
Department of Plastic Surgery, Royal Free London NHS Foundation Trust, Pond Street, London NW3 2QG, UK
Reconstructive Surgery & Regenerative Medicine Group, Institute of Life Science, Swansea University Medical School, Singleton Park, Swansea SA2 8PP, UK
Welsh Centre for Burns & Plastic Surgery, ABMU Health Board, Heol Maes Egwlys, Swansea SA6 6NL, UK

Ash Mosahebi
Department of Plastic Surgery, Royal Free London NHS Foundation Trust, Pond Street, London NW3 2QG, UK

Catherine A. Thornton
Reconstructive Surgery & Regenerative Medicine Group, Institute of Life Science, Swansea University Medical School, Singleton Park, Swansea SA2 8PP, UK

Iain S. Whitaker
Reconstructive Surgery & Regenerative Medicine Group, Institute of Life Science, Swansea University Medical School, Singleton Park, Swansea SA2 8PP, UK
Welsh Centre for Burns & Plastic Surgery, ABMU Health Board, Heol Maes Egwlys, Swansea SA6 6NL, UK

Edward Malins and Remzi Becer
Polymer Chemistry Laboratory, School of Engineering and Materials Science, Queen Mary University of London, Mile End Road, London E1 4NS, UK

Alexander M. Seifalian
Director/Professor Nanotechnology & Regenerative Medicine, NanoReg Med Ltd., The London Bio Science Innovation Centre, London NW1 0NH, UK

Hari Sharan Adhikari
Department of Chemistry, Western Region Campus, Institute of Engineering, Tribhuvan University, Pokhara, Nepal

Paras Nath Yadav
Central Department of Chemistry, Tribhuvan University, Kathmandu, Nepal

Yong Y. Peng, Veronica Glattauer and John A. M. Ramshaw
CSIRO Manufacturing, Bayview Avenue, Clayton, VIC 3169, Australia

Anuoluwa Abimbola Akinsiku, Kolawole Oluseyi Ajanaku, Olayinka Oyewale Ajani, Joseph Adebisi O. Olugbuyiro and Tolutope Oluwasegun Siyanbola
Department of Chemistry, Covenant University, PMB 1023, Ota, Ogun State, Nigeria

Enock Olugbenga Dare
Department of Chemistry, Federal University of Agriculture, PMB 2240, Alabata Road, Abeokuta, Nigeria

Oluwaseun Ejilude
Department of Medical and Parasitology, Sacred Heart Hospitals, Lantoro, Abeokuta, Nigeria

Moses Eterigho Emetere
Department of Physics, CovenantUniversity, PMB 1023, Ota, Ogun State, Nigeria
Department of Mechanical Engineering Science, University of Johannesburg, Auckland Park Kingsway Campus, Johannesburg 2006, South Africa

Watcharaphong Chaemsawang and Phanphen Wattanaarsakit
Department of Pharmaceutics and Industrial Pharmacy, Faculty of Pharmaceutical Sciences, Chulalongkorn University, Bangkok, Thailand

Weerapong Prasongchean
Department of Biochemistry and Microbiology, Faculty of Pharmaceutical Sciences, Chulalongkorn University, Bangkok, Thailand

Konstantinos I. Papadopoulos
THAI StemLife Co., Ltd., Bangkok,Thailand

Suchada Sukrong
Department of Pharmacognosy and Pharmaceutical Botany, Faculty of Pharmaceutical Sciences, Chulalongkorn University, Bangkok, Thailand

W. John Kao
Chemistry and Biology Centre, Li Ka Shing Faculty of Medicine and Faculty of Engineering,The University of Hong Kong, Hong Kong SAR, Hong Kong

Matthias A. König and Hans-Peter Simmen
Department of Traumatology, University Hospital Zurich, Zurich, Switzerland

Oliver P. Gautschi
Département de Neurosciences Cliniques, Geneva University Hospital, Geneva, Switzerland

Luis Filgueira
School of Anatomy and Human Biology, University ofWestern Australia, Perth,WA, Australia

Dieter Cadosch
Department of General and Trauma Surgery, Triemlispital, Zurich, Switzerland

Roohollah Sharifi
Department of Endodontics, School of Dentistry, Kermanshah University of Medical Sciences, Kermanshah, Iran

Davood Almasi
School of Dentistry, Kermanshah University of Medical Sciences, Kermanshah, Iran

Izman Bin Sudin
Department of Manufacturing and Industrial Engineering, Faculty of Mechanical Engineering, Universiti Teknologi Malaysia, 81310 Skudai, Johor, Malaysia

Mohammed Rafiq Abdul Kadir, Maliheh Sadeghi and Fatemeh Roozbahani
Medical Devices & Technology Group (MEDITEG), Faculty of Biosciences and Medical Engineering, Universiti Teknologi Malaysia, 81310 Johor Bahru, Johor, Malaysia

Ladan Jamshidy
Department of Prosthodontics, School of Dentistry, Kermanshah University of Medical Sciences, Kermanshah, Iran

Seyed Mojtaba Amiri
Department of Biostatistics and Epidemiology, School of Health, Kermanshah University of Medical Sciences, Kermanshah, Iran

Hamid Reza Mozaffari
Department of Oral and Maxillofacial Medicine, School of Dentistry, Kermanshah University of Medical Sciences, Kermanshah, Iran
Medical Biology Research Center, Kermanshah University of Medical Sciences, Kermanshah, Iran

Nida Iqbal
Bio-Medical Engineering Center, University of Engineering and Technology (UET), Lahore Kala Shah Kaku (KSK) Campus, Pakistan

Mohammed Yunus and Mohammad S. Alsoufi
Department of Mechanical Engineering, College of Engineering and Islamic Architecture, Umm Al-Qura University, Al-Abdiah, Makkah 24231, Saudi Arabia

Afaf AL-Haddad and Zeti A. Che Ab Aziz
Department of Restorative Dentistry, Faculty of Dentistry, University of Malaya, 50603 Kuala Lumpur, Malaysia

Index

www.ingramcontent.com/pod-product-compliance
Lightning Source LLC
Chambersburg PA
CBHW061258190326
41458CB00011B/3710